Electrons in Disordered Metals and at Metallic Surfaces

NATO ADVANCED STUDY INSTITUTES SERIES

A series of edited volumes comprising multifaceted studies of contemporary scientific issues by some of the best scientific minds in the world, assembled in cooperation with NATO Scientific Affairs Division.

Series B: Physics

RECENT VOLUMES IN THIS SERIES

Volume 33 – Microscopic Structure and Dynamics of Liquids
edited by J. Dupuy and A. J. Dianoux

Volume 34 – Path Integrals and Their Applications in Quantum, Statistical, and Solid State Physics
edited by George J. Papadopoulos and J. T. Devreese

Volume 35 – Correlation Functions and Quasiparticle Interactions in Condensed Matter
edited by J. Woods Halley

Volume 36 – Strongly Coupled Plasmas
edited by Gabor Kalman

Volume 37 – Coherence in Spectroscopy and Modern Physics
edited by F. T. Arecchi, R. Bonifacio, and M. O. Scully

Volume 38 – Theoretical Methods in Medium-Energy and Heavy-Ion Physics
edited by K. W. McVoy and W. A. Friedman

Volume 39 – Hadron Structure and Lepton–Hadron Interactions – *Cargèse* 1977
edited by Maurice Lévy, Jean-Louis Basdevant, David Speiser, Jacques Weyers, Raymond Gastmans, and Jean Zinn-Justin

Volume 40 – Kinetics of Ion–Molecule Reactions
edited by Pierre Ausloos

Volume 41 – Fiber and Integrated Optics
edited by D. B. Ostrowsky

Volume 42 – Electrons in Disordered Metals and at Metallic Surfaces
edited by P. Phariseau, B. L. Györffy, and L. Scheire

Volume 43 – Recent Advances in Group Theory and Their Application to Spectroscopy
edited by John C. Donini

This series is published by an international board of publishers in conjunction with NATO Scientific Affairs Division

A Life Sciences
B Physics — Plenum Publishing Corporation, London and New York

C Mathematical and Physical Sciences — D. Reidel Publishing Company, Dordrecht and Boston

D Behavioral and Social Sciences — Sijthoff International Publishing Company, Leiden

Electrons in Disordered Metals and at Metallic Surfaces

Edited by

P. Phariseau

Seminarie voor Theoretische Vaste Stof-en Lage Energie Kernfysica
Rijksuniversiteit-Gent
Ghent, Belgium

B.L.Györffy

H. H. Wills Physics Laboratory
University of Bristol
Bristol, United Kingdom

and

L. Scheire

Seminarie voor Theoretische Vaste Stof-en Lage Energie Kernfysica
Rijksuniversiteit-Gent
Ghent, Belgium

PLENUM PRESS • NEW YORK AND LONDON
Published in coöperation with NATO Scientific Affairs Division

Library of Congress Cataloging in Publication Data

Nato Advanced Study Institute on Electrons in Disordered Metals and at Metallic Surfaces, State University of Ghent, 1978.

Electrons in disordered metals and at metallic surfaces.

(NATO advanced study institutes series: Series B, Physics; v. 42)
"Proceedings of the NATO Advanced Study Institute on Electrons in Disordered Metals and at Metallic Surfaces, held at the State University of Ghent, Ghent, Belgium, August 28–September 9, 1978."
Includes index.
1. Electronic structure–Congresses. 2. Order-disorder alloys–Congresses. 3. Metals–Surfaces–Congresses. 4. Surface phenomena–Congresses. I. Phariseau, P. II. Gyorffy, B. L. III. Scheire, L. IV. North Atlantic Treaty Organization. V. Title. VI. Series.
QC176.8.E4N338 1978 530.4'1 79-12372
ISBN-13: 978-1-4684-3502-3 e-ISBN-13: 978-1-4684-3500-9
DOI: 10.1007/ 978-1-4684-3500-9

Lectures presented at the NATO Advanced Study Institute on Electrons in Disordered Metals and at Metallic Surfaces, held at the State University of Ghent, Ghent, Belgium, August 28–September 9, 1978.

Softcover reprint of the hardcover 1st edition 1979
A Division of Plenum Publishing Corporation
227 West 17th Street, New York, N.Y. 10011

Preface

We present here the transcripts of lectures and talks which were delivered at the NATO ADVANCED STUDY INSTITUTE "Electrons in Disordered Metals and at Metallic Surfaces" held at the State University of Ghent, Belgium between August 28 and September 9, 1978.

The aim of these lectures was to highlight some of the current progress in our understanding of the degenerate electron 'liquid' in an external field which is neither uniform nor periodic. This theme brought together such topics as the electronic structure at metallic surfaces and in random metallic alloys, liquid metals and metallic glasses. As is the case in connection with infinite ordered crystals, the central issues to be discussed were the nature of the electronic spectra, the stability of the various phases and the occurrence of such phenomena as magnetism and superconductivity.

In the theoretical lectures the emphasis was on detailed realistic calculations based, more or less, on the density functional approach to the problem of the inhomogeneous electron liquid. However, where such calculations were not available, as in the case of magnetism in random alloys and that of metallic glasses, simpler phenomenological models were used.

The theoretical discussions were balanced by reviews of the most promising experimental techniques. Here the stress was on results and their relevance to the fundamental theory. Moreover, the attention had centered on those experiments which probe the electronic structure in the greatest detail.

While the individual contributions are selfcontained accounts of the relevant topics, and no effort has been made to standardize the notations all through the text, cross references are frequent and each is written with evident awareness of the unity of the subject. It is hoped that by bringing together a variety of efforts to deal with the same underlying problem, namely the lack of crystalline symmetry, they would illuminate each other. Furthermore, the juxtaposition was also intended to call attention to the interesting variety of phenomena such broken symmetry can give rise to.

Unfortunately Professor Soven, Dr. Pendry and Dr. Bergmann, whose lectures contributed much to making the summerschool a balanced discussion of the subjects at hand, were unable to prepare their lecture notes for publication. Nevertheless, we have included a short summary of their contributions together with useful lists of suggested readings at the end of this volume (p.553).

The Advanced Study Institute was financially sponsored by the NATO Scientific Affairs Division (Brussels, Belgium). Co-sponsors were the National Science Foundation (Washington, D.C., U.S.A.), the Department of Higher Education and Scientific Research of the Ministry of National Education and Culture (Brussels, Belgium), and the Faculty of Sciences of the State University of Ghent. In particular we are indebted to Dr. T. Kester of the NATO Scientific Affairs Division, Prof. Dr. J. Hoste, President of the University of Ghent and Prof. Dr. R. Mertens, Dean of the Faculty of Sciences.

We are grateful to all lecturers for their most valuable contribution and their collaboration in preparing the manuscripts. Thanks are also due to the members of the International Advisory Board : F. Abelès (Paris, France), S. Berko (Waltham, Mass., U.S.A.), W. Dekeyser (Ghent, Belgium), H. Ehrenreich (Cambridge, Mass., U.S.A.), J.S. Faulkner (Oak Ridge, Tenn., U.S.A.), V. Heine (Cambridge, U.K.), S. Lundqvist (Gothenburg, Sweden) and W. Plummer (Philadelphia, Pa., U.S.A.).

The Institute itself could not have been realized without the enormous enthusiasm of all participants and lecturers and without the untiring efforts of our co-workers Mr. R. Rotthier and Mr. P. Van Steenberge at the "Seminarie voor Theoretische Vaste Stof- en Lage Energie Kernfysica". Also, Mrs. A. Goossens-De Paepe's help in typing the manuscripts is gratefully acknowledged.

P. Phariseau

B.L. Györffy

L. Scheire

Ghent and Bristol, December 1978

Contents

THE DENSITY FUNCTIONAL THEORY OF METALLIC SURFACES

O. Gunnarsson

Institut für Festkörperforschung der Kernforschungsanlage Jülich, D-5170 Jülich, F.R.G.
Institute of Theoretical Physics
S-412 96 Göteborg, Sweden

1. INTRODUCTION

Most calculations on clean surfaces and surfaces with chemisorbed atoms or molecules can be divided into two classes. In the first, one constructs a model Hamiltonian, for example for chemisorption systems the Anderson model |1| is often used. Usually the model is fairly simple and well suited to give a conceptual understanding of important features of the system as well as an indication of the importance of many-body effects|2|. In the second class one uses the density-functional (DF) formalism |3,4| or the Xα method |5|, which can be considered as a special case of the DF formalism. In this approach, ground-state properties are obtained by solving a Hartree-like equation (Eq. (10) below)

$$\{-\frac{\hbar^2}{2m}\nabla^2 + v_{eff}(r)\}\,\psi_\nu(r) = \varepsilon_\nu\,\psi_\nu(r) \qquad (10)$$

The relative simplicity of this equation makes it possible to use a fairly detailed and specific description of the system, without obtaining an unmanageable problem. The crucial quantity in Eq. (10) is $v_{eff}(r)$ which, in principle, contains all many-body effects. As a simple approximation for $v_{eff}(r)$, the so-called local density (LD) approximation, has been found to give generally good results, the method has become very popular. A detailed discussion of the DF formalism and the LD approximation is given in section 2.

The DF formalism has been applied by Lang and Kohn to the so-called planar uniform background model of a surface. They calculated properties such as the charge density, work function and surface energy and obtained good agreement with experiment (in the case of the surface energy the model needed to be refined slightly). Recently, models have been developed which take the atomic structure of the surface more explicitly into account. However, the simpler calculations of Lang and Kohn are still of great importance for our understanding of the surfaces of simple metals, and are described in section 3.

The planar uniform-background model has been used by Lang and Williams and by Gunnarsson, Hjelmberg and Lundqvist to describe chemisorption of atoms on simple metals. In section 4 selected results for H, Li, Si, Cl and Na chemisorbed on Al, Mg and Na are discussed.

2. THE DENSITY FUNCTIONAL FORMALISM

2.1. Basic Theorems

The DF formalism is based on two papers by Hohenberg and Kohn |3| and Kohn and Sham |4|. The basic quantity in this theory is the electron density $n(\mathbf{r})$, and we will first show the relation between the density and other ground-state properties. Assume that N interacting electrons are moving in an external potential $v(\mathbf{r})$, for instance the potential of the nuclei of the system. The Hamiltonian is

$$H = T + U + V \quad ,$$

where T is the kinetic energy operator, U is the electron-electron interaction term and V is the external potential operator corresponding to the potential $v(\mathbf{r})$. In principle, the corresponding density $n(\mathbf{r})$ can be calculated. Thus for a given potential $v(\mathbf{r})$ the density is uniquely determined. The converse statement is less trivial, but has been proven in the following way |3|: Assume that the same density is obtained for a different potential $v'(\mathbf{r})$ which differs from $v(\mathbf{r})$ by more than a trivial constant. The ground-state ψ' for the potential $v'(\mathbf{r})$ is different from the ground-state ψ for the potential $v(\mathbf{r})$, since they satisfy different Schrödinger equations. If the ground-state is nondegenerate |6|, the expectation value of the Hamiltonian has its lowest value for the exact ground-state wave function and

$$\begin{aligned} E' &= \langle\psi'|H'|\psi'\rangle < \langle\psi|H'|\psi\rangle \\ &= \langle\psi|H+V'-V|\psi\rangle = E + \int \left[v'(\mathbf{r})-v(\mathbf{r})\right] n(\mathbf{r})\, d^3r \end{aligned} \qquad (1)$$

However, the primed and unprimed quantities can be interchanged giving

$$E < E' + \int \left[v(\mathbf{r}) - v'(\mathbf{r})\right] n(\mathbf{r})\, d^3r \qquad (2)$$

Adding (1) and (2) we obtain

$$E + E' < E + E'$$

which disproves our assumption that there are two potentials $v(\mathbf{r})$ and $v'(\mathbf{r})$ (differing by more than a constant) which give the same density $n(\mathbf{r})$. Thus there exists only one $v(\mathbf{r})$ which gives rise to the density $n(\mathbf{r})$, and $v(\mathbf{r})$ can be considered a functional of $n(\mathbf{r})$ |8|. Once $v(\mathbf{r})$ is known ψ can, in principle, be calculated and all ground-state properties can be determined. It follows that all ground-state properties are functionals of the density, which is one of the basic results.

Examples of such functionals are the kinetic and electron-electron interaction energies and we call **their sum F[n]**

$$F[n] = \langle\psi |T + U| \psi\rangle \tag{3}$$

For a given potential $v(\mathbf{r})$ we define

$$E_v[n] = \int v(\mathbf{r})\, n(\mathbf{r})\, d^3r + F[n] \tag{4}$$

For the correct ground-state density $n(\mathbf{r})$ the functional $E_v[n]$ is equal to the ground-state energy E. This is actually the lowest value $E_v[n']$ can obtain for any density $n'(\mathbf{r})$ having the correct number of electrons

$$\int n'(\mathbf{r})\, d^3r = N \tag{5}$$

This variational principle is proven as follows |3|: Consider a state ψ' which is not the ground-state of H. Then

$$\begin{aligned} E_v[n] &= \langle\psi|H|\psi\rangle < \langle\psi'|H|\psi'\rangle \\ &= \langle\psi'|V+T+U|\psi'\rangle = \int v(\mathbf{r})\, n'(\mathbf{r})\, d^3r + F[n'] = E_v[n'] \end{aligned} \tag{6}$$

where we have used the definition (3).

To summarize, we have shown that there exists a universal (of $v(\mathbf{r})$ independent) functional $F[n]$ which gives the total energy via Eq. (4) and that the functional $F_v[n]$ in Eq. (4) has its minimum for the correct ground-state density. This gives us a general method for calculating ground-state properties: (a) Find an approximate functional $F[n]$ and (b) minimize $E_v[n]$.

An example of this approach is the Thomas-Fermi method |9|, in which it is assumed that the electrons are independent and that the kinetic energy density is the same as for a homogeneous medium. Then we obtain

$$F[n] = \frac{e^2}{2} \int \frac{n(\mathbf{r})\, n(\mathbf{r}')}{|\mathbf{r}-\mathbf{r}'|}\, d^3r\, d^3r' + C \int [n(\mathbf{r})]^{5/3}\, d^3r \quad ,$$

where C is a numerical constant. To minimize $E_v[n]$, we use the corresponding Euler equation which in this case is the well-known Thomas-Fermi equation |9|.

2.2. Derivation of a Hartree-like Equation

Although the Thomas-Fermi method gives a qualitative picture of, for instance, an atom, it is too crude for detailed quantitative calculations. Actually, to find good approximations for the functional $F[n]$ is very difficult and the method introduced by Kohn and Sham |4| is therefore of great importance. They realized that it is possible to separate (numerically) large contributions to $F|n|$ which can be treated exactly, so that only the smaller reminder has to be treated approximately. Thus they used the partitioning

$$E_v[n] = \int v(\mathbf{r})\, n(\mathbf{r})\, d^3r + \frac{e^2}{2} \int \frac{n(\mathbf{r})\, n(\mathbf{r}')}{|\mathbf{r}-\mathbf{r}'|}\, d^3r\, d^3r' + T_s[n] + E_{xc}[n] \tag{7}$$

The second term is the electrostatic interaction energy and the third is the kinetic energy of <u>noninteracting</u> electrons with the density $n(\mathbf{r})$. The final term is the exchange correlation energy, which would be zero if the electrons were noninteracting. For interacting electrons it contains all the many-body effects. The density is now varied under the constraint

$$\int \delta n(\mathbf{r})\, d^3r = 0$$

to find the minimum of $E_v[n]$. We obtain the Euler equation

$$\int \delta n(\mathbf{r}) \left\{ \frac{\delta T_s[n]}{\delta n(\mathbf{r})} + v_{eff}(\mathbf{r}) \right\} d^3r = 0 \tag{8}$$

with

$$v_{eff}(\mathbf{r}) = v(\mathbf{r}) + e^2 \int \frac{n(\mathbf{r}')}{|r-r'|}\, d^3r' + \frac{\delta E_{xc}[n]}{\delta n(\mathbf{r})} \tag{9}$$

Kohn and Sham observed that Eq. (8) is identical to the equation for noninteracting electrons moving in the potential $v_{eff}(\mathbf{r})$. Therefore we can solve Eq. (8) by using the Hartree method

$$\{-\frac{\hbar^2}{2m} \nabla^2 + v_{eff}(\mathbf{r})\}\, \psi_\nu(\mathbf{r}) = \varepsilon_\nu\, \psi_\nu(\mathbf{r}) \tag{10}$$

$$n(\mathbf{r}) = \sum_{\nu=1}^{N} |\psi_\nu(\mathbf{r})|^2 \tag{11}$$

For a given functional $E_{xc}[n]$ Eqs. (9)-(11) are solved by a self-consistent approach; we guess a density which, via Eq. (9), gives an effective potential and, via Eqs. (10)-(11), a new density.

It should be emphasized that this is a method for calculating ground-state properties. The energy eigenvalues ε_ν and eigenfunctions $\psi_\nu(\underset{\sim}{r})$ of Eq. (10) have not been shown to be excitation energies and wave functions, respectively, but are only auxiliary quantities obtained in the calculation of the density. The quantities that can be calculated are the electron density, the total energy and properties which can be derived from these quantities. For a chemisorbed atom or molecule, for example, we can obtain the binding energy, the equilibrium geometry (i.e. adsorption site and separation distance), the vibration frequencies (if the Born-Oppenheimer approximation is used), the activation energy for diffusion and the dipole moment.

The equation (10) is a one-body equation, i.e. the normal electron-electron interaction term

$$\sum_{i \neq j} \frac{e^2}{|\underset{\sim}{r}_i - \underset{\sim}{r}_j|}$$

does not enter explicitly. This is the main reason to the simplicity of this approach. Furthermore, the potential is local, in contrast to, for example, the Hartree-Fock potential

$$V_{HF}\ \psi(\underset{\sim}{r}) = \int V_{HF}(\underset{\sim}{r},\underset{\sim}{r}')\ \psi(\underset{\sim}{r}')\ d^3r' \tag{12}$$

Actually, if the so-called local density approximation (see below) is used for $E_{xc}[n]$ Eqs. (9)-(11) are not more difficult to solve than the Hartree equations.

The importance of the method of Kohn and Sham can be illustrated by some typical numbers. Fig. 1 shows the partitioning of the valence contribution to the total energy for a manganese atom. The kinetic energy, T_s, the electrostatic interaction between the valence electrons and the core, E_{vc}, and the electrostatic interaction between the valence electrons, E_{vv}, are all treated exactly. Only the exchange energy, E_x, and the correlation energy, E_c, (not shown in the figure, but substantially smaller than E_x) require approximation. Note the contrast to attempts of approximating the functional $F[n]$ directly, e.g., the Thomas-Fermi method, in which case also the large term T_s is treated approximately.

The numbers in Fig. 1 do not mean, of course, that exchange and correlation effects can be neglected. On the contrary, for the calculation of many properties good approximations for E_x and E_c are needed.

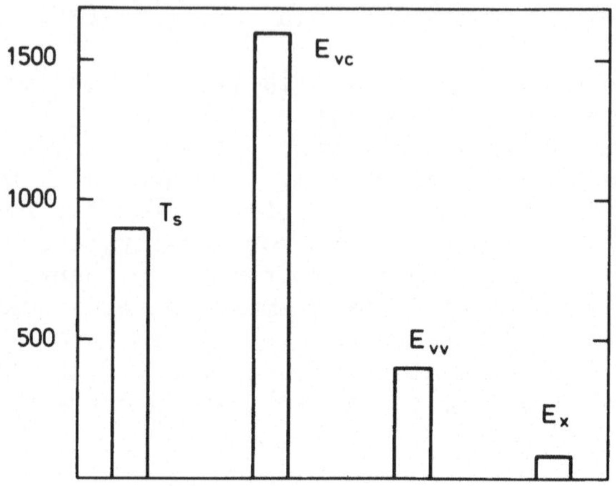

Fig. 1: The kinetic energy of the valence electrons (T_s), the core-valence (E_{vc}) and valence-valence (E_{vv}) electrostatic interaction energies and the valence exchange energy (E_x) for a manganese atom |18|. All energies are given in eV.

2.3. The Local Density Approximation

The most commonly used approximation is the local density (LD) approximation, which assumes that the electron density variations are spatially slow. Consider an electron at a point r in space. If the density varies little over some typical distance, say a few inverse Fermi wave vectors, this electron "sees" an essentially homogeneous medium. Then we can associate an exchange-correlation energy to this electron which is the same as for an electron in a homogeneous medium with the density $n(\underset{\sim}{r})$. For the total exchange-correlation energy we obtain |3,4|

$$E_{xc}[n] \approx \int n(\underset{\sim}{r})\ \varepsilon_{xc}(n(\underset{\sim}{r}))\ d^3r \qquad (13)$$

where $\varepsilon_{xc}(n(\underset{\sim}{r}))$ is the exchange-correlation energy per particle for a homogeneous system with the density $n(\underset{\sim}{r})$. Inserting this approximation in Eq. (9) gives us the exchange-correlation contribution to the effective potential

$$v_{xc}(\underset{\sim}{r}) \equiv \frac{\delta E_{xc}[n]}{\delta n(\underset{\sim}{r})} \quad \frac{\partial}{\partial n(\underset{\sim}{r})} \{n(\underset{\sim}{r})\ \varepsilon_{xc}(n(\underset{\sim}{r}))\} \qquad (14)$$

This potential is not only local in the sense mentioned above (Eq. (12)) but it has also a local density dependence, i.e. $v_{xc}(\underset{\sim}{r})$ depends on the density at the point $\underset{\sim}{r}$ only.

The LD approximation can be extended to spin-polarized systems. The basic quantities are now the density of spin-up and spin-down electrons, $n_+(\underset{\sim}{r})$ and $n_-(\underset{\sim}{r})$, respectively. We obtain |10,11,7| the local spin density (LSD) approximation

$$E_{xc}[n_+,n_-] \approx \int n(\underset{\sim}{r})\ \varepsilon_x(n_+(\underset{\sim}{r}),\ n_-(\underset{\sim}{r}))\ d^3r \qquad (15)$$

with obvious notation. It should be noted that Eq. (15) is not rotationally invariant in spin space.

If correlation effects are neglected in Eq. (13), we get

$$E_{xc}[n] \approx \int n(\underset{\sim}{r})\ \varepsilon_x(n(\underset{\sim}{r}))\ d^3r = \frac{3}{2}\left(\frac{3}{\pi}\right)^{1/3} \int [n(\underset{\sim}{r})]^{4/3}\ d^3r \qquad (16)$$

where ε_x is the exchange energy per particle of a homogeneous system. The corresponding potential (Eq. (14)) is

$$v_{xc}(n(\underset{\sim}{r})) = \mu_x(r_s) = -1.22/r_s \ \text{Ry}$$

where r_s is given by

$$\frac{4\pi}{3}(r_s a_o)^3 = \frac{1}{n(\underset{\sim}{r})}$$

Results for ε_{xc} and v_{xc} including correlation effects have been given in the literature |12,10,7|. For instance, in the absence of spin polarization we have |12|

$$v_{xc}(n(\underset{\sim}{r})) = \beta(r_s)\ \mu_x(r_s)$$

$$\beta(r_s) = 1 + 0.0368\ r_s\ \ln\left(1 + \frac{21}{r_s}\right)$$

The function $\beta(r_s)$ describes correlation effects. It has a fairly weak dependence on r_s, as is shown in Fig. 2.

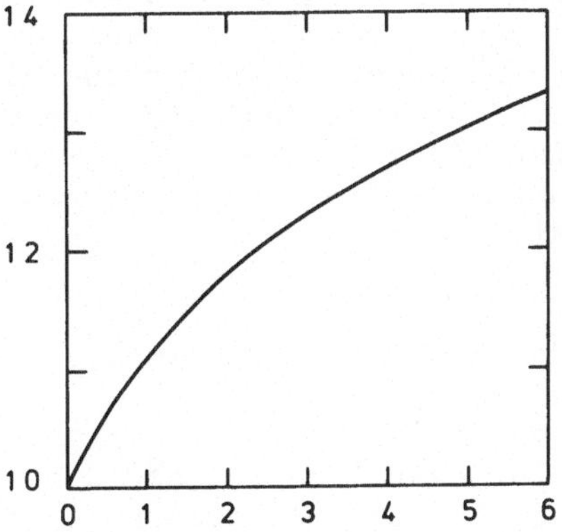

Fig. 2 : The function $\beta(r_s)$ which describes the effects of correlation on the exchange-correlation potential v_{xc}. If correlation is neglected $\beta(r_s) \equiv 1$ and in the Xα approximation $\beta(r_s)$ is replaced by a constant $3\alpha/2$ (after Ref. |12|).

A slightly different approximation, the so-called Xα approximation, was proposed by Slater |5| who, in essence, multiplied Eq. (16) by a constant 3 / 2 and obtained

$$E_{xc}[n] \approx \frac{3}{2} \alpha \int n(\underset{\sim}{r})\ \varepsilon_x(n(\underset{\sim}{r}))\ d^3r$$

This approach gives the so-called Xα potential

$$v_{X\alpha}(n(\underset{\sim}{r})) = \frac{3}{2} \alpha\ \mu_x(r_s)$$

The case α=2/3 is sometimes referred to as the Dirac-Gáspár-Kohn-Sham |13,4| potential, and α=1 gives the Slater potential. Schwarz has proposed the use of an atom-dependent value of α which for most atoms is in the range 0.70-0.75 |14|. If these values of α are used, the Xα and LD approximations give fairly similar results for systems without spin-polarization. However, the spin-dependent version of the Xα approximation gives a stronger spin-dependence than the LSD approximation |15,16| and it overestimates, for instance, the tendency to ferromagnetism for transition metals.

As discussed below, it is not at all clear a priori that approximations (13)-(15) are valid for the systems of interest. It is therefore of great importance that these approximations have now been tested for a large number of systems such as atoms |17,7,18|, small molecules |15,19,20|, simple and transition metals |21-23| and compounds |24|. As examples we show in Fig. 3 the ionization potential of a large number of atoms and in Fig. 4 the binding energy, the equilibrium separation and the vibration frequency of diatomic molecules. The molecular results are discussed in detail by R.O. Jones in this volume.

Generally good agreement with experiment is obtained. For instance, the ionization energy of atoms is typically 1/2 eV in error, the binding energy of small molecules is correct within 2 eV and the cohesive energy of metals within 1/2-1 eV. The errors for the separation distance is typically of the order 1/10 atomic unit. Usually the results are somewhat worse for the 3d series than for other series. This has been discussed extensively by Harris and Jones for atoms and diatomic molecules |25|. However, the magnetic properties of the 3d metals at zero temperatures seem to be described well |16,21-23|.

2.4. Nonlocal Functionals

In the LD approximation $\varepsilon_{xc}(\underset{\sim}{r})$ and $v_{xc}(\underset{\sim}{r})$ depend only on the "local" density $n(\underset{\sim}{r})$. In general there should be a dependence on the density at all points in the neighbourhood of $\underset{\sim}{r}$. Two such approximations were proposed in the original papers |3,4|. The LD approximation can be viewed as the lowest order term in a gradient expansion. Including the next term we obtain |3,4|

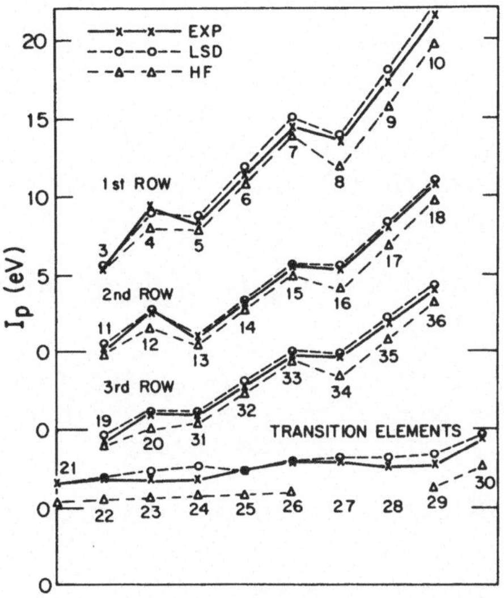

Fig. 3 : The first ionization potential of atoms in the LSD and Hartree-Fock (HF) approximations compared with experiment. The numbers show the atomic numbers of the atoms considered. The zero of energy is shifted 5 eV, 10 eV and 15 eV for the second row, the third row and the transition element series, respectively (after Ref. |18|).

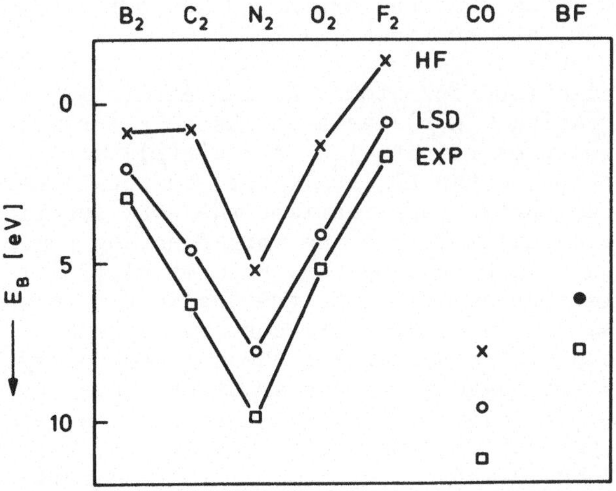

Fig. 4 : The binding energy of first-row molecules in the LSD and HF approximations compared with experiment. The LSD results are calculated in the LMTO method |19| using a limited basis set and larger binding energies would be obtained if the basis set were complete (after Ref. |19|).

$$E_{xc}[n] \approx \int n(\underset{\sim}{r})\, \varepsilon_{xc}(n(\underset{\sim}{r}))\, d^3r + \int C(n(\underset{\sim}{r}))\, \frac{|\nabla n(\underset{\sim}{r})|^2}{[n(\underset{\sim}{r})]^{4/3}}\, d^3r \ , \quad (18)$$

where $C(n(\underset{\sim}{r}))$ is a function |26| of the density. It is also possible to sum all terms which are of second order in the density variations, giving |3,4,27|

$$E_{xc}[n] \approx \int n(\underset{\sim}{r})\varepsilon_{xc}(n(\underset{\sim}{r}))d^3r - \frac{1}{4} \iint K_{xc}(\underset{\sim}{r}-\underset{\sim}{r}',n)(n(\underset{\sim}{r})-n(\underset{\sim}{r}'))^2 d^3r d^3r' \quad (19)$$

where $K_{xc}(\underset{\sim}{r}-\underset{\sim}{r}',n)$ is related to the dielectric function of a homogeneous medium with the density n. The density argument n is in general not uniquely specified by the theory.

The functional (19) is exact in the limit of weak density variations

$$n(\underset{\sim}{r}) = n_o + \Delta n(\underset{\sim}{r}) \quad (20)$$

with

$$|\Delta n(\underset{\sim}{r})| << n_o$$

In this limit it is therefore possible to compare the gradient expansion (18) and the LD approximation (13). It is questionable whether the gradient expansion improves the LD approximation in this limit for density variations having wave lengths typical for most systems |28,29|. In the more realistic situation of strong density variations the gradient expansion has primarily been applied to atoms |30| and surfaces |31-34|.

In the calculations for atoms, Herman et al |30| obtained improvements by treating $C(n(\underset{\sim}{r}))$ as an adjustable parameter. However, if the first principles results |26| for $C(n(\underset{\sim}{r}))$ are applied to atoms the correction to the LD result for the total energy has the wrong sign. In the surface applications the work function and surface energy were calculated. For the work function numerically very different results have been obtained. Rose et al |33| found an almost negligible correction due to the gradient term while Lau and Kohn |31| obtained a substantial contribution which made the agreement with experiment worse. For the surface energy Lau and Kohn |31| found worse agreement when gradient terms are included, while Rose et al |33| claimed improved agreement. As the experimental results for the surface energy are fairly uncertain Perdew et al |35| instead considered a simple model for the surface, where the contributions from different wave vectors were considered. The results for the surface energy were compared with the ones of the gradient expansion and Perdew et al concluded that the gradient expansion gives too large corrections and is inappropriate for surfaces.

The expansion to second order in the density variations (19) is exact for densities of the type (20) in the limit $|\Delta n|/n \to 0$. In this case there is no restriction on how rapid the variations can be spatially. The formula is derived for a density of the type (20) and the density argument in the kernel $K(\mathbf{r}-\mathbf{r}',n)$ is then n_o. However, for realistic density distributions with strong density variations, the theory does not prescribe the choice of density argument, and various possibilities have been proposed such as (i) $n((\mathbf{r}+\mathbf{r}')/2)$ and (ii) $(n(\mathbf{r})+n(\mathbf{r}'))/2$. For a density of the type (20) these two choices give identical results but for a more realistic density they differ crucially. For instance, the energy of an atom and the surface energy in the jellium model are finite if (ii) is used but infinite if (i) is applied |29|. These two choices of density argument are by no means the only plausible ones and therefore it seems that (19) can give results over a wide range, making the approach less useful.

2.5. Justification of the LD Approximation

The LD approximation and the gradient expansion have been derived under the assumption that the density is slowly varying. This can be expressed as |3|

$$\frac{1}{k_F(n(\mathbf{r}))}\left|\frac{\nabla n(\mathbf{r})}{n(\mathbf{r})}\right| << 1 \quad , \tag{21}$$

where $k_F(n(\mathbf{r}))$ is the Fermi wave vector of a homogeneous system with the density $n(\mathbf{r})$. Equation (21) requires that the relative density variation over the distance $k_F(n(\mathbf{r}))^{-1}$ is much smaller than unity. This variation |29| is shown for bulk copper in Fig. 5a and for a surface in the planar uniform background model (see below) in Fig. 5b. These figures show that (21) is of the order of unity or larger,

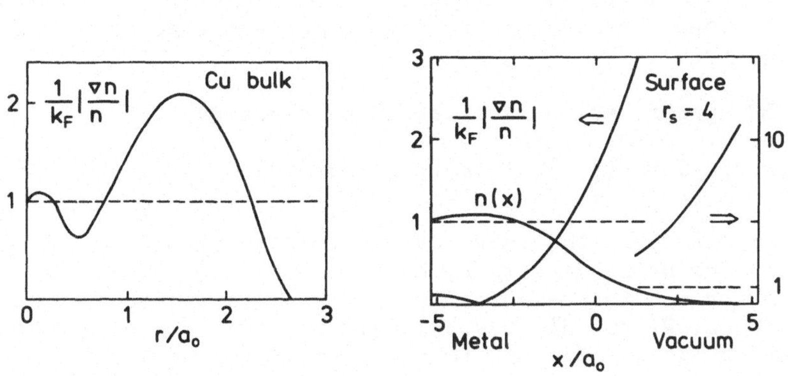

Fig. 5: The relative density variation (see Eq. (21)) for bulk copper as a function of the distance from the closest nucleus and for the planar uniform-background model of a surface. n(x) is the electron density (after Ref. |29|).

and the assumption under which these approximations have been derived is therefore invalid. Similarly, the condition $|\Delta n(\mathbf{r})| << n_o$ (Eq. (20)) for the second order expansion (19) is violated for most systems of interest. Nevertheless the LD approximation has been found to give good results, as discussed above. This is particularly surprising since the formally better gradient expansion does not seem to improve the LD approximation and the second order expansion, containing a partial infinite gradient summation, is worse for the cases investigated |29|.

To discuss this apparent contradiction, we use an exact formula |7,36| for $E_{xc}[n]$

$$E_{xc}[n] = \int d^3r \int d^3r' \frac{n_{xc}(\mathbf{r}, \mathbf{r}-\mathbf{r}')}{|\mathbf{r}-\mathbf{r}'|} \tag{22}$$

where

$$n_{xc}(\mathbf{r}, \mathbf{r}-\mathbf{r}') = n(\mathbf{r}') \int_0^{e^2} d\lambda \, [g(\mathbf{r},\mathbf{r}';\lambda) - 1] \tag{23}$$

The pair correlation function $g(\mathbf{r},\mathbf{r}';\lambda)$ is the exact one calculated for a system with the physical density $n(\mathbf{r})$ but with the electron-electron interaction reduced to $\lambda/|\mathbf{r}_i-\mathbf{r}_j|$. Formula (22) is obtained by using the Pauli trick of differenting the interaction energy with respect to λ and then integrating.

The factor $|g(\mathbf{r},\mathbf{r}';\lambda)-1|$ describes how the relative probability of finding two electrons close to each other is reduced due to the Pauli principle and the Coulomb interaction, i.e. exchange and correlation effects. If there is an electron at the point $\mathbf{r}$ the quantity $n_{xc}(\mathbf{r},\mathbf{r}')$ essentially describes how, on the average, the number of electrons at a point $\mathbf{r}'$ is reduced relative to an independent particle picture. We can view this as a hole in the electron density accompanying the electron. The interaction between the electron and the hole gives rise to the exchange-correlation energy (Eq. (22)).

The LD approximation is obtained by assuming

$$n_{xc}(\mathbf{r},\mathbf{r}'-\mathbf{r}) = n(\mathbf{r}) \int_0^{e^2} d\lambda \, [g_h(\mathbf{r}-\mathbf{r}';\lambda;n(\mathbf{r})) - 1] \tag{24}$$

where $g_h(\mathbf{r}-\mathbf{r}';\lambda;n)$ is the pair correlation function of a homogeneous medium with the density n. We note that once we have made an assumption for n_{xc} the calculation of $E_{xc}[n]$ is a simple electrostatic problem. We now discuss this electrostatic problem to see how sensitive the results are to the approximations for n_{xc}.

Using the variable substitution $\mathbf{R}=\mathbf{r}-\mathbf{r}'$ Eq. (22) becomes

$$E_{xc}[n] = \int d^3r \, n(\mathbf{r}) \int_0^{\infty} R \, dR \int d\Omega \, n_{xc}(\mathbf{r},\mathbf{R}) \tag{25}$$

Thus $E_{xc}[n]$ only depends on the spherical average |7|

$$n_{xc}^{s.a.}(\underset{\sim}{r},R) = \frac{1}{4\pi} \int_{|\underset{\sim}{r}'|=R} d^3r' \; n_{xc}(\underset{\sim}{r},\underset{\sim}{r}')$$

This is very important as in the LD approximation (24) the hole is assumed spherically symmetric. Equation (24) shows that neglecting the non-spherical components of the correct n_{xc} is of no importance as these integrate to zero. This is illustrated in Fig. 6, where we show the exchange-correlation hole for a nitrogen atom. We note that the exact and approximate holes are very different. Most of these differences are, however, due to the unimportant non-spher-

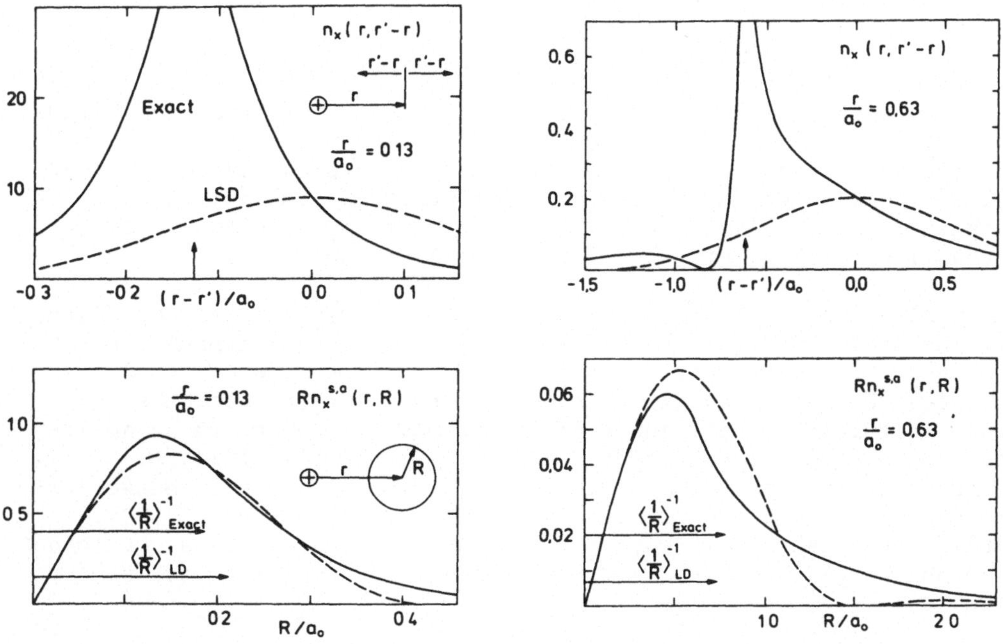

Fig. 6: The exact (full line) and LSD (broken line) exchange hole for a spin-up electron in a nitrogen atom for r=0.13 and 0.63 a.u. In the top figure the axis is through the nucleus and the point r, i.e. r' is parallel to r and r'-r=0 gives the electron position. The nuclear position is marked by an arrow. We note that the exact hole has a large weight at the nucleus while the LSD hole is centered at the electron. The lower part of the figure shows the spherical average around the electron position of the exchange correlation hole. The area under the curve is proportional to the exchange energy. The figure also shows the results for <1/R> defined in Eq. (27).

ical components and the exact and approximate spherical parts are fairly similar.

In addition, there is a sum rule |7| for the spherically symmetric part of the hole

$$4\pi \int_0^\infty R^2 \, dR \, n_{xc}^{s.a.}(\mathbf{r},R) = - e^2 \qquad (26)$$

stating that the hole should contain exactly one electron. This sum rule is fulfilled in the LD approximation and it leads to a systematic partial cancellation of positive and negative errors, as illustrated in Fig. 6. The conclusion is that E_{xc} is very insensitive to the shape and the detailed form of the hole if the sum rule (26) is satisfied.

Using the sum rule, we can alternatively view the integral

$$\int d^3R \, \frac{n_{xc}(\mathbf{r},\mathbf{R})}{|\mathbf{R}|} \equiv -e^2 < \frac{1}{R} > \qquad (27)$$

as a weighted average of R^{-1}, where $n_{xc}(\mathbf{r},\mathbf{R})$ is a normalized weight function |7|. The quantity 1/<1/R> can be considered as the "size" of the hole. The analysis above shows that we actually do not need to know the very complicated function $g(\mathbf{r},\mathbf{r}';\lambda)$ but only have to make a reasonable assumption for <1/R>. Fig. 6 illustrates how insensitive this quantity is to the details of the hole.

Although one could find approximate $n_{xc}(\mathbf{r},\mathbf{r}')$ which would lead to the gradient expansion (18) and the second order expression (19), no such expressions have been proposed and there are no natural choices. Definite statements about whether, for example, the sum rule is fulfilled can therefore not be made. However, in an application to copper metal using the density argument $((\mathbf{r}+\mathbf{r}')/2)$ it was found that most of the 29 electrons contribute to the exchange-correlation energy of an electron at the Wigner-Seitz radius |29|. This unphysical result suggests that the failure of the second order expression (19) results from not properly incorporating the sum rule (26).

The justification of the LSD approximation therefore does not lie in considering it as the first term in a rapidly converging gradient expansion. As we have seen the second order expression (19), which is formally obtained by a partial infinite summation of gradient terms, can lead to divergent results in cases for which the LSD approximation gives good results. The LSD approximation should rather be considered as an approximation in its own right, which fulfils the sum rule (26) and is a sensible way of estimating the hole "size" (27).

2.6. Excitation Energies

The scheme we have discussed so far has been derived for the ground-state. The eigenvalues ε_ν (Eq. (10)) have not been shown to be excitation energies and the calculation of these energies has a much less firm theoretical basis than the ground-state calculations. In some cases the calculations are theoretically justified but in most cases the support is more an empirical one. In this section, some of the methods will be reviewed and the support - theoretical or empirical - will be discussed.

There are basically two classes of methods. In the first class the ground-state scheme is applied to both the excited state and the ground state and the energy difference is calculated. This approach is often called the ΔSCF method, since we use the energy difference between two self-consistent (field) (SCF) calculations. The second class of methods uses the Dyson's equation and the fact that the self-energy is a functional of the density.

As an example of the first class we consider the calculation of the ionization potential, I_p, of an atom. We have

$$I_p = E_{N-1} - E_N \quad , \tag{28}$$

where E_N and E_{N-1} are the ground-state energies of the neutral atom and the ion, respectively. Thus the use of the ground-state scheme for the calculation E_N and E_{N-1} is fully justified. In Fig. 3 we show that the results obtained using Eq. (28) are in good agreement with experiment.

There are few cases for which the excitation energy can be expressed as the difference between two ground-state energies. However, the DF formalism has been extended also to a certain class of excited states, namely the lowest state for each set of quantum numbers |7|. Consider, e.g., the excitation 1s→2p in a hydrogen atom. We have

$$\Delta E(1s\rightarrow 2p) = E(2p) - E(1s) \quad , \tag{29}$$

with obvious notation. As the 2p state is the lowest state with p symmetry, the ground-state scheme can be used to calculate both E(2p) and E(1s).

The exchange-correlation functional in this generalized DF formalism depends in principle on the quantum numbers of the state |7|, but since the correct functional is not known the LSD approximation has often been used for all states |7,19|. Recently, Ziegler and Rauk |37| and von Barth |38| have shown that the LSD approximation works well for states for which the correct symmetry can be obtained using one single determinant. For other states they proposed a new method and obtained generally good results.

For small systems, such as atoms and small molecules, quite a few states can be treated with the methods described above. For larger systems, in particular infinite systems, this is not the case and we are forced to give up the restriction of only applying the formalism to the lowest state of each symmetry. It should be emphasized that this has not been justified theoretically. Nevertheless the results obtained are in many cases in good agreement with experiment.

For localized excitations, we can use a method applied by Zunger and Freeman |24| to a LiF crystal. For this system excitonic like excitations are important and the interaction between the excited electron and the hole left behind cannot be neglected. We first note that for the crystal in its ground-state all the solutions ψ_ν of Eq. (10) are Bloch type states, extending over the whole system. We can create an excitation by leaving one of the ψ_ν in the valence band empty and instead filling a level in the conduction band. Then both the particle and the hole state are extended and the strong electron-hole interaction typical for excitons is not present. To treat exciton-like excitations, Zunger and Freeman therefore introduced a large unit cell containing 8 or 16 atoms. For instance, to describe an exciton with a hole in the fluorine 2p-like band and an electron in the lithium 2s-like band, they prepared an initial potential by having one electron less on one of the F atoms in the unit cell, the missing electron being on one of the neighbouring Li atoms. The corresponding (non self-consistent) potential was iterated to self-consistency under the restriction that the F 2p band and the Li 2s band have one hole and one electron per unit cell, respectively. Although the scheme allows the initially localized perturbation to spread out, the calculation showed that in this case the excitation remains localized. The excitation energy is obtained as the difference between the total energy of the excited system and the system in its ground state. A similar method was applied to calculate ionization energies. The results of Zunger and Freeman are in good agreement with experiment.

In the methods discussed above the excitation energy is the difference between the total energies of two ground-state calculations. This difference is usually much smaller than the total energy and a high numerical accuracy is therefore required. The transition state technique introduced by Slater |5| is therefore very useful. We describe this technique by essentially following Williams et al |39|, and to be specific we consider the ionization energy of an atom. Set $E(n_1,n_2,\ldots,n_N)$ be the total energy of the atom with n_i electrons in the orbital i. The ionization energy is then given by

$$I_p = E(n_1-1,n_2,\ldots,n_N) - E(n_1,n_2,\ldots n_N) \qquad (30)$$

if the electron is removed from orbital 1. Noting that $E(n_1,n_2,\ldots,n_N)$ can be defined also when $\{n_i\}$ are noninteger numbers we obtain

$$I_p = \int_0^1 dn \frac{\partial E(n_1-n, n_2, \ldots, n_N)}{\partial n} \quad (31)$$

However,

$$\frac{\partial E(n_1-n, n_2, \ldots, n_N)}{\partial n} = -\varepsilon_1 \quad (32)$$

where ε_1 is the energy eigenvalue for orbital 1 |40,5| giving

$$I_p = -\int_0^1 dn\, \varepsilon_1(n_1-n, n_2, \ldots, n_N) \quad (33)$$

We note that ε_i depends on the occupancy $\{n_i\}$. Using the "midpoint formula" for integrals we find

$$I_p \approx -\varepsilon_1(n_1-\tfrac{1}{2}, n_2, \ldots, n_N) \quad , \quad (34)$$

i.e. we have to perform a self-consistent calculation for a system where half an electron has been removed. The arguments above are trivially generalized to the case in which an electron is excited from an orbital i to orbital j and the excitation energy is then given by

$$\varepsilon_j(n_1,\ldots,n_i-\tfrac{1}{2},\ldots,n_j+\tfrac{1}{2},\ldots) - \varepsilon_i(n_1,\ldots,n_i-\tfrac{1}{2},\ldots,n_j+\tfrac{1}{2},\ldots) \quad (35)$$

The main advantage of this approach is that the eigenvalues ε_i and ε_j are usually numerically much smaller than the total energy and the numerical accuracy required is thus lower.

We now discuss the second class of methods for calculating excitation energies, in which the Dyson's equation |41| is the starting point. This approach is based on a paper by Sham and Kohn |42|. The excitation energies are given by the eigenvalues E_k of the Dyson's equation

$$\{-\frac{\hbar^2}{2m}\nabla^2+V(\underset{\sim}{r})\}\phi_k(\underset{\sim}{r}) + \int \Sigma(\underset{\sim}{r},\underset{\sim}{r}';E_k)\phi_k(\underset{\sim}{r}')d^3r' = E_k\,\phi_k(\underset{\sim}{r}) \quad (36)$$

where $V(\underset{\sim}{r})$ is the electrostatic potential from the electronic and nuclear charge and $\Sigma(\underset{\sim}{r},\underset{\sim}{r}';E_k)$ is the self-energy operator |41|. Sham and Kohn |42| observed that the self-energy is a ground-state property and therefore a functional of the density. The self-energy is a nonlocal complex and energy-dependent quantity, which in general is very difficult to calculate. In addition, Eq. (36) is complicated to solve even when Σ is known. Therefore, two approximations are normally introduced.

Firstly, the self-energy is replaced by the self-energy of a homogeneous medium. Hedin and Lundqvist |12| advocate the approximation

$$\Sigma(\underset{\sim}{r},\underset{\sim}{r}';E) \approx \Sigma_h(\underset{\sim}{r}-\underset{\sim}{r}',E-\mu+\mu_h(n(\underset{\sim}{r}_o)),n(\underset{\sim}{r}_o)) \qquad (37)$$

where Σ_h is the self-energy of a homogeneous system with the density $n(\underset{\sim}{r}_o)$, with $\underset{\sim}{r}_o=(\underset{\sim}{r}+\underset{\sim}{r}')/2$. The quantities μ and $\mu_h(n)$ are the chemical potentials of the system under consideration and a homogeneous system, respectively. This is the equivalent of the LD approximation and Eq. (37) is exact in the limit of slowly varying density. Although Σ_h is fairly well known, the solution of Eq. (36) is still complicated by the non locality of Σ_h. Therefore a WKB-type approximation can be introduced and we write |42|

$$\phi_k(r) \approx A(\underset{\sim}{r}) \exp\{i\, \underset{\sim}{p}(\underset{\sim}{r})\cdot\underset{\sim}{r}\} \qquad (38)$$

If the $\underset{\sim}{r}$ dependence of $A(\underset{\sim}{r})$ and $\underset{\sim}{p}(\underset{\sim}{r})$ is small over the non locality range of Σ_h |43| we obtain

$$\int \Sigma_h(\underset{\sim}{r}-\underset{\sim}{r}',E-\mu+\mu_h(n(\underset{\sim}{r}_o));n(\underset{\sim}{r}_o))\phi_k(\underset{\sim}{r}')d^3r' \qquad (39)$$

$$\approx \Sigma_h(p(\underset{\sim}{r}),E-\mu+\mu_h(n(\underset{\sim}{r}_o)),n(\underset{\sim}{r})\ \phi_k(\underset{\sim}{r})$$

For the calculation of $p(\underset{\sim}{r})$ in Eq. (39) we use the prescription |42,12|

$$E(p,n) = E_k - \mu + \mu_h(n) \quad , \qquad (40)$$

where $E(p,n)$ is the energy of a quasi-particle of momentum p in a homogeneous medium with the density n. This gives us a potential for the calculation of excitation energies

$$V_{xc}(r) = \Sigma_h(p(\underset{\sim}{r});E(p(\underset{\sim}{r}),n(\underset{\sim}{r}));n(\underset{\sim}{r})) \quad , \qquad (41)$$

and the problem has been reduced to the solution of a Hartree-like equation

$$\{-\frac{h^2}{2m}\nabla^2 + V(\underset{\sim}{r}) + V_{xc}(\underset{\sim}{r})\}\ \phi_k(\underset{\sim}{r}) = E_k\ \phi_k(\underset{\sim}{r}) \qquad (42)$$

However, it should be emphasized that the approximations (37) and (39) have been introduced and that the range of validity of Eq. (42) is not clear.

Close to the Fermi wave vector V_{xc} (Eq. (41)) has a fairly weak momentum dependence. If $p(\underset{\sim}{r})$ is replaced by the Fermi wave vector the ground-state potential in the LD approximation is recovered. If this additional approximation is used the excitation energies E_k of Eq. (41) are identical to the eigenvalues ε_ν in the ground-state scheme. This approach is used in almost all band calculations and for metals it gives fair agreement with experiment, although discrepancies are known. For instance, the d band tends to be too high relative to the sp band for, e.g., Cu |44|, Zn |45| and Pd |21|. On the other hand, for systems with localized states, e.g. atoms the eigenvalues differ substantially from the excitation energies. For instance, for the ionization potential of atoms this procedure gives errors of the order 2-10 eV, while the proper method of using Eq. (28) is correct within about 1/2 eV. One reason for this large discrepancy is that there is no Koopman's theorem in the DF formalism. If, for example, the energy difference $E_{N-1}-E_N$ in Eq. (28) is calculated by removing one electron from the N electron system keeping all orbitals frozen, the result can be quite different from the corresponding eigenvalue. The two procedures give the same result only if the electron is removed from an extended orbital. An empirical relation between the degree of localization, measured by the width of the band, and the deviation between the eigenvalue and the excitation energy is suggested by the following numbers |46|: For broad free-electron like bands the errors are small, while the deviation is about 0.5 eV, 1.5-2 eV and 6 eV for the 5 eV broad d band of Pd |21|, the 1.5 eV broad Zn d band |45| and the sharp d level of a Zn atom, respectively.

The difficulties that may arise from using the eigenvalues can be illustrated for the case of hydrogen atom chemisorbed on an aluminum surface. For the hydrogen atom far from the surface, we find a self-consistent solution with a, say, spin-up electron on the H atom. This electron is located in a very narrow resonance about 7.6 eV below vacuum, which is only half the energy required to remove an electron from the H atom. To use the transition state technique (34) or the ΔSCF approach (28) gives the same result, since all the states involved are extended. Thus none of the methods seem to work in this limit. As the atom is brought closer to the surface the resonance is broadened and the interpretation of the eigenvalues as excitation energies becomes more credible. To obtain a better understanding of this problem the cases in which the atom is far inside the metal and in the surface region have been investigated by Vinter |47| and by Hjelmberg |48|, respectively.

Vinter |47| considered a proton in a homogeneous electron gas and calculated the change in the density of states due to the proton. He used the Sham-Kohn method (Eq. (36)) together with approximation (37). However, he avoided the approximation (39). In Fig. 7, his results are compared with the eigenvalue spectrum obtained by Nørskov |49|, using the eigenvalues of the Kohn-Sham equation (10). The results are qualitatively similar and in particular the split-off state below the bottom of the band has almost the same position in the two calculations.

Hjelmberg |48| performed a calculation for a proton slightly inside the surface of a sodium substrate in the jellium model. For the proton 3 a.u. inside the jellium edge, there is a split-off state below the bottom of the band. In a transition state calcula-

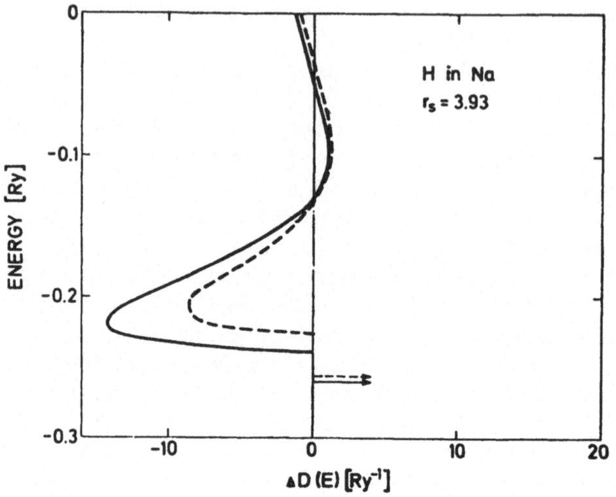

Fig. 7 : The induced density of states ΔD(E) for a proton in an electron gas with the same density as sodium. The full curve shows the eigenvalue spectrum of Nørskov |49| and the dashed curve the results of Eqs. (36),(37). All energies are measured relative the chemical potential (after Ref. |47|).

tion this level was shifted 0.6 eV. Provided that the transition state calculation gives a reasonable estimate of the excitation energy, this small shift suggests that the eigenvalue spectrum is relevant for the case of interest, namely the adatom being in the surface region.

3. CLEAN SURFACES OF SIMPLE METALS

Most early attempts to describe surfaces of simple metals used the so-called planar uniform-background model |50-56|, in which the ions are smeared out to give a positive background. In this section we shall focus on the work by Lang and Kohn |53-55|, which has been discussed in a review by Lang |56|. Recently developed methods which introduce the atomic structure non perturbatively |57-59| fall outside the scope of this section.

3.1. The Planar Uniform-Background Model

It is well known that bulk properties of simple metals can be well described using pseudopotentials |60|. As the pseudopotentials are relatively weak, they may be neglected entirely and replaced by a uniform positive background, which serves the purpose of keeping the system neutral. This so-called electron gas (or jellium) model gives a qualitative description of many simple metals. To apply it to a surface, we must decide where to cut the positive background. Sometimes the corrugated-background model is used, in which the geometrical arrangement of the surface atoms is imitated by cutting the positive background along a corrugated surface. Mostly, however, this cut is made along a plane surface, giving the uniform-background model. In this model, the system is translationally invariant parallel to the surface (the y and z directions) and the positive background charge density $n_+(x)$ is constant up to the surface (x=0) and then zero, i.e.

$$n_+(x) = \begin{cases} \bar{n} & x \leq 0 \\ 0 & x > 0 \end{cases} \tag{44}$$

The density $\bar{n}$ is given by the average electron density of the bulk metal. Usually the parameter r_s, defined by

$$\frac{4\pi}{3}(r_s a_o)^3 = \frac{1}{\bar{n}} \quad , \tag{45}$$

(a_o is the Bohr radius), is used to describe the metal. We note that $r_s a_o$ is the radius of a sphere which on the average contains one electron.

This model gives rise to an essentially one-dimensional problem. The effective potential in the DF formalism (Eq. (9)) consists of an electrostatic potential $\phi(x)$ from the positive background and the electrons and an exchange-correlation potential $v_{xc}(x)$ (Eq. (14)). The general shape of $\phi(x)$ and $v_{eff}(x)$ is shown schematically in Fig. 8. As discussed earlier, the exchange-correlation potential

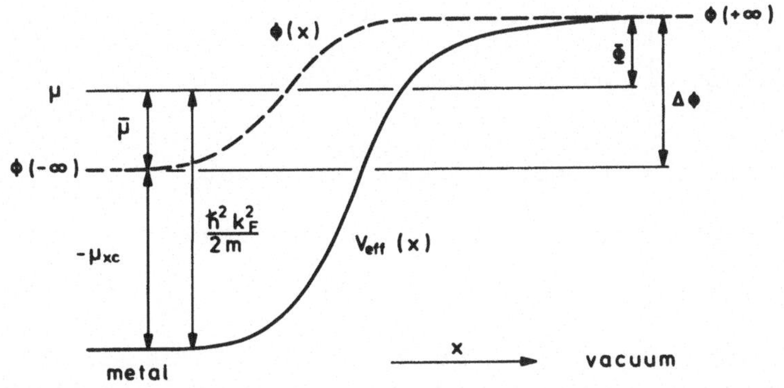

Fig. 8 : Schematic representation of the electrostatic (ϕ) and effective (v_{eff}) potentials in the planar uniform-background model.

$v_{xc}(\underset{\sim}{r})$ can be considered as the interaction between the electron and its surrounding exchange-correlation hole. Deep inside the surface the hole is, in this model, spherically symmetric and centered around the electron. For an electron closer to the surface the hole is distorted and has most of its weight on the metal side of the electron. Finally, if an electron is outside the surface (on the vacuum side), the hole is left behind and localized to the surface. Obviously the interaction between the electron and the hole is small in this limit and should have a 1/x dependence. Thus as the electron moves through the surface $|v_{xc}|$ is reduced and for large distances it goes over to the image type potential $v_{xc}(x) \approx -1/(4x)$ a.u. Since the effective potential is above the Fermi energy μ for large positive x, the wave functions and the density decay exponentially outside the surface. Deep inside the metal the density should go to a constant, although, due to the abrupt change in potential in the surface region, there are Friedel oscillations which decay only as x^{-2} when $x \to -\infty$.

This model has been solved |62| using the Thomas-Fermi |9| approximation, which reproduces the general features of the charge density. However, it fails in several respects and gives, for example, a vanishing work function and a negative surface energy. The Friedel oscillations inside the surface are also absent. The difficulties in the Thomas-Fermi model were removed in part by Smith |51|, who introduced a gradient correction to the kinetic energy expression and included exchange and correlation effects within the LD approximation. He obtained results for the work function in fair agreement with experiment, while the surface energy came out too small.

3.2. Self-Consistent Wave-Mechanical Solution

Instead of using a Thomas-Fermi like approximation for the kinetic energy as in the previous section, Bennett and Duke |52| and Lang and Kohn |53-55| treated the kinetic energy exactly by using the scheme of Kohn and Sham described in section 2.2. (Eqs. (9)-(11)). Except for the LD approximation (13) for exchange and correlation, the work of Lang and Kohn gives an essentially exact solution of the planar uniform-background model.

In this approach, one has to solve Eq. (10)

$$\{-\frac{\hbar^2}{2m}\nabla^2 + v_{eff}(\underset{\sim}{r})\}\,\psi_\nu(\underset{\sim}{r}) = \varepsilon_\nu\,\psi_\nu(\underset{\sim}{r}) \qquad (10)$$

Due to the translational invariance parallel to the surface the effective potential depends only on the coordinate x normal to the surface, and the solutions can be written

$$\psi_{k,k_y,k_z}(\underset{\sim}{r}) = \psi_k(x)\,\exp\,[i(k_y\,y + k_z\,z)] \qquad (46)$$

The function $\psi_k(x)$ is calculated numerically. Far inside the metal the potential is constant and the solution takes the form $(x \to -\infty)$

$$\psi_k(x) = \sin\,[kx - \gamma(k)] \quad , \tag{47}$$

where $\gamma(o)=0$. Putting the Fermi energy equal to zero, the eigenvalues are given by

$$\varepsilon_{k,k_y,k_z} = \frac{\hbar^2}{2m}(k^2 + k_y^2 + k_z^2 - k_F^2) \tag{48}$$

and Eq. (10) takes the form

$$\{-\frac{\hbar^2}{2m}\frac{d^2}{dx^2} + v_{eff}(x)\}\,\psi_\nu(x) = \frac{\hbar^2}{2m}(k^2-k_F^2)\,\psi_k(x) \tag{49}$$

The electron density is

$$n(x) = \frac{1}{\pi^2}\int_o^{k_F}(k_F^2-k^2)\,|\psi_k(x)|^2\,dk \tag{50}$$

The effective potential is determined by the density (Eq. (9),(13)) and Eqs. (9),(13),(49),(50) are solved self-consistently. We note that improved convergence can be obtained by using a method of Arponen et al |63|.

In Fig. 9 we show the results obtained by Lang and Kohn for the charge density. The results are calculated for r_s=2,4 and 6, which are close to the r_s values of Al, Na and Cs, respectively. On the vacuum side the charge density decays exponentially as it should.

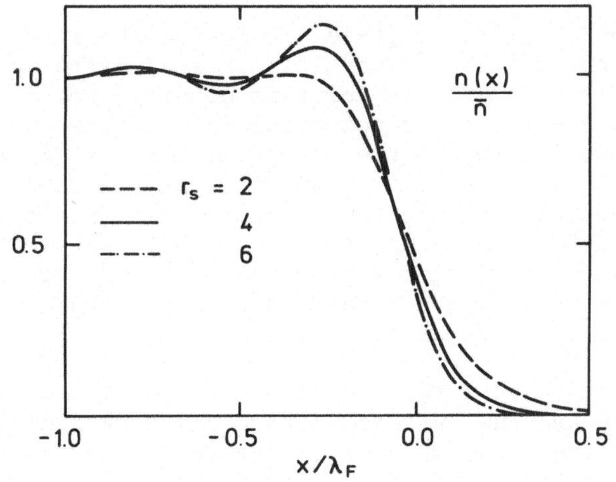

Fig. 9 : The charge density for r_s=2,4, and 6 in units of the bulk density. The distance is measured in units of the Fermi wavelength $\lambda_F=2\pi/k_F \approx 3.3\,r_s a_o$.

Inside the metal surface, Friedel oscillations of the form

$$A \cos(2k_F x - 2\gamma(k_F)) \,/\, x^2 \quad , \tag{51}$$

are present as expected. The amplitude increases rapidly with r_s.

The results for the electrostatic and effective potentials are shown in Fig. 10. We note that, for $r_s=4$, the electrostatic potential has a fairly weak variation and that most of the surface

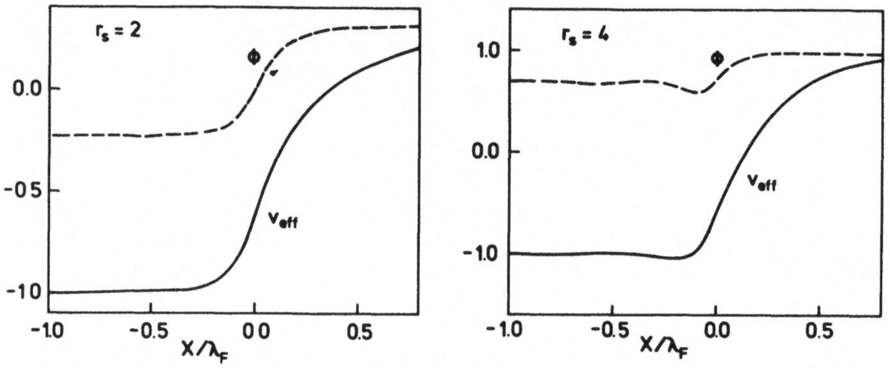

Fig. 10 : The electrostatic (ϕ) and effective (v_{eff}) potentials for $r_s=2$ and $r_s=4$. The potentials are given in units of the Fermi energy which is 0.92 Ry and 0.23 Ry for $r_s=2$ and $r_s=4$, respectively. The Fermi level is set equal to zero.

potential barrier ($v_{eff}(\infty)-v_{eff}(-\infty)$) is due to the strong variation of the exchange-correlation potential. This illustrates the great importance of many-body effects for this density. For $r_s=2$ the variation in the electrostatic potential is stronger and it gives a substantial contribution to the potential barrier. As mentioned earlier $v_{xc}(x)$ should decay as $-1/(4x)$ for large values of x, while it has an exponential decay in the LD approximation. To test the effects of this deficiency, Lang and Kohn |53| performed a check calculation in which they used a potential with the correct asymptotic form for intermediate and large distances. They found the effects on the density to be small.

We shall next discuss two measurable properties of the surface, the work function and the surface energy. The work function Φ is the minimum work that must be done to remove an electron from the crystal. Clearly an electron at the Fermi energy should be removed and Fig. 10 shows that

$$\Phi = \phi(+\infty) - \mu \tag{52}$$

Since the electrostatic potential is different outside the different faces of a crystal, the work function is more rigorously defined as the energy required to remove an electron a distance from the surface which is large compared with atomic dimensions but small compared with the geometrical extent of the surface. Defining the electrostatic barrier

$$\Delta\phi = \phi(+\infty) - \phi(-\infty) \tag{53}$$

we obtain (see Fig. 8)

$$\Phi = \Delta\phi - \bar{\mu} \quad , \tag{54}$$

where $\bar{\mu}$ is the chemical potential of a homogeneous system with density $\bar{n}$. We have used the fact that

$$\mu = \bar{\mu} + \phi(-\infty) \tag{55}$$

We note that $\bar{\mu}$ can be portioned into a kinetic part, $h^2k_F^2/2m$, and an exchange-correlation part, μ_{xc},

$$\bar{\mu} = \frac{\hbar^2 k_F^2}{2m} + \mu_{xc} \tag{56}$$

The results for the work function Φ and its two components $\Delta\phi$ and $\bar{\mu}$ are shown in Fig. 11. For large values of r_s (e.g., K, Rb and Cs) the electrostatic barrier $\Delta\phi$ is almost negligible, while it is

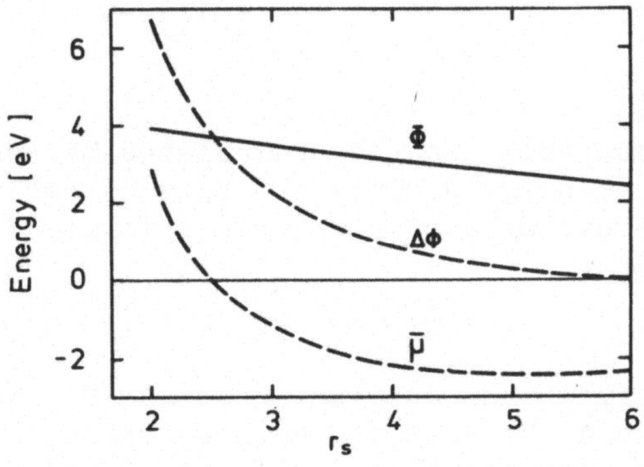

Fig. 11 : The work function (Φ) and its two components, the electrostatic barrier ($\Delta\phi$) and the chemical potential ($\bar{\mu}$).

large for the high density metals (e.g., Al with r_s=2). While $\Delta\phi$ and $\bar{\mu}$ each vary strongly with r_s, the work function has a fairly weak dependence on r_s. In Fig. 12 the calculated work function is compared with experiment and found to agree fairly well.

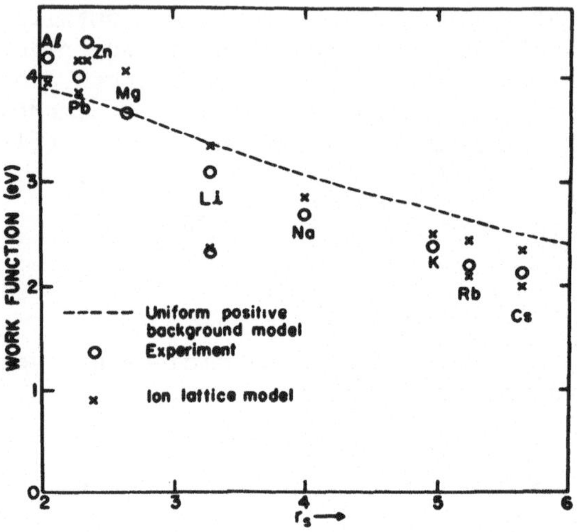

Fig. 12 : The theoretical work function in the planar uniform-background model (broken curve) and with pseudopotential corrections (crosses) compared with experiment. For Li, two experimental results are shown and for some elements two theoretical values are presented, corresponding to different values which have been proposed for r_c in the Ashcroft pseudopotential. The experimental results are for polycrystalline samples and the theoretical results with pseudopotential corrections represent the average of the results for low index faces (after Ref. |54|).

The second measurable quantity calculated by Lang and Kohn |53| was the surface energy σ. This quantity is defined as the energy required, per unit area of new surface formed, to split the crystal into two parts,

$$\sigma = \frac{1}{2A}(E - 2E') \quad , \tag{57}$$

where E is the energy of the original crystal, E' is the energy of one of the parts and A the area of the new surface created on one of the parts. It is convenient to divide the surface energy into kinetic, exchange-correlation and electrostatic contributions

$$\sigma = \sigma_s + \sigma_{xc} + \sigma_{es} \quad , \tag{58}$$

which are shown in Table 1. The kinetic energy part σ_s is always negative, as the electron can spill out beyond the jellium edge when the crystal is cut. Thus the electrons have a larger volume available and can lower their kinetic energy, according to the Heisenberg uncertainty principle. Since the kinetic energy increases rapidly with the density, this effect becomes very important for large densities (small r_s).

Table 1 : The surface energy σ and its components in ergs/cm^2 for the planar uniform-background model.

r_s	σ_s	σ_{xc}	σ_{es}	σ
2.0	- 5600	3260	1330	- 1010
2.5	- 1850	1460	430	40
3.0	- 720	750	170	200
3.5	- 320	430	80	190
4.0	- 145	265	40	160
4.5	- 70	165	25	120
5.0	- 30	115	15	100
5.5	- 15	80	10	75
6.0	- 5	55	10	60

The spilling out of electrons also leads to a reduction of the magnitude of the exchange-correlation energy due to the reduced density in the surface region. In the unsplit crystal the electrons move in a relatively high density and the exchange-correlation energy, resulting from the electrons being kept apart by many-body effects, is large. In the low density surface region of the split crystal the electrons are relatively far apart even in an independent particle description, and the reduction of the interaction energy due to many-body effects is smaller. Finally, the spilling out of electrons also costs electrostatic energy as some of the electrons are farther away from the positive background. The large values of σ_{xc} shows that Thomas-Fermi or Hartree calculations of the surface energy are inappropriate.

Table 1 shows that the surface energy becomes negative for small r_s. That means that energy is gained by splitting the crystal, i.e. the crystal is instable, according to the calculation. Since the exactly calculated kinetic energy dominates for small r_s, this failure cannot be due to the LD approximation but must result from the model.

3.3. Pseudopotential Corrections

To improve the model Lang and Kohn |53| introduced pseudopotentials and treated them in first order perturbation theory, i.e. the electron density was assumed not to change. Lang and Kohn used the Ashcroft pseudopotential |64|

$$v_{ps}(\mathbf{r}) = \begin{cases} 0 & |\mathbf{r}| \leq r_c \\ -\frac{Z}{r} & |\mathbf{r}| > r_c \end{cases} \tag{59}$$

where Z is the ionic charge and r_c is a cut-off radius which has been determined for each metal to reproduce bulk properties. These pseudopotentials are introduced in a three-dimensional lattice and replace the positive background in the planar uniform-background model. Thus the perturbation is

$$\delta v(\mathbf{r}) = \sum_{\mathbf{R}} v_{ps}(\mathbf{r}-\mathbf{R}) - v_+(\mathbf{r}) \tag{60}$$

where $\{\mathbf{R}\}$ are the lattice sites of the system and $v_+(\mathbf{r})$ is the potential from the positive background, which must be subtracted.

We first discuss the effects on the surface energy. As the electron density is assumed to remain unchanged, the kinetic and exchange-correlation energies do not change and the change of the surface energy $\delta\sigma$ is due entirely to the electrostatic term. Lang and Kohn |53| calculated this term in two steps (see Fig. 13). In the first step, the crystal is divided into two, keeping the electron density uniform up to the surface. This contribution is called

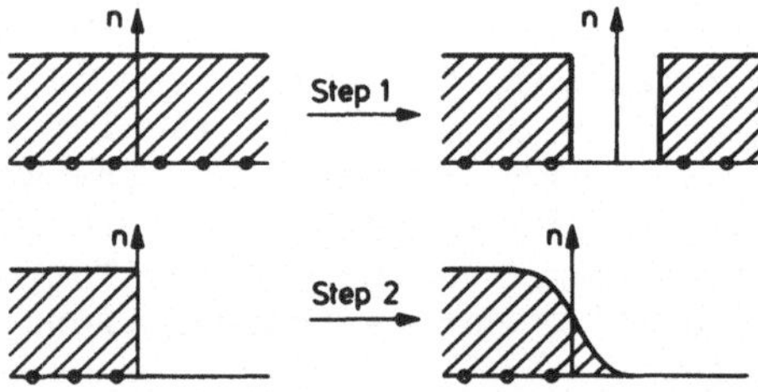

Fig. 13 : The calculation of the pseudopotential contribution to the surface energy divided into two steps.

the classical cleavage energy $\delta\sigma_{cl}$. In step 2, the electron density is allowed to relax from the step form to the profile of the planar uniform-background model. The pseudopotential contribution in this step is called $\delta\sigma_{ps}$.

Step 1 requires no energy in the planar uniform-background model. In this step the pseudopotentials may be replaced by point charge potentials, provided that the cleavage plane does not intersect any of the cores. Then we obtain the situation illustrated in Fig. 14, where the positive point charges are situated in a uniform neutralizing background. The interaction energy between the two

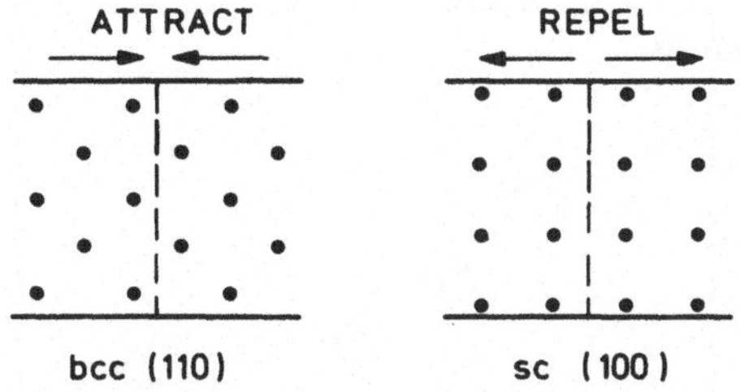

Fig. 14 : The interaction between the two parts of a crystal consisting of point ions and a neutralizing background. The cleavage plane is indicated by the dashed line and the dots show the ions in a plane perpendicular to the cleavage plane. The figure shows that the interaction is attractive for a bcc(110) surface and repulsive for a sc(100) (simple cubic) surface.

halves can be obtained from classical electrostatics and depends on the crystal structure and the cleavage plane. In step 2, the pseudopotential contribution is given by

$$\delta\sigma_{ps} = \int_{-\infty}^{\infty} \delta v_{av}(x) \left[n(x) - n_+(x)\right] dx \qquad (61)$$

where $\delta v_{av}(x)$ is the average of $\delta v_{ps}(\mathbf{r})$ over the y-z plane.

Since very little data is available for the surface energy of simple metals in the solid phase, Lang and Kohn compared their results with the surface energy of liquid metals. These experimental results were extrapolated to zero temperature, a procedure which introduces fairly large uncertainties. To obtain the best possible representation of a surface, Lang and Kohn performed calculations for the densely packed faces of a fcc and a bcc lattice. These results are compared with experiment in Fig. 15. We note that the calculated surface energies are positive for all metals. In most cases, the agreement with experiment is very good and only in the case of Pb (not shown in the figure) was a serious discrepancy obtained.

To introduce the pseudopotential corrections to the work function, we note that Eq. (52) can be written as

$$\Phi = \left[\phi(\infty) + E_{N-1}\right] - E_N \quad , \qquad (62)$$

i.e. the energy difference between the original N particle system and a system with one electron far outside the surface at the potential $\phi(\infty)$ and N-1 electrons in the metal. As the electron far

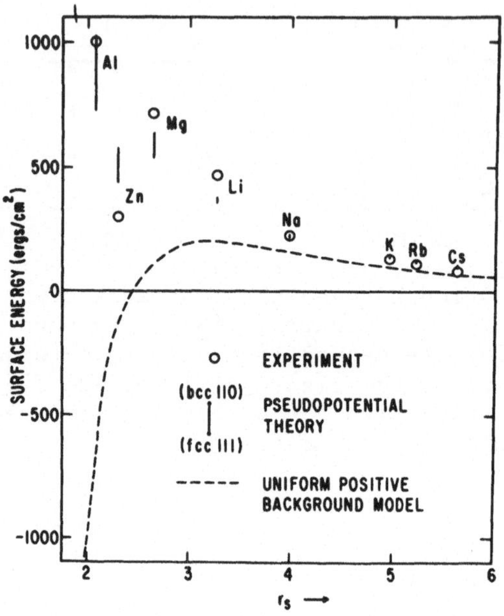

Fig. 15 : The surface energy in the planar uniform-background model (broken curve) and with pseudopotential corrections (vertical lines) compared with liquid-metal surface tensions extrapolated to zero temperature. The lowest point of each vertical line gives the results calculated for a fcc lattice and the highest point for a bcc lattice. The most densely packed plane is considered in both cases. For the alkali metals the lines are contracted almost to a point (after Ref. |53|).

outside the surface is not influenced by the pseudopotentials, standard perturbation theory gives the first order correction

$$\delta\Phi = \int \delta v(\underset{\sim}{r})\, n'_N(\underset{\sim}{r})\, d^3r - \int \delta v(\underset{\sim}{r})\, n_N(\underset{\sim}{r})\, d^3r \equiv \equiv \int \delta v(\underset{\sim}{r})\, n_\sigma(\underset{\sim}{r})\, d^3r \qquad (63)$$

where $n_N(\underset{\sim}{r})$ and $n'_N(\underset{\sim}{r})$ are the densities of the ground-state and the state for which one electron has been removed, respectively. The difference $n_\sigma(\underset{\sim}{r})$ is constant in the y-z plane and therefore only the average of $\delta v(\underset{\sim}{r})$ over this plane enters the calculation.

To discuss the variation of the work from one crystal face to another, we assume for a moment that the pseudopotentials are replaced by point ion potentials. According to the arguments above, the charge of the ions in a certain plane can be distributed over the plane, and the amount of charge on each plane is proportional to the spacing d of the planes. Together with the potential $-v_+(r)$, this gives rise to a potential oscillating between zero and a negative minimum which is proportional to d^2. If the point ion potentials are replaced by pseudopotentials, the average potential is shifted upwards by an amount which is the same for all crystal faces. Thus the average of $\delta v(r)$ is lower for a face with d large, i.e., a closed packed surface. As $n_\sigma(\underset{\sim}{r})$ integrates to -1, we would expect the largest work function for the close packed face. This argument is not completely rigorous |56|, since the averaging is done with the rather complicated weight function $n_\sigma(r)$, but in most cases Lang and Kohn |54| found results in accordance with it. Their results are shown in Fig. 12 and compared with experimental results for polycrystalline samples. The agreement is generally very good.

As the pseudopotential perturbation is fairly strong and has a very large effect on the surface energy, Perdew and Monnier |65| have tried to go beyond first order perturbation theory and introduced the so-called variational self-consistent method. Since the average of the perturbation $\delta v(\underset{\sim}{r})$ has a fairly large value inside the metal and is zero outside, they introduced an additional potential

$$V(x) = C\,\Theta(-x) \tag{64}$$

in the planar uniform-background model. As this potential is y and z independent, it does not cause any numerical complication. This gives a density which depends on the parameter C and the surface energy can be calculated to first order in

$$\delta v(\underset{\sim}{r}) - C\,\Theta(-x)\ , \tag{65}$$

where $\delta v(\underset{\sim}{r})$ is defined in Eq. (60). The constant C was varied to minimize the surface energy, giving a perturbation (65) which was in general weaker than $\delta v(\underset{\sim}{r})$. In a second calculation, the potential (64) was replaced by

$$V(x) = \langle\delta v\rangle_{av}\,\Theta(-x+X)\ , \tag{66}$$

where $\langle\delta v\rangle_{av}$ is the average of $\delta v(r)$ and X is a variational parameter and the surface energy was minimized as a function of X. The lowest value for the surface energy obtained from either (64) or (66) is shown in Table 2 and compared with the results of Lang and Kohn |53| and with experiment. In all cases but Pb the agreement between the two theoretical calculations is fairly good, giving a support for the first order perturbation treatment used by Lang and Kohn. The two theoretical calculations show a substantial disagreement for Pb, however which is tetravalent and has a relatively strong pseudopotential, and in this case the variational treatment brings theory into fair agreement with experiment. The method of Perdew and Monnier gives a face dependent density profile, which in chemisorption calculations gives more reasonable separation distances.

Table 2 : Comparison of the surface energy in erg/cm^2 calculated by Lang and Kohn (σ_{LK}) and by Perdew and Monnier (σ_{PM}) with the experimental results (σ_{exp}) given in Ref. |53|. The calculated results are obtained for a fcc (111) and a bcc (110) surface.

Metal	σ_{LK}	σ_{PM}	σ_{exp}
Al	730 - 915	643 - 870	1000
Pb	1140 - 1400	365 - 820	620
Zn	341 - 410	304 - 338	300
Mg	544 - 612	542 - 595	720
Li	353 - 375	330 - 358	480
Na	223 - 229	223 - 227	230
K	136 - 139	135 - 137	150
Rb	122 - 122	110 - 108	120
Cs	104 - 103	88 - 85	90

This discussion has left out several interesting aspects; for instance, a discussion of the surface energy in terms of the surface modes |66|, sum rules for surface properties |67| and the position of the image plane |55|.

4. CHEMISORPTION ON SIMPLE METALS

By chemisorption is meant the adsorption of an atom or a molecule on a surface due to the formation of a chemical bond with a binding energy larger than about 1/2 eV. Various aspects of the problem have been discussed in recent review articles |68|. Here we limit ourselves to calculations where the substrate is described by the planar uniform-background model |69|. In this model, the ions are replaced by a uniform positive background $n_+(x)$, as in section 3,

$$n_+(x) = \begin{cases} \bar{n} & x \leq 0 \\ 0 & x > 0 \end{cases} \qquad (44)$$

where the substrate density $\bar{n}$ is specified by the parameters r_s (see Eq. (45)). The adsorbed atom is described by a positive charge Ze at x=d, and the electrons of the substrate and the adsorbed atom (adatom) are allowed to move freely in the potential of this charge. The only free parameters are r_s and Z specifying the substrate and the adatom, respectively.

This model is difficult to solve due to its complicated geometry. The translational invariance of the substrate perpendicular to the surface is broken by the surface and the adatom breaks the invariance parallel to the surface. There remains only the cylinder symmetry around the adatom normal to the surface. The first calculation was performed by Smith, Ying and Kohn |70|. They used an extension of the Thomas-Fermi method, applied earlier by Smith |51| to the clean surface, and treated the perturbation from the adatom in linear response. These approximations have later been found to be unsatisfactory |71,72| and therefore more accurate methods have been developed by Lang and Williams |72-75| and by Gunnarsson, Hjelmberg and Lundqvist |71,76-79,48|, who solved the model without introducing any essential approximations, other than the LD approximation. As in the case of the clean surface, the substrate pseudopotentials have also been introduced and treated in first order perturbation theory |76,79,75|. For the details of these methods the reader is referred to the original papers. We shall discuss below selected results which have been obtained.

4.1. Adsorbate Induced Density of States

We define the local density of states

$$n(\underset{\sim}{r},\varepsilon) = \sum_{\nu} |\psi_{\nu}(\underset{\sim}{r})|^2 \, \delta(\varepsilon-\varepsilon_{\nu}) \tag{67}$$

where $\psi_{\nu}(\underset{\sim}{r})$ and ε_{ν} are the eigenfunctions and eigenvalues, respectively, of Eq. (10). Integration over the energy gives the density

$$n(\underset{\sim}{r}) = \int_{-\infty}^{\varepsilon_F} n(\underset{\sim}{r},\varepsilon) \, d\varepsilon = \sum_{\nu=1}^{N} |\psi_{\nu}(\underset{\sim}{r})|^2 \quad , \tag{68}$$

where ε_F is the Fermi energy, and integration over space gives the density of states

$$n(\varepsilon) = \int d^3r \; n(\underset{\sim}{r},\varepsilon) = \sum_{\nu} \delta(\varepsilon-\varepsilon_{\nu}) \tag{69}$$

The importance of $n(\underset{\sim}{r},\varepsilon)$ is twofold. Via Eq. (68) it gives us information about the density and the bonding properties, and from Eq. (69) we obtain the density of states, which is often used to discuss excitation properties (see the discussion in section 2.6 for the extent to which this approach is justified). Further we define the induced local density of states, i.e. the change due to the presence of the adatom,

$$\delta n(\underset{\sim}{r},\varepsilon) = n(\underset{\sim}{r},\varepsilon) - n_o(\underset{\sim}{r},\varepsilon) \tag{70}$$

where $n(\underset{\sim}{r},\varepsilon)$ and $n_o(\underset{\sim}{r},\varepsilon)$ are calculated with and without the adatom, respectively. Similar definitions are introduced for $\delta n(\underset{\sim}{r})$ and $\delta n(\varepsilon)$.

For large separation distances d the substrate-adatom interaction is negligible and $n(\varepsilon)$ is simply the sum of the bare surface and free atom densities of states. Thus $\delta n(\varepsilon)$ has just a number of δ-functions corresponding to the discrete atomic levels. For smaller values of d the adatom has a stronger interaction with the substrate, and the discrete levels are broadened to resonances, which are characterized by their widths and positions.

Fig. 16 shows how the width varies with the distance d for a hydrogen atom chemisorbed on a r_s=2 surface. As d is decreased the larger overlap between the wave functions of the bare surface and the H atom increases the coupling, leading to a broader resonance (d=1.1 a_o and 0). Simultaneously, the resonance moves down in en-

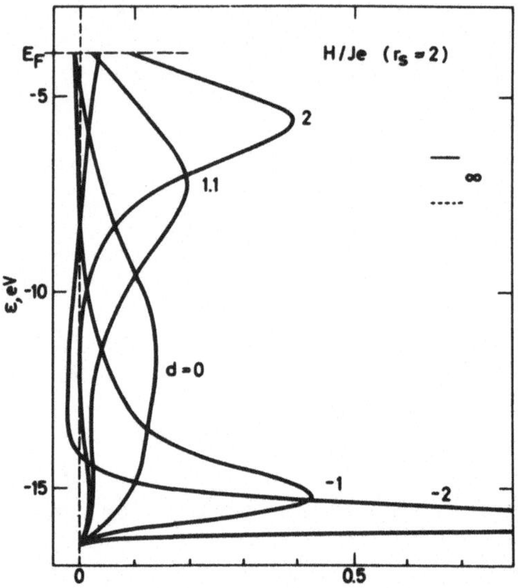

Fig. 16 : The induced density of states $\delta n(\varepsilon)$ in eV^{-1} (horizontal axis) for a hydrogen atom chemisorbed on a substrate (r_s=2) for values of the distance to the positive background edge d=-2,-1,0, 1.1 and 2 a_o. The energy eigenvalues of a free hydrogen atom (∞) are given for the spin-polarized (dashed line) and the spin-independent (full line) calculations. The bottom of the substrate band lies at about -16.4 eV (after Ref. |78|).

ergy and for d=-1 a_o and -2 a_o it is close to the bottom of the substrate band, where the substrate density of states is low and each state has a small amplitude in the surface region. This reduces the coupling as illustrated by the narrow resonance for d=-1 a_o and d=-2 a_o. This has also been discussed by Muscat and Newns |83| using a one-parameter model.

Fig. 17 shows the variation of the resonance position as a function of distance for hydrogen on different substrates. The resonance tends to follow the effective potential and inside the surface it is very close to the bottom of the substrate band. This

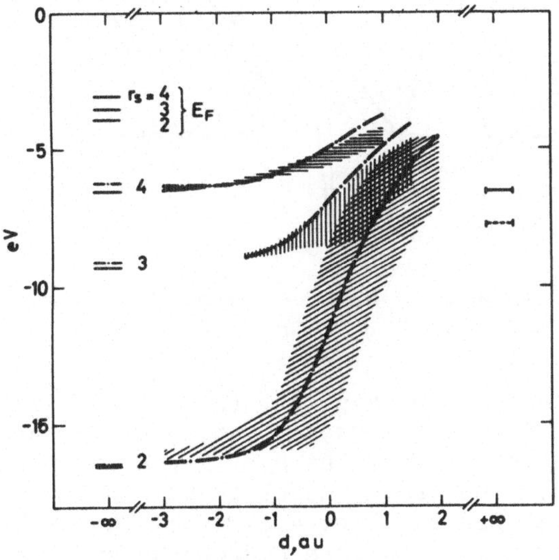

Fig. 17 : The resonance (shaded area) position in the induced density of states as a function of the distance from the edge of the positive background for three values of the substrate density (r_s=2,3,4). The dash-dotted lines show the corresponding effective potentials for the free surfaces. Inside the bulk (d=$-\infty$), sharp levels are split off below the bottom of the bands and for d=∞ the results for the free H atom, according to a spin-polarized calculation (dashed curve) and a spin-independent (full curve) calculation, are shown (after Ref. |78|).

happens although the substrate density ($\sim r_s^{-3}$) varies by a factor of 8 and the position of the bottom of the band relative the vacuum zero varies by a factor of 3. A similar effect was first observed |73| when the resonance position of different adatoms on a r_s=2 substrate was studied. The 3s and 3p levels of Si and the 3p level of Cl were found to have fairly constant energy separations to the effective potential as a function of the separation distance |73|.

The spatial character of $\delta n(\mathbf{r},\varepsilon)$ varies strongly with the energy, as is illustrated in Fig. 18 for H on a r_s=2 substrate |77|. Close to the bottom of the band (ε=-14.8 eV), $\delta n(\mathbf{r},\varepsilon)$ has a very smooth variation and extends deep into the substrate. At the bottom of the resonance, $\delta n(\mathbf{r},\varepsilon)$ has bonding character, i.e. charge is built up in the region between the surface and the adatom. Note the fairly smooth variation corresponding to a low kinetic energy.

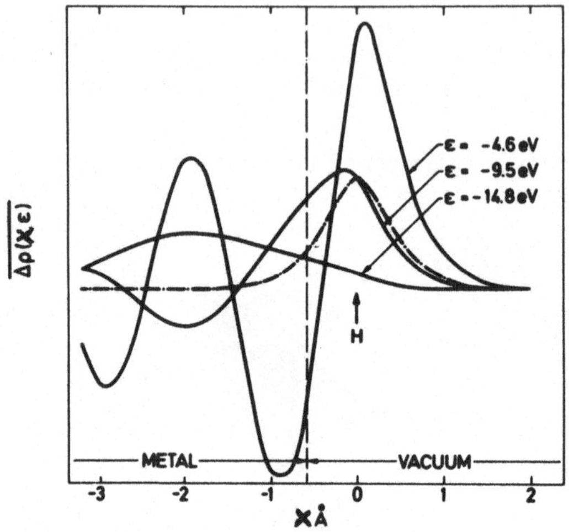

Fig. 18 : The local density of states $\delta n(\underset{\sim}{r},\varepsilon)$ for a hydrogen atom at $d=1.1\ a_o$ on a $r_s=2$ surface integrated over a plane parallel to the surface and a distance x Å from the hydrogen atom, $\Delta\rho(x,\varepsilon) = \int rdr \int d\phi\ \delta n(x,r,\phi,\varepsilon)$. The vacuum level is at $\varepsilon=0$ and the H resonance is located at $\varepsilon=-7$ eV and is about 4 eV wide. The dashed curve gives the results for $\Delta\rho(x,\varepsilon_H)$ for a free hydrogen atom. The vertical scale is linear and the same for all curves and the curves are normalized according to $\int\Delta\rho(x,\varepsilon)dx=1$ (after Ref. |77|).

At the top of the resonance ($\varepsilon\approx 4.6$), $\delta n(\underset{\sim}{r},\varepsilon)$ has anti-bonding character and charge is removed from the bond region and accumulated on the vacuum (high potential) side of the adatom. The bonding anti-bonding behaviour of the resonance is also illustrated by the results for Si on Al |75| in Fig. 19, which shows the behaviour of the s and p resonances of Si.

4.2. The Charge Density and the Effective Potential

Lang and Williams |73| have compared chemisorption on Al of Li, Si and Cl, which are examples of cationic, covalent and anionic chemisorption, respectively. Fig. 20 shows the results for the density of states. As expected a low-lying resonance is obtained for Cl, which is the most electronegative of the three. The resonance is below the Fermi energy, and is filled, leading to a charge transfer to the Cl atom. This is shown in Fig. 20, where the contours for the total charge density deflect away from the atom. For the electropositive Li the resonance is almost entirely above the Fermi energy and essentially empty, yielding a charge transfer from the Li atom. The charge is removed from the vacuum side of the Li atom and accumumlated in the bond region between the atom and the

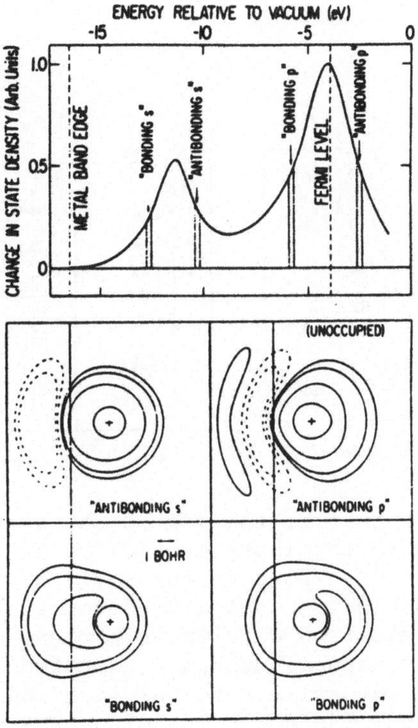

Fig. 19 : The upper part of the figure shows the density of states $\delta n(\varepsilon)$ for Si chemisorbed on a $r_s=2$ substrate. The two peaks correspond to the 3s and 3p atomic states. The lower part of the figure shows the density contours (contours of $\delta n(\underset{\sim}{r},\varepsilon)$) associated with the four shaded regions in the density of states curve. Solid lines correspond to positive values, dashed lines to negative values (after Ref. |75|).

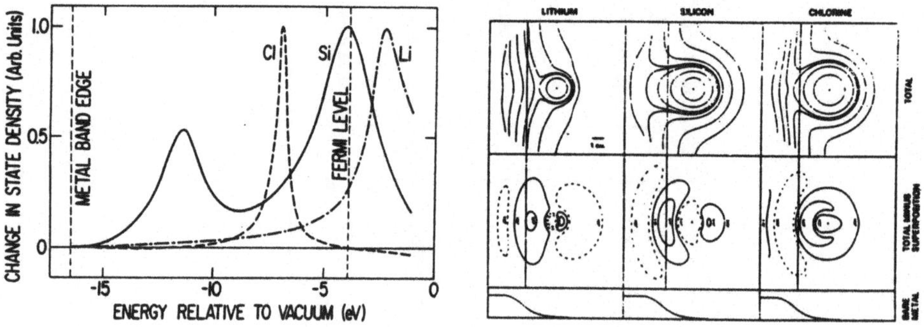

Fig. 20 : The left-hand part of the figure shows the change in the density of states due to chemisorption of Cl, Si and Li on a $r_s=2$ substrate. The curves correspond to the equilibrium separation. The right-hand part shows electron-density contours. The upper row shows contours of constant electron density in a plane normal to the metal surface containing the adatom nucleus (indicated by a dot). The metal is to the left and the solid vertical line indicates the edge of the positive background. The row below gives the total density minus the superposition of atomic and bare-metal densities and the bottom row the bare-metal electron-density profile (after Ref. |75|).

surface. For Si, the p-like (highest) resonance is at the Fermi energy and only partially occupied. Thus the bonding orbitals in Fig. 19 are occupied and the anti-bonding are empty, giving rise to a strongly directional bond. This is illustrated by the rather non spherical charge density contours around the Si atom.

Fig. 21 shows the charge density induced by a hydrogen atom on two different substrates with $r_s=2$ and $r_s=4$. The figure illustrates how rapidly the charge density is screened, although there remain

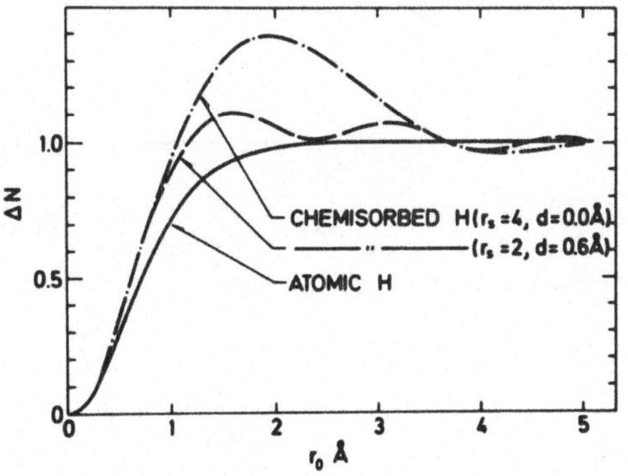

Fig. 21 : The electron-density $\delta n(\underset{\sim}{r})$ induced by an adsorbed hydrogen atom on two different substrates ($r_s=2$ and $r_s=4$), compared with the electron-density of a free hydrogen atom. The figure shows the amount of charge ΔN inside a sphere with radius r_o centered on the adatom. $\Delta N(r_o) = \int_0^{\infty} r^2 \, dr \, d\Omega \, \delta n(\underset{\sim}{r})$ (after Ref. |77|)

long-ranged Friedel oscillations in the density which decay as

$$\frac{1}{r^3} \cos(2k_F r + \alpha) \quad , \qquad (71)$$

where the Fermi momentum k_F is inversely proportional to r_s. Thus the wave length of these oscillations is doubled when r_s is increased from 2 to 4, as illustrated in Fig. 21.

A comparison of the induced charge density with the density of a free hydrogen atom (see Fig. 21) shows that the former is enhanced close to the nucleus. This contraction of the density towards the nucleus has been discussed extensively by Ruedenberg for molecules |81| and is similar to the bonding mechanism in alkali metals discussed by Wigner and Seitz |82|. Due to the overlap between the hydrogen 1s function and the metal states one can construct smooth bonding-type wave functions with low kinetic energy. Therefore, the wave functions can contract towards the proton and gain potential energy from the interaction with the nucleus, without producing an excessive increase in kinetic energy. Thus the interaction between the induced charge and the nucleus increases from

-27 eV for a free hydrogen atom to -31 eV for hydrogen chemisorbed on a r_s=2 substrate. This gain of 4 eV should be compared with the cost in kinetic energy of about 2.5 eV and the binding energy of 1.6 eV, and it plays an important role in the binding of the hydrogen atom |76|.

Fig. 22 shows the effective potential around a hydrogen atom chemisorbed on a r_s=2.0 substrate for d=1 a_o and d=0, which is typical of the equilibrium positions found when pseudopotential corrections are included. It is striking that the potential barrier

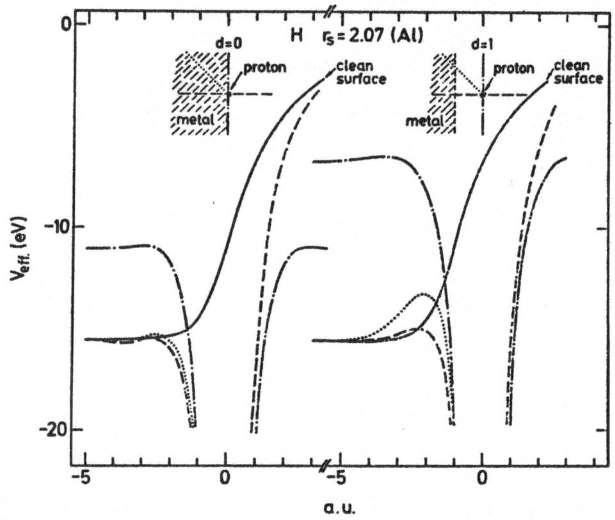

Fig. 22 : The effective potential $v_{eff}(r)$ for H chemisorbed on a r_s=2.07 substrate for d=0 and d=1 a_o. The curves show v_{eff} as a function of the distance from the proton along lines normal (---), parallel (-•-•-) and at an angle of 45^0 to the surface. The full line shows the potential of the bare metal surface (after Ref. |48|).

between the metal and the hydrogen atom is very small, contrary to popular belief. Since the resonance position is at about -8 eV (d=1 a_o) and -12 eV (d=0), electron states at the resonance energy have an extended character and can mix with the substrate states.

4.3. The Binding Energy

The total energy of a chemisorption system as a function of the adatom position is of great interest, giving a prediction of the equilibrium geometry and vibration frequencies. Although such predictions are unreliable without the inclusion of pseudopotentials, we shall show results for the uniform-background model, as these results show some trends which should be valid also for other models.

Fig. 23 shows the total energy minus the energy of the bare surface and the free atom, all calculated using the LSD approximation. The binding energy E_B is given by the minimum of the curves

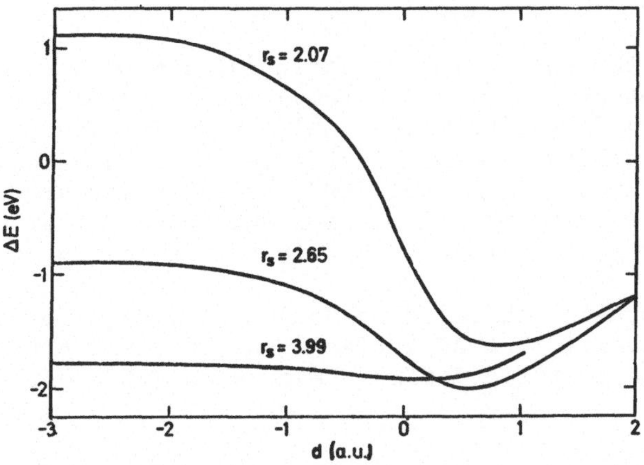

Fig. 23 : The energy change ΔE(d) upon chemisorption as a function of distance for r_s=2.07 (Al), 2.65 (Mg) and 3.99 (Na) (after Ref. |48|).

and takes the values 1.6, 2.0 and 1.9 eV for the r_s values 2.07 (Al), 2.65 (Mg) and 3.99 (Na), respectively. This small variation (≈25 per cent) in the binding energy is remarkable, considering that the density is varied by almost a factor of 8. We also note that the curves are very similar on the vacuum side of the minimum, say for d>1, while they differ strongly on the metal side. The main reason for the rapid rise of the curve for r_s=2 is an increase in the kinetic energy. For instance, if the adatom is moved from d≈2 a.u. to d=-1 a.u., the kinetic energy increases by about 10 eV |76|. This can be understood in the following way. For the hydrogen atom at 2 a.u., the electrons have the volume of both the substrate and the adatom available. For d=-1, on the other hand, the adatom is mainly inside the surface, leading to a reduction of the volume with low potential in which the electrons can move. According to Heisenberg's uncertainty principle this gives rise to a higher kinetic energy. This "kinetic energy repulsion" is an important factor for the equilibrium separation d on a substrate with a small r_s, while the effect is much weaker for larger values of r_s.

4.4. Pseudopotential Corrections

So far the substrate has been characterized by one single parameter, r_s, and no distinction has been made between chemisorption on different surfaces of the same substrate or different positions relative to the substrate lattice on the same surface. To give a more detailed description the ion pseudopotentials must be included, which has been done using the same methods as for the clean surface, namely first-order perturbation theory and the variational self-consistent method of Perdew and Monnier |65|.

In the former method, the correction to the total energy is

$$\int \delta v(\underset{\sim}{r})\, \Delta n(\underset{\sim}{r})\, d^3r \quad , \tag{71}$$

where $\delta v(\underset{\sim}{r})$ is the change in the potential due to the introduction of the pseudopotentials (see Eq. (60)) and $\Delta n(\underset{\sim}{r})$ is the induced electron density $\delta n(\underset{\sim}{r})$ (see Eqs. (68),(69)) plus the proton charge. This method is called model b. In the second method, called model c, the clean surface results of Perdew and Monnier are used in the planar uniform-background calculation and in the first order correction (71) $\delta v(\underset{\sim}{r})$ is replaced by $\delta v(\underset{\sim}{r})-c\Theta(-x)$ as in Eq. (65).

Typical results, obtained by Hjelmberg |79|, are shown in Fig. 24 for H on two different Al surfaces, Al(100) and Al(110). The figure shows the dramatic difference between different adsorption

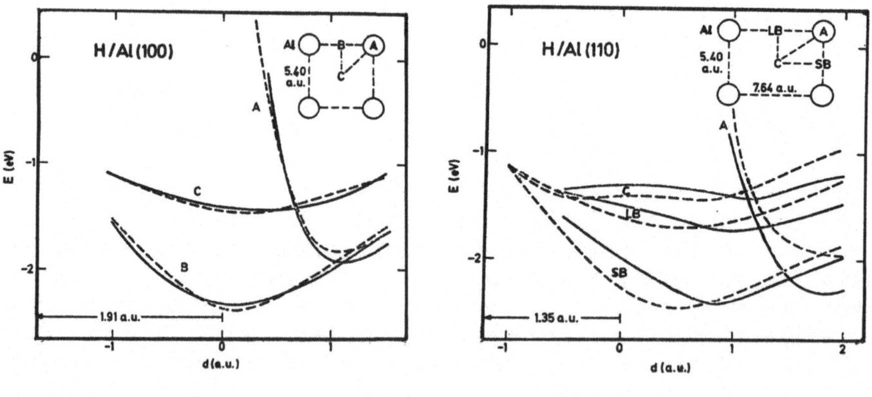

Fig. 24 : Calculated energy curves for H chemisorbed in different configurations on a (100) and a (110) surface of Al. The dashed curves have been obtained in model b and the full curve in model c (see text below Eq. (71)). The distance from the positive background edge to the first plane of Al ions is given in the lower left corner for the two surfaces. In the upper right corner is a figure of the unit cell for the surface, defining the different configurations (after Ref. |79|).

sites. The A curve (atop site) shows the energy variation when the adatom moves towards one of the substrate atoms. For d 1, the hydrogen atom starts to penetrate the core of the closest substrate atom, leading to a strong repulsion and a rapid rise of the energy curve. This core repulsion is a general effect for adsorption in the A position and is important for the equilibrium position |76|. For the bridge (B) site, the penetration of the core occurs for much smaller values of d and for the centered (C) site, the hydrogen atom can move all the way to the first ion plane without penetrating the core. For the B and C sites, therefore, the equilibrium positions are closer to the surface and the "kinetic energy repulsion", discussed in section 4.3. is important for the rise of the energy curve for small d.

Model c predicts the binding energies 2.3, 2.4 and 1.9 eV for the Al(100), Al(110) and Al(111), respectively |79|. Thus the energy gain from the chemisorption of two hydrogen atoms is comparable to or less than the binding energy of a hydrogen molecule. Therefore, if diffusion can take place, chemisorbed hydrogen atoms will recombine and form a molecule and perhaps leave the surface. By moving an H atom along different paths on an Al(100) surface, the activation energy for diffusion is estimated to be 0.1-0.2 eV, corresponding roughly to a temperature of 30-50 K |76|. According to the model, therefore, the observation of chemisorbed atomic hydrogen requires a substrate temperature below this value. This is not inconsistent with the failure to observing H on Al at room temperature |83|.

Fig. 25 shows results obtained by Hjelmberg |79| for H chemisorbed on a Na(110) surface. For $d \lesssim 0$, the C and LB curves are

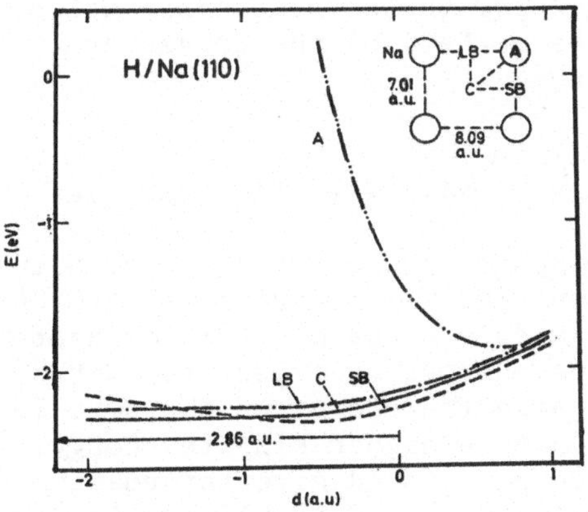

Fig. 25 : Calculated energy curves for H chemisorbed on a (110) surface of Na. The notation is the same as in Fig. 24 (after Ref. |79|).

almost flat and the theory predicts absorption of an H atom, with a binding energy of 2.4 eV. This is in contrast to the results for H on Al for which the theory predicts an activation energy for absorption to be 1.2-1.4 eV. For the A-site, the model predicts a Na-H separation of 3.5 a_o, which is close to the separation (3.57 a_o) in a Na H molecule.

Due to the lack of experimental results, it is not possible to make definite statements about the accuracy of these calculations. The essential approximations are the LD approximation and the approximate treatment of the pseudopotentials. The difference in Fig. 24 between the two methods of including the pseudopotentials indicates the uncertainty in the treatment. For instance, the difference in binding energy is less than 0.1 eV for Al(100) and Al (111) and 0.4 eV for Al(110) |79|. The results obtained are in general reasonable. For chemisorbtion on Al the activiation energy for diffusion into the substrate is large (1.2-1.4 eV) while it is small for Mg (0.4 eV) and Na (0) |79|. This is consistent with the existence of magnesium and sodium hydrides, while aluminum hydride does not form. The calculated formation energies are 2.6 eV (MgH) and 2.4 eV(NaH) which is in fair agreement with the experimental results 2.8 eV(MgH) and 3.0 eV (NaH) |79|. The discrepancy should at least partly be due to relaxation and structural changes of the lattice, not accounted for in the theory. The predicted equilibrium positions give distances to the closest substrate atom(s) which are usually not very far from the separation in the corresponding diatomic molecule |79|. For an aluminum surface, the predictions are 3.4 a_o (Al(100)), 3.5 a_o(Al(110)) and 2.9-3.4 a_o (Al(111)) compared with 3.1 a_o for the Al-H molecule and for the Mg(0001) surface the calculated separation is 4.0 a_o compared to 3.3 a_o for an MgH molecule. As already mentioned, the bond length (3.5 a_o) for H in an atop position on Na(110) is in good agreement with the result (3.57 a_o) for the corresponding molecule. For the other chemisorption sites on Na(110) the curves are to flat to predict a separation distance.

4.5 Extra-Atomic Relaxation Energies

A commonly used method to probe the state of a chemisorbed atom is X-ray photoemission (XPS) from core levels (ESCA), in which X-rays are used to emit a core electron from the adatom. If the adatom has a net negative charge, we expect the corresponding electrostatic potential to raise the energy of the core level, and the emitted electron should have a higher kinetic energy. Thus the energy of the emitted electrons should give information about the charging of the adatom. This highly simplified picture is complicated by, amongst other things, relaxation effects. The kinetic energy is given by

$$T = h\omega + E_N - E_{N-1} \quad , \qquad (72)$$

where $\hbar\omega$ is the photon energy and E_N and E_{N-1} are the energies of the system with and without the core electron, respectively. The kinetic energy is influenced strongly by the rearrangement of the N-1 particle system to obtain its lowest possible energy in the presence of a core-hole, i.e. relaxation effects. We shall not discuss the possibility that the N-1 particle system may be in an excited state, which gives rise to the so-called shake-up structure. Relaxation has been discussed by Lang and Williams |84| for atoms chemisorbed on simple metals and we follow their approach.

According to the transition state technique (see section 2.6.) the ionization energy is (Eq. (34))

$$I_p = - \varepsilon_i \; (n_i = 1/2) \; , \tag{73}$$

where ε_i is the energy eigenvalue of the core orbital with 1/2 electron removed. This energy is compared with the corresponding free atom result, which gives the extra-atom shift Δ

$$\Delta = \varepsilon_i \; (\text{chemisorbed atom, } n_i = 1/2) - \varepsilon_i \; (\text{free atom, } n_i = 1/2) \tag{74}$$

where Δ, which in principle depends on i, contains all the effects due to chemisorption. The change in the eigenvalue is due to a change of the effective potential in the core region upon chemisorption. As this change of the potential is mainly due to changes outside the core, it has a weak variation over the core region. Thus the shift of deep core levels can be approximated by the shift of the potential at the nucleus, which is almost entirely due to the electrostatic potential and we obtain

$$\Delta = \delta\phi(0, n_i = 1/2) \quad , \tag{75}$$

where $\delta\phi(0, n_i=1/2)$ is the electrostatic potential difference between a chemisorbed and free atom, each having 1/2 core electron.

This shift is divided into a chemical shift Δ_c, which depends on the properties of the system before the core hole is created, and a relaxation shift Δ_r, which depends on the response of the system to the creation of the core hole. Thus we have

$$\Delta_c = \delta\phi \; (0, n_i = 1) \tag{76}$$

$$\Delta_r = \delta\phi \; (0, n_i = 1/2) - \delta\phi \; (0, n_i = 1) \tag{77}$$

The quantity Δ_c describes factors such as charge transfer to the adatom and the effect of the strongly varying potential of the surface. When the core hole is created the valence orbital contracts to take advantage of the more attractive potential, giving the intra-atomic relaxation energy. For the chemisorbed atom there is also a rearrangement of the charge density outside the adatom and we define the extra-atomic screening charge

$$n(r;\ \text{extra-atomic}) = |n(r,\ n_i{=}0)-n(r,\ n_i{=}1)|_{\text{chemisorbed}} - |n(r,\ n_i{=}0)-n(r,\ n_i{=}1)|_{\text{free}} \tag{78}$$

Thus we have

$$\delta\phi(0,\ n_i{=}0) - \delta\phi(0,\ n_i{=}1) = e^2 \int \frac{n(\underset{\sim}{r},\ \text{extra-atomic})}{|\underset{\sim}{r}|} d^3r \tag{79}$$

Assuming linear response Eqs. (77),(79) gives |84|

$$\Delta_r = \frac{e^2}{2} \int \frac{n(r,\ \text{extra-atomic})}{|\underset{\sim}{r}|} d^3r \tag{80}$$

This very useful formula is essentially the DF analogy of the formula derived by Hedin and Johansson |85| in the Hartree-Fock formalism.

Lang and Williams |84| have calculated n(r; extra-atomic) for O, Na, Si and Cl chemisorbed on a r_s=2 substrate, using the methods described earlier. The density of states $\delta n(\varepsilon)$ (Eqs. (69),(70)) for Na are shown in Fig. 26. Before the creation of the core hole the 3s-derived resonance is essentially empty, while the attractive core hole pulls the resonance down so that it becomes partially

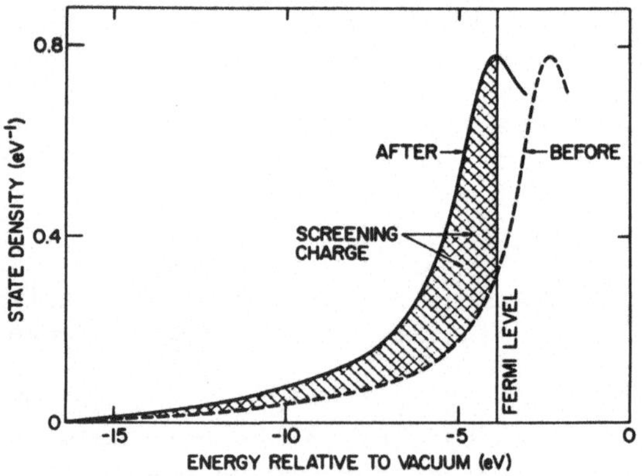

Fig. 26 : The induced density of states for Na chemisorbed on a r_s=2 substrate before and after a 2s core electron is removed (after Ref. |84|).

filled. This suggests the so-called excited atom approximation, in which the screening charge is obtained by subtracting the charge of an ion (free atom with a core hole) from that of the excited atom, for which a core electron has been transferred to the lowest

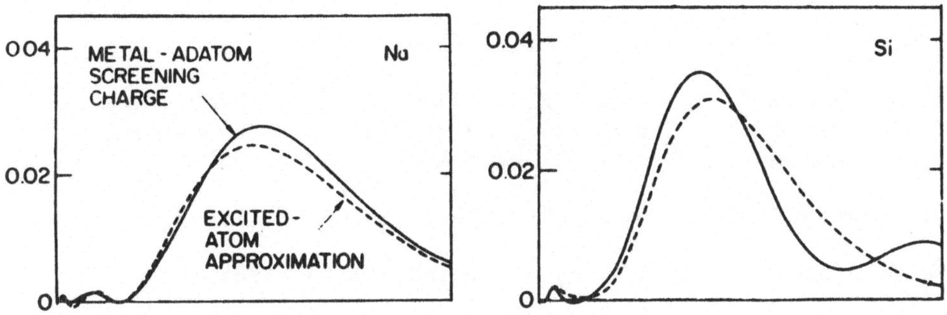

Fig. 27 : Radial distribution of screening charge for Na and Si chemisorbed on a r_s=2 substrate. The full curve shows the screening charge of the chemisorption system and the dashed curve is obtained in the excited atom approximation (after Ref. |84|).

unoccupied valence orbital. In Fig. 27, n($\underset{\sim}{r}$, extra-atomic) is compared with the screening charge in this approximation for Na and Si with a 2s hole. The good agreement shows that for these systems the metal substrate essentially acts as a reservoir, providing an electron which occupies a valence level of the adatom. The excited atom model was proposed by Shirley et al |86|. The results of Lang and Williams |84| show that it is a good approximation for adatoms which have a partially empty valence level close to the Fermi energy. However, it should not be applied to systems such as O or Cl on a r_s=2 substrate, for which the valence resonance is filled.

The excited atom approximation gives a very simple way of calculating relaxation energies |86,84| for, e.g. chemisorption systems. The extra-atomic screening charge is obtained from two atomic calculations, which via Eq. (80) gives the relaxation energy. This method has also been applied successfully to the calculation of relaxation energies of bulk transition metals |87|. The excited atom approximation can be used to construct simple models of chemisorption systems. For instance, it has been used to discuss the shake-up structure of the XPS spectra from core levels |88|.

ACKNOWLEDGEMENT

The author is grateful to R.O. Jones for useful discussions and a critical reading of the manuscript.

REFERENCES

|1| D.M. Newns, Phys. Rev. 178, 1123 (1969).

|2| K. Schönhammer, Phys. Rev. B13, 4336 (1976); W. Brenig and K. Schönhammer, Z. Physik 267, 201 (1974); K. Schönhammer, V. Hartung and W. Brenig, Z. Physik B22, 143 (1975).

|3| P. Hohenberg and W. Kohn, Phys. Rev. 136, B864 (1964).

|4| W. Kohn and L.J. Sham, Phys. Rev. 140, A1133 (1965).

|5| J.C. Slater, Quantum Theory of Molecules and Solids, Vol. IV, (McGraw-Hill, New York, 1974).

|6| If the ground state is degenerate we can limit ourselves to states of a given symmetry |7|. In this restricted set of states, the state with the lowest energy is non-degenerate.

|7| O. Gunnarsson and B.I. Lundqvist, Phys. Rev. B13, 4274 (1976).

|8| It has been shown that there is at most one $v(\mathbf{r})$ which gives the density $n(\mathbf{r})$. However, there might be densities $n(\mathbf{r})$ which cannot be obtained for any external potential $v(\mathbf{r})$ |3|. Thus the functional giving $v(\mathbf{r})$ may not be defined for all densities. In the following the functional is assumed to exist for any physically reasonable density.

|9| L.H. Thomas, Proc. Camb. Philos. Soc. 23, 542 (1926); E. Fermi, R.c.Acc. Lincei 6, 602 (1927).

|10| U. von Barth and L. Hedin, J. Phys. C 5, 1629 (1972).

|11| A.K. Rajagopal and J. Callaway, Phys. Rev. B7, 1912 (1973).

|12| L. Hedin and B.I. Lundqvist, J. Phys. C 4, 2064 (1971).

|13| P.A.M. Dirac, Proc. Camb. Philos. Soc. 26, 376 (1930); R. Gáspár, Acta Phys. Hung. 3, 263 (1954).

|14| K. Schwarz, Phys. Rev. B5, 2466 (1972).

|15| O. Gunnarsson, P. Johansson, S. Lundqvist and B.I. Lundqvist, Int. J. Quantum Chem. 9 S, 83 (1975).

|16| O. Gunnarsson, J. Phys. F 6, 587 (1976).

|17| B.Y. Tong and L.J. Sham, Phys. Rev. 144, 1 (1966).

|18| O. Gunnarsson, J. Appl. Phys. 49, 1399 (1978).

|19| O. Gunnarsson, J. Harris and R.O. Jones, Phys. Rev. B15, 3027 (1977); J. Chem. Phys. 67, 3970 (1977); J. Harris and R.O. Jones, J. Chem. Phys. 68, 1190 (1978) and to be published.

|20| E.J. Baerends and P. Ros, to be published.

|21| U.K. Poulsen, J. Kollâr and O.K. Andersen, J. Phys. F 6, L241 (1976); O.K. Andersen, J. Madsen, U.K. Poulsen, O. Jepsen and J. Kollâr, Physica 86B, 249 (1977).

|22| J.F. Janak and A.R. Williams, Phys. Rev. B14, 4199 (1976); V.L. Moruzzi, A.R. Williams and J.F. Janak, Phys. Rev. B15, 2854 (1977); J.F. Janak, Phys. Rev. B16, 255 (1977) and to be published.

|23| J. Callaway and C.S. Wang, Physica 91B, 337 (1977); C.S. Wang and J. Callaway, Phys. Rev. B15, 298 (1977); S.H. Vosko, J.P. Perdew and A.H.McDonald, Phys. Rev. Lett. 35, 1725 (1975); A.H. McDonald, K.L. Liu and S.H. Vosko, Phys. Rev. B16, 777 (1977).

|24| A. Zunger and A.J. Freeman, Phys. Rev. B15, 5049 (1977); B16, 906 (1977); B16, 2901 (1977); B17, 2030 (1978).

|25| J. Harris and R.O. Jones, J. Chem. Phys. 68, 3316 (1978) and to be published. See also R.O. Jones in this volume.

|26| D.J.W. Geldart and M. Rasolt, Phys. Rev. B13, 1477 (1976).

|27| L.J. Sham, Phys. Rev. B7, 4357 (1973).

|28| D.J.W. Geldart, M. Rasolt and R. Taylor, Solid State Commun. 10, 279 (1972).

|29| O. Gunnarsson, M. Jonson and B.I. Lundqvist, to be published.

|30| F. Herman, J.P. van Dyke and I.B. Ortenberger, Phys. Rev. Lett. 22, 807 (1969).

|31| K.H. Lau and W. Kohn, J. Phys. Chem. Solids 37, 99 (1976).

|32| J.S.Y. Wang and M. Rasolt, Phys. Rev. B13, 5330 (1976); M. Rasolt, J.S.Y. Wang and L.M. Kahn, Phys. Rev. B15, 580 (1977).

|33| J.H. Rose, H.B. Shore, D.J.W.Geldart and M. Rasolt, Solid State Commun. 19, 619 (1976).

|34| A.K. Gupta and K.S. Singwi, Phys. Rev. B15, 1801 (1977).

|35| J.P. Perdew, D.C. Langreth and V. Sahni, Phys. Rev. Lett. 38, 1030 (1977).

|36| J.P. Perdew and D.C. Langreth, Solid State Commun. 24, 765 (1977).

|37| T. Ziegler and A. Rauk, Theoret. Chim. Acta 43, 261 (1977).

|38| U. von Barth, to be published.

|39| A.R. Williams, R.A. de Groot and C.B. Sommers, J. Chem. Phys. 63, 628 (1975).

|40| M. Ross, Phys. Rev. 179, 612 (1969) ; J.C. Slater, J.B. Mann, T.M. Wilson and J.H. Wood, Phys. Rev. 184, 672 (1969); J.F. Janak, to be published.

|41| L. Hedin and S. Lundqvist, Solid State Physics, edited by H. Ehrenreich, F. Seitz and D. Turnbull, 23, 1 (1969).

|42| L.J. Sham and W. Kohn, Phys. Rev. 145, 561 (1966).

|43| This approximation has been discussed by M. Rasolt and S.H. Vosko, Phys. Rev. Lett. 32, 297 (1974); Phys. Rev. B10, 4195 (1974); S.B. Nickerson and S.H. Vosko, Phys. Rev. B14, 4399 (1977), and references therein.

|44| J.F. Janak, A.R. Williams and V.L. Moruzzi, Phys. Rev. Lett. 6, 4367 (1972); Phys. Rev. B11, 1522 (1975).

|45| K.Y.Yu, J.N. Miller, P. Chye, W.E. Spicer, N.D. Lang and A.R. Williams, Phys. Rev. B14, 1446 (1976).

|46| A.R. Williams, private communication.

|47| B. Vinter, Phys. Rev. B17, 2429 (1978).

|48| H. Hjelmberg, Physica Scripta (in print).

|49| J. Nørskov, Solid State Commun. 24, 691 (1977).

|50| J. Bardeen, Phys. Rev. 49, 653 (1936).

|51| J. Smith, Phys. Rev. 181, 522 (1969).

|52| A.J. Bennett and C.B. Duke, in Structure and Chemistry of Solid Surfaces, edited by G.A. Somorjai (Wiley, New York 1969).

|53| N.D. Lang and W. Kohn, Phys. Rev. B1, 4555 (1970).

|54| N.D. Lang and W. Kohn, Phys. Rev. B3, 1215 (1971).

|55| N.D. Lang and W. Kohn, Phys. Rev. B7, 3541 (1973).

|56| See also references in N.D. Lang, Solid State Physics, edited by H. Ehrenreich, F. Seitz and D. Turnbull, 28, 225 (1973).

|57| J.A. Appelbaum and D.R. Hamann, Phys. Rev. B6, 2166 (1972).

|58| G.P. Alldredge and L. Kleinman, Phys. Rev. B10, 559 (1974).

|59| O. Jepsen, J. Madsen and O.K. Andersen, Phys. Rev. B18, 605 (1978); S.G. Louie, Phys. Rev. Lett. 40, 1525 (1978).

|60| W. Harrison, Pseudopotentials in the theory of metals, (Benjamin, New York, 1966); V. Heine and D. Weaire, Solid State Physics 24, 249 (1970); V. Heine, Solid State Physics 24, 1 (1970).

|61| R. Smoluckowski, Phys. Rev. 60, 661 (1941).

|62| J. Frenkel, Z. Physik 51, 532 (1928).

|63| J. Arponen, P. Hautojärvi, R. Nieminen and E. Pajanne, J. Phys. F 3, 2092 (1973); M. Manninen, R. Nieminen, P. Hautojärvi and J. Arponen, Phys. Rev. B12, 4012 (1975); R.M. Nieminen, J. Phys. F 7, 375 (1977).

|64| N.W. Ashcroft, Phys. Lett. 23, 48 (1966).

|65| R. Monnier and J.P. Perdew, Phys. Rev. B17, 2595 (1978); J.P. Perdew and R. Monnier, Phys. Rev. Lett. 37, 1286 (1976).

|66| See, e.g., the recent review by D.C. Langreth, Comments Solid State Phys. 8, 129 (1978) and references therein.

|67| See, e.g., H.F. Budd and J. Vannimenus, Phys. Rev. Lett. 31, 1218 (1973); J. Vannimenus and H.F. Budd, Solid State Commun. 15, 1739 (1974); A. Sugiyama, J. Phys. Soc. Japan 15, 965 (1965); D.C. Langreth, Phys. Rev. B5, 2842 (1972); B11, 2155 (1975); P. Feibelman, Phys. Rev. B3, 220 (1971); J.A. Appelbaum and E.I. Blount, Phys. Rev. B8, 483 (1973); J.E. Inglesfield and E. Wikborg, Solid State Commun. 15, 1727 (1974).

|68| T.B. Grimley, in Electronic Structure and Reactivity of Metal Surfaces, edited by E.G. Derouane and A.A. Lucas (Plenum, New York, 1976) p. 35; J.R. Schrieffer, J. Vac. Sci. Technol. 9, 561 (1972)1 13, 335 (1976); and in Dynamic Aspects of Surface Physics, edited by F.O. Goodman (Editrice Compositori,

Bologna, 1974); R. Gomer, in Solid State Physics, Vol. 30, p. 93; J.W. Gadzuk, in Surface Physics of Materials, edited by J.M. Blakely (Academic, New York, 1975), Vol. II, p. 339; B.I. Lundqvist, H. Hjelmberg and O. Gunnarsson, in Photoemission from Surfaces, edited by B. Feuerbacher, B. Fitton and R.F. Willis, (Wiley, New York, 1978).

|69| The numerous calculations using a cluster of atoms to describe the system are not discussed due to space limitations. See, for example, R.P. Messmer, C.W. Tuckler, Jr. and K.H. Johnson, Surf. Sci. 42, 341 (1974); I.P. Batra and O. Robaux, Surf. Sci. 49, 653 (1975); J. Harris and G.S. Painter, Phys. Rev. Lett. 36, 151 (1976); P.J. Jennings, G.S. Painter and R.O. Jones, Surf. Sci. 61, 255 (1976); N. Rösch and D. Menzel, Chem. Phys. 13, 243 (1976); C.H. Li and J.W.D. Connolly, Surf. Sci. 65, 700 (1977); R.P. Messmer and D.R. Salahub, Phys. Rev. B16, 3415 (1977), and references therein. See also R.O. Jones, this volume.

|70| J.R. Smith, S.C. Ying and W. Kohn, Phys. Rev. Lett. 30, 610 (1973); Phys. Rev. B11, 1483 (1975); L.M. Kahn and S.C. Ying, Solid State Commun. 16, 799 (1975). See also H.B. Huntington, L.A. Turk and W.W. White III, Surf. Sci. 48, 187 (1975).

|71| O. Gunnarsson and H. Hjelmberg, Physica Scripta 11, 97 (1975).

|72| N.D. Lang and A.R. Williams, Phys. Rev. Lett. 34, 531 (1975).

|73| N.D. Lang and A.R. Williams, Phys. Rev. Lett. 37, 212 (1976).

|74| K.Y. Yu, J.N. Miller, W.E. Spicer, N.D. Lang and A.R. Williams, Phys. Rev. B14, 1446 (1976).

|75| N.D. Lang and A.R. Williams, to be published.

|76| O. Gunnarsson, H. Hjelmberg and B.I. Lundqvist, Phys. Rev. Lett. 37, 292 (1976).

|77| O. Gunnarsson, H. Hjelmberg and B.I. Lundqvist, Surf. Sci. 63, 348 (1977).

|78| H. Hjelmberg, O. Gunnarsson and B.I. Lundqvist, Surf. Sci. 68, 158 (1977).

|79| H. Hjelmberg, submitted to Surf. Sci.

|80| J.P. Muscat and D.M. Newns, Phys. Lett. 60A, 348 (1977); 61A, 481 (1977).

|81| K. Ruedenberg, Rev. Mod. Phys. 34, 326 (1962); Localization and Delocalization in Quantum Chemistry, (Reidel, Dordrecht-

Holland, 1975), Vol. i, 223; M.J. Feinberg and K. Ruedenberg, J. Chem. Phys. 54, 1495 (1971).

|82| See, e.g., F. Seitz, The Modern Theory of Solids, (McGraw-Hill, New York, 1940), p. 352.

|83| S.A. Flodström, L.G. Petersson and S.B.M. Hagström, J. Vac. Sci. Technol. 13, 280 (1976).

|84| N.D. Lang and A.R. Williams, Phys. Rev. B16, 2408 (1977).

|85| L. Hedin and A. Johansson, J. Phys. B2, 1336 (1969).

|86| D.A. Shirley, Chem. Phys. Lett. 16, 220 (1972); L. Ley, S.P. Kowalczyk, F.R. McFeely, R.A. Pollak and D.A. Shirley, Phys. Rev. B8, 2392 (1973).

|87| A.R. Williams and N.D. Lang, Phys. Rev. Lett. 40, 954 (1978).

|88| K. Schönhammer and O. Gunnarsson, Solid State Commun. 23, 691 (1977); 26, 399 (1978); O. Gunnarsson and K. Schönhammer, Solid State Commun. 26, 147 (1978); Surf. Sci., to be published.

DENSITY FUNCTIONAL CALCULATIONS FOR ATOMIC CLUSTERS

R.O. Jones

Institut für Festkörperforschung der Kernforschungsanlage Jülich
D-5170 Jülich, F.R.G.

I. INTRODUCTION

The density functional approach to the calculation of electronic properties of surfaces and amorphous materials has played a central role in the lectures of this school. In this approach it is possible, within a single-particle picture, to derive equations which determine the total energy of an interacting system of electrons in an external field. An approximation to the exchange-correlation energy functional is unavoidable, but can be justified to a very considerable extent /1/. Perhaps less evident is the relationship between atomic clusters and surfaces or disordered systems, and the motivation for performing cluster calculations in the context of chemisorption will be discussed in Sec. II. In Sec. III, we discuss the solution of the density functional equations for clusters using the linear muffin-tin orbital method of Andersen /2/, and note some of the advantages of the scheme. Before examining larger clusters, it is essential to perform detailed calculations for small molecules, in order to test the ability of the density functional method to describe chemical bonding in cases where extensive data, both theoretical and experimental, are available. In Sec. IV, therefore, we discuss the bonding in some s- and sp-bonded diatomic molecules. There are pronounced trends which correlate well with the form of the valence functions of the constituent atoms and shed light on the use of pseudopotentials in density functional theory. We conclude (Sec. V) with a discussion of energies in atoms of transition elements, and note some difficulties in the use of a local approximation for the exchange-correlation energy density.

II. CLUSTER CALCULATIONS AND CHEMISORPTION

Some of the most important problems in surface physics involve the interaction between an adsorbed atom and a surface or between atoms in the neighbourhood of a surface. These situations are illustrated in Fig. 1 for the case of an ideal unreconstructed surface. As noted above, the density functional approach provides a

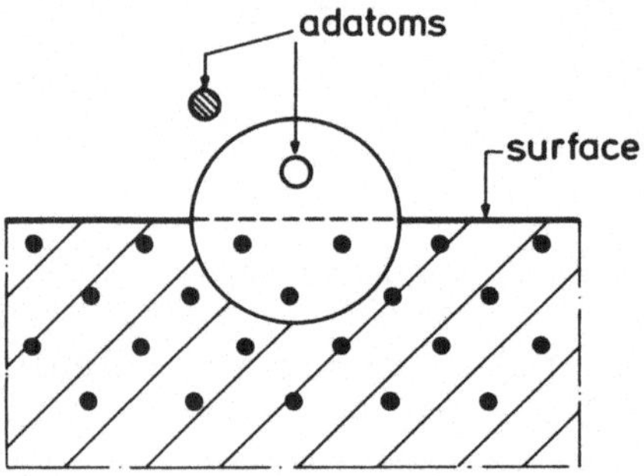

Fig. 1 : Two adatoms in the neighbourhood of a surface. The cluster approach focuses on the adatom and near neighbours (encircled). An example of the embedding approach would be to model the hatched region by jellium.

means of determining the total energy and the density of a system of electrons. In the context of surface physics, this would allow us in principle to calculate some very interesting quantities. For example, determination of the total energy as a function of adsorbate position would yield the equilibrium position of the adsorbate, the frequency of vibration around that position and the activation energy for diffusion on the surface. Knowledge of the same quantity for more than one adatom would allow the determination of reaction paths. The charge density is a necessary input into the calculation of quantities such as dipole moments and work functions.

The solution of the density function equations for chemisorption of a single atom on an extended surface has generally been limited to substrates represented by a uniform positive background /3/ or such a model supplemented by a weak pseudopotential /4/. These calculations indicate, however, that the main perturbation to the density is localized to within a few Bohr radii of the adatom and lends support to the local picture of chemical bonding used by chemists. In the cluster approach to chemisorption /5/, we follow this picture and represent the effect of the surface by a few near neighbours of the adsorbed atom. The model could be improved by increasing the cluster size until the binding energies are independent of the number of atoms or by embedding the cluster to simulate the boundary conditions more realistically /6/. For example, the hatched region in Fig. 1 could be replaced by a jellium half-space. At present, progress on such extensions has not been great and most work has focused on free clusters, typified by that encircled in Fig. 1.

The problem of calculating electronic charge distributions and energies of atomic clusters has confronted molecular physicists and chemists for decades. Before discussing the application of the density functional method, we should examine why the traditional ab initio methods of quantum chemistry are inappropriate for present purposes. In principle, it is possible to include correlation effects in the configuration interaction scheme, where the wave function of the <u>interacting</u> system is expressed as a linear combination of "configurations" of occupied spin-orbitals. The number of such determinantal functions is very large, however, even in small molecules. For example, the O_2 molecule has four valence orbitals ($3\sigma_g$, $1\pi_u$, $1\pi_g$, $3\sigma_u$) which could hold a total of twelve electrons. Since there are only eight valence electrons, the total number of determinantal functions is 12!/8!4!=495, and there are 142 multiplets. Although it is possible to eliminate many configurations because they contribute negligibly to the molecular state in question, for example on symmetry grounds, the convergence problems for larger molecules are prohibitive /8/. The density functional method requires the solution of a problem for a system of non-interacting particles, for which the wave function arises from a single configuration /9/. The numerical advantages are obvious.

III. DENSITY FUNCTIONAL EQUATIONS AND MUFFIN-TIN ORBITALS

In the density functional scheme /1/, the total energy E of a system of electrons in an external potential, V^{ext}, is a functional of the electron density $n(\underline{r})$. The ground state energy minimizes the functional and can be found by solving the one-particle problem

$$\left[-\frac{1}{2}\nabla^2 + V^{eff}(\underline{r}) - \varepsilon_n\right]\psi_n = 0 \quad . \qquad (1)$$

The eigenfunctions, ψ_n, determine the density

$$n(\underline{r}) = \sum_n f_n |\psi_n|^2 \quad , \qquad (2)$$

where the occupation numbers f_n must be compatible with the symmetry of the state in question. The effective potential, $V^{eff}(\underline{r})$, is given by

$$V^{eff}(\underline{r}) = V^{ext}(\underline{r}) + \frac{1}{2}\int d\underline{r}' \frac{n(\underline{r}')}{|\underline{r}-\underline{r}'|} + \frac{\delta E^{xc}[n]}{\delta n(\underline{r})} \qquad (3)$$

and the total energy

$$E[n] = T_o[n] + \int d\underline{r}\, n(\underline{r})\, V^{ext}(\underline{r}) + \frac{1}{2}\int d\underline{r} \int d\underline{r}' \frac{n(\underline{r})n(\underline{r}')}{|\underline{r}-\underline{r}'|} + E^{xc}[n] \qquad (4)$$

$T_o[n]$ is the kinetic energy of a system of non-interacting particles of density n, and the only term for which an approximation is necessary is the exchange-correlation energy functional, E^{xc}. For this we use the local spin-density (LSD) functional approximation /10/.

To solve the single-particle equations (1),(3), we use the linear muffin-tin orbital (LMTO) method /2/, developed originally for band theory. In close-packed materials, the one-electron potential is generally close to a muffin-tin form, i.e. spherically symmetric inside spheres centred on the atoms and constant between them. This approximation is illustrated in the cluster case in Fig. 2. In the ith sphere (or muffin-tin) centred at $\underline{r}_i^o$, solutions of (1) can be found, $\phi_{\ell i}(E_\nu,|r-r_i^o|)$, for each partial wave ℓ and energy E_ν. Andersen /2/ showed that it is possible to form a linear combination of $\phi_{\ell i}$ and its derivative with respect to energy, $\partial\phi_{\ell i}/\partial E$ for $E=E_\nu$, which match smoothly onto appropriate interstitial functions, i.e. spherical Bessel functions with kinetic energy parameter, κ, given by $\kappa^2=E-V_o$. Moreover, it is possible to construct linear combinations of these orbitals (LMTO), such that the eigenvalues obtained for a muffin-tin potential are correct to order $(E-E_\nu)^4$ and the eigenfunctions are correct to order $(E-E_\nu)^2$, where E_ν is the trial energy. The linear eigenvalue problem which results has much greater numerical efficiency than methods such as the KKR or scattered-wave method /11/.

In open systems such as small molecules, the muffin-tin approximation is inadequate for determining total energies. Danese /12/ has shown, for example, that the total energy of the C_2 mole-

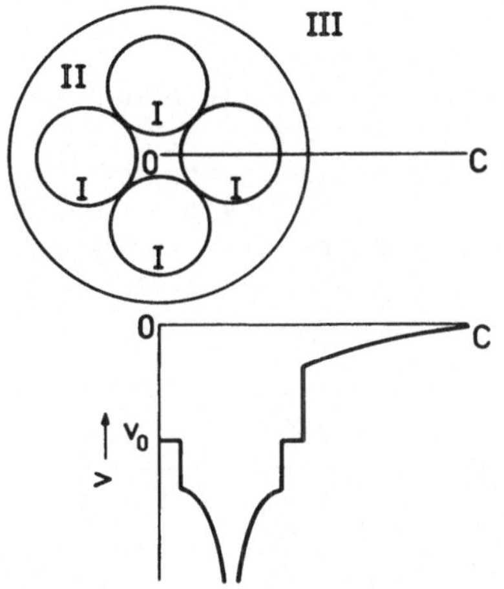

Fig. 2 : Muffin-tin approximation for a cluster. The potential is shown for the section OC. V_o is the interstitial potential or muffin-tin zero.

cule shows no minimum as a function of internuclear separation in a muffin-tin calculation, whereas its experimental binding energy is 6.25 eV. For our purposes, the LMTO method has the striking advantage that the inclusion of non-spherical potential terms is straightforward. For a non muffin-tin potential, the κ-values are free parameters and allow variational estimates of the energy.

In our calculations, therefore, we expand solutions of (1) as

$$\psi_n(\underline{r}) = \sum_{Li} C^n_{Li}\ \chi^i_L(\kappa,\underline{r}_i) \quad ; \quad L \equiv (\ell,m)\ ,\ \underline{r}_i \equiv \underline{r}-\underline{r}^o_i \tag{5}$$

and diagonalize the corresponding secular matrix

$$M^{ij}_{LL'} \equiv \langle\chi^i_L| - \frac{1}{2}\nabla^2 + V^{eff}(\underline{r}) - \varepsilon|\chi^j_{L'}\rangle \quad . \tag{6}$$

In evaluating the matrix elements and the total energy, the full three-dimensional potential is used. The calculation is carried out self-consistently and the total energy is minimized with respect to κ.

IV. BONDING IN ALKALI AND GROUP IV DIMERS

The alkali dimers Li_2, Na_2, K_2, Rb_2 and Cs_2 have a single σ-bond, as in the case of the hydrogen molecule /13/, and have been studied extensively over the last fifty years. In Fig. 3, we show calculated spectroscopic constants for these molecules and

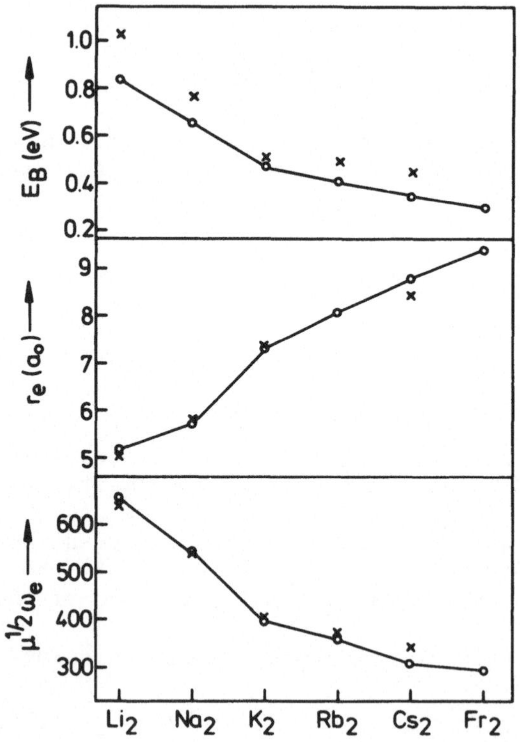

Fig. 3 : Binding energies (E_B), equilibrium internuclear separations (r_e) and weighted vibration frequencies ($\mu^{1/2}\ \omega_e$, where μ is the reduced mass in atomic weight units and ω_e is in cm^{-1}) for the alkali dimers. Experimental values (crosses), density functional results (circles). See Ref. 14.

the corresponding measured values. The agreement is very satisfactory and comparison with accurate MC-SCF calculations for Li_2 and Na_2 is also good /14/. It is significant that there is a break in binding energy trends between Na_2 and K_2, and we return to this below. For the moment, we note that the break correlates with the extent of the tails of the valence wave functions in the constituent atoms (Fig. 4).

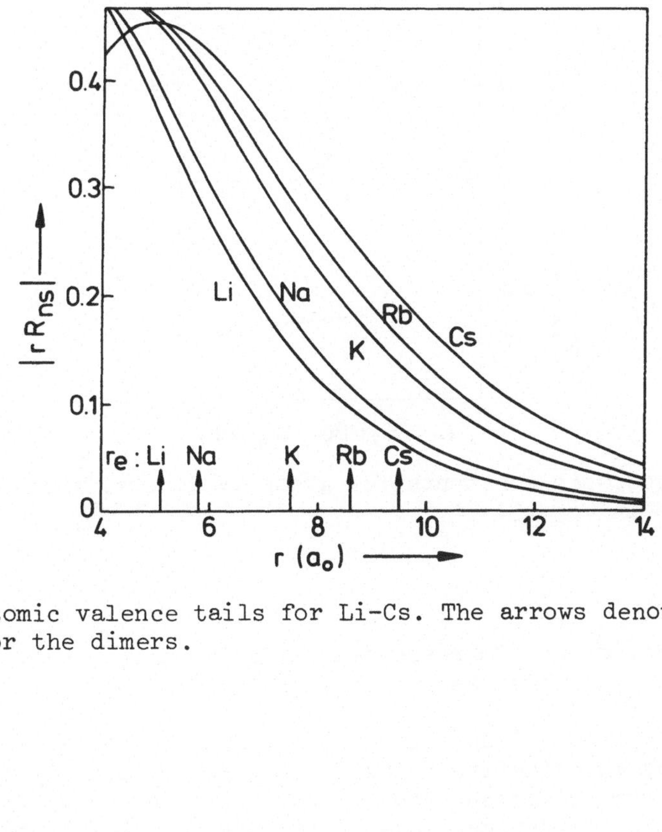

Fig. 4 : Atomic valence tails for Li-Cs. The arrows denote $r_e/2$ calculated for the dimers.

Bonding trends are also apparent in the recent work on group IV dimers (C_2, Si_2, Ge_2, Sn_2, Pb_2) /15/. In each case, we have performed calculations for a variety of low-lying states and find that the ground state is $^3\Sigma_g^-$, except for C_2, where the experimental ground state is $^1\Sigma_g^+$. In Fig. 5, we show spectroscopic constants for the $^3\Sigma_g^-$ state of each molecule. Although the experimental data is much less extensive than for alkali dimers, the calculated results show pronounced discontinuities similar to those

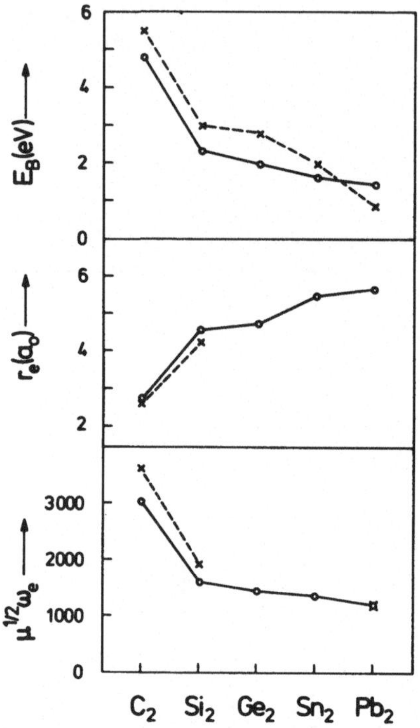

Fig. 5: Spectroscopic constants for group IV dimers. The notation is the same as in Fig. 3.

observed in that group. Corresponding effects on the valence function tails are also to be seen in Fig. 6.

The pseudopotential approach, in which the effects of the core are replaced by an effective potential, is an obvious way to study similar atoms from different rows of the periodic table, and Austin and Heine /16/ applied these ideas to trends in atomic radii and s-p splittings in alkali and noble metals. With increasing atomic number, there is a general increase in atomic radius and a reduction in the ionization energy. Superposed on these trends, however, they noted that the introduction of a new sub-shell in the core (e.g. a p-shell on going from Li to Na or a d-shell in going from Si to Ge) leads to a more attractive potential and a smaller increase in atomic radius. This is the case if the additional level is weakly bound and contributes to the core potential at radii larger than

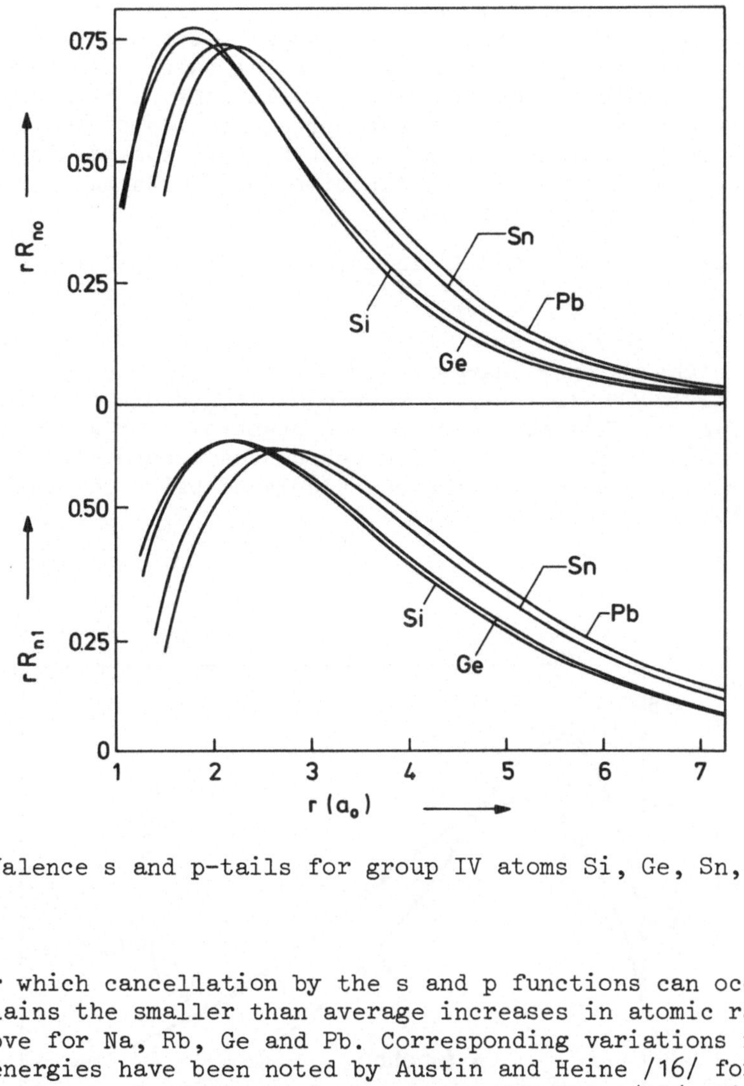

Fig. 6 : Valence s and p-tails for group IV atoms Si, Ge, Sn, Pb.

those for which cancellation by the s and p functions can occur. This explains the smaller than average increases in atomic radii noted above for Na, Rb, Ge and Pb. Corresponding variations in ionization energies have been noted by Austin and Heine /16/ for the alkalis and are also present in the group IV atoms /15/. Since the binding trends in group IA and group IV dimers correlate rather well with the atomic radii (or the tails of the valence functions) and these trends can be understood in terms of pseudopotentials, it is natural to examine the role of pseudopotentials in density functional theory.

The pseudopotential method has been used widely to calculate electronic properties and charge densities. In such calculations /17/, the total density in Eq. (2)-(4) is replaced by the pseudodensity, obtained from smooth, nodeless valence orbitals, and V^{ext} by a pseudopotential V^{ext}_{ps}, which simulates all effects of the nuclei and the core electrons. As can be seen by inspecting these equations, it is not at all obvious that a pseudopotential - often obtained by fitting optical band gaps in the bulk - can be used in the density functional context. The correlations we have noted above suggest, however, that an appropriate choice of pseudopotential would be to maximize the agreement between the pseudo-wavefunction and the atomic valence function outside the core region. For Si, the pseudopotential /18/

$$V_{ps}^{ext}(r) = \frac{-Z_v}{r} \frac{\{1-\exp(-\lambda r)\}}{\{1+\exp\lambda(r_c-r)\}} + V_o \exp[\lambda(r_c-r)-1]\ \Theta(r_c-r) \tag{7}$$

where Z_v is the valence charge (4), λ=14 a.u., r_c=0.98 a.u. and V_o=0.2 Ry gives the fit between pseudofunction and atomic function shown in Fig. 7. This choice of pseudopotential also gives a good fit to the Si^+ ionic functions and reproduces accurately the ionization and s-p promotion energies calculated using the full functional /18/.

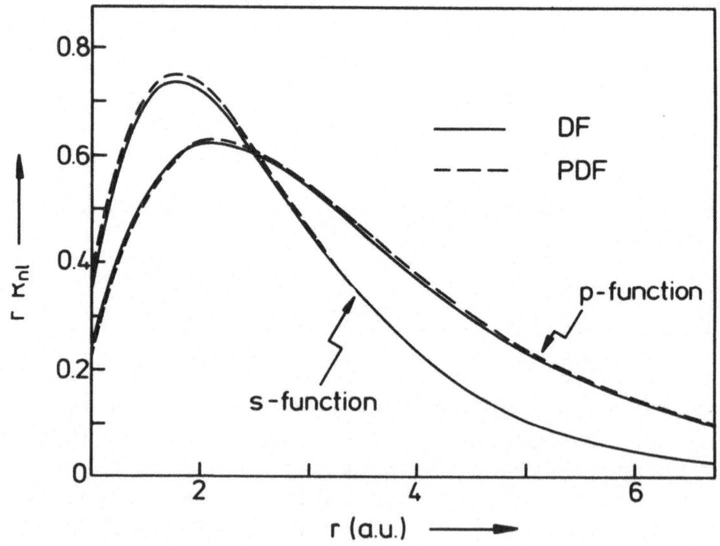

Fig. 7 : Tails of valence eigenfunctions for the Si atom calculated using density function (DF, full curve) and pseudodensity functional (PDF, broken curve).

Calculation of the binding energy using the PDF with this pseudopotential shows remarkably good agreement with the DF results, as shown in Fig. 8. For comparison, we show the results using the pseudopotential of Appelbaum and Hamann /19/ ((d) in Fig. 8). The

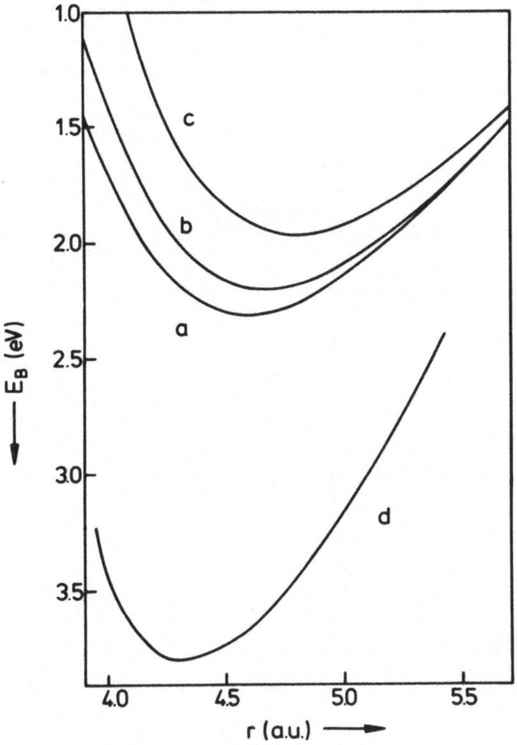

Fig. 8 : Binding energy curves for the ground state of Si_2: (a) DF; (b) PDF with $\lambda=14$, $r_c=0.98$, $V_o=0.2$ Ry; (c) PDF with $\lambda=12$, $r_c=1.07$, $V_o=0$; (d) PDF with core potential of Appelbaum and Haman (Ref. 19).

discrepancy in this case shows that the pseudopotential must be chosen with great care. The charge transfer or density difference plots which result from the DF and PDF schemes is shown in Fig. 9 for the ground state. Differences outside the core are unavoidable, since the fit to the valence functions is not perfect, but it is clear that the differences are minor in this region. Calculations have also been carried out for Sn_2 and Pb_2 with similar results /15/.

This comparison between the results of pseudopotential and all-electron calculations strongly supports the use of pseudopoten-

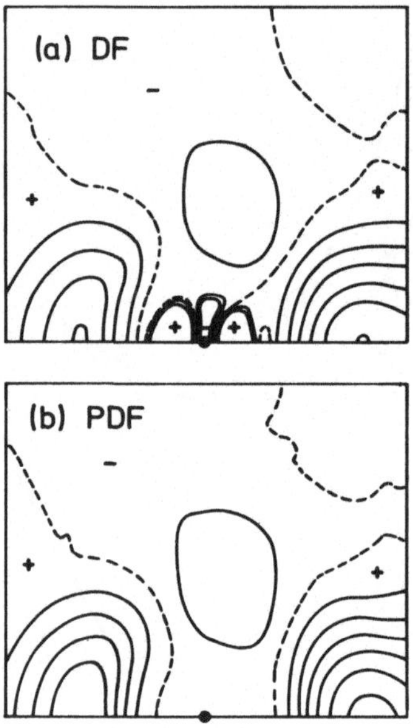

Fig. 9 : Density difference plot for Si_2: (a) density functional, (b) pseudodensity functional, for r_e=5.4 a_o. The zero contour is broken and the interval is 0.04 electrons a_o^{-3}.

tials in energy calculations, provided that the pseudopotential is chosen correctly. It should be noted, however, that the PDF and DF calculations take comparable amounts of computer time and the difference between the larger total energies in the latter (molecule and constituent atoms) can be determined with essentially comparable accuracy.

V. LOCAL SPIN-DENSITY FUNCTIONAL AND TRANSITION ELEMENT ATOMS

The previous sections have dealt with bonding involving predominantly s and p-electrons, where spin-polarization effects play a secondary role. This is not true in systems containing d-electrons and the use of the local spin density (LSD) function is more open to question. A discussion of our calculations of transition metal dimers is outside the scope of this lecture, but it is interesting to study calculations for transition atoms, where extensive experimental data are available. Calculations of energy differences in atoms have proved to be of great value in studying trends in molecular binding energies and cohesive energies and in assessing the accuracy of the LSD approximation.

In Fig. 10, we show the differences between the total energies for the $d^{n-1}s^1$ and $d^{n-2}s^2$ configurations for iron-series atoms /20/. Similar trends are obtained for the 4d-series, though d-occupancy is generally more favourable. Also shown are results of relativistic Hartree-Fock calculations /21/ of the same energy difference. It is

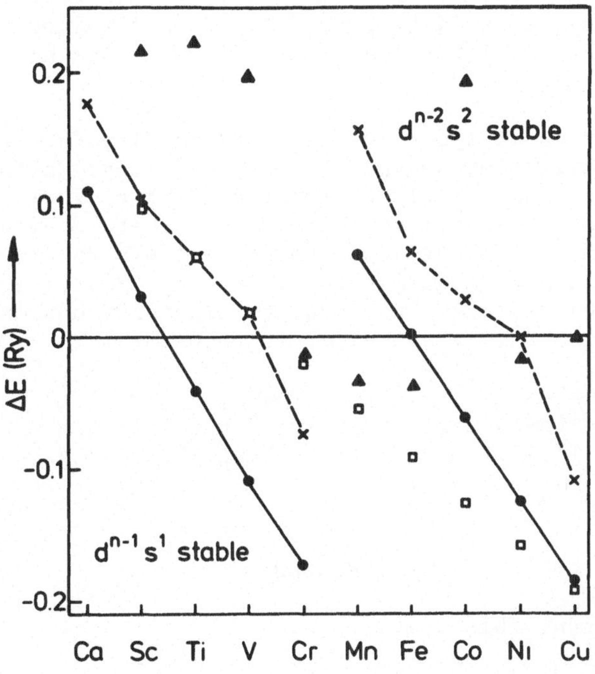

Fig. 10 : Energy differences, ΔE, for atoms of 3d-transition series. Squares: non spin-polarized results, full circles: LSD results, crosses: experimental values, triangles: relativistic Hartree-Fock values (Ref. 21). For more information, see Ref. 20.

evident that the LSD approach describes the trends across the series rather well, favouring d-occupancy by consistently 1 eV in both series. The break in the centre of the series, arising from the flipping of a d-spin in the presence of s-spins, is given very accurately by the LSD functional.

The tendency of the functional to favour d-occupancy is particularly interesting in the context of energy differences between bulk and constituent atoms (cohesive energy) or a molecule and its constituent atoms (binding energy). In cases where both bulk or molecular and the atomic states have essentially the same d-character, the errors due to the incorrect description of s-d transfer should nearly cancel. An example is the cohesive energy of 4d transition elements, where excellent agreement with experiment has been obtained /22/. In the 3d series, on the other hand, most of the atoms have s^2 ground states, while the bulk state is generally closer to s^1. This leads to substantially greater errors in the cohesive energies, as evident from Ref. 22. A detailed discussion of similar effects in low-lying states of 3d-dimers has been given elsewhere /23/.

Our experience in applying the local density functional method to atoms and small molecules has been encouraging. Calculated binding energies are consistently better than Hartree-Fock and comparable with extensive ab initio calculations where available. The extension to non-local or ℓ-dependent potentials should remove some of the systematic discrepancies and we are optimistic about the prospects for the calculation of larger clusters, not only in the context of chemisorption, but also for problems involving isolated defects.

The work reviewed here was carried out together with John Harris (Refs. 9,14,15,18,20,23). We thank Olle Gunnarsson for helpful discussions.

REFERENCES

/1/ O. Gunnarsson, this volume.

/2/ O.K. Andersen and R.G. Woolley, Mol. Phys. 26, 905 (1973); O.K. Andersen, Phys. Rev. B12, 3060 (1975).

/3/ N.D. Lang and A.R. Williams, Phys. Rev. Lett. 37, 212 (1976).

/4/ O. Gunnarsson, H. Hjelmberg and B.I. Lundqvist, Phys. Rev. Lett. 37, 292 (1976), Surface Sci. 63, 348 (1977).

/5/ For early calculations in which the emphasis is on eigenvalues rather than the total energy, see I.P. Batra and O. Robaux, Surface Sci. 49, 653 (1975) and J. Harris and G.S. Painter, Phys. Rev. Lett. 36, 151 (1976).

/6/ Discussions of embedding have been given by T.B. Grimley and C. Pisani, J. Phys. C 7, 2831 (1974); E.A. Hyman, Phys. Rev. B11, 3739 (1975); O. Gunnarsson and H. Hjelmberg, Physica Scripta 11, 97 (1975).

/7/ There are numerous parametric schemes, such as the extended Hückel method, which have been applied in the present context. They lie outside the scope of the present discussion.

/8/ A discussion of methods for including correlation effects in molecular calculations has been given by A.C. Wahl and G. Das, Advan. Quantum Chem. 6, 261 (1972). Chemisorption cluster calculations which include dominant correlation effects have been given by, for example, S.P. Walch and W.A. Goddard III, Surface Sci. 72, 645 (1978).

/9/ J. Harris and R.O. Jones, Phys. Rev. (to be published).

/10/ For more details, see O. Gunnarsson, J. Harris and R.O. Jones, Phys. Rev. B15, 3025 (1977). The spin-density functional of O. Gunnarsson and B.I. Lundqvist, Phys. Rev. B13, 4274 (1976) is used throughout.

/11/ See J.C. Slater, Quantum Theory of Molecules and Solids, Vol. IV (McGraw-Hill, New York, 1974).

/12/ J.B. Danese, J. Chem. Phys. 61, 3071 (1974).

/13/ For density functional calculations of H_2, see O. Gunnarsson and P. Johansson, Int. J. Quantum Chem. 10, 307 (1976) and Gunnarsson, Harris and Jones (Ref. 10).

/14/ For more details, see J. Harris and R.O. Jones, J. Chem. Phys. 68, 1190 (1978).

/15/ J. Harris and R.O. Jones, J. Chem. Phys. (to be published).

/16/ B.J. Austin and V. Heine, J. Chem. Phys. 45, 928 (1966).

/17/ See, for example, J.A. Appelbaum and D.R. Hamann, Rev. Mod. Phys. 48, 479 (1976); H. Wendel and R.M. Martin, Phys. Rev. Lett. 40, 950 (1978).

/18/ J. Harris and R. O. Jones, Phys. Rev. Lett. 41, 191 (1978).

/19/ J.A. Appelbaum and D.R. Hamann, Phys. Rev. B8, 1777 (1973).

/20/ J. Harris and R.O. Jones, J. Chem. Phys. 68, 3316 (1978) and to be published.

/21/ T. Kagawa, Phys. Rev. A12, 2245 (1975).

/22/ V.L. Moruzzi, A.R. Williams and J.F. Janak, Phys. Rev. B15, 2854 (1977).

/23/ J. Harris and R.O. Jones, J. Chem. Phys. (to be published).

IMPURITY KNIGHT SHIFT AND ELECTRIC FIELD GRADIENTS AT Al NUCLEI IN DILUTE SUBSTITUTIONAL Al-Li ALLOYS

M. Manninen

Res. Inst. for Theor. Phys., Univ. of Helsinki
00170 Helsinki 17, FINLAND

R. Monnier

Laboratorium für Festkörperphysik, EHT-Z
8093 Hönggerberg, Switzerland

After a brief summary of the necessary notions of nuclear magnetic resonance, we present results for the contact spin-densities at a substitutional Li nucleus, and for the electric field gradient caused by the impurity at the first three nearest neighbour host nuclei in a dilute Al-Li alloy. The displaced charge- and spin-densities around the impurity are computed self-consistently using the spin-density functional formalism and including the main effect of the discreteness of the lattice. The latter is seen to play an essential role in determining the phase of the Friedel oscillations (and therefore the electric field gradient) and the impurity Knight shift. Due to the presence of polarizable bound states, the conventional formula for the Knight shift $K \sim \langle |\Psi_{\vec{k}}(0)|^2 \rangle_{FS}$ is found to be inadequate.

I. INTRODUCTION

This lecture is concerned with the effect of a small concentration of substitutional lithium impurities on the nuclear magnetic resonance (NMR)[1] properties of aluminium metal.

If the concentration of impurities is not too small (say a few percent), one will observe a signal associated with the Li nuclei themselves. This extra absorption will occur when the local magnetic field at the impurity, H(0) , satisfies the resonance condition :

$$2\pi.\nu = \gamma_i \, H(0) \tag{1}$$

where ν is the frequency of the microwave field and γ_i is the gyromagnetic ratio for the impurity nucleus.

H (0) differs from the applied static magnetic field by a hyperfine contribution ΔH^i_{hf} which, at a site of cubic symmetry and in the absence of spin-orbit coupling, is given by the Fermi contact term:

$$\Delta H^i_{hf} = -(8\pi/3)\ \mu_B\ (n^+(0) - n^-(0)) \tag{2}$$

where μ_B is the Bohr magneton and $n^+(0)$ ($n^-(0)$) is the contact density of electrons with spin parallel (antiparallel) to the applied field. A more conventional, though less general* expression for the hyperfine field can be written down:

$$\Delta H^i_{hf} = (8\pi/3)\ <|\Psi_k(0)|^2>_{FS}\ V\cdot\chi_m\cdot H \tag{3}$$

Here χ_m is the spin susceptibility per unit volume of the host metal, $<|\Psi_{\vec{k}}(0)|^2>_{FS}$ is the square of the wave-function at the impurity nucleus, averaged over all Bloch states $\vec{k}$ at the Fermi surface, V is the normalization volume for $\Psi_{\vec{k}}(\vec{r})$, and H is the applied magnetic field.

Experimentally, the hyperfine field is obtained as the difference (at constant microwave frequency) between H^i_m, the value of H at which resonant absorption by the impurity occurs in the metal, and H^i_d, the corresponding value for the same nucleus in an insulating diamagnetic salt **. The <u>impurity Knight shift</u> is defined as the ratio:

$$K^i_m = \frac{H^i_d - H^i_m}{H^i_m} \cong \frac{\Delta H^i_{hf}}{H^i_m} \tag{4}$$

With our expressions for the hyperfine field this becomes

---

* The derivation of eq. (3) assumes that the polarizing magnetic field only affects the relative population of up and down spin electrons, without modifying the corresponding wave-functions, which are taken to be the same as in the absence of field. This is not strictly true since the two population will not experience the same exchange potential any more. The approximation is particularly severe in the presence of localized bound states of s-like symmetry, whose contribution to the hyperfine field (≡core polarization) can be quite large.

** A small extra correction arises from the change in chemical environment (≡chemical shift).

$$K_m^i = (8\pi/3) < |\Psi_{\vec{k}}(0)|^2 >_{FS} V \chi_m \tag{5a}$$

and

$$K_m^i = (8\pi/3) \frac{n^+(0) - n^-(0)}{n_0^+ - n_0^-} \chi_m \tag{5b}$$

where in (5b) we have used the definition of the bulk magnetization

$$M_z = -\mu_B (n_0^+ - n_0^-) = \chi_m H \tag{6}$$

In section 4 we shall present results for K_{Al}^{Li} computed from equations (5a) and (5b), using densities and wave-functions at the Fermi surface obtained from a selfconsistent density functional calculation. Note that use of eq. (5b) requires a spin-polarized treatment, whereas with (5a) the computations can be done for the unpolarized case.

Next let us study how the NMR signal from the solvent nuclei themselves is affected by the presence of the impurities. Due to the oscillating spin density induced by the latter, the hyperfine field felt by a close-by host nucleus will be modified with respect to its value in the pure metal. This change will depend on the distance of the host atom from the impurity, so that one expects a series of weak satellites to appear besides the main absorption peak, each of them corresponding to a given shell of neighbours. In aluminium based non-magnetic dilute alloys, these lines cannot be resolved and produce only a broadening and a slight shift of the main peak.

This small effect is completely overwhelmed by the change in the absorption line-shape associated with the interaction of the electric quadrupole moment of the host nuclei with the electric field gradient (EFG) caused by the impurity at the corresponding lattice site *. Defining the z-axis of a coordinate system to be along the line connecting the impurity (placed at the origin) with the nucleus of interest (placed at $\vec{r}_n$), the EFG is given by

$$eq(\vec{r}_n) \equiv \left. \frac{\partial^2 V}{\partial z^2} \right|_{\vec{r} = \vec{r}_n} \tag{7}$$

* In the pure metal there is no quadrupolar coupling because each nucleus is at a site of cubic symmetry, where the EFG vanishes identically as a result of Laplace's equation (the EFG is assumed to arise from charges outside of the nucleus).

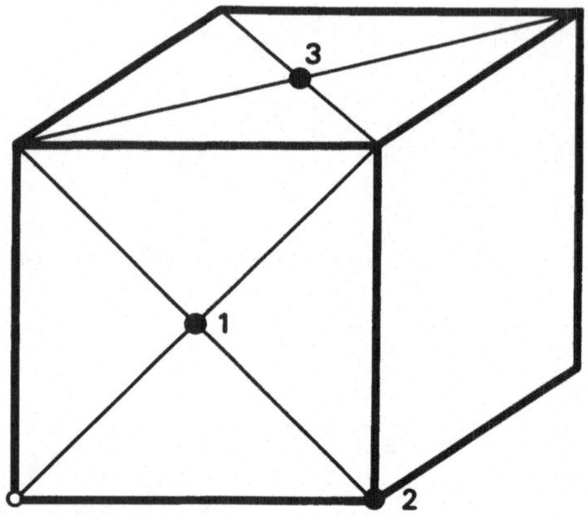

Figure 1: Crystal structure of aluminium containing a single impurity denoted by the open circle. The numbers indicate the shell of neighbours to which the host atoms belong.

where $V(\vec{r}_n)$ is the electrostatic potential at $\vec{r}_n$. With an appropriate choice of the other two axes, the tensor of second derivatives of V can be made diagonal. One then defines an asymmetry parameter

$$\eta(\vec{r}_n) = (1/eq(\vec{r}_n)) \left(\frac{\partial^2 V}{\partial x^2} - \frac{\partial^2 V}{\partial y^2} \right)\Bigg|_{\vec{r} = \vec{r}_n} \quad (8)$$

which measures the deviations from cylindrical symmetry about the z-axis. If the latter is a n-fold rotation or rotation-reflexion axis with $n>2$, the asymmetry parameter vanishes. Fig. 1 shows the position of the first three nearest neighbours to the impurity in the fcc structure. According to the above criterion one expects $\eta \neq 0$ for 1 and 3, and $\eta = 0$ for 2.

To first order, the coupling of the EFG to the electric quadrupole moment eQ of a nucleus with spin $I>1/2$ produces a splitting of the absorption line into $2 \cdot I$ components, shifted with respect to the unperturbed resonance by an amount

$$\Delta\nu(m \leftrightarrow m-1) = \frac{1}{2} (m-\frac{1}{2}) \nu_Q \{3 \cos^2\theta - 1 - \eta \sin^2\theta \cos 2\phi\} \quad (9)$$

where (θ,ϕ) specifies the orientation of the applied field in the above-defined coordinate system, m and m-1 are the azimuthal quantum numbers for the two levels between which the transition occurs, and ν_Q is the so-called quadrupolar frequency

$$\nu_Q(\vec{r}_n) = \frac{3eq(\vec{r}_n)eQ}{2I(2I-1)h} \qquad (10)$$

Equation (9) shows that for a nucleus with half-integer spin, the transition ($1/2 \leftrightarrow -1/2$) is unaffected by the quadrupolar coupling. The experimental absorption curve consists in a complicated superposition of such ($2\cdot I$)-tuplets, and we refer the reader to the specialized literature[1-4] for more details about the methods used to extract NMR parameters from the observed spectra.

In the limit of vanishingly small asymmetry, the impurity affects the frequency shift (eq. (9)) only through the associated electric field gradient $e\cdot q(\vec{r}_n)$. In section 4 we shall present results for the EFG at the first three nearest neighbours obtained in that limit * for AlLi, using a theory due to Kohn and Vosko[5] (KV). These authors write the EFG distribution around an isolated impurity as

$$q(\vec{r}) = (8\pi/3)\ \alpha(\vec{k}_F)\ \delta n(r) \qquad (11)$$

where $\delta n(r)$ is the excess density which the impurity would induce if placed in a uniform electron gas of density equal to that of the host, and $\alpha(\vec{k}_F)$ is the so-called Bloch (or core) enhancement factor which accounts for the effects of band structure and core-orthogonalization, and only involves properties of the host metal. The latter has been computed by Holtham and Jena[6] for Al, using an asymptotic formula, i.e. valid in principle only at large distance from the impurity, given by KV. We shall use their values in conjunction with our density functional results for $\delta n(r)$ to estimate the EFG's.

II. SPIN-DENSITY FUNCTIONAL FORMALISM

The spin-density functional (SDF) method is a generalization[7-9] in order to include spin-dependent effects, of the theory of the inhomogeneous electron gas developped by Hohenberg, Kohn and Sham (HKS)[10,11]. Due to the additional degree of freedom introduced with the spin, the basic variable of the theory is now the 2 x 2 spin-density matrix $\overleftrightarrow{s}(\vec{r})$ defined by

$$s_{\alpha\beta} = \langle\Phi|\psi^+_\beta(\vec{r})\ \psi_d(\vec{r})|\Phi\rangle \qquad (12)$$

---

* This involves no approximation for the next-nearest neighbours. For the nearest neighbours, experimentally observed anisotropies range from 3 percent in AlGe, AlGa, AlCd, AlSi to 37 percent in AlSn (see ref. 4).

where $|\Phi\rangle$ is the ground state wave function and $\psi^+_\beta(\psi_\alpha)$ is a creation (annihilation) operator for an electron with spin $\beta(\alpha)$. The total density of electrons $n(\vec{r})$, which used to be the basic variable in the HKS theory, is obtained as the trace of $\vec{s}$ over spin states. In the presence of a polarizing magnetic field H_o, which fixes the direction of the quantization axis, the spin-density matrix becomes diagonal, and the ground state of the system is completely specified by the two spin-densities parallel and antiparallel to the applied field, introduced in the preceding section. Another commonly used pair of variables is the total density, $n(\vec{r}) = n_+(\vec{r}) + n_-(\vec{r})$, and the local polarization $\xi(\vec{r}) = (n_+(\vec{r})-n_-(\vec{r}))/n(\vec{r})$. For electrons in simple metals (Pauli paramagnets) $\xi(\vec{r})$ is always negative with our definition. In terms of the above variables, the ground state energy of a metal containing a single impurity with atomic number Z_i can be written

$$E[n;\xi] = T_s[n;\xi] + E_{xc}[n;\xi] + \frac{e^2}{2}\int d^3r d^3r' \frac{n(\vec{r})n(\vec{r}')}{|\vec{r}-\vec{r}'|}$$

$$+ \int d^3r\, w_i(|\vec{r}|)n(\vec{r}) + \int d^3r \sum_{\vec{\ell}\neq\vec{0}} w_m(|\vec{r}-\vec{\ell}|)n(\vec{r})$$

$$+ \sum_{\vec{\ell}\neq\vec{0}} \frac{(Z_i e)(Z_m e)}{|\vec{\ell}|} + \frac{1}{2}\sum_{\vec{\ell},\vec{\ell}'}{}' \frac{(Z_m e)^2}{|\vec{\ell}-\vec{\ell}'|} \qquad (13)$$

where T_s is the kinetic energy of an assembly of non-interacting electrons with density n and polarization ξ, E_{xc} is the exchange correlation energy of the interacting system and the third term on the rhs of (13) is the Hartree electrostatic energy. The electrons interact with the impurity via a local potential w_i to be specified later, and the electron-host ion interaction is described by the local pseudopotential w_m, which consists in a long-range attractive part and a short-range repulsive part w_R

$$w_m(r) = -\frac{Z_m e^2}{r} + w_R(r) \qquad (14)$$

The last two terms in (13) describe the Coulomb interaction between the ions distributed over sites $\vec{\ell}$, and the prime over the last sum means that the terms $\ell,\ell' = 0$ and $\ell = \ell'$ are omitted.

At large distances from the impurity the polarization $\xi(r)$ goes to its bulk value ξ_o, which can be thought of as being maintained by the stabilizing external magnetic field H_o:

$$\xi_o = -(\chi_m H_o / \mu_B n_o) = (n_o^+ - n_o^-)/n_o \tag{15}$$

where $n_o = n_o^+ + n_o^-$ is the average density of valence electrons in the pure host.

For macroscopic systems, the individual Coulomb terms in (13) are badly behaved, and it is advantageous to add and subtract a fictitious neutralizing positive background $n^B(\vec{r})$, which results in the following expression for the ground state energy:

$$E[n;\xi] = T_s[n;\xi] + E_{xc}[n;\xi] + \frac{1}{2}\int d^3r\, \phi_n(\vec{r})[n(\vec{r}) - n^B(\vec{r})]$$

$$+ \int d^3r\, w_i(r)\, n(\vec{r}) + \int d^3r\, \delta v(\vec{r})[n(\vec{r}) - n^B(\vec{r})]$$

$$+ \int d^3r \sum_{\vec{\ell}\neq o} w_R(|\vec{r}-\vec{\ell}|)n^B(\vec{r}) + \frac{e^2}{2}\iint d^3r\, d^3r' \frac{n^B(\vec{r})n^B(\vec{r}')}{|\vec{r} - \vec{r}'|}$$

$$- \int d^3r \sum_{\vec{\ell}\neq o} \frac{Z_m e^2}{|\vec{r}-\vec{\ell}|} n^B(\vec{r}) + \sum_{\vec{\ell}\neq o} \frac{(Z_i e)(Z_m e)}{|\vec{\ell}|} + \frac{1}{2}\sum_{\vec{\ell},\vec{\ell}'}{}' \frac{(Z_m e)^2}{|\vec{\ell}-\vec{\ell}'|} \tag{16}$$

where

$$\phi_n(\vec{r}) = e^2 \int d^3r' \frac{n(\vec{r}')-n^B(\vec{r}')}{|\vec{r} - \vec{r}'|} \tag{17}$$

and

$$\delta v(\vec{r}) = \sum_{\vec{\ell}\neq o} w_m(|\vec{r} - \vec{\ell}|) + e^2 \int d^3r' \frac{n^B(\vec{r}')}{|\vec{r} - \vec{r}'|} \tag{18}$$

Since the impurity is substitutional, i.e. occupies a vacancy in the pure host, a natural choice for $n^B(r)$ is:

$$n^B(\vec{r}) = n_o\, \theta(r - R_{WS}) \tag{19}$$

where R_{WS} is the host Wigner-Seitz radius and θ is the step function.

The functional (16) satisfies a variational principle according to which the ground state density and polarization (or equivalently the ground state spin-densities) are the ones which minimize $E[n;\xi]$ under the subsidiary condition that the number of electrons is kept constant. As a consequence of this property, the ground state spin-densities can be obtained from a selfconsistent solution of the following set of coupled Schrödinger-like equations:

$$[-(\hbar^2/2m)\ \nabla^2+\phi_n(\vec{r})+w_i(\vec{r})+v^\sigma_{xc}(\vec{r})+\delta v(\vec{r})]\,\psi^\sigma_i(r) = \varepsilon^\sigma_i\ \psi^\sigma_i(\vec{r}), \qquad (20)$$

by summing the modulus squared of the auxiliary single-particle wave functions over all occupied levels:

$$n^\sigma(\vec{r}) = \sum_i^{occ} |\psi^\sigma_i(\vec{r})|^2 \qquad (21)$$

The exchange-correlation potential $v^\sigma_{xc}(\vec{r})$ is equal to the functional derivative of $E_{xc}[n;\xi]$ $(= E_{xc}(n^+;n^-))$, with respect to n^σ. Since the exchange-correlation energy is not known exactly, it is replaced in practice by its value in the local spin-density approximation (LSDA):

$$E_{xc}[\,n;\xi\,] \cong \int d^3r\ n(\vec{r})\ \varepsilon_{xc}\ (n(\vec{r});\xi(\vec{r})) \qquad (22)$$

where ε_{xc} is the exchange-correlation energy per particle in an electron gas of uniform density and polarization*. In the LSDA, $v_{xc}(r)$ is approximated by the exchange-correlation contribution to the chemical potential

$$v^\sigma_{xc}(\vec{r}) = \mu^\sigma_{xc}(n(\vec{r});\xi(r)) = \frac{\partial(n(\vec{r})\ \varepsilon_{xc}(n(\vec{r});\xi(\vec{r}))}{\partial n^\sigma(\vec{r})} \qquad (23)$$

Parametrized forms for $\varepsilon_{xc}(n;\xi)$ and $\mu^\sigma_{xc}(n;\xi)$ can be found in ref. 9.

To make equations (20) and (21) operational, we still have to define $w_i(r)$ and $w_R(r)$. Since we are interested in the actual spin-densities at the impurity site, it is out of the question to describe the electron-impurity interaction by a pseudopotential. In fact for small values of the impurity charge Z_i, there is no

---

* Note that with this approximation the rigorous proof of the minimal property on $E[n;\xi]$ is lost.

difficulty in using for $w_i(r)$ the bare Coulomb interaction between the nucleus and the electron

$$w_i(r) = - \frac{Z_i e^2}{r} \tag{24}$$

For large values of Z_i, the above potential leads to bound states which are strongly localized and therefore poorly described by the LSDA. To get around that difficulty, Bryant and Mahan[12] have recently proposed a "hybrid"-type of approach in which so-called self-interaction terms which are overestimated in the LSDA are subtracted in an ad hoc way (see ref. 12 for more details).

The ground state properties of many simple metals are well described[13,14] by the Ashcroft empty-core pseudopotential[15], for which $w_R(r)$ cancels $-Z_m e^2/r$ exactly within a distance r_c from the nucleus

$$w_R(r) = \frac{Z_m e^2}{r} \theta(r_c - r) \tag{25}$$

The model works particularly well for aluminium for which the relevant parameters are $Z_m = 3$ and $r_c = 1.12$ a.u. The radius of the Wigner-Seitz sphere for Al is 2.99 a.u. (1 a.u. = 0.529 A).

III. APPROXIMATE TREATMENTS

The exact solution of the single-particle equations (20) requires a full three-dimensional calculation, due to the presence of the term $\delta v(\vec{r})$ in the effective one-body potential

$$V^{\sigma}_{eff}(\vec{r}) = \phi_n(\vec{r}) - \frac{Z_i e^2}{r} + \mu^{\sigma}_{xc}(n(\vec{r});\xi(\vec{r})) + \delta v(\vec{r}) \tag{26}$$

A common practice, based on the experience that in the pure host, band structure effects (i.e. effects due to the discreteness of the ionic distribution) are small, has been to leave out the term $\delta v(\vec{r})$ from the equations, leading to the jellium model for the substitutional impurity[12,16]. With this approximation, the effective potential is spherically symmetric, so that the problem is reduced to one dimension *.

In the jellium model one assumes that $\delta v(\vec{r})$ (which is characteristic of the the defected crystal) is a small perturbation, by appealing to a property of the perfect host (weakness of band-structure effects). Such a line of thought is of course incorrect: In the pure host, inhomogeneities in the electron density ** are only due to the oscillations of the periodic lattice potential around its average, irrespective of the value of the latter (a shift of the total potential by a constant does not affect the wave functions and the density). In the defected crystal, the situations is completely different. By virtue of Gauss' law, $\delta v(\vec{r})$ vanishes identically in the Wigner-Seitz sphere centered at the origin, whereas far away from the impurity it becomes equal to the lattice potential *** of the pure host, whose average is in general different from zero. If the latter is large, the fact that the average lattice potential close to and far away from the impurity is different can have a substantial effect on the density and therefore on all ground state properties. As an illustration let us see what happens in the case of aluminium. The average lattice potential is best calculated by assuming that the crystal is made up of neutral Wigner-Seitz spheres. Then the potential vanishes at the boundary of each sphere, and one only needs to calculate its average in one of them, with the result

$$\langle \delta v_{WS} \rangle_{av} = \frac{-3}{10} (Z_m/R_{WS}) \left[1-5(r_c/R_{WS})^2 \right] \tag{27}$$

in atomic units. With the parameters for Al: $\langle \delta v_{WS} \rangle_{av} = -0.09$ a.u. (- 2.44 eV). This is close to one fourth of the free electron Fermi energy ($E_F = 11.69$ eV), which is the natural energy scale for the

---

* Expressions for the single-particle equations and spin-densities in jellium model for the impurity can be found in refs. 17 and 18, where the hyperfine field at an interstitial positive muon is calculated (note that for an interstitial impurity, the fictitious (≡jellium)background takes the simple from $n^B(\vec{r}) = n_o$).

** Strictly, the electron pseudo-density, since we describe the host ions by a pseudopotential.

*** We always include the neutralizing background in our definition of the lattice potential.

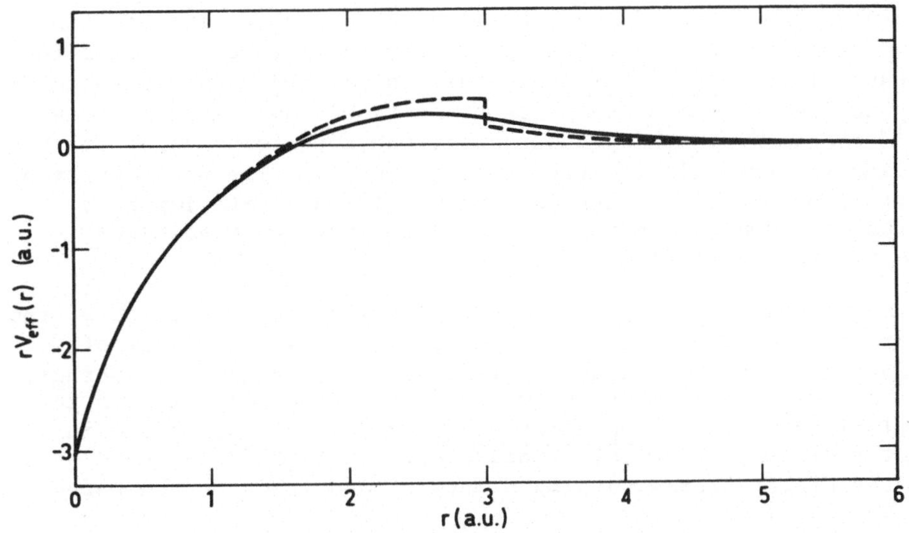

Figure 2: Selfconsistent effective potential for the jellium model (continuous line) and in the variational-selfconsistent approach (dashed line)

problem, and cannot be ignored in a realistic calculation. Fortunately it turns out that this effect of the discreteness of the lattice can be incorporated in the selfconsistent calculation without any loss in computational simplicity: with the above model for the crystal, $\delta v(\vec{r})$ differs from the potential for the perfect lattice only by the fact that one sphere is missing at the origin. Therefore we can account for the change in average lattice potential when moving away from the impurity, by subtracting $\langle \delta v_{WS} \rangle_{av}$ for that sphere, i.e. by adding to the effective one-body potential of the jellium model the (spherically symmetric) term

$$V_{Vac}(\vec{r}) = - \langle \delta v_{WS} \rangle_{av} \, \theta(R_{WS} - r) \tag{28}$$

What is left of $\delta v(\vec{r})$ is now a perturbation of zero average.

Of course the true crystal is not made up of Wigner-Seitz spheres, and the average of the pure host lattice potential may differ somewhat from $\langle \delta v_{WS} \rangle_{av}$. We can account for that by appealing to the variational principle. That is we replace the coefficient of the step function in (28) by a set of constants $\{C\}$. To each constant will correspond a selfconsistent density (the jellium density corresponds to C = 0). According to the variational principle, the "best" density will be the one that minimizes the ground state energy (13), fixing thereby the optimum value of the constant. This variational-selfconsistent method has been successfully applied to the calculation of surface[14] and vacancy formation[19] energies. In the present case, the minimizing value of the constant is virtually identical with $-\langle \delta v_{WS} \rangle_{av}$. The selfconsistent effective potentials in the jellium model and in its improved version are compared in figure 2 for the case where no magnetic field is present ($\xi(\vec{r}) \equiv 0$).

Up to now we have replaced $\delta v(\vec{r})$ by a constant outside of the Wigner-Seitz sphere centered at the origin. The next level of sophistication is to replace it instead by its spherical average, i.e. to take into account the variations of the potential in the radial direction. This spherical-solid model was introduced in a slightly different form by Almbladh and von Barth[20] in their study of X-ray singulatities in simple metals, and leads to the potential show in fig. 3.

Since the radial potential oscillations occur both in the perfect and in the defected crystal, the evaluation of the impurity-induced (spin) density $\delta n^{(\sigma)}(r)$ now requires a selfconsistent calculation for both situations. In order to simplify the computations, it is convenient to introduce a cut-off radius ρ, beyond which the

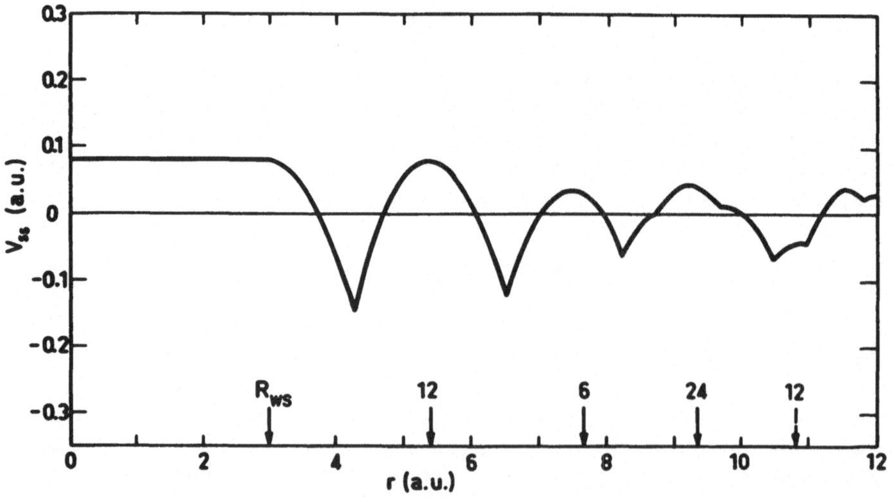

Figure 3: Spherical-solid model potential for a vacancy in aluminium. The Al ions are described by an Ashcroft empty-core pseudopotential, with core radius equal to 1.12 a.u.

potential is set equal to a constant, i.e.

$$V_{SS}(r) \equiv V_{SS}(\rho) \quad , \quad r \geqslant \rho \tag{30}$$

Clearly ρ must be chosen large enough for the results to be cut-off independent. In the present context we are only interested in what happens up to the third nearest neighbours to the impurity, and our calculations show that a cut-off radius of 10 a.u. will do.

IV. RESULTS AND DISCUSSION

We have performed two sets of calculations. One in the absence of external magnetic field (i.e. with the bulk polarization ξ equal to zero), from which we have determined the value at the origin of the Fermi level wave-function squared $<|\Psi_{\vec{k}}(0)|^2>_{FS}$, and the impurity-induced density. The latter is displayed in figure 4 for the three approximate models discussed in the preceding section. One observes that the introduction of discrete lattice effects strongly alters the phase of the Friedel oscillations. Going from the variational-selfconsistent to the spherical-solid model mainly changes their amplitude. Using the above densities in the Kohn-Vosko formula (eq. (11)), we have computed the values of the electric field gradient at the first three nearest neighbours to the impurity. These are shown in table 1. Compared with what is observed in other dilute aluminium alloys [*,4], the values we obtain for the EFG at the first nearest neighbour are strikingly small. This could be partly due to our use of an asymptotic formula for describing what happens at a relatively short distance from the impurity. However, in view of the success with which this approach has met in explaining the EFG induced by an interstitial positive muon at nearest-neighbour copper nuclei[21] ($r_{\mu^+-Cu} \simeq 3.4$ a.u.), we are fairly confident that the effect is real. Finally we should mention that the EFG's obtained from the spherical-solid model should be taken with a grain of salt: our approach assumes that by subtracting the density for the perfect lattice from the one in presence of the impurity, we account for the effect of the latter alone. This ignores possible interferences between impurity- and bandstructure-induced density fluctuations, which can arise because of the presence of the oscillating potential $V_{SS}(r)$- If these interference effects were to be large, the separation of the EFG into a Bloch enhancement factor (which should take care of all bandstructure effects) and the response of a homogeneous electron gas would be inconsistent for the spherical-solid model.

★ The reported measurements typically yield values arount 10^{14} cgs esu for the EFG at the first nearest neighbour.

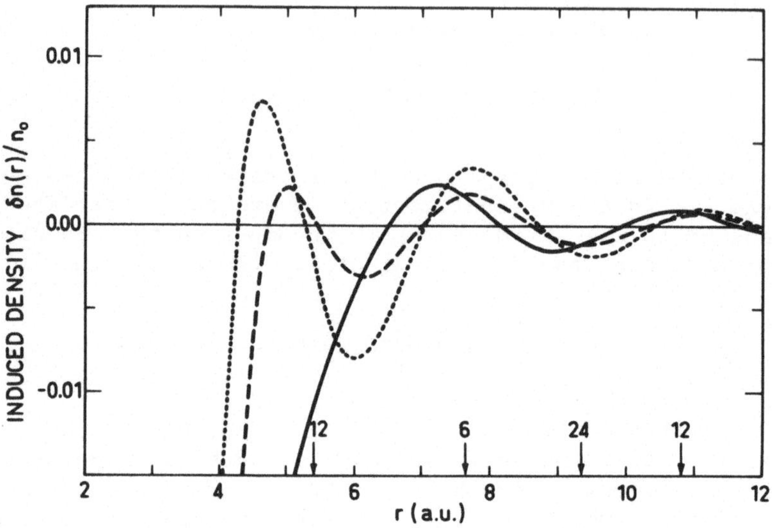

Figure 4: Selfconsistent impurity-induced density around a substitutional Li nucleus in Al. Solid line: jellium model, dashed line: variational-selfconsistent approach, dotted line: spherical solid model. The position of the first four nearest neighbour shells is also shown, together with the number of ions in each shell.

Our spin-polarized calculations of the impurity Knight shift (eq. (5b)) were done with the bulk polarization ξ_o equal to -0.05. While this corresponds to a stabilizing external field of the order of 50 MegaGauss(!), the use of a smaller value for ξ_o would result in a serious loss in numerical accuracy. Fortunately, the ratio $(n^+(0) - n^-(0))/(n_o^+ - n_o^-)$ which appears in eq. (5b) is independent of $|\xi_o|$ as long as the latter does not exceed 0.5[17,18,22], so that our results should be reliable for laboratory-size fields as well. Results for the impurity Knight shifts obtained from equations (5a) and (5b) are compared in table 2. The Knight shifts deduced from the spin-polarized calculation are seen to be systematically smaller than the ones obtained from the conventional treatment. As already mentioned in the introduction, this is a consequence of the formation of bound states (one for each spin) of s-like symmetry. Due to the paramagnetic response of the electron gas, the contact density at the nucleus will contain a small excess of down-spin conduction electrons, so that it becomes energetically more favourable to bind an electron with spin parallel to the applied field (i.e. $E_B^+<E_B^-<0$ and $|\Psi_B^+(0)|^2>|\Psi_B^-(0)|^2$). This makes a contribution to the hyperfine field which is opposite to the one from the conduction electrons,

Table 1. : Calculated electric field gradients at host neighbour sites in a dilute Al-Li alloy.
A: jellium model
B: variational-selfconsistent approach
C: spherical-solid model
The Bloch enhancement factors $\alpha(\vec{k}_F)$ are from P.M. Holtham and P. Jena, J. Phys. F 5, 1649 (1975)

Neighbour	Position r (a.u.)	$\alpha(\vec{k}_F)$	EFG (10^{13} egs esu) A	B	C
first	5.338	6.4	-5.5	0.2	-1.3
second	7.619	6.2	0.8	0.8	1.5
third	9.332	8.9	-0.8	-0.7	-1.2

Table 2.: Impurity Knight shifts (in parts per million) for a dilute Al-Li alloy.

A, B and C: as in Table 1.

χ_m is the exchange-correlation-enhanced Pauli susceptibility corresponding to the Gunnarsson-Lundqvist exchange-correlation potential used in these calculations.

$\chi_m = 1.38\ \chi_{Pauli} = 1.73\ 10^{-6}$ cgs volume units

For comparison, the Knight shift in pure Li is 260 ppm.

	A	B	C
$(8\pi/3)\langle|\Psi_{\vec{k}}(0)|^2\rangle_{FS}\ V\chi_m$	203	168	144
$(8\pi/3)\ \dfrac{n^+(0) - n^-(0)}{n_o^+ - n_o^-}\ \chi_m$	170	128	99

thereby reducing the Knight shift. A similar phenomenon was observed by Munjal and Petzinger in their study of muonic Knight shifts in copper and alkali metals[18]. Finally, note the large change in the magnitude of the shifts when the discreteness of the host lattice is taken into account.

In conclusion, we hope that the detailed results we have presented for the impurity Knight shift and the electric field gradients in a dilute Al-Li alloy will encourage an experimental investigation of the NMR properties of this system. The latter should provide an ideal testing ground for the theory, since lattice distorsion effects, which are difficult to treat from first principles and form a major source of uncertainty in the interpretation of the experimental results, can be neglected[23].

Part of this work was supported by the Swiss National Science Foundation.

REFERENCES

1. The following textbooks give a clear account of the theory of nuclear magnetic resonance:

   A. Abragam, The Principles of Nuclear Magnetism, Clarendon Press, Oxford (1961)

   Charles P. Slichter, Principles of Magnetic Resonance, Harper & Row, New York (1963)

   For a complete review of NMR studies in metals and alloys with an exhaustive list of references up to 1977 see:

   G.C. Carter, L.H. Bennett and D.J. Kahan, Metallic Shifts in NMR, Volume 20 of Progress in Materials Science, Pergamon, Oxford (1977)

2. L. Jørgensen, R. Nevald, D. Ll. Williams, J. Phys. F 1, 972 (1971)
3. J.A.R. Stiles and D. Ll. Williams, J. Phys. F 4, 2297 (1974)
4. M. Minier and S. Ho Dung, J. Phys. F 7, 503 (1977)
5. W. Kohn and S.H. Vosko, Phys. Rev. 119, 912 (1960)
6. P.M. Holtham and P. Jena, J. Phys. F 5, 1649 (1975)
7. U. von Barth and L. Hedin, J. Phys. C 5, 1629 (1972)
8. A.K. Rajagopal and J. Callaway, Phys. Rev. B 7, 1912 (1973)
9. O. Gunnarsson and B.I. Lundqvist, Phys. Rev. B 13, 4274 (1976)
10. P. Hohenberg and W. Kohn, Phys. Rev. 136, B864 (1964)
11. W. Kohn and L.J. Sham, Phys. Rev. 140, A1133 (1965)
12. G.W. Bryant and G.D. Mahan, Phys. Rev. B 17, 1744 (1978)
13. N.W. Ashcroft and D.C. Langreth, Phys. Rev. 155, 682 (1967)
14. R. Monnier and J.P. Perdew, Phys. Rev. B 17, 2595 (1978)
15. N.W. Ashcroft, Phys. Lett. 23, 48 (1966)
16. M. Manninen, P. Hautojärvi and R. Nieminen, Solid State Comm. 23, 795 (1977)
17. P. Jena, K.S. Singwi and R. Nieminen, Phys. Rev. B 17, 301 (1978)

18. K.G. Petzinger and R. Munjal, Phys. Rev. B 15, 1560 (1977), R. Munjal and K.G. Petzinger, in Hyperfine Interactions 4, 301 (1978)

19. R.M. Nieminen, J. Nucl. Mat. 69-70, 633 (1978), M. Manninen and R.M. Nieminen, J. Phys. F (to appear)

20. C.O. Almblach and U. von Barth, Phys. Rev. B 13, 3307 (1976)

21. P. Jena, S.G. Das un K.S. Singwi, Phys. Rev. Lett. 40, 264 (1978)

22. M. Manninen (unpublished)

23. W.B. Pearson, Handbook of Lattice Spacings and Structures of Metals and Alloys (Pergamon, New York, 1964) p. 346.

FIRST PRINCIPLES BAND THEORY FOR RANDOM METALLIC ALLOYS*

B.L. Györffy †

Brookhaven National Laboratory, Upton, N.Y. 11973, U.S.A.

G.M. Stocks

Oak Ridge National Laboratory, Oak Ridge, Tenn. 37830, U.S.A.

PROLOGUE

Two years ago at an Advanced Study Institute in Gent, we had given five lectures on the same topics as treated here. Since then the subject developed considerably and we have written an entirely new set of lecture notes. On that previous occasion we have concentrated on the formal structure of the theory. Here the focus is on results and applications. In order to make the present notes self-contained, we did not refrain from covering the same ground when it was necessary. However, a significant part of the earlier material had to be cut. The interested reader may find these in Györffy and Stocks (1976) "Electrons in Finite and Infinite Structures", Ed. P. Phariseau and L. Scheire.

I. INTRODUCTION

(1) The Band Theory Program for Random Alloys

In these lectures we will describe and advocate a first-principles approach to the problem of electronic states in random substitutional metallic alloys. By "first principles" we do not mean that the theory is in any way exact. We merely wish to indicate that our methodology is to follow the first principles as best as

---

* Research supported by the Division of Basic Energy Sciences, Department of Energy, under Contract No. EY-76-C-02-0016.

† Permanent Address : H.H. Wills Physics Laboratory, University of Bristol, Bristol BS8 1TL, U.K.

we can (i.e. make many approximations) without introducing phenomenological parameters in an essential way. Namely we do not want to
fit constants in our theory to experiments. Nor do we aim to explain
the band structure of the random alloys from the energy bands of the
constituent pure metals. On the contrary, we are going to consider
the problem of constructing crystal potentials and solving the appropriate Schrödinger equation for an alloy at a given concentration,
with the experimentally observed crystal structure. Before a useful
discussion of the motivation behind this strategy can be given we
must digress to explore briefly what it entails.

The modern <u>band theory program for pure metals</u> consists of
three parts :

(a) An effective one electron crystal potential is constructed, usually in the form of potential wells centered on the lattice sites :

$$V(\underline{r}) = \sum_i v(\underline{r} - \underline{R}_i) \qquad \text{I-1}$$

(b) The corresponding Schrödinger equation :

$$(-\nabla^2 + \sum_i v(\underline{r} - \underline{R}_i))\psi_{\underline{k}}(\underline{r}) = \varepsilon_{\underline{k}}\psi_{\underline{k}}(\underline{r}) \qquad \text{I-2}$$

is solved for the energy eigenvalues $\varepsilon_{\underline{k}}$ and the eigenfunctions $\psi_{\underline{k}}$.

(c) Using the $\psi_{\underline{k}}$'s and $\varepsilon_{\underline{k}}$'s obtained above, physical observables
such as the Fermi surface, cohesive energy, photoelectric cross
section are calculated.

Steps (b) and (c) are done numerically and more or less exactly. In this way the potential function $V(\underline{r})$ acquires a central
significance in the theory. In case of disagreement with experiments
apart from computational difficulties, it is the potential function
which is found inadequate. What lends credence to this approach is
the fact that nowadays Eq. I-2 has a proper many-body foundation.
As explained elsewhere at this summer school [Gunnarsson (1978)]
it no longer has to be viewed as a one-electron theory from which
most of the exchange and all of the correlation effects have been
left out. Rather, it is an effective one-electron theory which for
many (but by no means all) purposes takes full account of the many-
body interactions. To what extent this is the case depends on the
method of constructing the crystal potential. (Obviously the many-
body problem is never solved exactly). In the language of the density functional theory of the inhomogeneous electron gas $V(\underline{r},[\rho])$ is
a functional of the charge distribution $\rho(r)$. Eq. I-2 has to be
solved self-consistently. This means that one must start with a
guess for $\rho_i(\underline{r})$, construct the crystal potential $V(\underline{r},[\rho_i])$, solve
Eq. I-2 for $\varepsilon_{\underline{k}}$ and $\psi_{\underline{k}}$, construct the new charge distribution

$$\rho_n(\underline{r}) = e \sum_{\underline{k}}^{\varepsilon_{\underline{k}}<\varepsilon_F} |\psi_{\underline{k}}(\underline{r})|^2$$

and repeat the process starting with $\rho_n(\underline{r})$ until convergence. For the Local Density Approximation $V_{LD}(\underline{r}_i, [\bar{\rho}])$ [Kohn and Sham (1965)] which treats exchange and correlation on equal footing (all be it approximately) this program can be readily carried through. [Tong (1972)].

Thus, the study of the electronic structure of solids naturally falls into two parts. On the one hand, one works on the conventional many-body aspects of the problem, namely at devising better schemes for the crystal potential. On the other hand, using the computational techniques of band theory, one endeavors to make increasingly more detailed contact between the crystal potential and experiments. For a most readable example of current thinking in the former vain, see Gunnarsson and Lundqvist (1976). For the latter see Callaway (1974) or Ziman (1971).

There is another, less elevated, reason for keeping the crystal potential in the center of our thinking about electronic structure. It turns out that it is a relatively simple function : (usually) spherically symmetric potential wells centered at regular lattice sites. At the same time it gives rise to a bewildering complexity of energy bands and wavefunctions which, in turn, describe a rich variety of physical phenomena observed as highly structured data. Hence, the potential function is not only useful in making contact with our most general theory of the interacting many-electron system but also a most efficient tool for correlating our understanding of diverse manifestations of the electronic structure.

Unfortunately, there is no room here even to list the most conspicuous advances in our understanding of ordered solids brought about by the above program in recent years. As examples peruse the cohesive energy calculations for metals with Z up to 50 by Moruzzi, Williams and Janak (1977), the zero temperature magnetization calculation of Wang and Callaway (1977) or the calculation of the superconducting transition temperatures for the 4d transition metals by Butler (1977).

The purpose of this summer school is to report on progress in formulating the band theory approach to metals without full crystalline symmetry. In particular, these lectures will be concerned with the extent to which this program can be implemented for random alloys. Having scratched the background, let us now see what is involved at first in principle and then in practice.

Similarly to the case of pure metals, the band theory program for random alloys will consist of three parts :

(a) As in the case of ordered systems one must begin by constructing an alloy potential $V(\underline{r})$. This will again consist of potential wells centered on the lattice sites. However, in general, each will be different.

$$V_{alloy}(\underline{r}) = \sum_i v_i(\underline{r} - \underline{R}_i) \qquad \text{I-3}$$

To simplify matters, let us assume that there are only two kinds of wells described by the potential functions $v_A(\underline{r})$ and $v_B(\underline{r})$. Then, we may write

$$v_i(\underline{r} - \underline{R}_i) = \xi_i \, v_A(\underline{r} - \underline{R}_i) + (1 - \xi_i) v_B(\underline{r} - \underline{R}_i) \qquad \text{I-4}$$

where the occupation variable $\xi_i = 1$ if the site at $\underline{R}_i$ is occupied by an A well and $\xi_i = 0$ if the well at $\underline{R}_i$ is B type. A configuration is specified if the values of all the occupation variables $\{\xi_i\}$ are given.

(b) Having constructed a crystal potential for a given configuration, $\{\xi_i\}$, we must now solve the Schrödinger equation

$$(-\nabla^2 + \sum_i v_i(\underline{r} - \underline{R}_i))\psi_n(\underline{r}, \xi_1 \ldots \xi_N) = \varepsilon_n(\xi_1 \ldots \xi_N)\psi_n(\underline{r}, \xi_1 \ldots \xi_N) \qquad \text{I-5}$$

where we have indicated explicitly that the eigenfunctions $\psi_n(\underline{r}, \xi_1, \ldots \xi_N)$ and eigenvalues $\varepsilon_n(\xi_1 \ldots \xi_N)$ depend on the configuration.

(c) Using ψ_n and ε_n we then have to calculate "observables" e.g., the matrix elements of some operator O, $O_{n,n'}(\xi_1, \ldots \xi_N) = \langle \psi_n | O | \psi_{n'} \rangle$. Fortunately, we do not have to stop here. Clearly to obtain $\psi_n(\underline{r}, \xi_1, \ldots \xi_N)$ and $\varepsilon_n(\xi_1, \ldots \xi_N)$ would be an impossible task. Since our experiment measures only averages as a final step we must calculate the average of $O_{n,n'}(\xi_1, \ldots \xi_N)$ over all possible configurations :

$$\langle O_{n,n'} \rangle = \sum_{\{\xi\}} P(\{\xi\}) \, \langle \psi_n(\{\xi\}) | O | \psi_{n'}(\{\xi\}) \rangle \qquad \text{I-6}$$

where $P(\{\xi\})$ is the probability that a given configuration $\{\xi\}$ occurs. (Normally we will assume that the sites are uncorrelated, e.g., $P(\{\xi\}) = \prod_i^N P(\xi_i)$ and take $P(\xi_i) = c\xi_i + (1 - c)(1 - \xi_i)$ where c is the concentration of the A type wells).

Evidently, this task is much more complicated than the corresponding program for ordered systems. The surprising thing is not the way it is done, but that it is done at all. Nevertheless, a number of groups (A. Bansil, H. Ehrenreich, and L. Schwartz in Boston ; An-Ban Chen and A. Sher in Virginia, and us : Györffy, Stocks and Temmerman in Bristol, U.K. and Oak Ridge, Tenn., U.S.A.

together with Giuliano and Ruggeri in Messina, Italy and Pindor in Warsaw, Poland) have been at it for the past several years. The attraction is that if a reliable route through a maze of possible simplifications can be found the problem would separate into two parts : the construction of the alloy potential and the solving for the various configurationally averaged quantities. Note that both problems are extremely difficult ; the former has to do with the allusive question of local charge rearrangements in a solid and involves the microscopic understanding of such phenomenological concepts as charge transfer, relative electronegativity and covalency, and the latter is concerned with studying the eigenvalue spectrum of random Hamiltonians. While a theory of the electronic structure in random alloys implies a coherent treatment of both of these problems it would clearly be a great help to be able to discuss them one at a time. The alloy potential is the conceptual tool which allows this separation without distorting either aspect unduly.

Before one knew what to do about a Schrödinger equation with a random potential there was little incentive to think about the question of alloy potentials. Our main aim in these lectures is to demonstrate that we now have sufficiently reliable methods for going from a realistic random potential to experimental observables to warrant a more serious effort towards a theory of it. That is to say, within the view presented here the construction of adequate alloy potentials becomes the heart of the alloy problem.

Hopefully, the burden of these general remarks will become clearer as we consider specific instances of the theory at work. It will also be easier to compare the present approach to various alternatives later when more of the details in the overall picture are filled in. However, before we can get more specific, we must deal with some conceptual matters regarding the treatment of disorder.

(2) The Coherent Potential Approximation

The theory of eigenvalue problems which correspond to random Hamiltonians has a vast literature. Of this only the advances brought about by the discovery of the Coherent Potential Approximation (CPA) [Soven (1967), Taylor (1968), Onodera (1968)] concern us here. For a good review of this major development, see Elliott et al (1974) and Ehrenreich et al (1976). Without attempting to do justice to this very interesting subject we shall now sketch the main outline of the argument.

To do this, it is sufficient to consider a problem which is much simpler than that of a quantum mechanical electron propagating among random scattering centers. In fact, the main idea behind the CPA can be illustrated by studying a classical current through an inhomogeneous conductor. This is a very old problem which was first discussed by Maxwell (1873).

Consider a conductor which, on a macroscopic scale, consists of a random mixture of two metals : one with conductivity σ_1 and the other with σ_2. A particular configuration is depicted in Fig.1.

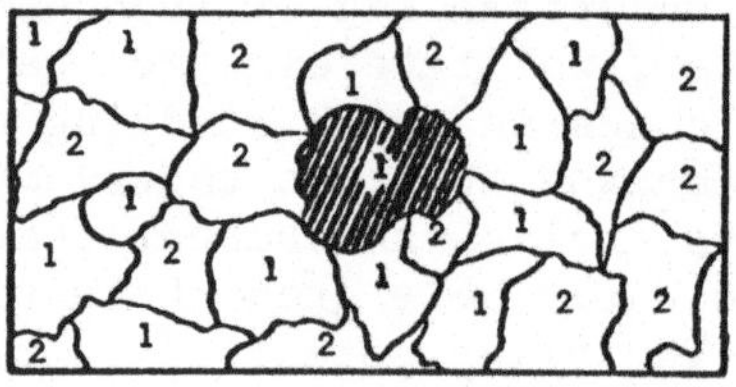

Fig. 1 : The shaded crystal of type 1 is surrounded by crystals of both types, which are imagined to be replaced by a single medium of uniform conductivity.

A steady-state current flowing through this system is described by the constitutive equation

$$\underline{j}(\underline{r}) = \sigma(\underline{r})\ \underline{E}(\underline{r}) \qquad \text{I-7}$$

where $\underline{j}(\underline{r})$ is a time-independent current density, $\sigma(\underline{r})$ is the conductivity (σ_1 in metal 1 and σ_2 in metal 2) and $\underline{E}(\underline{r})$ is the electric field at the position $\underline{r}$. Since it is impossible to determine the exact configuration of the sample, we would like to know the effective conductivity such that

$$\langle \underline{j} \rangle = \sigma_{eff} \langle \underline{E} \rangle \qquad \text{I-8}$$

Evidently, the simplest thing to do is to assume that $\langle \underline{j} \rangle = \langle \sigma(\underline{r})\ \underline{E}(\underline{r}) \rangle \cong \bar{\sigma} \langle \underline{E} \rangle$ where the averaged conductivity $\bar{\sigma}$ is defined in terms of the volume fractions $c_1 = \Omega_1/\Omega$ and $c_2 = \Omega_2/\Omega$ for metal 1 and 2 respectively as

$$\bar{\sigma} = c_1\sigma_1 + c_2\sigma_2 \qquad \text{I-9}$$

For our random alloy Schrödinger equation this approximation corresponds to averaging the crystal potential and solving the Schrödinger equation appropriate to

$$\bar{V}(\underline{r}) = \langle \sum_i v_i(\underline{r} - \underline{R}_i) \rangle = \sum_i (c v_A(\underline{r} - \underline{R}_i) + (1 - c) v_B(\underline{r} - \underline{R}_i)).$$

This is the Virtual Crystal Approximation (VCA) introduced into the electronic problem by Nordheim (1931) and Muto (1938). Since $\bar{V}(\underline{r})$ has the periodicity of the lattice the corresponding Schrödinger equation can be solved by the standard methods of band theory. For an application of this method to Cu_cZn_{1-c} see Sommers, Amar, and Johnson (1966).

It is not hard to see that Eqs. I-8 and I-9 are good approximations only if σ_1 is close to σ_2. The trouble is that in general $\langle\sigma(\underline{r})E(\underline{r})\rangle \neq \bar{\sigma}\langle\underline{E}\rangle$. An external field will create a steady current but charges will pile up at the boundaries of regions with different conductivities and this will create additional fields which will make $\underline{E}(\underline{r})$ random and not independent of σ at $\underline{r}$. In fact, one can easily show using perturbation theory [Herring (1960)] that for small $\sigma_1 - \sigma_2/\bar{\sigma}$

$$\begin{aligned}\sigma_{eff} &= \bar{\sigma}\left(1 - \frac{1}{3}\frac{\langle(\sigma - \bar{\sigma})^2\rangle}{\bar{\sigma}^2}\right) \\ &= \bar{\sigma}\left(1 - \frac{1}{3}c_1c_2\frac{(\sigma_1 - \sigma_2)^2}{\bar{\sigma}^2}\right)\end{aligned} \qquad \text{I-10}$$

Clearly, the analogue to this approach in the alloy problem is appropriate to simple metals (see the lectures by R. Evans) where perturbation theory can be used with confidence. However, we are interested in transition metal alloys as well. For these the crystal potential is not a small perturbation. Moreover, even the difference $v_A - v_B$ can be large enough so that an expansion in powers of $v_A - v_B$ is of no use. In our simple problem of an inhomogeneous conductor this corresponds to the situation where $(\sigma_1 - \sigma_2)/\bar{\sigma}$ is not a small parameter. To treat this case something more subtle than perturbation theory is required.

As was mentioned before already Maxwell (1873) has dealt with this problem. However, we shall give here the Effective Medium Theory of Bruggeman (1935). The basic idea of this approach is to introduce an effective medium with a conductivity σ_m and attempt to determine σ_m by the requirement that σ_m should describe the average behavior of the random system.

To do this consider the effect of a spherical inclusion of the metal 1 in the effective medium on a steady current. The configuration we have in mind is illustrated in Fig. 2. Since there are no time dependent magnetic fields around, $\underline{\nabla} \times \underline{E}(\underline{r}) = \underline{0}$ and $\underline{E}$ may be represented by a scalar potential such that $\underline{E} = -\underline{\nabla}\Phi$. From charge conservation ($\dot{\rho}(\underline{r}) = 0$) we also have that $\underline{\nabla}.\underline{j}(\underline{r}) = 0$. Hence

$$\underline{\nabla}.\underline{j}(\underline{r}) = \underline{\nabla}.(\sigma(\underline{r})\underline{E}(\underline{r})) = \begin{cases} -\sigma_m\nabla^2\Phi(\underline{r}) = 0 \text{ for } \underline{r} \text{ in the medium} \\ -\sigma_1\nabla^2\Phi(\underline{r}) = 0 \text{ for } \underline{r} \text{ in the inclusion}\end{cases} \qquad \text{I-11}$$

Since the current across the boundary between the two regions must

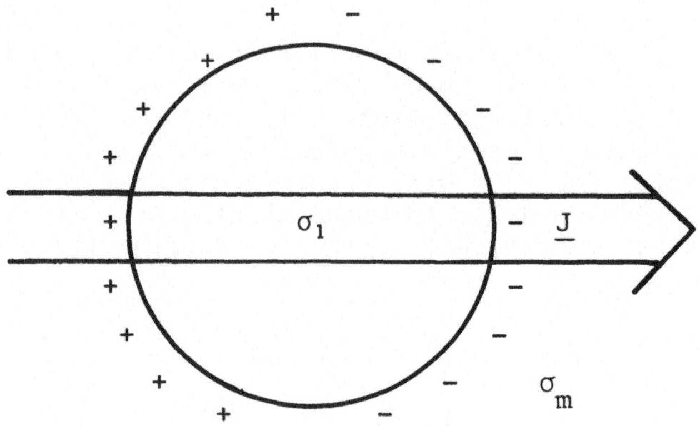

Fig. 2 : The current and medium configuration corresponding to the effective medium calculation discussed in the text. Outside the central sphere is the effective medium with conductivity σ_m. The inside is filled with the component whose conductivity is σ_1. $\underline{J}$ is the measuring current.

be continuous the boundary conditions are [Landau and Lifshitz (1960)]

$$\sigma_1 E^{\perp}_{in}(\underline{s}) = \sigma_m E^{\perp}_{out}(\underline{s})$$

$$E^{\parallel}_{in}(\underline{s}) = E^{\parallel}_{out}(\underline{s}) \qquad \text{I-12}$$

where $\underline{s}$ is a position vector on the boundary surface, the subscripts "in" and "out" are self-explanatory, and the superscripts "$\perp$" and "$\parallel$" denote components perpendicular and parallel to the surface at $\underline{s}$.

Formally, Eq. I-11 together with I-12 is a simple electrostatic problem which can be solved by the usual means: since $\Phi(\underline{r})$ must be a solution to Laplace's equation we may write

$$\Phi_{in}(\underline{r}) = \sum_{\ell=0}^{\infty} A_\ell r^\ell P_\ell(\cos\theta)$$

and

$$\Phi_{out}(\underline{r}) = \sum_{\ell=0}^{\infty} [B_\ell r^\ell + C_\ell r^{-(\ell+1)}] P_\ell(\cos\theta).$$

If $\underline{E}_0$ is an applied field then

$$\lim_{r\to\infty} \Phi_{out}(\underline{r}) = -zE_0 = -r\cos\theta\, E_0$$

This limit together with the boundary conditions gives the coefficients as

$$A_\ell = B_\ell = C_\ell = 0 \text{ for } \ell \neq 1, A_1 = -\frac{3\sigma_m}{\sigma_1+2\sigma_m}E_0,\; B_1 = -E_0$$

and

$$C_1 = a^3\frac{\sigma_1-\sigma_m}{\sigma_1+2\sigma_m}E_0$$

Outside the inclusion we find that

$$\underline{E}_{outside} = \underline{E}_0 - a^3\frac{\sigma_1-\sigma_m}{\sigma_1+2\sigma_m}\underline{\nabla}\left(\frac{\underline{r}\cdot\underline{E}_0}{r^3}\right) \qquad \text{I-13}$$

Thus the presence of the inclusion induced a dipole field which must be added to the applied field $\underline{E}_0$ (see Fig. 2). Clearly if the inclusion was of metal 2 a similar extra dipole field would be induced with σ_2 replacing σ_1 everywhere. Imagine now the situation in which the material within the inclusion is metal 1 with the probability c_1 and metal 2 with probability c_2. If we then choose the effective medium conductivity by requiring that the induced dipole field is zero on the average, e.g., σ_m is to be determined by the condition that

$$c_1\frac{\sigma_1-\sigma_m}{\sigma_1+2\sigma_m} + c_2\frac{\sigma_2-\sigma_m}{\sigma_2+2\sigma_m} = 0 \qquad \text{I-14}$$

then $\langle\underline{E}_{outside}\rangle = \underline{E}_0$. Evidently, this means that any region of our effective medium may be replaced by metal 1 and 2 in a random fashion and, on the average, the fields outside that region will be the same as if all the sample consisted of the medium. Thus we conclude that we have found an effective medium which gives a good account of the average behavior of our random mixture. Clearly, $\langle\underline{j}\rangle = \sigma_m\underline{E}_0$ and therefore the effective conductivity $\sigma_{eff} = \sigma_m$.

Eq. I-14 is only a quadratic equation. Its positive solution is

$$\sigma_m = \frac{1}{4}\left(\gamma + (\gamma^2 + 8\sigma_1\sigma_2)^{1/2}\right) \qquad \text{I-15}$$

where $\gamma = (3c_1 - 1)\sigma_1 + (3c_2 - 1)\sigma_2$. This result is illustrated in Fig. 3 for the cases where one of the constituents has either vanishingly small conductivity (insulator) or vanishingly small resistivity (superconductor).

It turns out that this is a very good overall theory of inhomogeneous conductors covering a range of conductivities from

10^{-10} 1/Ωcm (tefalon) to >10^6 1/Ωcm (Cu at T = 300). From the point of view of our present concern its most interesting feature is that unlike the VC approximation it predicts that when the volume fraction of the good conductor is reduced to 1/3 the mixture ceases to conduct. This metal insulator transition can be thought of as a phase transition and our EMT theory correctly predicts its existence. However, the details are wrong near the transition which is called the percolation threshold. More sophisticated approaches to the percolation problem [Kirkpatrick (1973), Stephen (1978), Wegner (1976)] show that the EMT is the analogue of the mean field theory in statistical mechanics of second order phase transitions. Thus, it is a good theory away from the percolation limit. This is illustrated in Fig.4 for a resistance network which is a slightly more complicated model than the one we have considered here.

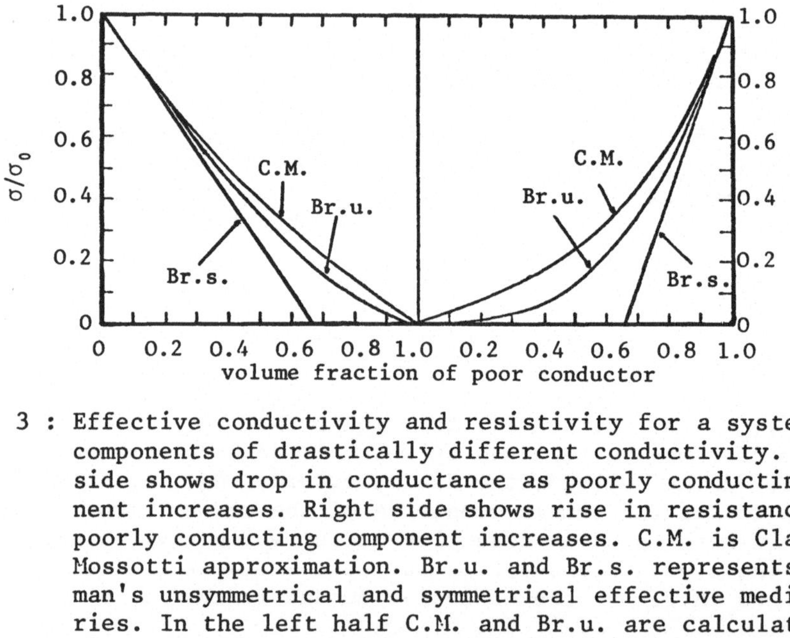

Fig. 3 : Effective conductivity and resistivity for a system with components of drastically different conductivity. Left side shows drop in conductance as poorly conducting component increases. Right side shows rise in resistance as poorly conducting component increases. C.M. is Clausius-Mossotti approximation. Br.u. and Br.s. represents Bruggeman's unsymmetrical and symmetrical effective medium theories. In the left half C.M. and Br.u. are calculated on the assumption that poor conductor is introduced into a continuous background of good conductor. The right hand side assumes that the poor conductor is the continuous material.

As you might have guessed it by now the EMT can be adapted for treating an electron in a random potential such as described by Eqs.I-3 and I-4. More generally, the basic philosophy of the EMT can be used to discuss the averaged eigenvalue spectra of random Hamiltonians describing a wide class of elementary excitations such as phonons, magnons, excitons, as well as electrons. Nowadays, the

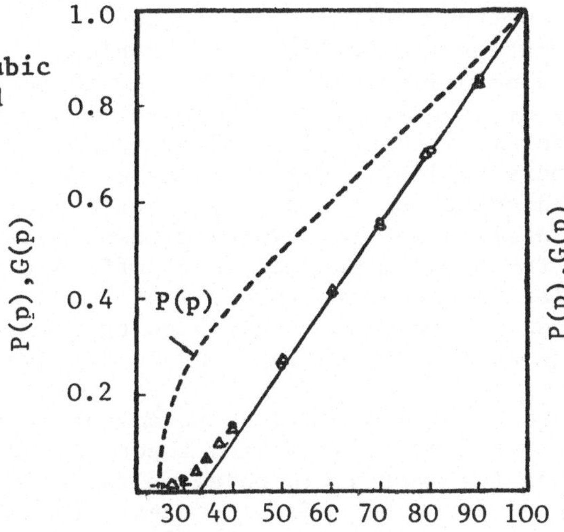

Fig. 4 : Percolation probability, P(p) (dashed line), and conductance, G(p) (data points), for bond percolation on 3D simple cubic network. G(p) is normalized to unity at p=1. The networks studied ranged in size from 15 x 15 x 15 to 25 x 25 x 25. The solid line indicates the prediction of the effective medium theory. The arrow indicates the position of the percolation threshold p_c.

resulting approximation schemes are called the Coherent Potential Approximation (CPA)[Elliott et al (1974)] . Interestingly, in almost all cases the CPA was derived in each field independently and the connection with methods used for other kinds of problems was noted later.

In the context of electrons in random substitutional alloys the CPA was discovered by P. Soven (1967). Since then it has been implemented for a variety of simple model Hamiltonians usually of the tight binding form. It was shown to be exact in the impurity limits ($c \to 0$, $c \to 1$) and in the limits of very weak and very strong scattering. Moreover, it was found to extrapolate very well between these extremes [Ehrenreich and Schwartz (1976)] . The analogue of the percolation threshold in this problem is localization [Anderson (1958), Thouless (1974)] . It turns out that this is a subtle problem and the CPA misses it completely. However, this is of no great consequence for us since the alloys we have in mind are good metals at all concentrations. A more serious drawback is that in the usual formulation of the CPA the short-range order is left out of the theory. This aspect will eventually need to be remedied at the least in connection with magnetism in random alloys (see Dr. Edwards's lectures).

As should be evident from the foregoing discussion, the CPA is a fairly general strategy for dealing with problems involving random Hamiltonians and not tied to specific forms these can take. That is to say, it may be used to discuss a model with a single tight-binding band, random site energies and non-random overlap integrals as well as more realistic descriptions of the alloy Hamiltonian. Moreover, the adequacy of the CPA does not appear to depend on whether the random Hamiltonian is simple or complicated. However, the implementation of the program becomes more difficult as the basic random Hamiltonian becomes more sophisticated. For instance, no one as yet has solved in the CPA a tight-binding Hamiltonian with random site energies, random overlaps and random hybridization.

Thus we conclude that the CPA would be an adequate treatment of the disorder in our first principles band theory program if it could be implemented for the muffin-tin model of the alloy potential described by I-3 and I-4. The general success with model crystal potentials, which consist of non-overlapping muffin-tin wells, for pure metals ensures that we have a reasonably realistic description of the problem. The large body of work on simple models using the CPA guarantees that we have a good mean field theory for treating the disorder.

Though the guidelines are clear it is nevertheless a non-trivial matter, except in the simplest case, to formulate the CPA for each new problem. Solving the fundamental equations for the effective medium (e.g., the coherent potential), the analogue of Eq. I-14, is frequently even more daunting.

It is with some surprise that we report in the next two lectures that both of these tasks can be readily performed for our Schrödinger equation given in Eq. I-5.

In summary, let us list again the steps in our "first principles" program for random alloys.

(a) Construction of the random alloy potential in the form of non-overlapping potential wells.

(b) Solving the fundamental equation of CPA for the effective medium which describes the average behavior of an electron in the presence of the random potential constructed in (a).

(c) Calculation of configurationally averaged observables using the effective medium determined in (b).

Our main concern for the rest of these lectures will be to demonstrate that steps (b) and (c) can be done with computational efforts not particularly large on the scale of modern band theoretical calculations. Unfortunately, on (a) we have very little new to say, except to point out that since (b) and (c) can now be done, it becomes one of the central issues in alloy physics. For the specific alloy systems we shall be studying we do have reasonable alloy potentials. However, these have been constructed by ad-hoc procedures applicable only to these relatively simple systems (no charge transfer). It is probably fair to say that at this point there is no general method which is known to work. Our view is that the development of such method will involve a great deal of trial and error.

II. A MULTIPLE SCATTERING THEORY APPROACH TO THE PROBLEM OF ELECTRONIC STATES IN RANDOM ALLOYS (KKR-CPA)

After the rather general discussion of the previous lecture, let us now turn to the detailed description of the theory. First we discuss, briefly, some general features of the crystal potential. Then we shall introduce the language of scattering theory using the single scatterer as an example. Next we shall consider an electron multiple scattering from an infinite array of scattering centers and derive the Korringa-Kohn-Rostoker (KKR) method of band theory for pure metals. Finally, we shall formulate the CPA for a random alloy potential which consists of non-overlapping muffin-tin potential wells.

(1) The Crystal Potential in the Form of Non-Overlapping Muffin-Tin Wells

In order not to loose sight completely of the physical problem we are trying to elucidate during the somewhat formal development that will follow, we shall now examine briefly the physical origin of the crystal potential. For the purposes of a qualitative discussion it is useful to think of it as arising from the overlap of atomic potentials centered on the lattice sites. By atomic potential we mean some self-consistent potential seen by an electron in a neutral atom. It is due to the nuclear charge and all the other electrons. (This potential is readily available to us from atomic physics. Hartree-Fock-Slater calculations of it are tabulated in the widely available Herman-Skillman tables [see Loucks (1967)]). The bound states of this potential constitute the electronic structure of the neutral atom and the extent to which they are filled (to obtain neutrality) defines the position of the atom in the periodic table. Clearly, it is spherically symmetric and as $r \to \infty$, it goes to an energy which we call the atomic zero.

Let us now consider an infinite array of such potential wells centered on regular lattice sites which are very far apart. The potentials will hardly overlap in spite of their 1/r tail and the potential hills separating the large negative wells will reach the atomic zero V_{AZ}. Now let us bring these atoms together to form a solid by reducing the lattice spacing. In the interstitial space one is adding the negative tails of contributions from each site. Thus the tops of the hills between the wells rapidly sink. In fact, for the observed lattice spacings they will be below the energies of some of the occupied states of the free atom potentials. When this happens the relevant electrons will flow from site to site like water. (Some of the electrons in the lower states will only be able to tunnel to other sites. Though this will, in principle, broaden their states into bands these core bands will be very narrow and a tunneling event will be very infrequent). The potential function at a site will no longer be spherically symmetric. It will have the point group symmetry of the lattice. However, in a metal its angular variation will be small compared with the rapid rise as one moves away from the nucleus at the center in any radial direction. Let us therefore imagine that the potential function is spherically averaged within the largest sphere that can be drawn about each site within the corresponding unit cell. Since the potential function also varies slowly in the space left outside these inscribed spheres, let us set it equal to a constant V_{MTZ} which is equal to the average of the potential in the region between the cell boundaries and the inscribed sphere. This construction is shown in Fig. 5. Because the array of spherically symmetric potential wells separated by a constant plateau resembles the button of a muffin-tin pan (encountered in kitchens of the English-speaking world) they are called muffin-tin potentials. The interstitial constant V_{MTZ} is called the muffin-tin zero and the radius of the inscribed sphere is referred to as the muffin-tin radius r_M. For one component cubic systems $r_M = a/2$ where a is the lattice parameter.

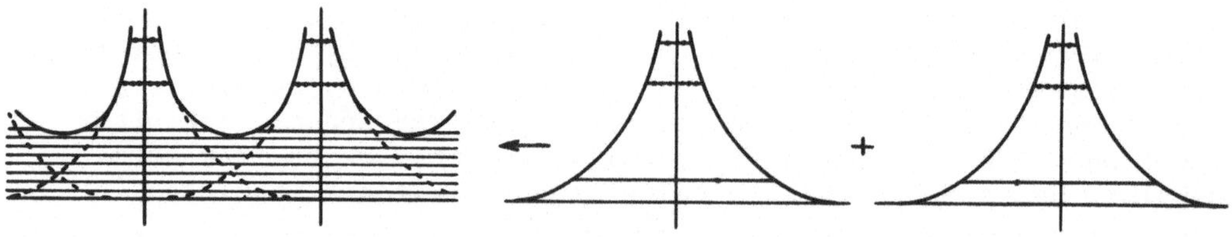

Fig. 5 : A schematic illustration of the construction of the muffin-tin potential for Na. The solid dots in the lower two levels represent the bound core electrons. The dot in the upper level is the "conduction" electron. The shaded area is the conduction band in the metal.

As is evident from the way it was constructed V_{MTZ} lies well below V_{AZ}. Physical intuition also tells us that the bottom of the conduction band will be somewhere near V_{MTZ} and the filled part of the band will be within 1 Ry or so above V_{MTZ}. For important technical reasons in solving the Schrödinger equation we shall take

V_{MTZ} as the origin of the energy scale. Hence, most of the energy bands will be at positive energies. (Of course they will be bound states of the solid and therefore they will be at negative energies with respect to V_{AZ}). The most natural language for describing the solutions of a Schrödinger equation at positive energies is that of scattering theory. This will be the subject of the next two sections.

It is useful to note at this stage that although each potential well is made up of the atomic potential from that site and the tails of potentials hanging over from the neighboring sites (typically 5-6 nearest neighbor shells contribute) the dominant contribution still comes from the atomic potential. Thus, while V_{MTZ} is a typical solid state concept in the sense that it is mainly determined by the density of sites, the potential wells carry a great deal of information about the properties of the atoms which make up a solid. If we would take V_{AZ} as our zero of energy as in a tight-binding calculation this information would be most usefully described by giving the atomic energies and wavefunctions. Having taken V_{MTZ} as our energy zero will make the attributes of the atomic species which constitutes the solid show up in the scattering properties of the potential wells. For instance, in energy regions where the atom had a bound state we expect strong scattering. We shall be more explicit about this remark in the next section. Here we merely wanted to point out that while the language has changed due to our choice of the energy zero, the relevance of atomic bound states to the energy bands of the solid and the corresponding intuitive understanding remain the same.

(2) Description of a Single Scatterer

For a more complete account of the scattering theory relevant to our present concern, you are referred to the review article by Lloyd and Smith (1972). Here, we shall present only the bear essentials.

Consider a single muffin-tin potential shown in Fig.6. Since the potential is spherically symmetric the solution to the Schrödinger equation

$$[-\nabla^2 + v(r)]\ \psi(\underline{r}) = \varepsilon\psi(\underline{r}) \qquad \text{II-1}$$

can be written in the form

$$\psi(\underline{r};\varepsilon) = \sum_L a_L(\varepsilon) R_\ell(r;\varepsilon) Y_L(\hat{r}) \qquad \text{II-2}$$

where $Y_L(\hat{r})$ is a spherical harmonic with polar and azimuthal quantum numbers $(\ell,m) \equiv L$ [Lloyd and Smith (1972)] and $R_\ell(r;\varepsilon)$ is the solution of the radial Schrödinger equation

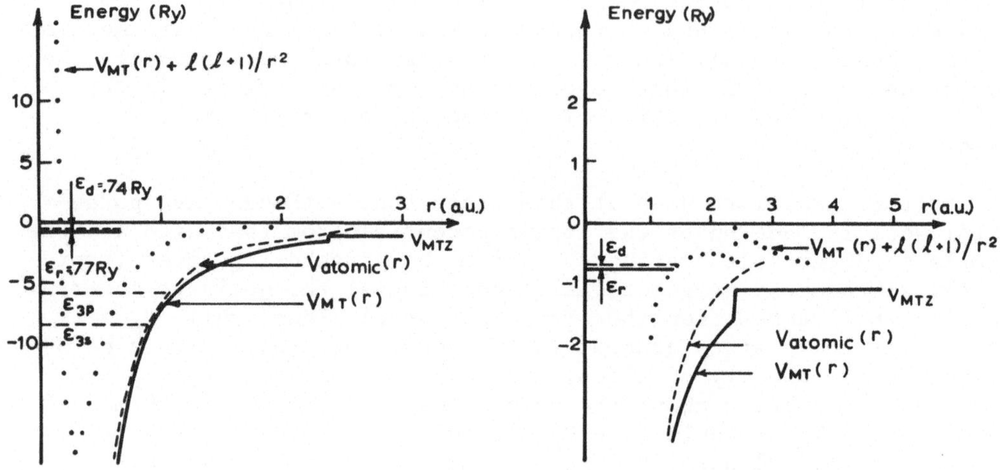

Fig. 6 : The muffin-tin potential $v_{MT}(r)$, the atomic potential $v_{atomic}(r)$ and the sum of $v_{MT}(r)$ and the centrifugal term $\ell(\ell+1)/r^2$ for a Cu crystal. Fig. 6b is the same as 6a but on an enlarged scale to show what happens at r_{MT}.

$$\left[-\frac{1}{r}\frac{d^2}{dr^2}r + \frac{\ell(\ell+1)}{r^2} + v(r)\right] R_\ell(r;\varepsilon) = \varepsilon R_\ell(r;\varepsilon) \qquad \text{II-3}$$

For $r > r_{MT}$ Eq. II-3 is the radial Schrödinger equation for free electrons, whose two linearly independent solutions are the spherical Bessel and Neumann functions $j_\ell(\sqrt{\varepsilon}r)$ and $n_\ell(\sqrt{\varepsilon}r)$ respectively. Thus for $r > r_{MT}$, $R_\ell(r;\varepsilon)$ is a linear combination of j_ℓ and n_ℓ. Therefore, without loss of generality we may define $R_\ell(r;\varepsilon)$ as those solutions of Eq. II-3 which satisfy the boundary condition that

$$R_\ell(r;\varepsilon) = \cos\delta_\ell j_\ell(\sqrt{\varepsilon}r) - \sin\delta_\ell n_\ell(\sqrt{\varepsilon}r), \quad r > r_{MT} \qquad \text{II-4}$$

Inside the muffin-tin sphere $R_\ell(r;\varepsilon)$ is that solution of Eq. II-3 which is regular at the origin and at $r = r_{MT}$ its logarithmic derivative

$$\gamma_\ell(\varepsilon) = R_\ell^{-1}\frac{d}{dr}R_\ell\Big|_{r = r_{MT}}$$

must be equal to the logarithmic derivative of $R_\ell(r;\varepsilon)$ on the outside given in Eq. II-4. From this condition it follows that

$$\delta_\ell(\varepsilon) = \cot^{-1}\left[\frac{\sqrt{\varepsilon}\; n_\ell'(\sqrt{\varepsilon}r_{MT}) - \gamma_\ell(\varepsilon)n_\ell(\sqrt{\varepsilon}r_{MT})}{\sqrt{\varepsilon}\; j_\ell'(\sqrt{\varepsilon}r_{MT}) - \gamma_\ell(\varepsilon)j_\ell(\sqrt{\varepsilon}r_{MT})}\right] \qquad \text{II-5}$$

Note that asymptotically

$$\lim_{r\to\infty} R_\ell(r;\varepsilon) = \frac{1}{\sqrt{\varepsilon}r}\sin\left[\sqrt{\varepsilon}r - \frac{\ell\pi}{2} + \delta_\ell(\varepsilon)\right] \qquad \text{II-6}$$

differs from the free electron solution only by a shift of phase $\delta_\ell(\varepsilon)$. Hence, $\delta_\ell(\varepsilon)$ is usually referred to as the phase shift which, as indicated, is a function of the energy.

The connection between the phase shifts and scattering theory can be seen if we recall that in scattering theory one seeks the solution of Eq. II-1 such that $\psi_{\underline{k}}(\underline{r})$ is an eigenfunction with energy $\varepsilon = k^2$ and asymptotically it can be written as the superposition of an incident plane wave and an outgoing spherical wave, e.g., [Lloyd and Smith (1972)]

$$\psi_{\underline{k}}(\underline{r}) = e^{i\underline{k}\cdot\underline{r}} + f(k^2,\theta)\,\frac{e^{ikr}}{kr}, \qquad \text{II-7}$$

where θ is the angle between the vectors $\underline{k}$ and $\underline{r}$. Setting $\varepsilon = k^2$ in Eq. II-2 and requiring that $\psi(\underline{r};k^2)$ has the asymptotic form given in Eq. II-7 we find

$$a_L(k^2) = 4\pi i^\ell e^{i\delta_\ell} Y_L^\star(\hat{k})$$

Furthermore

$$\psi_{\underline{k}}(\underline{r}) = 4\pi\sum_L i^\ell e^{i\delta_\ell} R_\ell(r;k^2)Y_L(\hat{r})Y_L^\star(\hat{k}) \qquad \text{II-8}$$

and the scattering amplitude $f(k^2;\theta)$ is given by

$$f(\varepsilon;\theta) = \sum_\ell (2\ell+1)\; f_\ell(\varepsilon)\; P_\ell(\cos\theta) \qquad \text{II-9}$$

where

$$f_\ell(\varepsilon) = \sin\delta_\ell(\varepsilon)e^{i\delta_\ell(\varepsilon)} \qquad \text{II-10}$$

Using the definition that the total scattering cross section is given by $\sigma = \frac{2\pi}{\varepsilon}\int_{-1}^{1} d(\cos\theta)\,|f(\theta)|^2$ one can also show that

$$\sigma = \frac{4\pi}{\varepsilon}\sum_\ell (2\ell+1)\;\sin^2\delta_\ell(\varepsilon) \qquad \text{II-11}$$

Thus, a phase shift near $\pi/2$ means very strong scattering while δ_ℓ near 0 or π means weak scattering.

As is evident from Eqs. II-7, II-9, and II-10 the phase shifts completely determine the scattering properties of the potential function. To put it another way, the effect of the potential within the muffin-tin sphere on the wavefunction outside this region is completely determined by the δ_ℓ's. Thus the phase shifts are a very efficient description of the potential function for problems where we are interested in the wavefunctions only outside the scattering region. This is particularly so in the case of muffin-tin potentials one encounters in most metals. It turns out that except for the f-band metals, $\delta_\ell(\varepsilon)$ is practically zero for $\ell > 2$ in the energy range of the conduction band.

As will be seen later the energy bands of a pure metal depend on the muffin-tin potential only through the phase shifts. Thus, three simple functions of energy and the crystal structure determine the band structure (but not the wavefunctions) completely. Clearly, this is a much more concise description than that afforded by an even halfway realistic tight-binding Hamiltonian with many overlap integrals.

The phase shifts for a Na and a Cu muffin-tin potential are shown in Fig. 7. Note that the sodium muffin-tin potential, like a pseudopotential, is a weak scatterer in spite of the fact that it is sufficiently strong to bind all the core electrons (20). On the other hand, the Cu potential is a very strong scatterer at energies near .49 Ry above the muffin-tin zero since $\delta_2(\varepsilon)$ passes through $\pi/2$.

A phase shift rising from 0 to π within a small energy interval Γ about ε_r where it is equal to $\pi/2$ is called a scattering resonance. It is customary to refer to ε_r and Γ as the position and the width of the resonance and parametrize the corresponding scattering amplitude by the Breit-Wigner form

$$f(\varepsilon) = \frac{\Gamma}{(\varepsilon-\varepsilon_r) + i\Gamma} \qquad \text{II-12}$$

Alternatively, one can write $\tan \delta(\varepsilon) = \Gamma(\varepsilon-\varepsilon_r)^{-1}$. Clearly, the contribution of the resonance to the scattering cross section is a Lorentzian peak $\sigma \sim \Gamma[(\varepsilon-\varepsilon_r)^2 + \Gamma^2]^{-1}$ which describes the rise and fall of the scattering strength of the potential as ε passes through ε_r.

An elegant result in scattering theory is that the time spent by an electron at the scattering center, the Wigner delay time τ_w, is given by $\tau_w = 2\hbar\, \partial/\partial\varepsilon\, \delta_\ell(\varepsilon)$. [For a very enlightening discussion see Faulkner (1977)]. Thus an electron with an energy near the resonance will spend a long time at the scattering center. As can be seen in Fig. 7 the Cu d-band straddles the resonant energy where

$$\delta_2^{Cu}(\varepsilon_r) = \frac{\pi}{2}$$

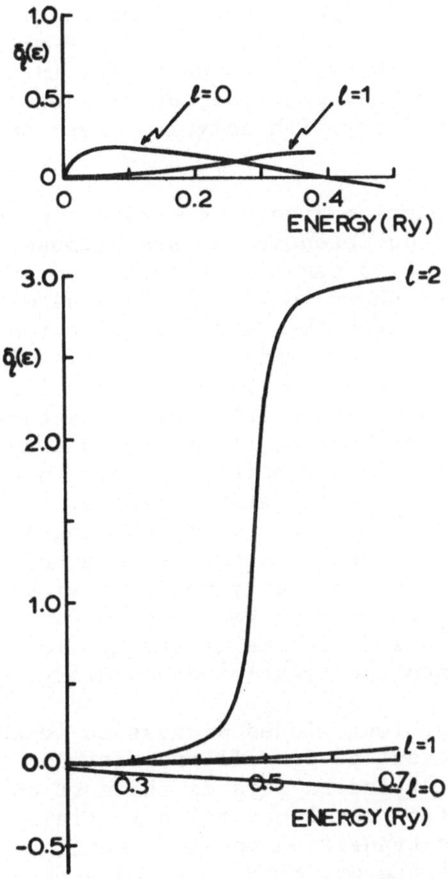

Fig. 7 : The phase shifts $\delta_\ell(\varepsilon)$ for the muffin-tin potential wells of Na (upper) and Cu (lower) constructed for the observed lattice spacing.

Thus, an electron with energy in the d-band will spend a long time at each site as it travels through the lattice. Moreover, while at a site its wavefunction will have predominantly d-character about that site. Clearly this is the same physical picture as provided by the tight-binding theory where one assumes that a d-electron in a narrow d-band, like that of Cu, is localized in an atomic d-state most of the time and hops to another site only occasionally. The deceptively different language is due to the fact that, as mentioned earlier, in the tight-binding picture one deals with bound states at negative energies which decay exponentially in space, while in scattering theory we work at positive energies where the scattered wave travels off to infinity with only its phase shifted by the interaction at a site.

To make this connection even more explicit we note that the $\ell = 2$ resonance in the Cu potential arises because the potential $v(r) + \ell(\ell+1)/r^2$ in Eq. II-3 can almost form a bound state behind the centrifugal barrier shown in Fig. 6. Because we are at positive energies this state is metastable since an electron can always tunnel out. In fact the tunneling probability gives the width of the resonance. The long Wigner delay time is merely a reflection of the fact that the electron can be trapped into this metastable bound state. Let us now remove the neighboring atoms around a scattering center. From the way the muffin-tin potential has been constructed it is clear that V_{MTZ} will rise to the atomic zero of the energy scale but the potential will remain relatively unchanged (we are removing the tails of the atomic potentials on the neighboring sites). Consequently, at the same absolute energy where previously we had a resonance we will now have a real bound state whose wavefunction decays exponentially in space. This is the atomic bound state whose wavefunction enters into the tight-binding theory.

From this somewhat long winded discussion it should be clear that the muffin-tin potential for all transition metals should have an $\ell = 2$ resonance. Indeed this is the most characteristic feature of the transition metal crystal potentials. Furthermore, all the characteristic transition metal properties like high resistivity, large cohesive energy, high superconducting transition temperature, metallic magnetism, etc. can be fairly directly attributed to the presence of this d-resonance.

Having discovered the scattering resonance in the muffin-tin potential of transition metals it is now clear why we cannot use perturbation theory to describe their band structure. In fact there is a theorem in scattering theory which says that the Born series does not converge at the resonance and therefore no finite order perturbation theory has any meaning. Thus, for alloys involving transition metals one is forced to consider some non-perturbative method of treating disorder as a matter of principle.

Let us now turn to the discussion of the Green's function and the t-matrix in the language of scattering theory.

Almost by definition the Green's function of Eq. II-1 is

$$G(\underline{r},\underline{r}';\varepsilon) = \sum_n \frac{\psi_n^\star(\underline{r})\psi_n(\underline{r}')}{\varepsilon - \varepsilon_n + i\eta} \qquad \text{II-13}$$

where the $\psi_n(\underline{r})$'s form an orthonormal complete set and are eigenfunctions of the Hamiltonian $H = -\nabla^2 + v(\underline{r})$. It is a fairly complicated matter but it can be shown that the functions $\psi_{\underline{k}}(\underline{r})$ in Eq. II-8 for all $\underline{k}$ form a complete set provided the potential $v(\underline{r})$ has no bound states. However, they must be treated with care since their norm, like that of plane waves, is infinite e.g., $\langle\psi_{\underline{k}}|\psi_{\underline{k}'}\rangle = (2\pi)^3\delta(\underline{k}-\underline{k}')$. Nevertheless, in the interest of getting on, allow us to assert that they can be used to evaluate Eq. II-13 for positive energies. This gives us

$$G(\underline{r},\underline{r}';\varepsilon) = \int \frac{d^3k}{(2\pi)^3} \frac{\psi_{\underline{k}}^\star(\underline{r})\psi_{\underline{k}}(\underline{r}')}{\varepsilon - k^2 + i\eta} \qquad \text{II-14}$$

and

$$\mathrm{Im}G(\underline{r},\underline{r}';\varepsilon) = -\sqrt{\varepsilon}\sum_L R_\ell(r;\varepsilon)R_\ell(r';\varepsilon)Y_L^\star(\hat{r})Y_L(\hat{r}') \qquad \text{II-15}$$

where we have made use of Eq. II-8 and that $\varepsilon = k^2$.If there was no potential at the origin, that is to say for free electrons, Eq. II-15 would read

$$\mathrm{Im}G_0(\underline{r},\underline{r}';\varepsilon) = -\sqrt{\varepsilon}\sum_L j_\ell(\sqrt{\varepsilon}r)j_\ell(\sqrt{\varepsilon}r')Y_L^\star(\hat{r})Y_L(\hat{r}') \qquad \text{II-16}$$

and hence the change in the density of states due to the introduction of the potential is

$$\Delta n(\varepsilon) = n(\varepsilon) - n_0(\varepsilon) = \frac{\sqrt{\varepsilon}}{\pi}\sum_\ell (2\ell+1)\int_0^\infty dr\, r^2[R_\ell^2(r;\varepsilon) - j_\ell^2(\sqrt{\varepsilon}r)]$$

$$= \frac{1}{\pi}\sum_\ell (2\ell+1)\frac{\partial}{\partial\varepsilon}\delta_\ell(\varepsilon) \qquad \text{II-17}$$

where the details of the calculation are given in [Lloyd and Smith (1972)].

This is the famous Friedel sum and it shows how a phase shift varying rapidly with energy can give rise to a large density of states near the resonance energy. Using the phase shift for Cu we show $\Delta n(\varepsilon)$ in Fig. 8. Of course the density of states curve in Fig.8

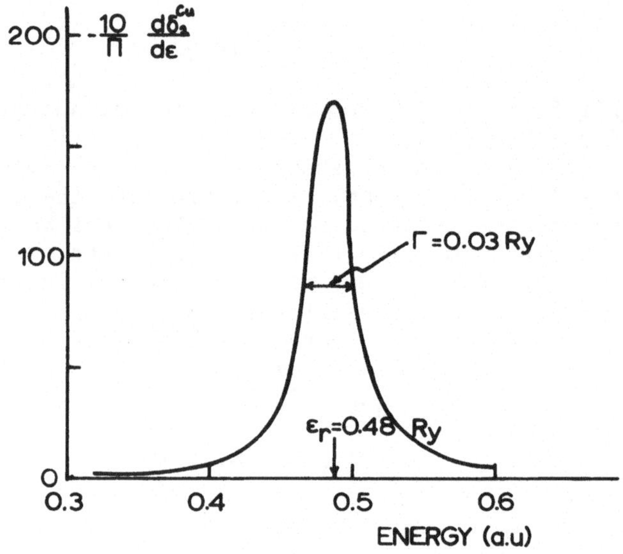

Fig. 8 : The Friedel sum contribution to the density of states for the Cu muffin-tin potential discussed in the text.

is not to be confused with the d-band density of states in Cu. Here, Δn is due to a single Cu muffin-tin potential in a flat potential background. Its width is the width of the d-resonance Γ and not the width of the d-band. However, when scattering from more potentials is included (the multiple scattering corrections), this resonance peak will broaden out into the d-band. All the same, it remains true that a sharp resonance will give rise to a narrow d-band and a broad resonance to a broad one. The point of this short discussion was to give you an inkling of how resonance scattering is related to large densities of states. Obviously, in the scattering theory parlance, it is this connection which replaces the small-overlap-integral, flat-band, high density of states argument of tight-binding theory.

As an alternative to constructing the Green's function from

the eigenfunctions of Eq. II-1 as in Eq. II-13, we could start with the integral equation

$$G(\underline{r},\underline{r}';\varepsilon) = G_0(\underline{r},\underline{r}';\varepsilon) + \int d^3r_1\, G_0(\underline{r},\underline{r}_1;\varepsilon)v(\underline{r}_1)G(\underline{r}_1,\underline{r}';\varepsilon) \qquad \text{II-18}$$

which follows from Eq. II-1 and Eq. II-13 if one notes that the free particle Green's function $G_0(\underline{r},\underline{r}';\varepsilon)$ is defined by the equation

$$(\nabla^2 + \varepsilon + i\eta)G_0(\underline{r} - \underline{r}';\varepsilon) = \delta(\underline{r} - \underline{r}') \qquad \text{II-19}$$

A way of solving Eq. II-18 is to define the t-matrix by the relation

$$G(\underline{r},\underline{r}';\varepsilon) = G_0(\underline{r} - \underline{r}';\varepsilon) + \int d^3r_1 \int d^3r_2 G_0(\underline{r} - \underline{r}_1;\varepsilon)t(\underline{r}_1,\underline{r}_2;\varepsilon)\cdot$$

$$\cdot\, G_0(\underline{r}_2 - \underline{r}';\varepsilon) \qquad \text{II-20}$$

and attempt to solve the integral equation

$$t(\underline{r},\underline{r}';\varepsilon) = v(\underline{r})\delta(\underline{r}-\underline{r}') + \int d^3r_1 v(\underline{r})G_0(\underline{r}-\underline{r}_1;\varepsilon)t(\underline{r}_1,\underline{r}';\varepsilon) \qquad \text{II-21}$$

which follows from Eqs. II-18 and II-20.

In its full generality Eq. II-21 is quite difficult to solve, however, the "on the energy shell" angular momentum components of $t(\underline{r},\underline{r}';\varepsilon)$ turn out to be simple quantities. To define these consider the matrix elements of the t-matrix in the plane wave representation

$$t(\underline{k},\underline{k}';\varepsilon) = \int d^3r_1 \int d^3r_2 e^{-i\underline{k}\cdot\underline{r}_1}\, t(\underline{r}_1,\underline{r}_2;\varepsilon)e^{i\underline{k}'\cdot\underline{r}_2} \qquad \text{II-22}$$

In general the above matrix elements are defined for all $\underline{k}$ and $\underline{k}'$, that is to say for arbitrary incoming and outgoing plane wave energies k^2 and k'^2 respectively. If $k^2 = k'^2 = \varepsilon$ we are describing elastic scattering and $t(\underline{k},\underline{k}';\varepsilon)$ is the probability amplitude that a plane wave $|\underline{k}\rangle$ scatters into the state $|\underline{k}'\rangle$. This situation is often referred to as being "on the energy shell". Using the expansion

$$e^{i\underline{k}\cdot\underline{r}} = 4\pi \sum_L i^\ell j_\ell(kr)Y_L^\star(\hat{k})Y_L(\hat{r})$$

it is easy to see that

$$t(\underline{k},\underline{k}';\varepsilon) = \sum_{L,L'} i^{-\ell+\ell'}\,(4\pi)^2\, Y_L(\hat{k})t_{L,L'}(\varepsilon)Y_{L'}^\star(\hat{k}') \qquad \text{II-23}$$

where the angular momentum components of the "on the energy shell" t-matrix $t_{L,L'}$ are defined as

$$t_{L,L'}(\varepsilon) = \int d^3r_1 \int d^3r_2 j_\ell(\sqrt{\varepsilon}r_1) Y_L^*(\hat{r}_1) t(\underline{r}_1,\underline{r}_2;\varepsilon) Y_{L'}(r_2) j_{\ell'}(\sqrt{\varepsilon}r_2) \qquad \text{II-24}$$

We note in passing that for a general potential we should be talking about a matrix $t_{L,L'}(\varepsilon)$. However, for a spherically symmetric potential v(r) the t-matrix is diagonal and it is sufficient to define $t_{LL'}(\varepsilon) = t_L(\varepsilon)\delta_{LL'}$. Given what we said about the meaning of $t(\underline{k},\underline{k}';\varepsilon)$ "on the energy shell" it should not come as a surprise that

$$t_L(\varepsilon) = -\frac{1}{\sqrt{\varepsilon}} f_\ell(\varepsilon) = -\frac{1}{\sqrt{\varepsilon}} \sin\delta_\ell \, e^{i\delta_\ell} \qquad \text{II-25}$$

This is shown in [Lloyd and Smith (1972)]. Following a somewhat imprecise practice we shall refer to both $t_L(\varepsilon)$ and $f_\ell(\varepsilon)$ as scattering amplitudes.

Using Eqs. II-15 and II-24 we can thus write

$$\mathrm{Im}G(\underline{r},\underline{r}';\varepsilon) = \sum_L \Delta_L^*(\underline{r};\varepsilon)\,\Delta_L(\underline{r}';\varepsilon)\,\mathrm{Im}t_L(\varepsilon) \qquad \text{II-26}$$

where

$$\Delta_L(\underline{r};\varepsilon) = -\sqrt{\varepsilon}\,\frac{R_\ell(r;\varepsilon)}{\sin\delta_\ell(\varepsilon)}\,Y_L(\hat{r}) \qquad \text{II-27}$$

It might come as a let down to end the presentation of a sequence of sophisticated results with this rather artificial expression. However, bear with us. In the next section we shall show how Eq. II-26 generalizes very nicely when we consider the imaginary part of the Green's function at a site in the presence of scattering centers at all other sites. Then the expression analogous to Eq. II-26 will be a non-trivial result.

In summary the density of states for a single scatterer is given in terms of the phase shifts in Eq. II-17. Physical observables which depend on the wavefunction as well as the eigenvalues can frequently be calculated from $\mathrm{Im}G(\underline{r},\underline{r}';\varepsilon)$. For such purposes Eq. II-26 can be used. Clearly, $\mathrm{Im}G(\underline{r},\underline{r}';\varepsilon)$ depends on the wavefunctions $\Delta_L(\underline{r};\varepsilon)$ as well as on the phase shifts.

As a final remark we note that in scattering theory we have a continuum of states even for one scatterer. This is evident from Fig. 8 or from the definition of the functions $\psi_{\underline{k}}(\underline{r})$. This continuity is due to the fact that we are working at positive energies. In the tight-binding formulation one is solving the Schrödinger equation at negative energies and therefore, in the case of the one atom problem, the spectrum is discrete. It is then a very subtle mathematical process by which this discrete spectrum turns into a continuum when the possibility of hopping from site to site all the way to infinity is introduced. No finite cluster will give a continuum. This is an important technical difference which must be borne in mind when approximations in the two different pictures are compared.

(3) Multiple Scattering Theory

Let us now consider the Schrödinger equation for the full crystal lattice with the appropriate muffin-tin potential on every site. As was the case in the previous section this can be done in one of three ways. One can solve the Schrödinger equation for the wave-functions. Equivalently, one can attempt solving Eq. II-18 with $v(\underline{r})$ replaced by the full crystal potential $V(\underline{r}) = \sum_i v(\underline{r} - \underline{R}_i)$. Finally, one can calculate the Green's function from the t-matrix for the full crystal potential defined as

$$G(\underline{r},\underline{r}';\varepsilon) = G_0(\underline{r} - \underline{r}';\varepsilon) + \int d^3r_1 \int d^3r_2\ G_0(\underline{r} - \underline{r}_1;\varepsilon)T(\underline{r}_1,\underline{r}_2;\varepsilon)\cdot G_0(\underline{r}_2 - \underline{r}';\varepsilon) \qquad \text{II-28}$$

In this case the problem is to find $T(\underline{r},\underline{r}';\varepsilon)$ by solving

$$T(\underline{r},\underline{r}';\varepsilon) = V(\underline{r})\delta(\underline{r} - \underline{r}') + \int d^3r_1 V(\underline{r})G_0(\underline{r} - \underline{r}_1;\varepsilon)T(\underline{r}_1,\underline{r}';\varepsilon) \qquad \text{II-29}$$

An efficient way of proceeding is to follow this last method.

To solve Eq. II-29 we introduce the scattering path operator [Györffy and Stott (1973)] by the relation

$$\tau^{ij}(\varepsilon) = v_i\delta_{ij} + \sum_k v_i\ G_0(\varepsilon)\ \tau^{kj}(\varepsilon) \qquad \text{II-30}$$

where we used the operator notation to simplify the algebra. Obviously, $v(\underline{r})\delta(\underline{r} - \underline{r}')$ is the matrix element $\langle\underline{r}|v|\underline{r}'\rangle$ and $G_0(\underline{r} - \underline{r}';\varepsilon) \equiv \langle\underline{r}|G_0(\varepsilon)|\underline{r}'\rangle$ etc. Furthermore, by taking similar matrix elements of Eq. II-29 we obtain an integral equation like Eq. II-30.

Summing over i and j in Eq. II-30 one obtains the operator version of Eq. II-29 if one identifies $\sum_{i,j}\tau^{ij}$ with T. Thus solving Eq. II-30 is equivalent to solving Eq. II-29.

By taking the term with k = i on the right hand side of Eq. II-30 over to the left and dividing both sides of the equation by the operator $1 - v_iG_0$ we can rewrite Eq. II-30 as

$$\tau^{ij} = t_i\delta_{ij} + \sum_{k\neq i} t_iG_0\tau^{kj} \qquad \text{II-31}$$

where the t-matrix operator t_i is defined as $(1 - v_iG_0)^{-1}v_i$. Its matrix elements $\langle\underline{r}|t_i(\varepsilon)|\underline{r}'\rangle$ are the t-matrix defined in Eq. II-20 with the exception that now the muffin-tin potential $v(\underline{r} - \underline{R}_i)$ is centred at $\underline{R}_i$ and not at the origin.

In our version of the multiple scattering theory [Lloyd and Smith (1972)] Eq. II-31 is the fundamental equation. It is a way of constructing a solution to the many scatterer problem namely $T(\underline{r},\underline{r}';\varepsilon) \equiv \langle\underline{r}|T(\varepsilon)|\underline{r}'\rangle$ from the knowledge of the solution to the individual scattering problems $\langle\underline{r}|t_i(\varepsilon)|\underline{r}'\rangle$. Its meaning becomes apparent if we note that quite generally the scattering solution to a Schrödinger equation can be written as $\psi_{\underline{k}} = \phi_{\underline{k}} + \Phi_{\underline{k}}$ where $\phi_{\underline{k}}$ is the incident wave and $\Phi_{\underline{k}}$ is the scattered wave as in II-7 and it can be shown that $|\Phi_{\underline{k}}\rangle = G_0 t|\phi_{\underline{k}}\rangle$. Thus, the t-matrix operator generates the scattered wave from the incident wave. More specifically t_i generates the scattered wave due to the potential $v(\underline{r} - \underline{R}_i)$ and T generates the scattered wave due to the whole crystal potential. A little further thought along these lines reveals that the scattering path operator τ^{ij}, when it operates on the wave incident at the site $\underline{R}_j$, gives the scattered waves emanating from the site at $\underline{R}_i$ and includes the effects of all the scattering in between. Evidently $\sum_{ij} \tau^{ij}$ takes the incident waves arriving at all the sites in the lattice, turns them into scattered waves emanating from all other sites and adds up all the scattered waves. This is precisely what T does, hence the equality. In these terms Eq. II-31 is merely a self consistency requirement on τ^{ij} : a wave incident at $\underline{R}_j$ becomes a scattered wave at $\underline{R}_k$, then propagates to $\underline{R}_i$ according to the free particle propagator G_0, to become an incident wave there and there to be turned into a scattered wave from $\underline{R}_i$ by t_i; the sum of all such processes must be equal to τ^{ij} again for $i \neq j$. For $i = j$ there is also a direct scattering term : t_i giving the scattered wave due to the direct scattering at $\underline{R}_i$ and $\sum_{k\neq i} t_i G_0 \tau^{kj}$ giving the contribution to the scattered wave which arises because the scattered particle has returned to the site $\underline{R}_i$, after multiple scattering at the other sites, there to be scattered according to t_i.

The importance of this formulation of the problem to the crystal potential which consists of non-overlapping muffin-tin wells is that for non-overlapping potentials the "on the energy shell" part of Eq. II-31 decouples from the rest and this gives rise to a particularly powerful tool for calculating the energy bands.

To see what is meant by this statement let us define, in analogy with Eq. II-24, the "on the energy shell" components of the matrix elements $\langle\underline{r}|\tau^{ij}(\varepsilon)|\underline{r}'\rangle \equiv \tau^{ij}(\underline{r},\underline{r}';\varepsilon)$ of the scattering path operator $\tau^{ij}_{LL'}(\varepsilon)$ as

$$\tau^{ij}_{LL'}(\varepsilon) = \int d^3r_i \int d^3r'_j Y^{\star}_L(\hat{r}_i) j_\ell(\sqrt{\varepsilon}r_i)\tau^{ij}(\underline{r}_i,\underline{r}'_j;\varepsilon) j_{\ell'}(\sqrt{\varepsilon}r'_j) Y_{L'}(\hat{r}'_j) \qquad \text{II-32}$$

where L is the angular momentum about the site $\underline{R}_i$ and L' is that about the site $\underline{R}_j$, $\underline{r}_i = \underline{r} - \underline{R}_i$ and $\underline{r}'_j = \underline{r}' - \underline{R}_j$. The use of these spatial variables is legitimate since it follows from the integral equation version of Eq. II-31 that $\tau^{ij}(\underline{r},\underline{r}';\varepsilon)$ depends on $\underline{r}$ and $\underline{r}'$ only through $\underline{r} - \underline{R}_i$ and $\underline{r}' - \underline{R}_j$.

We now want to find an expression for $\tau^{ij}_{LL'}(\varepsilon)$ in terms of $t_L(\varepsilon)$ from Eq. II-31. To do this, take the matrix elements of this equation between the states $|\underline{r}>$ and $|\underline{r}'>$ inserting the identity $I = \int d^3r|\underline{r}><\underline{r}|$ between operator products and then take the "on the energy shell" components of both sides of the equation by the operation defined in Eq. II-32. Recognizing that for $\underline{r}$ in the muffin-tin sphere surrounding $\underline{R}_i$ and $\underline{r}'$ in the muffin-tin sphere surrounding $\underline{R}_j$ with $i \neq j$

$$G_0(\underline{r} - \underline{r}';\varepsilon) = \sum_{LL'} Y^\star_L(\hat{r}_i)j_\ell(\sqrt{\varepsilon}r_i)G_{LL'}(\underline{R}_i - \underline{R}_j;\varepsilon)j_{\ell'}(\sqrt{\varepsilon}r_j)Y_{L'}(\hat{r}_j) \qquad \text{II-33}$$

where

$$G_{LL'}(\underline{R}_i - \underline{R}_j;\varepsilon) = 4\pi\sqrt{\varepsilon}\sum_{L''} C^{L''}_{LL'}\, i^{\ell''} h^+_{\ell''}(\sqrt{\varepsilon}|\underline{R}_i - \underline{R}_j|)Y_{L''}(\widehat{\underline{R}_i - \underline{R}_j}) \qquad \text{II-34}$$

with h^+_ℓ a Hankel function and $C^{L''}_{LL'}$ a Gaunt number defined by $C^{L''}_{LL'} = \int d\Omega Y_L(\Omega)Y_{L''}(\Omega)Y^\star_{L'}(\Omega)$ it is entirely straightforward to show that

$$\tau^{ij}_{LL'}(\varepsilon) = t_L(\varepsilon)\delta_{ij}\delta_{LL'} + \sum_{k\neq i,L''}\sum t_L(\varepsilon)G_{LL''}(\underline{R}_i - \underline{R}_k;\varepsilon)\tau^{kj}_{L''L'}(\varepsilon) \qquad \text{II-35}$$

This is, then, our fundamental multiple scattering equation "on the energy shell". It gives $\tau^{ij}_{LL'}$ in terms of $t_L(\varepsilon)$, that is to say in terms of the phase shifts, and the structure constants $G_{LL'}(\underline{R}_i - \underline{R}_j;\varepsilon)$ which do not depend on the potential function and are determined completely by the spatial arrangement of the scattering sites. Eq. II-35 is valid for any arrangement of potentials even if at each site we have a different scatterer [in this case we would have to replace $t_L(\varepsilon)$ by $t_{i,L}(\varepsilon)$]. Thus, it is a good starting point to discuss pure metals, liquids as well as random alloys. The only limitation on its validity is the requirement that the potential wells at the different sites may not overlap. If they are allowed to overlap we need an expression for $G_0(\underline{r},\underline{r}';\varepsilon)$ which is valid for $\underline{r} = \underline{r}'$ where Eq. II-33 cannot be used. The unfortunate consequence of overlapping potentials is that the "on the energy shell" part of the problem becomes coupled to the "off the energy shell" part and except for some simplified models the problem becomes intractable. The physical reason for the "off the energy shell" component coming in for overlapping potentials is easy to see. For non-overlapping potentials one elastic scattering process is completely over before the next one begins. So, you stay on the energy shell, and the only information about the potential which matters are the phase shifts. For overlapping scatterers before a scattering process is over (i.e. one is still within the scattering region) the next one begins because the tail of the next potential hangs into the scattering region of the first. Under these circumstances the inelastic virtual processes begin to play a role. Fortunately, we will not need the theory under such adverse circumstances.

Before closing this section we would like to introduce a notation which will simplify many manipulations later on in these lectures. Consider the space spanned by the abstract vectors $|i,L\rangle$ and regard $\tau^{ij}_{LL'}(\varepsilon)$ as matrix elements of the operator $\tau(\varepsilon)$ in that space. Namely write

$$\begin{aligned} \tau^{ij}_{LL'}(\varepsilon) &\equiv \langle i,L|\tau(\varepsilon)|j,L'\rangle \\ t_{iL}(\varepsilon) &\equiv \langle i,L|t(\varepsilon)|i,L\rangle \\ G_{LL'}(\underline{R}_i - \underline{R}_j;\varepsilon) &\equiv \langle i,L|G_0(\varepsilon)|j,L'\rangle \qquad i \neq j \\ &\equiv 0 \qquad i = j \end{aligned} \qquad \text{II-36}$$

and regard

$$\tau(\varepsilon) = t(\varepsilon) + t(\varepsilon)G_0(\varepsilon)\tau(\varepsilon) \qquad \text{II-37}$$

as the operator equivalent of Eq. II-35 in the site-angular momentum space spanned by $|i,L\rangle$.

It is interesting to note that

$$[t^{-1}(\varepsilon) - G_0(\varepsilon)]\tau(\varepsilon) = I \qquad \text{II-38}$$

and therefore

$$\sum_{k,L''} [t^{-1}_{i,L}(\varepsilon)\delta_{LL''}\delta_{ik} - G_{LL''}(\underline{R}_i - \underline{R}_k;\varepsilon)]\tau^{kj}_{L''L'}(\varepsilon) = \delta_{ij}\delta_{LL'} \qquad \text{II-39.}$$

For those of you with a background in the "tight-binding" approach to the band theory of transition metals it should be illuminating to note that Eq. II-39 is precisely of the form of the Schrödinger equation for the Green's function in the tight binding language. If we expand the wavefunction about each site in a complete set of atomic states $\phi_\mu(\underline{r})$ and write the Bloch function as

$$\psi_{\underline{k}}(\underline{r}) = \sum_{i,\mu} a^{\mu,i}_{\underline{k}}\phi_\mu(\underline{r} - \underline{R}_i)$$

then the Green's function in Eq. II-13 may be written as

$$G(\underline{r},\underline{r}';\varepsilon) = \sum_{\substack{ij,\\ \mu\mu'}} \phi^\star_\mu(\underline{r} - \underline{R}_i)G_{\mu\mu'}(i,j;\varepsilon)\phi_{\mu'}(\underline{r}' - \underline{R}_j) \qquad \text{II-40}$$

The coefficients $G_{\mu\mu'}(i,j;\varepsilon)$ are called the Green's function in the site representation and they are determined by the Schrödinger equation :

$$\sum_{\mu'',k} [(\varepsilon - \varepsilon_{\mu})\delta_{\mu\mu''}\delta_{ik} - J_{\mu\mu''}(i,k)]\, G_{\mu''\mu'}(k,j;\varepsilon) = \delta_{\mu\mu'}\delta_{ij} \qquad \text{II-41}$$

where ε_{μ}'s are the atomic energies corresponding to the complete set of states $\phi_{\mu}(\underline{r})$ and $J_{\mu\mu'}(i,j)$ are the overlap integrals given in terms of the wavefunctions ϕ_{μ} and the local atomic potential. In order to keep the notation simple we have neglected the overlap integrals which refer to more than two sites. For a more complete discussion of this theory you are referred to standard textbooks like that of Kittel (1976) or Callaway (1974).

We now note that Eq. II-39 is of the same form as Eq. II-41. $\tau^{ij}_{LL'}(\varepsilon)$ plays the role of the Green's function $G_{\mu\mu'}(i,j;\varepsilon)$ and $(\varepsilon - \varepsilon_{\mu})\delta_{ij}\delta_{\mu\mu'}$ is replaced by $t^{-1}_{i,L}(\varepsilon)\delta_{ij}\delta_{LL'}$. For resonant scatterers this connection is even closer since then $t^{-1}_{i,L}(\varepsilon) = \Gamma^{-1}(\varepsilon - \varepsilon_r) - i\sqrt{\varepsilon}$. However, as these equations stand, this connection is mainly formal. The significance of the connection has been explored by Hubbard (1967), Heine (1967) and Pettifor (1969).

In the present form it is noteworthy that the "overlap integral" $G_{\mu\mu'}(\underline{R}_i - \underline{R}_j;\varepsilon)$ is infinite ranged and energy dependent but is independent of the potential. This fact is of enormous significance from the point of view of the CPA since, as we shall see, it eliminates the necessity of dealing with off-diagonal randomness without making any approximations. (The approximation is made when the crystal potential is constructed but not in solving the Schrödinger equation).

(4) The KKR Band Theory for Pure Metals

If the t-matrix $t_L(\varepsilon)$ is the same on every site, $\tau^{ij}_{LL'}(\varepsilon)$ depends only on $\underline{R}_{i,j}$ and, for an infinite lattice, Eq. II-35 can be easily solved by the method of lattice Fourier transforms. Defining $\tau_{LL'}(\underline{q};\varepsilon)$ and $G_{LL'}(\underline{q};\varepsilon)$ by the relations

$$\tau_{LL'}(\underline{q};\varepsilon) = \frac{1}{N}\sum_{ij} e^{-i\underline{q}\cdot(\underline{R}_i - \underline{R}_j)}\tau^{ij}_{LL'}(\varepsilon) \qquad \text{II-42}$$

and

$$G_{LL'}(\underline{q};\varepsilon) = \frac{1}{N}\sum_{ij} e^{-i\underline{q}\cdot(\underline{R}_i - \underline{R}_j)} G_{LL'}(\underline{R}_i - \underline{R}_j;\varepsilon) \qquad \text{II-43}$$

we find

$$\tau_{LL'}(\underline{q};\varepsilon) = [\,(t^{-1}(\varepsilon) - G_0(\underline{q};\varepsilon))^{-1}]_{LL'} \qquad \text{II-44}$$

where t^{-1} is a diagonal matrix with diagonal elements $t_L^{-1}(\varepsilon)$ and $G_0(\underline{q};\varepsilon)$ stands for the matrix $G_{LL'}(\underline{q};\varepsilon)$.

As we have mentioned before T and therefore τ generates the scattered waves from the incident waves. Thus τ is like a response function. Where it diverges there will be a scattered wave even in the absence of an incident wave. Evidently, these will be the stationary states of the system , that is to say the energy bands. It follows from the rules of matrix inversion

$$\tau_{LL'}(\underline{q};\varepsilon) = \frac{1}{\left|\left|t^{-1}(\varepsilon)-G_0(\underline{q};\varepsilon)\right|\right|} \operatorname{cof}[t_L^{-1}(\varepsilon)\delta_{LL'} - G_{LL'}(\underline{q};\varepsilon)] \qquad \text{II-45}$$

that $\tau_{LL'}(\underline{q};\varepsilon)$ will diverge where the determinant

$$\left|\left|t_L^{-1}(\varepsilon)\delta_{LL'} - G_{LL'}(\underline{q};\varepsilon)\right|\right| = 0 \qquad \text{II-46}$$

This is the well known KKR condition. [Korringa (1947), Kohn and Rostoker (1954) ; see also : Segall and Ham (1968)] . It can yield the energy bands in two ways. For a given $\underline{q}$ one can search in ε and the energies at which Eq. II-46 is satisfied will be the eigenenergies $\varepsilon_{\underline{q},n}$ at that $\underline{q}$. Depending how far up in energy one is willing to go one will find all the bands on the way. This mode of operation is usually referred to as the "constant $\underline{q}$ search". Alternatively, one can do a constant energy search. [Faulkner, Davis and Joy (1967)] . Operating the KKR in this mode consists of fixing the energy and searching for all the zeros of the KKR determinant in the Brillouin zone. This is done direction by direction. Starting at the zone center the KKR determinant is evaluated on a given mesh out to the zone boundary and in this way all its zeros, in that direction, are determined. When this is done for a sufficiently large number of directions (561 in 1/48th of the Brillouin zone) the $\underline{q}$ coordinates of the zeros map out a constant energy surface. The volume in $\underline{q}$-space below this surface is the integrated density of states.

In fact the integrated density of states may be written as

$$N(\varepsilon) = N_0(\varepsilon) - \frac{1}{\pi\Omega_{BZ}} \operatorname{Im}\int_{BZ} d^3q \, \ln\left|\left|t_L^{-1}\delta_{LL'} - G_{LL'}(\underline{q};\varepsilon)\right|\right| \qquad \text{II-47}$$

which is the integral of the phase of the KKR determinant over the Brillouin zone (BZ) whose volume is Ω_{BZ} plus the free electron contribution N_0. It can be regarded as the generalization of the Friedel sum for an infinite lattice. For a careful discussion, see Faulkner (1977). However, that the Brillouin zone integral of the phase of the determinant $\left|\left|t_L^{-1}\delta_{LL'} - G_{LL'}(\underline{q};\varepsilon)\right|\right|$ is the integrated density of states can be seen without such derivation. Take it from us that for cubic systems the KKR matrix $t_L^{-1}\delta_{LL'} - G_{LL'}(\underline{q};\varepsilon)$ is real. Thus, its determinant when regarded as a complex number will have a phase which is an integer multiple of π. If this phase is defined as $\phi(\underline{q};\varepsilon) = \mathrm{Im}\ \ln\ ||t_L^{-1}\delta_{LL'} - G_{LL'}(\underline{q};\varepsilon)||$ it will change discontinuously by π every time the determinant passes through 0 with either $\underline{q}$ or ε changing. Thus by determining the phase difference between the bottom of the band and an arbitrary ε for a given $\underline{q}$ we have found a number which is π times the number of states below that energy. Hence, Eq. II-47 is manifestly the integrated density of states. To illustrate this 'counting' in Fig. 9 we have plotted the energy bands of pure Cu along the Δ direction and the corresponding phase as a function of energy for the zone center (Γ-point) and the zone boundary (X-point). The singlet Γ_1 gives rise to a change of phase of π while the triplet $\Gamma_{25'}$ and doublet Γ_{12} states give rise to changes in the phase of 3π and 2π respectively, thus accounting correctly for the degeneracy of the levels and in the process yielding at all energies the phase with respect to the bottom of the band.

Observe that the possibility of a constant energy search differentiates sharply between the scattering theory approach to solving the Schrödinger equation and the Rayleigh-Ritz type of methods for solving eigenvalue problems. As is evident from the foregoing discussion, using the KKR in the constant energy search mode allows us to determine the eigenstates at a given energy without knowing all the states below. This makes this method particularly efficient for accurate determination of the Fermi energy, Fermi surface, and the density of states at the Fermi energy. It will also facilitate the solution of the random alloy problem energy by energy.

Of course there is no mystery connected with the above remarks. The Green's function is determined by all the eigenvalues and eigenfunctions of the Hamiltonian. All these must be known to calculate the Green's function at a given energy if the construction in Eq. II-13 is used. However, in scattering theory we can calculate the Green's function at a given energy directly and the eigensolutions at other energies are not needed. The mathematical theorem which ensures the equivalence of these two approaches is the Mittag-Leffler (1884) theorem which relates the values of a meromorphic function to the positions of its poles.

It will amplify these remarks and will be useful later if we note that for $\underline{r}$ and $\underline{r}'$ within the same muffin-tin sphere at $\underline{R}_i$

$$\mathrm{Im}G(\underline{r},\underline{r}';\varepsilon) = \sum_{LL'} \Delta_L^{\star}(\underline{r};\varepsilon)\Delta_{L'}(\underline{r}';\varepsilon)\,\mathrm{Im}\ \tau_{LL'}^{ii}(\varepsilon) \qquad \text{II-48}$$

where $\Delta_L(\underline{r};\varepsilon)$ is defined in Eq. II-27. This expression is clearly the analogue of Eq. II-26 in the case of many scatterers and was shown to be valid for an arbitrary number and arrangement of non-overlapping scatterers by Györffy and Stott (1972) and also by

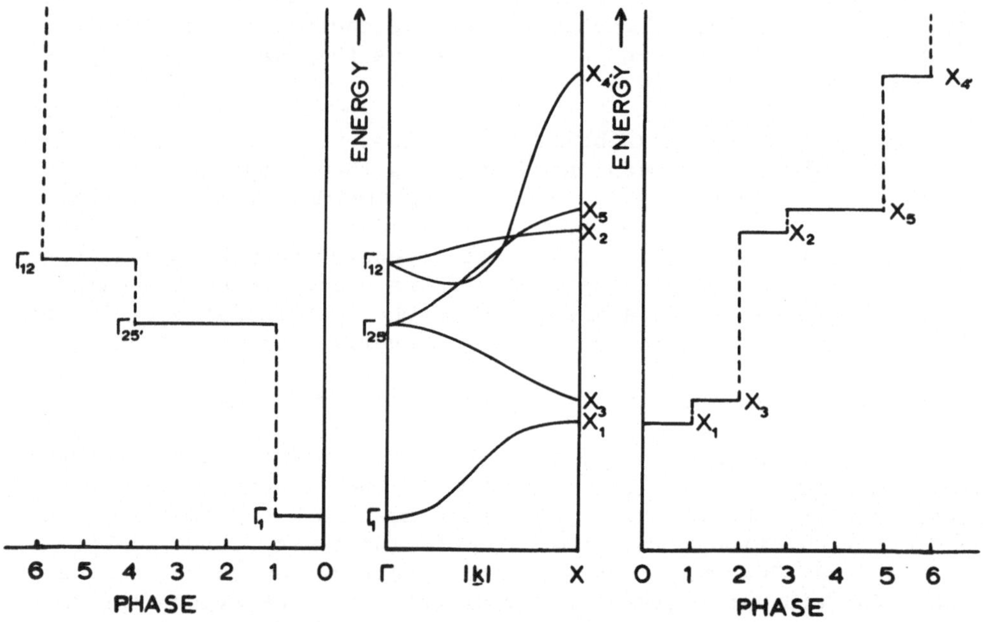

Fig. 9 : The energy bands of pure Cu in the Δ direction and the phase of the KKR determinant at the Γ and X points.

Beeby (1967) and Johnson (1966). For a finite group of scatterers or for a random system where the potential wells on different sites are different $\tau^{ii}_{LL'}(\varepsilon)$ is given by the solution of Eq. II-35. In the case of an ordered system of scatterers

$$\tau^{ii}_{LL'}(\varepsilon) = \frac{1}{\Omega_{BZ}} \int d^3q \; \tau_{LL'}(\underline{q};\varepsilon) \qquad \text{II-49}$$

Clearly, in going from Eq. II-26 to Eq. II-48 we have merely replaced the single site t-matrix on the energy shell $t_L(\varepsilon)$ with the multiple scattering t-matrix at the site in question $\tau^{ii}_{LL'}(\varepsilon)$ which is also on the energy shell. As we mentioned before $\tau^{ii}_{LL'}(\varepsilon)$ gives the scattered wave from the site at $\underline{R}_i$ and includes the contribution from all processes in which the electron scattered from the potential well at $\underline{R}_i$ scatters repeatedly at other sites and then returns to the site at $\underline{R}_i$ to be scattered there again. It is appealing to the intuition that the modification of the Green's function within the muffin-tin well at $\underline{R}_i$ due to the introduction of the other scattering centers only depends on the scattering properties, namely the phase shifts, at those new sites.

The Green's function contains much more information than the energy bands. One can calculate many useful observables from the Green's function within a muffin-tin well. This will be demonstrated in our final lecture where we shall discuss the soft X-ray spectra of random alloys.

Having developed the formal structure of the theory which will be used in the next section to discuss random alloys we would like to conclude this section by a comment on the alloy potential and the corresponding multiple scattering equations on the energy shell.

Consider a typical alloy configuration. If we were to construct the crystal potential all through the lattice by overlapping atomic potentials we would find that not all the muffin-tin potentials on the A sites are the same. Due to the fact that the atomic charge distributions at the neighboring sites will have tails which overhang into the muffin-tin sphere at the site in question the muffin-tin potential there will depend on the local environment. For most configurations this will vary from A site to A site. Of course, the same is true on the B sites. Consequently, at least in principle, we shall have to describe the "on the energy shell" scattering on A and B sites by sequences of scattering amplitudes $t^{\nu}_{A,L}(\varepsilon)$ and $t^{\nu}_{B,L}(\varepsilon)$ where the superscript ν labels all possible configurations within the first few nearest neighbor shells (as many shells as used in the construction of the potential). Elsewhere we have pointed out how this freedom in the theory can be used to describe short range order within the context of the CPA [Györffy and Stocks (1976)]. However, for now, we shall assume that all the A scattering amplitudes are the same and all the B scattering amplitudes are the same. Namely we set $t^{\nu}_{A,L} = t_{A,L}$ and $t^{\nu}_{B,L} = t_{B,L}$ for all ν. That this is a reasonable approximation can be seen from Fig. 6. Comparing the atomic potential and the muffin-tin potential it is evident that the effect of having neighbors is relatively small well within the muffin-tin radius. Furthermore, it is precisely this part of the potential which determines the position of the resonance. Since this is the most dominant aspect of our potential, as a first approximation, it is reasonable to neglect the slight variation in the potential due to the variation in the environment.

Let us then consider an alloy crystal potential which for each configuration is $v_A(\underline{r} - \underline{R}_i)$ on an A site and $v_B(\underline{r} - \underline{R}_j)$ on a B site. These potential functions may be constructed for the average environment and therefore depend on the concentration, the lattice spacing and the atomic charge distribution of the constituent atoms but are independent of the short range order. Under these circumstances the fundamental equations of multiple scattering on the energy shell are

$$\tau^{ij}_{LL'}(\varepsilon) = t_{i,L}(\varepsilon)\delta_{ij}\delta_{LL'} + \sum_{L''}\sum_{k\neq i} t_{i,L}(\varepsilon)G_{LL''}(\underline{R}_i - \underline{R}_k;\varepsilon)\,\tau^{kj}_{L''L'}(\varepsilon) \qquad \text{II-50}$$

where $t_{i,L} = t_{A,L}$ or $t_{B,L}$ as prescribed by the configuration in question.

Thus, the general alloy problem we want to solve consists of three parts : solution of Eq. II-50, the calculation of an interesting observable for a given configuration and averaging of the observable over the ensemble of all configurations consistent with a given concentration and any further information which is determined by the measuring process.

One of the simplest "observables" we shall want to calculate is the density of states. The real space analogue of Eq. II-47 is

$$N(\varepsilon)=N_0(\varepsilon) - \frac{1}{\pi N}\,\mathrm{Im}\,\ln\left|\left|t^{-1}_{i,L}(\varepsilon)\delta_{ij}\delta_{LL'}-G_{LL'}(\underline{R}_i-\underline{R}_j;\varepsilon)\right|\right| \qquad \text{II-51}$$

where the process of taking the determinant has to be carried out with respect to the site index as well as the angular momentum index L. For a recent discussion of this result see [Faulkner (1977)]. Taking the derivative of Eq.II-51 and averaging the results gives

$$\bar{n}(\varepsilon) = n_0(\varepsilon) - \frac{1}{\pi N}\Big[\mathrm{Im}\Big[\sum_{i,L}\Big(\frac{\partial}{\partial\varepsilon}t^{-1}_{A,L}\Big)\langle\tau^{ii}_{LL}\rangle_A + \sum_{i,L}\Big(\frac{\partial}{\partial\varepsilon}t^{-1}_{B,L}\Big)\langle\tau^{ii}_{LL}\rangle_B$$

$$- \sum_{i,j}\sum_{LL'}\Big(\frac{\partial}{\partial\varepsilon}G_{LL'}(\underline{R}_i - \underline{R}_j;\varepsilon)\Big)\langle\tau^{ij}_{LL'}\rangle\Big] \qquad \text{II-52}$$

where we have used the matrix identity $\ln\|\tau^{-1}\| = \mathrm{tr}\ \ln\tau^{-1}$ and the relation $(\tau^{-1})_{iL,jL'} = t^{-1}_{i,L}\delta_{ij}\delta_{LL'} - G_{LL'}(\underline{R}_i - \underline{R}_j;\varepsilon)$ which follows from Eq. II-39. By $\langle\tau^{ij}_{LL'}\rangle_A$ and $\langle\tau^{ij}_{LL'}\rangle_B$ we mean the average of $\tau^{ij}_{LL'}$ over all configurations which have an A and a B atom respectively at the site i. $\langle\tau^{ij}\rangle$ means an average without any restriction except that implied by the concentration. Thus the task of an alloy theory is to find $\langle\tau^{ij}_{LL'}\rangle_A$, $\langle\tau^{ij}_{LL'}\rangle_B$ and $\langle\tau^{ij}_{LL'}\rangle$.

(5) The KKR-CPA

We are now in the position to formulate the effective medium theory we have discussed in the introduction for our model alloy potential.

Let us begin by noting what the Virtual Crystal Approximation (VCA) means within our multiple scattering approach. Evidently, if we average the random crystal potential before solving the corresponding Schrödinger equation we have to associate with each scattering center an average potential $\overline{v}(\underline{r}) = cv_A(\underline{r}) + (1 - c)v_B(\underline{r})$. The scattering properties of this potential are described by the phase shifts $\delta^{VCA}_\ell(\varepsilon)$ constructed in the usual way from $\overline{v}(\underline{r})$. Then, the ordered array of scattering centers, each described by $f^{VCA}_\ell(\varepsilon) = \sin\delta^{VCA}_\ell \exp i\,\delta^{VCA}_\ell$ may be regarded as an effective medium. The behavior of an electron in this medium is proported to represent the averaged behavior of an electron in the random lattice. Since the scattering on each site is described by the same scattering amplitude the multiple scattering equations, Eq. II-39 can be readily solved and the eigenvalue spectrum of the effective medium is determined by the usual KKR formulae (see Section 4). The potential function $\overline{v}(\underline{r})$ changes smoothly from $v_A(\underline{r})$ to $v_B(\underline{r})$ as c goes from 0 to 1. The phase shifts will do the same. Thus, for a given concentration we obtain a band structure intermediate between that of the constituent metals. To complete the theory in this approximation one must, of course, fill up the bands according to the appropriate number of electrons per atom $Z_{eff} = cZ_A + (1 - c)\,Z_B$.

As we have mentioned before, the VCA is a good approximation only if the potential functions $v_A(\underline{r})$, $v_B(\underline{r})$ are very close. Under this circumstance $\overline{v}(\underline{r})$ does not change much with concentration and the bands only shift by small amounts without their shape being very much affected. Thus the VCA is just a fancy form of the rigid band model. Though in general unreliable, this model frequently works well as a qualitative guide. Nevertheless, as we shall show later, the VCA is apt to miss all the interesting alloying effects even if the general picture it gives is not grossly wrong [Giuliano et al (1977)].

A more sophisticated effective medium results if instead of

averaging the potential functions we average the scattering amplitudes on each site. We again obtain an ordered array of scattering centers each described by the effective scattering amplitudes $\bar{f}_\ell(\varepsilon) = cf_\ell^A(\varepsilon) + (1-c)f_\ell^B$. Evidently the multiple scattering equations (Eq. II-39) can again be solved by the method of lattice Fourier transform and apparently we are back to a KKR calculation for the effective lattice. This is the averaged t-matrix approximation first proposed by Korringa (1958). Since then it has been investigated by a number of workers ; see for instance Beeby (1964) and Soven (1966). More recently, it has been advocated as an adequate theory of the electronic structure for many alloys by Bansil, Schwartz and Ehrenreich (1975).

In spite of the similarity to the rather naive VCA theory, the ATA is a rather sophisticated approximation. A way of seeing this is to consider the Argand plot for the averaged scattering amplitude $\bar{f}_\ell = -\sqrt{\varepsilon}\bar{t}_\ell = cf_\ell^A + (1-c)f_\ell^B$. This is a plot of Im $\bar{f}_\ell$ vs. Re $\bar{f}_\ell$ for various values of the energy. As we shall show presently $\bar{f}_\ell$ describes an inelastic scatterer. In order to facilitate our discussion of this fact we shall now briefly digress to introduce some basic notions regarding such scatterers.

An elastic scatterer is always described by real phase shifts. From the definition of the scattering amplitude

$$f_\ell(\varepsilon) = \sin\delta_\ell e^{i\delta_\ell} = \frac{1}{2i}(e^{2i\delta_\ell} - 1) \qquad \text{II-53}$$

it follows that

$$(\text{Re } f_\ell)^2 + (\text{Im}f_\ell - \tfrac{1}{2})^2 = \frac{1}{4} \qquad \text{II-54}$$

Thus, for a resonant phase shift shown in Fig. 10a the Argand plot is a circle with a radius 1/2 centered at Re $f_\ell = 0$, Im$f_\ell = 1/2$ as shown in Fig. 10b. It is customary to refer to this circle as the unitarity circle. Note that f_ℓ as defined above automatically satisfies the optical theorem Im $f_\ell = |f_\ell|^2$.

On the other hand an inelastic scatterer is described by a complex phase shift, or equivalently it may be represented as

$$f_\ell(\varepsilon) = \frac{1}{2i}(\alpha_\ell(\varepsilon)e^{2i\delta_\ell(\varepsilon)} - 1) \qquad \text{II-55}$$

where $\delta_\ell(\varepsilon)$ is a real phase shift which describes the scattering power of the underlying potential and $\alpha_\ell(\varepsilon)$ is a real function of energy which, for an absorptive scatterer, is less than one. Recall that the optical theorem is the consequence of the conservation of the probability of finding the particle somewhere after the scattering process. In an inelastic process this conservation law is violated. It is easy to show that Im$f_\ell = |f_\ell|^2 - 1/4(1-\alpha_\ell^2)$.

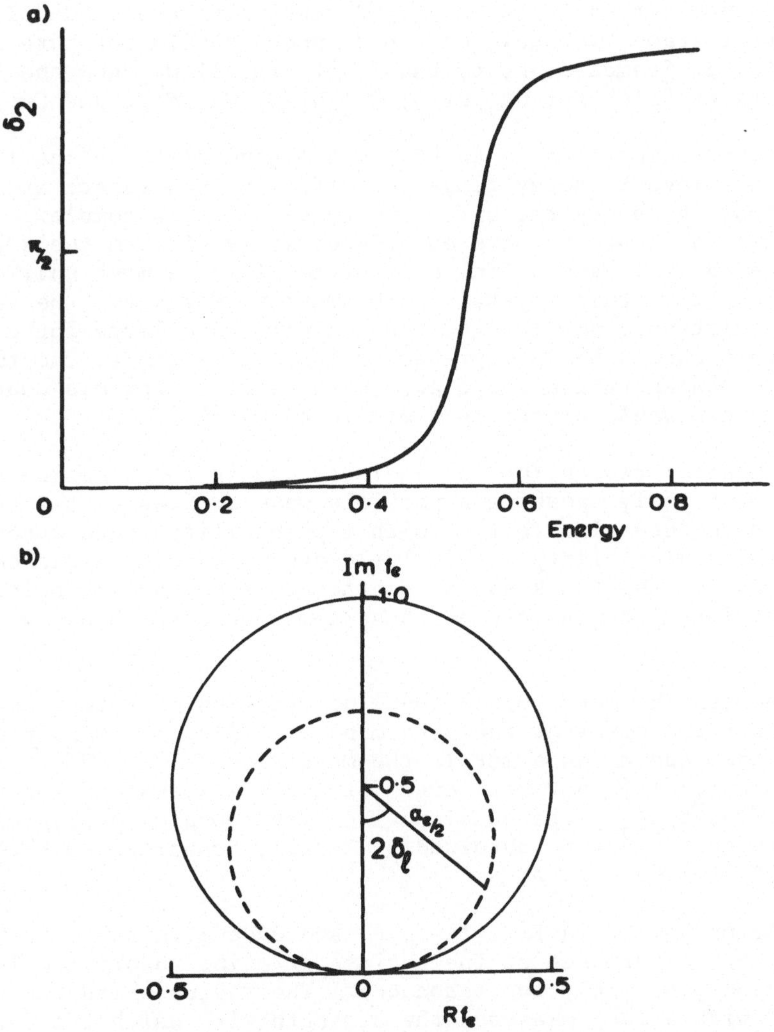

Fig. 10 : A generic resonant phase-shift. In b) the full line is the unitarity circle which is the Argand plot of the elastic scattering amplitude $\sin \delta_\ell(\varepsilon) e^{i\delta_\ell(\varepsilon)}$ for $\delta_\ell(\varepsilon)$ given in a). The dashed circle is the Argand plot for an inelastic scattering amplitude $[\alpha_\ell(\varepsilon) \exp 2i\delta_\ell(\varepsilon) - 1]/2i$ with $\alpha_\ell < 1$.

Hence, the probability that the particle was absorbed in the scattering process is $1/4(1 - \alpha_\ell^2)$. This Argand plot for a typical inelastic scattering amplitude is shown in Fig. 10b as marked.

Let us now return to the averaged scattering amplitude. In Fig. 11 upper frame, we show the $\ell = 2$ phase shifts for pure Ni and pure Cu. In frames a, b, c, and d, of Fig. 11 we show the Argand plots of $\overline{f}_2(\varepsilon)$ for $Cu_{90}Ni_{10}$, $Cu_{60}Ni_{40}$, $Cu_{40}Ni_{60}$ and $Cu_{10}Ni_{90}$.

The first point to note about these Argand plots is the fact that in the relevant energy range our effective scatterers are inelastic ; that is to say the effective phase shift is complex. This means that even though we have an ordered array of them the state with a given $\underline{q}$ will have a finite lifetime. This is most unlike the case in the virtual crystal approximation. Moreover, the inelasticity is relatively small for $c \sim 0$ and $c \sim 1$ and large for $c = .5$. Clearly, this should be interpreted as long liftetime at the two ends of the concentration range and short lifetime for mid concentration, as one would expect in a sensible theory.

The physical reason for the inelasticity is the randomness in the problem. Loosely speaking a particle wave arrives at a site and scatters there according to $t_{A,L}$ with a probability c and according to $t_{B,L}$ with a probability $1 - c$. The emerging wave is a random superposition of the two kinds of scattered waves. As a result of destructive interference between these components the beam is attenuated.

We recover the more conventional description of randomness if we search for the zeros of the determinant $||\overline{t}_L^{-1}\delta_{LL'} - G_{LL'}(\underline{q};\varepsilon)||$. Unlike in the case of pure metals the matrix $\overline{t}_L^{-1}\delta_{LL'} - G_{LL'}(\underline{q};\varepsilon)$ is not real even for cubic systems and for real $\underline{q}$ the eigenvalues $\varepsilon_{\underline{q},n}$ are complex. The first complex energy bands using this scheme were obtained by Bansil, Schwartz and Ehrenreich (1975) for $Cu_{.7}Ni_{.3}$.

The Argand plots in Fig. 11a,b,c, and d deserve some further comments. In the parlance of inelastic scattering theory two loops in the Argand plot imply two resonances. The radius which can be associated with a loop measures the strength with which the scattered particle is coupled into that resonance. A loop is shifted from the center of the unitarity circle because of background scattering due to other resonances. Clearly, all four Argand plots in Fig. 11 have two resonances : the one at lower energy is associated with the Cu resonance and the higher energy loop is due to the Ni resonance. For $c = .90$ the Cu resonance is more elastic and the coupling into this resonance is stronger than the coupling into the more severely damped Ni resonance. In this case we would expect the band structure to consist of fairly well defined Cu bands and smeared out Ni impurity bands. At the Ni rich end of the concentration

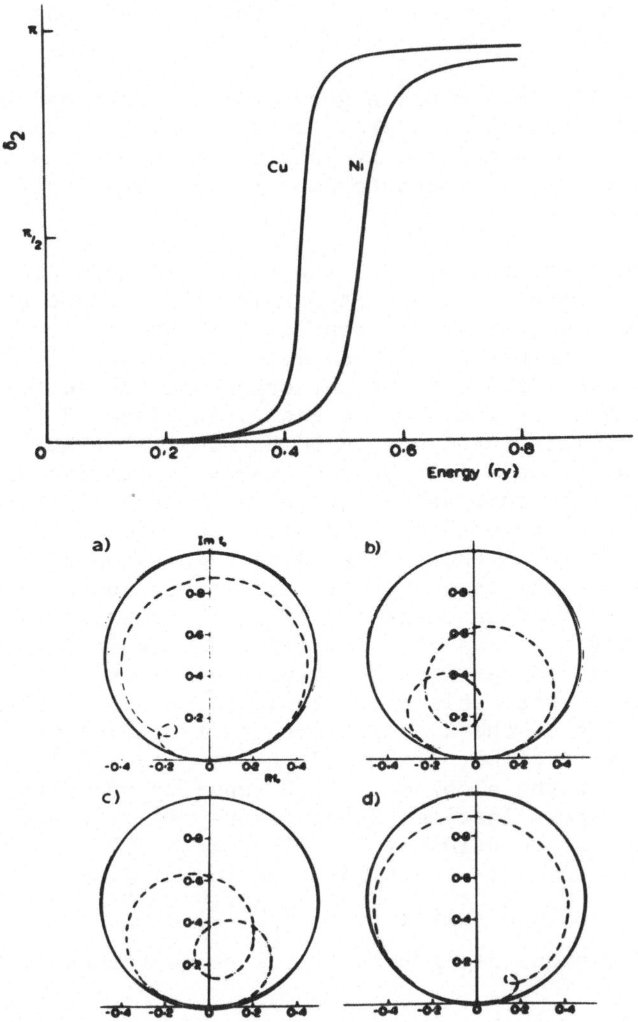

Fig. 11 : (upper frame). The $\ell = 2$ phase shifts used in our calculation to describe the scattering at Cu and Ni sites respectively. They are the same as in pure Cu and Ni, except here they are plotted with respect to an alloy V_{MTZ} whose choice is explained in the text. The Argand plots in a), b),c), and d) are the $\ell = 2$ averaged (ATA) scattering amplitudes $\bar{f}_2 = cf_2^{Cu} + (1-c)f_2^{Ni}$ for $Cu_{.9}Ni_{.1}$,$Cu_{.6}Ni_{.4}$, $Cu_{.4}Ni_{.6}$, $Cu_{.1}Ni_{.9}$.

range Fig.11d predicts the reverse situation. At $c \sim .5$ both sets of bands should be there but they will be severely damped. Of course, if the broadening is large enough no separate bands can be distinguished.

Evidently, all this makes a good deal of sense and suggests that the ATA is a reasonable first pass theory. Though we shall present evidence that it is inadequate in many respects, our more sophisticated calculations will tend to support the qualitative picture we have discussed above.

To make further headway with the theory we must calculate the averaged density of states. Unfortunately, this is not straightforward and at the moment the prescription for the density of states in the scattering theory ATA is somewhat controversial. (In the tight binding theory there is no corresponding ambiguity). The attractive thing to do would be to use $\bar{t}_L$ in place of $\bar{t}_L^{-1}$ and evaluate Eq. II-47. This turns out to be an easy calculation, however it is wrong. The trouble is that the energy derivative of $N(\varepsilon)$ in Eq. II-47 is not the same as the trace of the Green's function for an energy dependent potential. Unhappily the averaged t-matrix $\bar{t}_L(\varepsilon)$ corresponds to an energy dependent complex potential. Consequently, it is not the same approximation to evaluate Eq. II-47 using $\bar{t}_L(\varepsilon)$ or to find an expression for the trace of the Green's function directly in terms of $\bar{t}_L(\varepsilon)$ and evaluate it . Indeed, the first of these procedures does not give a sensible density of states. We believe that a reasonable thing to do is to evaluate Eq. II-52 (this is the trace of the averaged Green's function) by taking for $\langle\tau^{ij}_{LL'}\rangle$ the solution of Eq. II-50 and for $\langle\tau^{ii}_{LL'}\rangle_A$, $\langle\tau^{ii}_{LL'}\rangle_B$ use the solution for the problem of A, B impurity respectively in the averaged t-matrix lattice. Namely, take $\langle\bar{\tau}^{ii}_{LL'}\rangle_A = \bar{\tau}^{A,ii}_{LL'}$ where $\bar{\tau}^{A,ii}_{LL'}$ is the solution of Eq. II-50 with $t_{j,L} = \bar{t}_L + (t_{A,L} - \bar{t}_L)\delta_{ij}$ and for $\langle\bar{\tau}^{ii}_{LL'}\rangle_B = \bar{\tau}^{B,ii}_{LL'}$ where $\bar{\tau}^{B,ii}_{LL'}$ is defined similarly to $\bar{\tau}^{A,ii}_{LL'}$.
The first calculations along this line were performed by Bansil et al (1978).

Let us now return to our primary concern : the construction of an effective medium analogous to the one we have discussed, in connection with the inhomogeneous conductor, in the introduction. Having seen how the VCA and the ATA work within the framework of multiple scattering theory it is natural to seek the best possible effective medium in the form of an ordered lattice of scatterers each described by the same effective scattering amplitude $f_{C,L}(\varepsilon)$. The problem is, then, to find a suitable condition, analogous to Eq. I-14, which would determine $f_{C,L}(\varepsilon)$ in such a way that an electron would most nearly mistake the medium for the random lattice.

Notice that in this approach we bypass the problem of finding a "coherent potential" [a phrase due to Lax (1973)] to describe the effective medium by implicitly assuming that it is of the non-overlapping muffin-tin form and therefore all we want to know about it are the scattering amplitudes $f_{C,L}(\varepsilon)$.

The way to proceed is now fairly straightforward [Soven (1970), Shiba (1971), Györffy (1972)]. Consider an A impurity at $\underline{R}_0$ in the coherent potential lattice. The scattering from that site is described by $\tau^{A,00}_{LL'}(\varepsilon)$ which is the solution of Eq. II-50 with $t_{j,L} = t_{C,L} + (t_{A,L} - t_{C,L})\delta_{ij}$. Similarly the scattering from a B impurity at $\underline{R}_0$ is described by $\tau^{B,00}_{LL'}(\varepsilon)$. It is only a matter of some simple matrix algebra to show that

$$\tau^{A,00}_{LL'}(\varepsilon) = \left[\frac{1}{1-(t_A^{-1}-t_C^{-1})\tau^{C,ii}}\,\tau^{C,00}\right]_{LL'}$$

$$\tau^{B,00}_{LL'}(\varepsilon) = \left[\frac{1}{1-(t_B^{-1}-t_C^{-1})\tau^{C,ii}}\,\tau^{C,00}\right]_{LL'} \qquad \text{II-56}$$

where

$$\tau^{C,00}_{LL'}(\varepsilon) = \frac{1}{\Omega_{BZ}}\int d^3q\left[\frac{1}{t_C^{-1}-G(\underline{q};\varepsilon)}\right]_{LL'} \qquad \text{II-57}$$

The CPA condition that the averaged scattering from the site at $\underline{R}_0$ should be the same as in the pure coherent potential lattice now reads as

$$c\,\tau^{A,00}_{LL'}(\varepsilon) + (1-c)\,\tau^{B,00}_{LL'}(\varepsilon) = \tau^{C,00}_{LL'}(\varepsilon) \qquad \text{II-58}$$

Using Eq. II-56 it is a simple matter to rearrange this condition into the following more useful form [Györffy and Stocks (1974)].

$$f^{-1}_{C,L} = cf^{-1}_{A,L} + (1-c)\,f^{-1}_{B,L} + \sqrt{\varepsilon}(f^{-1}_{C,L} - f^{-1}_{A,L})\,\tau^{C,00}_{LL}\,(f^{-1}_{C,L} - f^{-1}_{B,L}) \qquad \text{II-59}$$

where we used f_L instead of $t_L=(-1/\sqrt{\varepsilon})f_L$ as a more convenient representation of the various scattering amplitudes.

This is then the fundamental equation for the effective scattering amplitude in the KKR-CPA. It determines $f_{C,L}(\varepsilon)$ in terms of the A site and B site scattering amplitudes $f_{A,L}(\varepsilon)$, $f_{B,L}(\varepsilon)$ and the crystal structure which fixes the structure constants $G_{LL'}(\underline{q};\varepsilon)$.

It is worth remembering that $f_{A,L}$ and $f_{B,L}$ need not be the same scattering amplitudes which determine the band structures of the pure metals A and B respectively. Normally, they would be constructed at each concentration according to one of the prescriptions mentioned in Sec. II-4, Eq. II-50. Thus, the KKR-CPA together with a prescription for the alloy potential is a first principles theory of the electronic states in random alloys in the same sense as one talks about first principles band theory for pure systems.

Obviously, there is no limitation to the applicability of the theory arising from the size and energy dependence of the scattering amplitudes. They can correspond to pure metals band structures with unequal band widths. Moreover, there can be any number of such bands overlapping and hybridizing in an arbitrary fashion. To put it another way, we are treating randomness in site energies, overlap integrals (arbitrary number of them) and hybridization coefficients on an equal footing. This is most unlike the situation with tight-binding model Hamiltonians where each of these features is a new, sometimes forbidding, complication.

As does the KKR, our method makes full use of the fact that for most metals only the first few phase shifts matter. Since $t_{C,L}(\varepsilon)$ will have the cubic symmetry of the crystal lattice for $\ell \leqslant 2$, we will have at the most four coupled equations to solve. Frequently, the alloying behavior is dominated by the difference between the $\ell = 2$ phase shifts on the A and B sites. Then, we shall have to solve only for an e_g and a t_{2g} scattering amplitude. Clearly $f_{C,L}$ as $\bar{f}_L$ in the previous section, will correspond to an inelastic scatterer. For physical reasons this will have to be an absorber and hence the effective scattering amplitudes $f_{C,L} = -\sqrt{\varepsilon} t_{C,L}$ will have to be within the unitarity circle. Thus, at each energy Eq. II-50 needs to be solved only for a limited set of complex numbers whose values are restricted, a priori, to a very limited part of the complex plane. As you shall see all these features play an important role in making the KKR-CPA a tractable proposition.

Model calculations and general theorems suggest that Eq. II-59 can always be solved by iteration. One assumes some value for $t_{C,L}$ and calculates $\tau_{LL'}^{C,00}$ numerically using Eq. II-57. A new value of $f_{C,L}$ is obtained by evaluating the right-hand side of Eq. II-59 for the assumed value of $f_{C,L}$. This process is continued until the new value is the same as the assumed value at the beginning of the last step. The difficult part of the calculation is the evaluation of the Brillouin zone integral in Eq. II-57 at each iteration.

Of course, finding $f_{C,L}$ is not enough. We must calculate observables. Consider, at first, the density of states. As in the case of the ATA we cannot use Eq. II-47 with t_L^{-1} replaced by $t_{C,L}^{-1}$ on account of the fact that $t_{C,L}$ corresponds to an energy dependent coherent potential. Thus, we must turn to Eq. II-52 which gives rigorously the trace of the averaged Green's function. Following the line of argument that leads to Eqs.II-56,57,58 we may make the substitutions

$$<\tau^{ii}_{LL'}>_A = \tau^{A,ii}_{LL'} \;;\; <\tau^{ii}_{LL'}>_B = \tau^{B,ii}_{LL'} \;;\; <\tau^{ij}_{LL'}> = \tau^{C,ij}_{LL'} \qquad \text{II-60}$$

Substituting these expressions into Eq. II-52 we can write our CPA prediction for the averaged density of states as

$$\overline{n}(\varepsilon) = n_0(\varepsilon) - \frac{1}{\pi N} \operatorname{Im} \{ \sum_{i,j} \sum_{L,L'} [(\frac{\partial}{\partial\varepsilon} t^{-1}_{C,L}) \delta_{LL'} \delta_{ij} - (\frac{\partial}{\partial\varepsilon} G_{LL'}(\underline{R}_i - \underline{R}_j;\varepsilon))] \tau^{C,ij}_{LL'} + \sum_{i,L} c(\frac{\partial}{\partial\varepsilon} t^{-1}_{A,L}) \tau^{A,ij}_{LL} + \sum_{i,L} (1-c)(\frac{\partial}{\partial\varepsilon} t^{-1}_{B,L}) \tau^{B,ij}_{LL} - \sum_{i,L} (\frac{\partial}{\partial\varepsilon} t^{-1}_{C,L}) \tau^{C,ii}_{LL} \} . \qquad \text{II-61}$$

The first term in this expression is immediately recognized as the energy derivative of the real space KKR-CPA determinant

$$|| t^{-1}_{C,L} \delta_{LL'} \delta_{ij} - G_{LL'}(\underline{R}_i - \underline{R}_j;\varepsilon) ||$$

The rest of the terms may be rearranged to yield

$$\overline{N}(\varepsilon) = N_0(\varepsilon) - \operatorname{Im} \frac{1}{\pi\Omega_{BZ}} \int_{BZ} d^3q \ln || t^{-1}_{C,L} \delta_{LL'} - G_{LL'}(\underline{q};\varepsilon) || + \operatorname{Im} \frac{c}{\pi} \ln \sum_L (\frac{t^{-1}_{B,L} - <t^{-1}_L>}{t^{-1}_{B,L} - t^{-1}_{C,L}}) + \operatorname{Im} \frac{1-c}{\pi} \ln \sum_L (\frac{t^{-1}_{A,L} - <t^{-1}_L>}{t^{-1}_{A,L} - t^{-1}_{C,L}}) \qquad \text{II-62}$$

The details of the algebra leading to Eq. II-62 are given in [(Győrffy and Stocks) 1976].

Note that the first term is just the integrated density of states for an ordered set of effective scatterers. The correction terms arise because the effective scattering amplitudes correspond to an energy dependent potential.

As mentioned before, the phase of the KKR determinant for a pure system consists of monotonically increasing steps as shown in Fig.9. For our effective scatterers these steps will be rounded off but, except for a limited energy range, the phase remains a monotonically increasing function. Such functions are relatively easy to integrate over the Brillouin zone. By contrast $\operatorname{Im}\tau^C(\underline{q};\varepsilon)$ will have sharp peaks and Re $\tau^C(\underline{q};\varepsilon)$ will behave as a rounded off 1/x singularity. For this reason, while Eq. II-57 is very hard to evaluate, the Brillouin zone integral in Eq. II-62 can be readily performed. The correction terms in Eq. II-62 do not require a Brillouin zone integration and they can be trivially evaluated once $t_{C,L}$ is known.

Another useful quantity to calculate is the Bloch wave spectral function which we shall now define. The usual spectral function is defined to be $\overline{A}(\underline{q};\varepsilon) = -(1/\pi)\mathrm{Im}\langle G(\underline{q},\underline{q};\varepsilon)\rangle$. This quantity is a set of reasonably well defined peaks at energies equal to the real part of the poles of $\overline{G}(\underline{q};\varepsilon)$ in the complex energy plane. Unfortunately it is not possible to calculate this spectral function from the knowledge of the "on the energy shell" t-matrices alone. However, we can define a similar quantity which carries roughly the same kind of information. For a pure system it makes sense to define

$$\overline{A}_B(\underline{q};\varepsilon) = \sum_n \delta(\varepsilon - \varepsilon_{\underline{q},n}) , \qquad \text{II-63}$$

where n is the band index and call it the Bloch spectral function. Evidently

$$\frac{1}{\Omega_{BZ}} \int_{BZ} d^3q\, \overline{A}_B(\underline{q};\varepsilon) = n(\varepsilon).$$

Thus $\overline{A}_B(\underline{q};\varepsilon)$ is the density of states per $\underline{q}$-point in the Brillouin zone. Therefore, in order to find the analogue of Eq. II-63 we must write Eq. II-62 as a single sum over the Brillouin zone. Using Eq. II-59 to write the correction terms in Eq. II-62 in terms of

$$\tau^{C,00}_{LL'} = \int_{BZ} d^3q\, \tau^C(\underline{q};\varepsilon)$$

and then differentiating the resulting equation with respect to ε leads to

$$\overline{A}_B(\underline{q};\varepsilon) = -\,\mathrm{Im}\,\frac{1}{\pi}\sum_{LL'}\frac{1}{\Omega_{BZ}}\cdot \qquad \text{II-64}$$

$$\cdot\left[\frac{\partial G_{LL'}}{\partial\varepsilon} - \frac{(t^{-1}_{A,L} - t^{-1}_{C,L})(\frac{\partial}{\partial\varepsilon}t^{-1}_{B,L}) - (t^{-1}_{B,L} - t^{-1}_{C,L})(\frac{\partial}{\partial\varepsilon}t^{-1}_{A,L})}{t^{-1}_{A,L} - t^{-1}_{B,L}}\delta_{LL'}\right]\tau^C_{L'L}(\underline{q};\varepsilon)$$

after some straightforward algebra.

By definition Eq. II-64 is the CPA Bloch spectral function. Its integral over the Brillouin zone is manifestly equal to the density of states and therefore has the same interpretation as Eq. II-63 for pure systems. Moreover, in the limit of no disorder e.g., $t_{A,L} = t_{B,L} = t_{C,L} = t_L$, Eq. II-64 reduces to

$$\bar{A}_B(\underline{q};\varepsilon) = \frac{-1}{\pi} \operatorname{Im} \frac{1}{\Omega_{BZ}} \frac{\partial}{\partial\varepsilon} \ln\left|\left| t_L^{-1}\delta_{LL'} - G_{LL'}(\underline{q};\varepsilon)\right|\right| \qquad \text{II-65}$$

By recognizing that the above formula is the energy derivative of the phase of the KKR-determinant which behaves like a sequence of step functions at the energy eigenvalues it is easy to see that Eq. II-65 is the same as Eq. II-63.

In more physical terms $\bar{A}_B(\underline{q};\varepsilon)$ may be interpreted as the probability that an electron with a Bloch vector $\underline{q}$ has an energy ε. Clearly, $\bar{A}_B(\underline{q};\varepsilon)$ is as close as we can get to looking at individual bands in the disordered systems.

We would like to note in passing that the interest in $\bar{A}_B(\underline{q};\varepsilon)$ is not entirely pedagogical. While $\bar{A}_B(\underline{q};\varepsilon)$, like the ε vs. $\underline{q}$ curves for pure systems, is not directly observable in any experiment, something very close to it is measured in angle resolved photoemission studies. In this experiment, which has become possible only recently, one measures not only the energy but also a part of the wavevectors of the photo-emitted electrons. Hence, if one is willing to disregard the effect of matrix elements and the complications introduced by the surface barrier one can interpret the spectra in terms of integrals of $\bar{A}_B(\underline{q};\varepsilon)$ along various directions in the Brillouin zone. Further discussion of this subject is given in Dr. Pendry's lectures. The averaged Bloch spectral function $\bar{A}_B(\underline{q};\varepsilon)$ is also a useful tool for analyzing the results of positron annihilation experiments and Compton profile studies. These experiments will be discussed by Professor Berko.

III. METHODS FOR SOLVING THE KKR-CPA EQUATIONS

In this lecture we shall discuss various ways of solving the fundamental equation of the KKR-CPA, namely Eq. II-59. Evidently this equation cannot be solved analytically. At the first sight even a numerical solution appears too difficult. Indeed, until a few years ago most workers in the field were resigned to having to do without a full solution and were busy devising various ATA-like alternatives [Bansil, Schwartz and Ehrenreich (1975), Györffy and Stocks (1974)]. The main burden of these lectures is to demonstrate that the pessimism was unwarranted since Eq. II-59 can be readily solved by presently available computational techniques.

The reason why the solution of Eq. II-59 appears so difficult at first sight is not hard to see. Evidently, during the iteration procedure we must evaluate the Brillouin zone integral in Eq. II-57 repeatedly. Note then that the integrand is the inverse of the KKR-matrix for an ordered set of effective scatterers and is, therefore, inversely proportional to the determinant

$$\left|\left| t_C^{-1} - G(\underline{q};\varepsilon)\right|\right|$$

(in the angular momentum space). For elastic scatterers this determinant goes to zero every time the eigenvalue corresponding to $\underline{q}$, namely $\varepsilon_{\underline{q}}$, is equal to ε. At these points the integrand has a pole. Mercifully, for inelastic scatterers this pole is not on the real axis and therefore the $\underline{q}$ integration will not have to go through actual singularities. Nevertheless, in most cases these poles will not be too far from the real axis and the integrand will have many sharp peaks in the imaginary part and rounded 1/x singularities in the real parts. Such rapidly varying integrands are notoriously difficult to integrate.

Under this circumstance it was natural that we sought to avoid doing the Brillouin zone integral altogether by attempting to solve Eq. II-59 in real space. This led us to an approximate method we shall call the cluster CPA (CCPA) [Stocks, Györffy, Giuliano and Ruggeri (1977)]. Next we considered a frontal attack which involved doing the Brillouin zone integrals but the structure constants were simplified to facilitate the calculation. This method will be referred to as the ASA-CPA [Temmerman, Györffy and Stocks (1978)]. Finally, it was realized that the energy dependence of $G_{LL'}(\underline{q};\varepsilon)$ can be eliminated from the problem by judicious fits of this matrix to polynomials in ε, then a complete solution was obtained [Stocks, Temmerman and Györffy (1978)]. We shall now briefly describe these methods in turn.

(1) The Cluster Method of Solving the KKR-CPA Equations

As an alternative to evaluating the Brillouin zone integral in Eq. II-57 we may attempt to find $\tau^{C,00}_{LL'}(\varepsilon)$ by solving the multiple scattering equation for $\tau^{C,ij}_{LL'}(\varepsilon)$ in real space, e.g.

$$\tau^{C,ij}_{LL'}(\varepsilon) = t_{C,L}\delta_{ij}\delta_{LL'} + \sum_{L''}\sum_{k\neq i} t_{C,L}\, G_{LL''}(\underline{R}_i - \underline{R}_k;\varepsilon)\tau^{C,kj}_{L''L'} \qquad \text{III-1}$$

By iterating this equation formally we find

$$\tau^{C,00}_{LL'} = t_{C,L}\delta_{LL'} + \sum_{i\neq 0}\sum_{L''} t_{C,L}G_{LL''}(\underline{R}_0 - \underline{R}_i)t_{C,L''}\, G_{L''L'}(\underline{R}_i - \underline{R}_0) + \dots \qquad \text{III-2}$$

It is interesting to note that if we keep only the first term in this series and substitute $\tau^{C,00}_{LL'}(\varepsilon) = t_{C,L}\delta_{LL'}$ into Eq. II-59 the solution is $f_{C,L} = \overline{f}_{C,L}$. That is to say, the ATA scattering amplitude may be thought of as an approximate solution to the CPA equation. The nature of the approximation is very suggestive. Clearly

$$\tau^{C,00}_{LL'} = t_{C,L}\delta_{LL'}$$

is the exact solution of the multiple scattering equation, Eq. III-1 for a "cluster" of scatterers containing one atom. Thus $t_{C,L} = \overline{t}_L$

for a cluster of one atom as it should be since in this limit both the ATA and the CPA are exact solutions to the problem. This suggests a way of going beyond the ATA toward the CPA. An obvious way to proceed further with the argument is to take instead of

$$\tau^{C,00}_{LL'}(\varepsilon) = t_{C,L}\delta_{LL'}$$

the solution of Eq. III-1 for a finite cluster surrounding the site at $\underline{R}_0$. We call the resulting approximation the cluster CPA. However, we stress that our CCPA is an approximation to the CPA equations and not, as another method by the same name, a means of going beyond the CPA [Nickel and Butler (1973)]. While less ambitious than those, our CCPA has all the correct analytic properties [Lloyd and Oglesby (1976)].

To calculate $\tau^{C,00}_{LL'}(\varepsilon)$ for a finite number of scatterers one merely has to invert the matrix

$$t_L^{-1}\delta_{ij}\delta_{LL'} - G_{LL'}(\underline{R}_i - \underline{R}_j;\varepsilon)$$

where the site indices i and j run from 0 tot N, the number of scatterers considered, and the angular momentum index L takes all possible values consistent with the largest ℓ for which the scattering amplitude $t_{C,L}$ is taken to be non-zero. For example in an fcc lattice an atom and its nearest neighbor shell is a cluster of 13 atoms. If we include a scattering amplitude with $\ell \leqslant 2$, L can take 9 different values (00,1-1, 10, 11, 2-2, 2-1, 20, 21, 22), hence one must invert a 117 by 117 matrix. This can be readily accomplished numerically. If we are willing to treat the s-p component of the effective scattering amplitude in the averaged t-matrix or virtual crystal approximation on the grounds that they are very much the same anyway, only the L-index corresponding to the 5d-states arises and the matrix to be inverted is only 65 by 65. In our actual calculations we shall always be using the d-block only.

With this in mind we proposed the following strategy for solving for the effective CPA scattering amplitude $t_{C,L}$. We solve Eq. II-59 using a finite cluster centred on the site at $\underline{R}_0$ to evaluate $\tau^{C,00}_{LL'}$ at each iteration. We then repeat the calculation using a larger cluster which includes the next nearest neighbor shell beyond the cluster in the previous calculation. This process will generate a set of scattering amplitudes, one for each cluster size. In principle this sequence will converge to the true solution of Eq. II-59. However, the physical idea that far away scatterers will have little effect on the very local CPA self-consistency between impurity scattering on a site and its environment suggests that the convergence will be rapid. Indeed, as we shall show presently, already the third nearest neighbor shell has practically no effect on $f_{C,L}$ for the systems we have considered.

In Fig. 12 we show $\mathrm{Im}f_{C,L}$ as a function of energy for 0, 1, and 3 nearest neighbor shell calculations for $Cu_{.87}Ni_{.13}$. Evidently $\mathrm{Im}f_{C,L}$ changes considerably from its ATA value (0 shell cluster) when the first nearest neighbor shell is introduced. As expected $f_{C,L}$ no longer corresponds to a spherically symmetric scatterer, e.g., it is different for different m. Since the site has cubic symmetry this m dependence is still degenerate and $f_{C,L}$ splits into only two components f_{C,e_g} and $f_{C,t_{2g}}$. When the next shell is introduced the t_{2g} component hardly changes but there is still quite a bit of movement in f_{C,e_g}. This can be understood easily. In an fcc lattice the e_g orbitals about a site ($x^2 - y^2$, $2z^2 - x^2 - y^2$) point towards next nearest neighbors. In a one nearest neighbor shell cluster the central atom sees no scatterers along its e_g orbital lobes. When the second shell is introduced there will be scatterers along these lobes. For this reason the CPA self-consistency yields two very different f_{C,e_g}'s. It is reassuring to note that the introduction of the third shell causes changes in both f_{C,e_g} and $f_{C,t_{2g}}$ which are small on the scale of changes in previous steps. Thus we assume that our scheme is reasonably well converged after two shells of neighbors have been included.

Note that, as expected on the grounds of our previous qualitative discussion of the likely shape the effective scattering amplitudes will have, the Ni impurity appears as an extra inelastic resonance loop in the Argand plot shown in Fig. 12. Corresponding to this weakly coupled resonance there is a broad but small amplitude impurity peak in the plot of $\mathrm{Im}f_{C,L}$ vs.ε. The position of this peak with respect to the Ni resonance in $\mathrm{Im}f_2^{Ni}(\varepsilon)$ and the corresponding feature in $\mathrm{Im}\,\overline{f}_2(\varepsilon)$ merits further discussion. This will be given later after we have described the other two methods for solving the KKR-CPA equation. For further applications of the CCPA, see [Giuliano, Ruggeri, Györffy and Stocks (1977)].

(2) <u>The Atomic Sphere Approximation to the KKR-CPA</u>

Noting that the shapes and relative position of the energy bands do not vary much if the muffin-tin zero V_{MTZ} is changed, Andersen (1973) approximated the KKR method for pure metals by adjusting V_{MTZ} so that the structure constants become energy independent (ASA - Atomic Sphere Approximation). We shall use the same method for simplifying the KKR-CPA. To motivate our procedure we shall now briefly describe the ASA scheme. Andersen has given a number of different derivations of this method. At the risk of obscuring its physical content, we shall present it in the form most appropriate to our purpose. In the KKR method one finds the energy bands by searching for the zeros of the KKR determinant. One seeks the solutions of the equation

$$||t_L^{-1}(\varepsilon;\kappa)\delta_{LL'} - G_{LL'}(\underline{q};\kappa)|| = 0 \qquad \text{III-3}$$

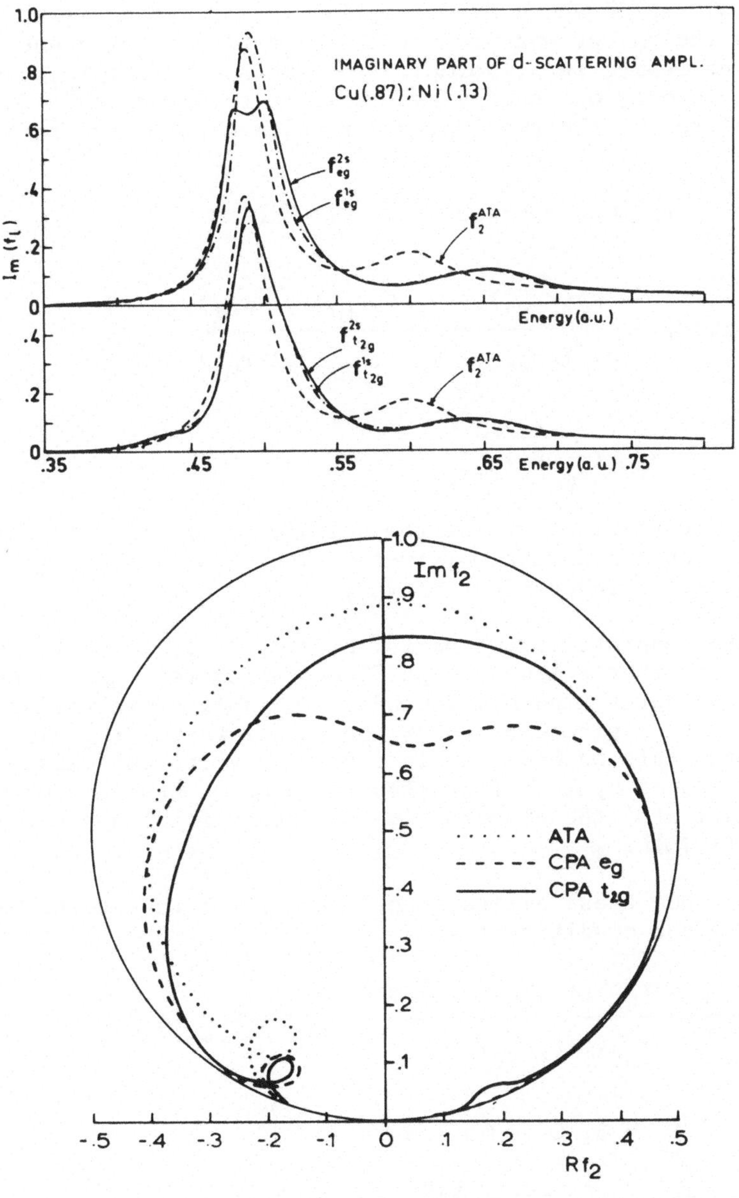

Fig. 12 : The imaginary part of the effective scattering amplitude $f_{C,L}(\varepsilon)$ for a 1 shell and 2 shell CPA showing the convergence discussed in the text. We also show the Argand plot of $f_{C,L}(\varepsilon)$ as obtained in the 2 shell calculation.

where $\kappa^2 = \varepsilon - V_{MTZ}$ is the kinetic energy of the electron in the interstitial region where the potential is a constant equal to V_{MTZ}, ε is the energy measured from some arbitrary reference point like the atomic zero. Normally ε is measured from V_{MTZ} and therefore there is only one energy available in the problem. To make the role played by the two energy variables ε and κ explicit we note that

$$t_L^{-1}(\varepsilon;\kappa) = -\kappa[\cot\,\delta_\ell(\varepsilon;\kappa) - i] \qquad \text{III-4}$$

and

$$\cot\delta_\ell(\varepsilon;\kappa) = \frac{\kappa n_\ell'(\kappa r_{MT}) - \gamma_\ell(\varepsilon, r_{MT}) n_\ell(\kappa r_{MT})}{\kappa j_\ell'(\kappa r_{MT}) - \gamma_\ell(\varepsilon, r_{MT}) j_\ell(\kappa r_{MT})} \qquad \text{III-5}$$

Since

$$\gamma_\ell = \left(\frac{1}{R_\ell} R_\ell'\right)_{r=r_{MT}}$$

is to be evaluated in terms of the solution $R_\ell(r;\varepsilon)$ of the Schrödinger equation inside the muffin-tin sphere, which is regular at the origin, it is independent of V_{MTZ}.

The ASA consists of two parts. Firstly, κ is taken to be zero. That is to say that somewhat superfluous parameter V_{MTZ} is set equal to the energy studied. Secondly, the muffin-tin radius r_{MT} in Eq. III-5 is replaced by the Wigner-Seitz radius r_{WS}. The rationale for this second step is provided by the analysis of the errors introduced by the first step in which we have neglected the propagation of the electron in the interstitial space [Andersen (1975)].

Having made these alterations in Eq. III-3 the energy bands are given by the condition that

$$\left|\left|\frac{C_\ell}{r_{WS}^{2\ell+1}} \frac{r_{WS}\gamma_\ell(\varepsilon, r_{WS}) + \ell + 1}{r_{WS}\gamma_\ell(\varepsilon, r_{WS}) - \ell} \delta_{LL'} - S_{LL'}(\underline{q})\right|\right| = 0 \qquad \text{III-6}$$

where $C_0 = 1$, $C_1 = 3$, $C_2 = 45$ and

$$S_{LL'}(\underline{q}) = \lim_{\kappa\to 0} [\kappa^{\ell+\ell'} G_{LL'}(\underline{q};\kappa^2)] \qquad \text{III-7}$$

The principal gain in going from Eq. III-3 to Eq. III-7 is that in the latter, i.e., in the ASA, the structure constants $S_{LL'}(\underline{q})$ are independent of the energy. Therefore, they can be calculated once and for all. In Eq. III-3 they have to be reevaluated at each energy. Clearly, a simplification of this sort will also speed up the KKR-CPA enormously. Taking the $\kappa \to 0$ limit simplifies the structure constants considerably. Nevertheless, one must still use the Ewald procedure to evaluate the lattice sums involved in their calculation.

Andersen (1973) and others have studied many fcc and hcp pure metals using the ASA scheme and found that the eigenvalues within about one Rydberg or so of the bottom of the conduction band always agreed to better than 10 mRy with exact calculations. For more open structures the agreement is less satisfactory. Errors as large as 50-80 mRy can occur even in bcc structures. Our calculations on pure Cu and pure Nb are also consistent with these findings. It is useful to note the $\kappa \to 0$ aspect of the ASA has a particularly appealing interpretation in the language of scattering theory. As is well known, for slow electrons, the scattering cross section is frequently well approximated by $4\pi a_0$ where a_0 is the scattering length defined by the

$$\lim_{\kappa\to 0} \frac{1}{\kappa} f_0(\kappa) \equiv a_0 .$$

By analogy, we also define the scattering "length" a_ℓ for $\ell > 0$ as

$$a_\ell \equiv \lim_{\kappa\to 0} \frac{1}{\kappa^{2\ell+1}} f_\ell(\kappa).$$

Obviously this will not have the dimension of length, hence the quotation marks around the word length. The low energy behavior of each partial wave amplitude $f_\ell(\kappa)$ is then described by the corresponding scattering "length" a_ℓ. When in the ASA we are setting V_{MTZ} equal to ε(i.e., $\kappa = 0$) we are in fact defining an energy dependent scattering "length"

$$a_\ell(\varepsilon;r_{MT}) = \lim_{\kappa\to 0} \frac{1}{\kappa^{2\ell+1}} f_\ell(\varepsilon,\kappa)$$

$$= \frac{r_{MT}^{2\ell+1}}{(2\ell+1)!!(2\ell-1)!!} \frac{r_{MT}\gamma_\ell(\varepsilon,r_{MT}) - \ell}{r_{MT}\gamma_\ell(\varepsilon,r_{MT}) + \ell + 1} \qquad \text{III-8}$$

which describes accurately, to lowest order in $\kappa = \sqrt{\varepsilon - V_{MTZ}}$ the scattering of those electrons which move slowly in the interstitial region. Thus, in terms of the rescaled scattering "length" a more transparent version of Eq. III-6 is

$$||a_\ell^{-1}(\varepsilon;r_{WS}) - S_{LL'}(\underline{q})|| = 0 \qquad \text{III-9}$$

which is to be compared with the full KKR equation in Eq. III-3.

To make use of the relative simplicity of the ASA in the context of random alloys we must approximate Eqs. II-59 and II-57 in the appropriate fashion. The argument that leads to setting κ equal to zero should be as valid here as in the case of Eq. III-3. Hence our first step is to take $\kappa \to 0$ in Eqs. II-59 and II-57. It is also reasonable to complete the approximation by replacing all quantities evaluated at r_{MT} by their equivalent at r_{WS}. In order to carry out the $\kappa \to 0$ process on Eqs. II-59 and II-57 we must have $t_{C,L}$ and therefore $f_{C,L}$ as functions of ε and κ. Thus we imagine that the effective scatterer corresponds to an energy dependent complex potential which, nevertheless, is of the muffin-tin form. We parameterize $f_{C,L}(\varepsilon;\kappa)$ in analogy with Eq. III-4, as

$$f^{-1}_{C,L}(\varepsilon;\kappa) \equiv \cot \delta^{C}_{\ell}(\varepsilon;\kappa) - i$$

$$= \left[\frac{\kappa n'_{\ell}(\kappa r_{MT}) - \gamma^{C}_{\ell}(\varepsilon,r_{MT})n_{\ell}(\kappa r_{MT})}{\kappa j'_{\ell}(\kappa r_{MT}) - \gamma^{C}_{\ell}(\varepsilon,r_{MT})j_{\ell}(\kappa r_{MT})}\right] - i \qquad \text{III-10}$$

where $\delta^{C}_{\ell}(\varepsilon;\kappa)$ is an effective complex phase shift and $\gamma^{C}_{\ell}(\varepsilon,r_{MT})$ is an effective logarithmic derivative. This is our new description of the effective scatterer. Clearly, the limit as κ goes to zero of $f^{-1}_{C,L}$ is now well defined. Since

$$t_{C,L}(\varepsilon;\kappa) = -\frac{1}{\kappa} f_{C,L}(\varepsilon;\kappa)$$

then

$$\lim_{\kappa\to 0} t^{-1}_{C,L}(\varepsilon;\kappa)$$

is also well defined.

Letting κ go to zero in Eq. III-10 and replacing r_{MT} by r_{WS} everywhere it is straightforward to derive the following equation for the effective CPA logarithmic derivatives

$$a^{-1}_{C,L}(\varepsilon) = c a^{-1}_{A,L}(\varepsilon) + (1-c) a^{-1}_{B,L}(\varepsilon) + \qquad \text{III-11}$$

$$+ \left[a^{-1}_{A,L}(\varepsilon) - a^{-1}_{C,L}(\varepsilon)\right] Z^{C,00}_{LL}(\varepsilon) \left[a^{-1}_{B,L}(\varepsilon) - a^{-1}_{C,L}(\varepsilon)\right]$$

where

$$Z^{C,00}_{LL'}(\varepsilon) \equiv \lim_{\kappa\to 0} \frac{1}{\kappa^{\ell+\ell'}} \tau^{C,00}_{LL'}(\varepsilon;\kappa) = \frac{1}{\Omega_{BZ}} \int d^3q \left[\frac{1}{a^{-1}_{C} - S(\underline{q})}\right]_{LL'} \qquad \text{III-12}$$

with the prescription that $a_C^{-1}(\varepsilon)$ and $S(\underline{q})$ are to be interpreted as matrices with elements $a_{C,L}^{-1}(\varepsilon)\delta_{LL'}$ and $S_{LL'}(\underline{q})$ respectively. The various scattering "lengths" $a_{C,L}$, $a_{A,L}$ and $a_{B,L}$ are related to the corresponding logarithmic derivatives γ_L^C, γ_L^A, γ_L^B by Eq. III-8.

These are the fundamental equations of the CPA in the Atomic Sphere Approximation. They determine the effective logarithmic derivatives $\gamma_\ell^C(\varepsilon, r_{WS})$ in terms of

$$\gamma_\ell^A(\varepsilon, r_{WS}),\ \gamma_\ell^B(\varepsilon, r_{WS})$$

and the structure constants $S_{LL'}(\underline{q})$.

Evidently Eqs. III-11 and III-12 still have to be solved by iteration and the need to do a Brillouin zone integral in Eq. III-12 at each step has not been eliminated. However, the energy independence of the structure constants $S_{LL'}(\underline{q})$ simplifies matters enormously. Note that once we have solved Eq. III-11 and III-12 for the effective logarithmic derivatives $\gamma_\ell^C(\varepsilon, r_{WS})$, we can construct the corresponding scattering amplitudes $f_{C,L}$ and use these to evaluate the full KKR-CPA formula for the averaged density of states and other quantities of physical interest. If we assume that the coherent potential corresponding to $\gamma_\ell^C(\varepsilon, r_{WS})$ is of the muffin-tin form and is equal to the alloy V_{MTZ} beyond r_{MT} out to the Wigner-Seitz radius r_{WS}, then $f_{C,L}$ may be obtained from Eq. III-10 with r_{MT} replaced by r_{WS} everywhere. Although it is a useful procedure to derive approximate expressions for the integrated density of states $N(\varepsilon)$ and the Bloch spectral function $\bar{A}(\underline{q}, \varepsilon)$ within the Atomic Sphere Approximation by letting $\kappa \to 0$ in Eq. II-62 and Eq. II-64 and thus evaluate all quantities of physical interest in the ASA [Gordon, Temmerman, Györffy and Stocks (1978)], in what follows we shall allways reconstruct $f_{C,L}$ from $\gamma_{C,L}^{ASA}$ and proceed with the full KKR-CPA formalism. This will facilitate the comparison between the various methods of solving Eq. II-59 and allow us to comment more easily on the general features of the complete theory.

For a detailed description of how Eq. III-11 and III-12 can be iterated to convergence see [Temmerman, Györffy and Stocks (1978)]. Here, we shall only sketch the most difficult aspect of the calculation, namely the repeated evaluation of the Brillouin zone integral in Eq. III-12.

As is indicated this is a three-dimensional volume integral over the full Brillouin zone. However, our lattice (fcc) has cubic symmetry and therefore it is sufficient to evaluate the integral in the irreducible 1/48-th of the Brillouin zone shown in Fig. 13a and multiply the result by 48. To do this we can use two somewhat different methods :

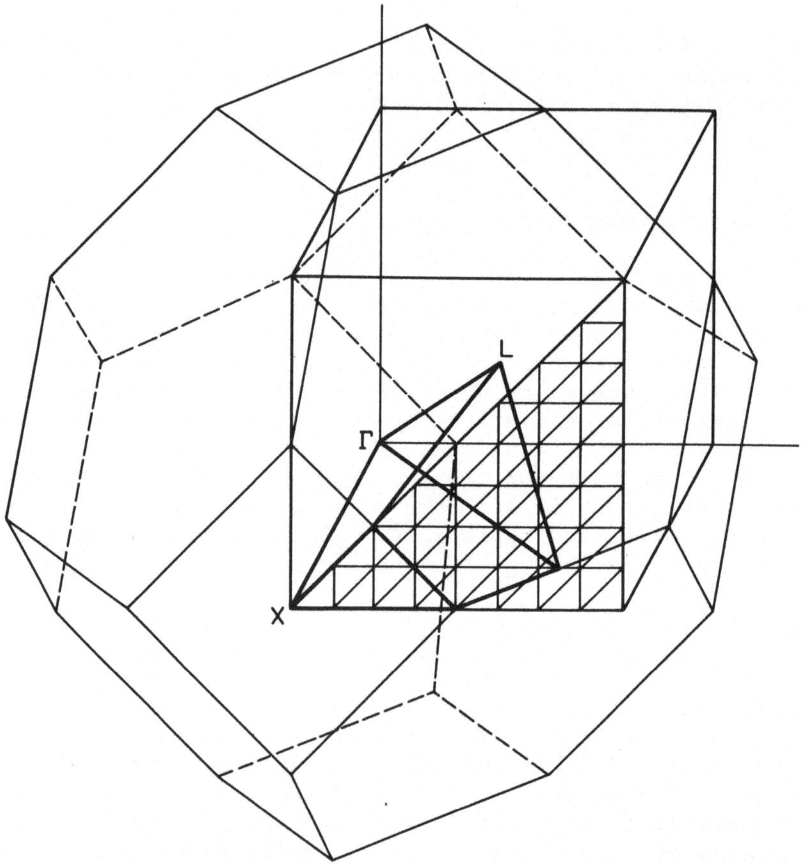

Fig. 13a : The irreducible 1/48-th of Brillouin zone for an fcc lattice. The tetrahedra discussed in integration method (b) of the text are those with apex at the Γ-point and a base comprising one of the small triangles shown in the figure. The integration is performed along the center of main line of each tetrahedron, the weight associated with each point on the line being taken as cross sectional area of the tetrahedra parallel to the Brillouin zone face.

(a) The simpler one is a very rapid technique for doing rough Brillouin zone integrals in cubic systems. As was noted by Bansil (1974) for integrals which have the full cubic symmetry a number of special directions emanating from the Γ-point can be found such that the sum of the line integrals along these directions out to the Brillouin zone boundary, gives exactly the contribution to the volume integral from a specified number of terms in the cubic harmonics expansion of the integrand. Since $\tau^{C,00}_{LL'}(\varepsilon)$ and therefore the Brillouin zone integral in Eq. III-12 have the full cubic symmetry this method is directly applicable to their evaluation. In most of our calculations we approximated the volume integral over the irreducible 1/48-th of the Brillouin zone by the sum of the line integrals along the set of 13 directions given by Bansil. This insured that we have treated the first 20 terms in the cubic harmonics expansion of the integrand correctly. For further details concerning the accuracy of the scheme the reader is referred to the original paper [Bansil (1975)] .

(b) To test the convergence of the above method we have also made use of an alternative. In the spirit of the constant energy search KKR method of Faulkner et al (1967) we have divided the 1/48-th of the Brillouin zone into small tetrahedra with one vertex at the Γ point and edges terminating at the Brillouin zone boundary. The volume integral was then evaluated by calculating the integrand along the center of mass line of these tetrahedra at a large number of points and multiplying it by the appropriate fraction of the volume. The final sum was performed in two steps ; first, we added up all contributions along a direction and then we summed over the directions (tetrahedra). Thus, again we reduced the volume integral to a sum of line integrals along directions emanating from the center of the Brillouin zone (the Γ point). The accuracy of this method can be consistently improved by constructing smaller and smaller tetrahedra, i.e., increasing the number of directions. The convenient numbers of directions are 36, 136, 561.

In both of these methods we have to evaluate line integrals along fixed directions. Since the structure constants are energy independent and do not vary too rapidly with $\underline{q}$ we found it worthwhile to fit polynomials to the $S_{LL'}(\underline{q})$ along the various directions used in the calculations and to store the polynomial coefficients. In subsequent calculations this procedure reduced the evaluation of the structure constants at any point along the selected directions to a small number of multiplications. Consequently, it was always a simpler matter to converge the one-dimensional integrals. Hence, the convergence of the volume integrals was a question of taking sufficiently a large number of directions. Clearly this procedure is most suitable in the case where one is dealing with roughly spherical energy surfaces centered on the Γ point. This would suggest that for energies in the middle of the d-band complex our integration scheme could become somewhat uncertain. However, due to the substantial smearing of all structure by the disorder in the problem, i.e., the inelasticity of $f_{C,L}(\varepsilon)$, this turned out not to be the case in our calculations.

Before describing the results of calculations obtained by the above method we wish to turn to ways of solving the full KKR-CPA equations.

(3) A Complete Solution of the KKR-CPA Equations

As it turns out, both approximate methods of the previous two sections for handling the KKR-CPA equations work well for the Ni_cCu_{1-c} system. This will be shown in the next section where we compare the CCPA and the ASA-CPA with the complete solution. However, we do not expect the same agreement in general. The CCPA is best at high energies where the wavelength of the electrons is small compared to the cluster size. But, even there, we do not expect it to be reliable in describing the finer details such as small but interesting features of the Fermi surface. The ASA-CPA is best at energies close to V_{MTZ} and appears to be useful only for such closed packed systems as fcc and hcp metals. Therefore, it is desirable to have a method which works in all circumstances and reliably reproduces the full content of the KKR-CPA theory. In this section, we introduce a method which answers this description.

It is a frontal attack on the problem. Namely, we solve Eq. II-57 and Eq. II-59 directly following the same route as in the case of ASA-CPA. The somewhat surprising thing is that this can be done with little extra computational effort.

We proceed by direct evaluation of Eq. II-59 for some initial guess of the $f_{C,L}$'s. On successive iterations of Eqs. II-59 and II-57 updated values of the $f_{C,L}$'s are obtained via the Newton-Ralphson method. The evaluation of Eq. II-57 is carried out by performing integrations along directions emanating from the Brillouin zone center. A summation is then made over directions. As before, the directions and weights are chosen either by the special directions method of Bansil (1975) and of Fehlner and Vosko (1976) or by a prism method based on the constant energy search pattern of Faulkner et al (1967).

Prior to the commencement of the iteration process the KKR structure constants $G_{LL'}(\underline{q};\varepsilon)$ are fitted to polynomials after a term

$$\sum_{\underline{K}_n} Z_{LL'}/(\varepsilon - |\underline{q} + \underline{K}_n|^2)$$

has been subtracted from each $G_{LL'}$ in order to remove the free electron singularities. The summation is over those reciprocal lattice vectors $\underline{K}_n$ which contribute a singularity along the $\underline{q}$-direc-

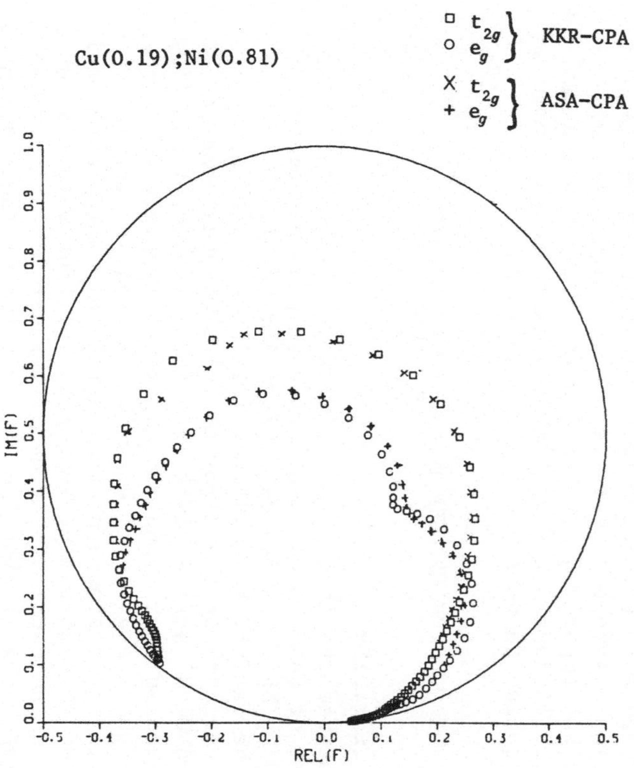

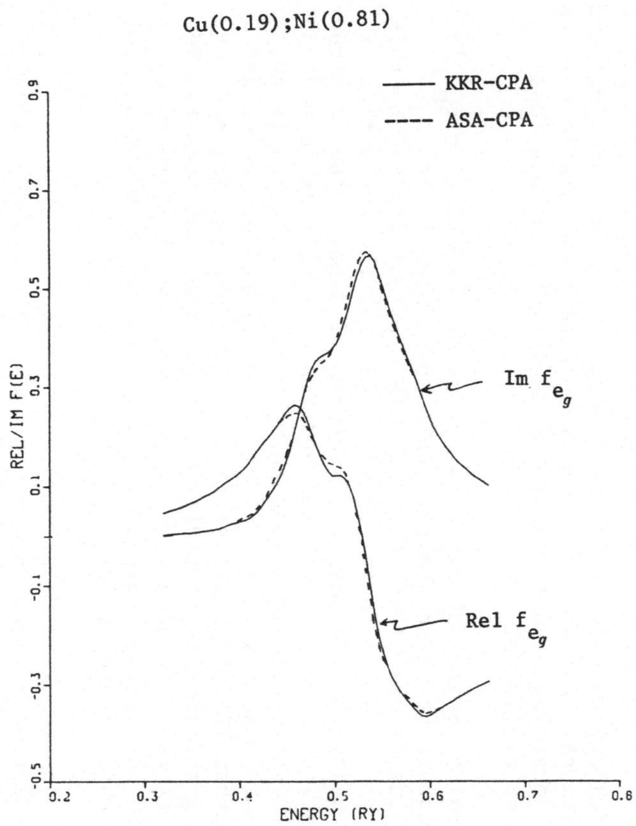

Fig. 13 b : Comparison between the e_g and t_{2g} scattering amplitudes as obtained in the Atomic Sphere Approximation (ASA) and in the full solution of the KKR-CPA.

tion under consideration. Evaluation of the $G_{LL'}(\underline{q};\varepsilon)$ for a particular energy and wavevector is thus reduced to a few additions and multiplications. From this point on the present calculations are no more time consuming than the approximate ASA-CPA calculations of Temmerman et al (1978).

Central to establishing the accuracy of the solution $f_{C,L}$'s is the accuracy to which the Brillouin zone integral in Eq. II-57 can be evaluated. For the Cu-Ni systems studied by this method [Stocks, Temmerman and Györffy (1978)] the evaluation of Eq. II-57 using the 13 special directions of Bansil, the 21 directions Gaussian formula of Fehlner and Vosko as well as 36 and 136 uniformly distributed directions shows that the self-consistent $f_{C,L}$'s based on the 13 special directions of Bansil are converged to better than 1% (often better than 0.2%). This is adequate for determining the density of states and Bloch spectral density. We feel that it would be necessary to go beyond the 13-direction method only for special purposes, e.g., accurate determination of the Fermi surface properties.

For a number of Cu-Ni systems, the solutions obtained in this way are essentially identical with those of the ASA-CPA calculation of Temmerman et al (1978). This is illustrated in Fig. 13b. The good agreement was expected since the Atomic Sphere Approximation is particularly good for describing the d-bands of fcc metals. However, it should be stressed that the present method removes all the restrictions imposed by the ASA on the systems which can be treated, whilst it does not significantly increase the computational effort required to perform the calculations.

IV. THE BAND STRUCTURE OF $Cu_c Ni_{1-c}$ ALLOYS

(1) The Density of States

Having described the theory we now wish to illustrate its use. Although a number of KKR-CPA calculations have been performed for the Ag_cPd_{1-c} and Mo_cNb_{1-c} alloys as well as for Cu_cNi_{1-c} we shall concentrate on the latter. As band theorists have tuned up their tools on Cu for decades, so are alloy calculations best tested on the well known Cu-Ni system. This is a good example of an alloy where the time-honored rigid band model gives a completely erraneous picture of the energy-level distribution and, at the same time, it is sufficiently simple so that a reliable alloy potential can be constructed. Consequently, even a calculation which is not self-consistent in the sense of the density functional theory can be expected to make contact with the most variety of thermodynamic, transport and, significantly, spectroscopic data which are available on their electronic properties. Thus, it is not surprising that efforts to understand the electronic structure of Cu_cNi_{1-c} have played an important role in the development of the theory of random alloys.

Before launching into a description of our results we wish to pause briefly to recall the evolution of our current understanding in this field. [For a more detailed account see Györffy and Stocks (1976)].

Nickel and copper, being neighbors in the periodic table, have very similar atomic charge distributions, except for the extra electron on Cu. An evidence for their similarity is that they form solid solutions for all concentrations. The crystal structure of pure Cu, pure Ni as well as all their alloys is fcc and the lattice parameter does not change very much with concentration. Naturally, for a long time it was assumed that on adding Ni to Cu nothing much will happen to the conduction band and the only effect of alloying will be a reduction of the Fermi energy ε_F due to the decrease in the average number of electrons per atom [Shimizu (1963), Kittel (1976); see however Friedel (1956)]. This is the simplest form of the rigid band model.

In Fig. 14 we show the density of states of Cu [Stocks, Williams and Faulkner (1971)]. Evidently, if the only effect of adding Ni is to reduce ε_F, initially $n(\varepsilon_F)$ will remain more or less constant and then will rise rapidly as we begin to empty the d-band. This would imply that the linear coefficient in the specific heat

$$\gamma = \frac{2}{3} \pi^2 k_B^2 \, n(\varepsilon_F)$$

will be roughly constant for small Ni concentrations up to a critical value and then it will show a steep increase. This is in fact what is observed.

That this picture is completely wrong was first shown by Seib and Spicer (1970) [see also Hüfner, Wertheim and Wernick (1973) and Shevchik and Penchina (1975)] who found that instead of the ε_F moving down with respect to the upper edge of the Cu d-band an extra Ni peak appeared in the density of states between the "Cu d-band" and ε_F. Such behavior could not be explained in terms of the rigid band model or its most sophisticated version, the VCA [Györffy and Stocks (1976)]. However, in the impurity limit, this is precisely what should be expected according to the virtual bound state idea of Friedel (1956). As we have mentioned earlier the CPA is exact in the impurity limit and therefore, for the appropriate values of the parameters in a given model Hamiltonian, it can give virtual bound states. However, it is a more general theory and it is also able to describe the evolution of such bound states with concentration. In fact the first realistic CPA calculation based on a tight binding model [Stocks, Williams and Faulkner (1971)] gave a good quantitative account of the photoemission experiments

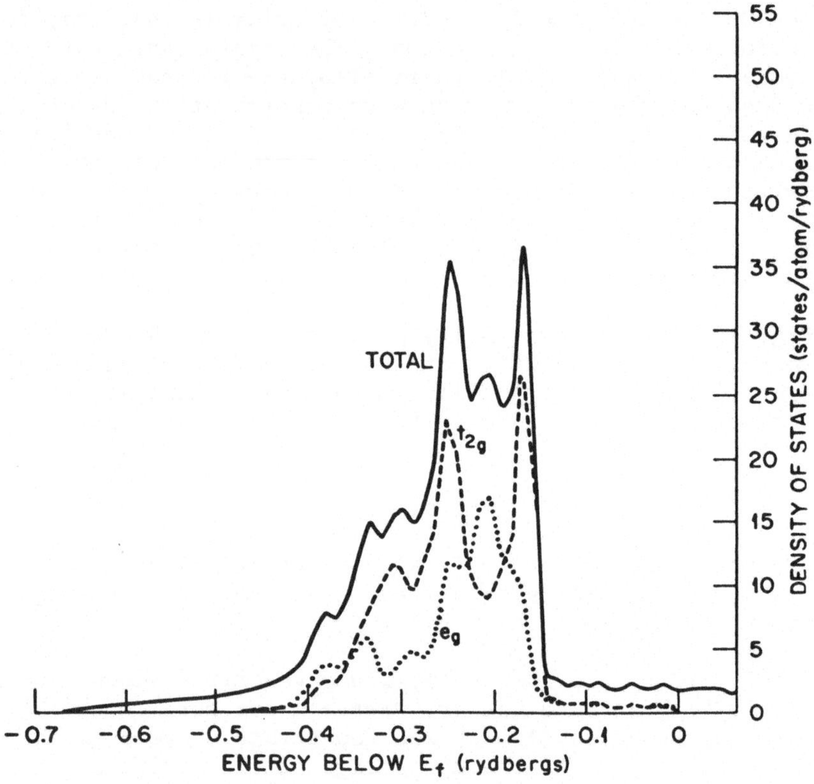

Fig. 14 : Density of states in pure Cu as in Stocks, Williams and Faulkner (1971).

on the Cu-rich alloys. This is illustrated in Fig. 15 where we compare the density of states obtained by SWF and the photoemission yield. Bearing in mind the uncertainties with regard to interpreting the photoemission spectra as a density of states (see the lectures by Dr. Pendry) the agreement is quite satisfactory. Indeed this early success of the CPA played an important role in establishing the CPA idea as a sound base on which to build an alloy theory.

The calculation of SWF was made possible by two simplifying facts :

(a) The potential functions around the Ni and the Cu sites in the alloys are practically the same as they were in the corresponding pure metals. This allowed these authors to determine the parameters of their model Hamiltonian by fitting to the band structures of the pure elements.

(b) Because the band widths of Cu and Ni are roughly the same ($\sim$ 2.3 eV) it was possible to neglect randomness in the overlap integrals and in the hybridization parameters and consider the relatively simple case where only the site energies vary from site to site.

Evidently, the first circumstance implies that we can construct a reliable alloy potential in the form necessary for KKR-CPA. The details of the recipe are given in Appendix 1. In Fig. 16 we show the relevant phase-shifts. Thus, we can carry out a KKR-CPA calculation on the same basis as the previous model calculation of SWF. Evidently, in such a calculation none of the approximations implied in (b) are made. The results for $Cu_{.77}Ni_{.23}$ are shown in Fig. 15. With regard to the relative positions and widths of the "Cu d-band" and the Ni-related impurity structure the agreement between the two calculations is gratifying (only if the Fermi energies are lined up, the absolute values of the density of states are directly comparable).

We emphasize that only the starting phase-shifts were the same in the two calculations, otherwise they proceeded along two very different routes. SWF used these phase-shifts to generate pure metal band structures to which they fitted their model. Both the form of this model and the various approximations which had to be made to render it tractable were arrived at by taking full advantage of the simplifying features of the Cu-Ni system. On the other hand, in the KKR-CPA we went from the phase-shifts directly to the alloy density of states without any approximation other than the CPA. Thus the quantitative agreement should be read in two ways.

(a) It should be taken as a proof that in carrying out the

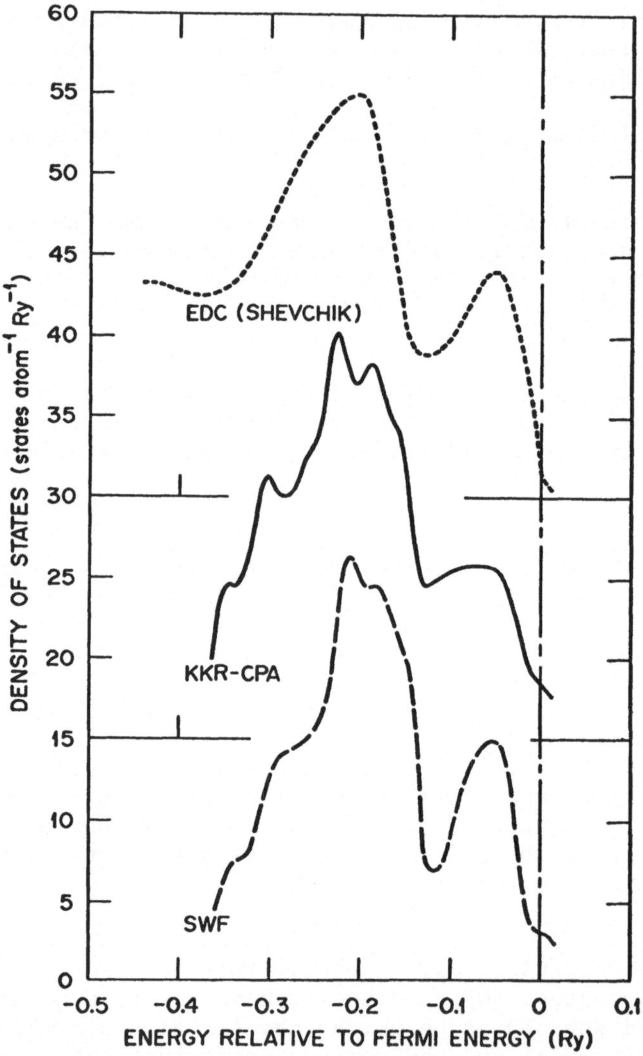

Fig. 15 : Comparison of the density of states as calculated in the KKR-CPA and the Nearly Free Electron-Tight-Binding - CPA with that deduced from photoemission experiments for $Cu_{.77}Ni_{.23}$.

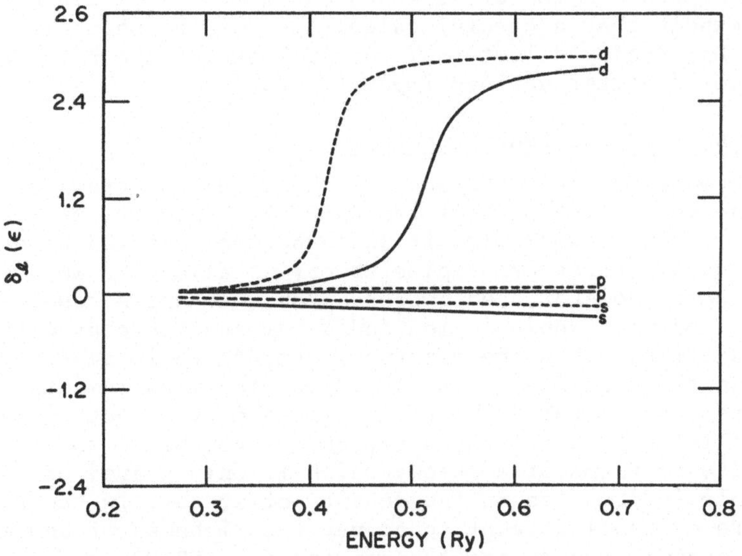

Fig. 16 : The s,p, and d phase-shifts which describe the scattering at the Ni sites (full line) and at the Cu sites (dashed lines) for our non-overlapping muffin-tin model Hamiltonian for Cu-Ni alloys.

KKR-CPA procedure we have indeed solved the problem correctly. The possibility of this check was the main reason for concentrating on the Cu-Ni systems as a first example to be treated in detail.

(b) It also shows that the long sequence of approximations by SWF retained all the essential physics in the problem.

Having examined the similarity between the calculation of SWF and the KKR-CPA it is worthwhile to note the differences. As can be seen in Fig. 15 in the Cu-related d-band region the KKR-CPA shows considerably more structure. This is the combined consequence of more realistic treatment of the band structure and a more general description of disorder in the KKR-CPA. That these extra features are also present in the experimental data gives us a hint of the enormously enhanced power of the new method. Given this good agreement at low energies we believe that the better agreement of the SWF results with the data is fortuitous. The impurity peak is less sharp in the KKR-CPA due to random hybridization. This effect was not included in the model calculation. Thus, tentatively, we attribute the sharpness in the experimental data to matrix element effects and expect that a careful calculation of the photoelectric yield (see Dr. Pendry's lectures) on the basis of the KKR-CPA would produce a better defined impurity peak.

(2) The Details of the Band Structure

If one measured only the specific heat and the transport coefficients one could live without the KKR-CPA. Moreover, with such scant information to work with it would be fool-hearted to set out to develop such a first principles theory of alloys as we are talking about here. However, such modern probes as angle resolved photoemission, positron annihilation and X-ray spectroscopy with synchroton radiation probes the electronic structure in vastly greater details than the classical physical properties measurements. Admittedly, these sophisticated experiments pose many as yet unresolved problems of interpretation. Even so, one cannot begin to discuss their results in terms of a theory which is only geared up to provide reliable information on the density of states. In the rest of this lecture we shall attempt to convey the richness of new details that becomes available during the course of a KKR-CPA calculation. We shall focus on those general features which could be detected by the experiments we have mentioned above.

Since the complete solution of the KKR-CPA equation and the ASA-CPA are almost indistinguishable for the Cu_cNi_{1-c} alloys, in the following we shall use the results of our more extensive ASA-CPA calculations.

In Fig. 17 we show the evolution of the $\ell = 2$ effective scattering amplitude $f_{C,L}$ and the ATA scattering amplitude $\overline{f}_2(\varepsilon)$ with addition of increasing amount of Ni. The dominant qualitative feature of these curves is the appearance and progressive strengthening of highly inelastic Ni impurity resonance above the host Cu resonance. Thus our main expectation based on the averaged scattering amplitudes (see Fig. 11) is borne out by the self-consistent calculation. However, looked at more closely, the CPA scattering amplitude differs from $\overline{f}_2(\varepsilon)$ in a number of interesting respects.

Evidently, the position of the Ni impurity resonance in Im $\overline{f}_2(\varepsilon)$ is almost an eV (.82 eV) below the corresponding peaks predicted by the CPA solutions. As can be easily seen the ATA gives the impurity resonances at the energy where the Ni site scattering amplitude has its resonance (.52 Ry). This is because $\overline{f}_2(\varepsilon)$ does not "know" about the band structure. On the other hand the CPA is

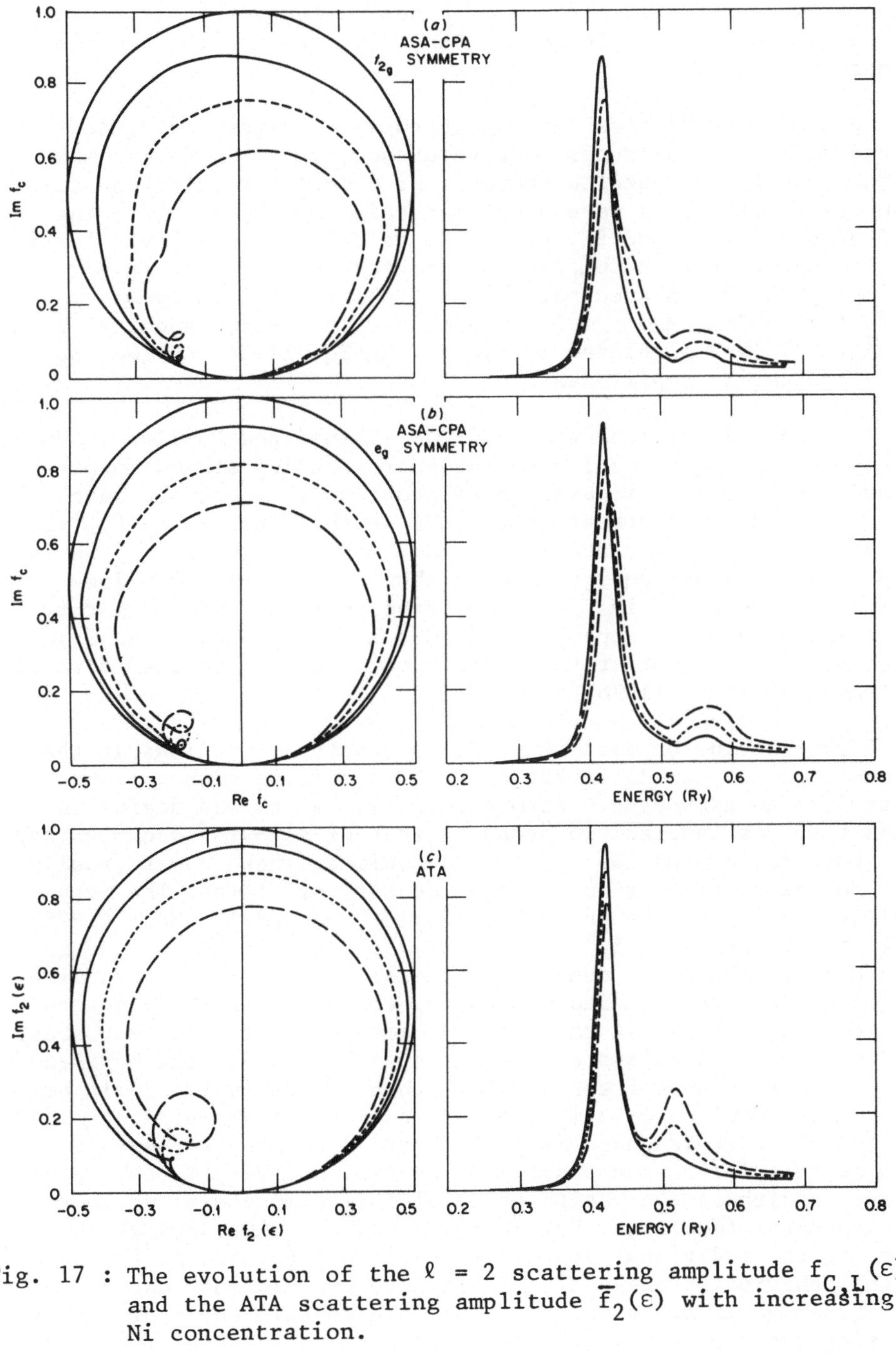

Fig. 17 : The evolution of the ℓ = 2 scattering amplitude $f_{C,L}(\varepsilon)$ and the ATA scattering amplitude $\overline{f}_2(\varepsilon)$ with increasing Ni concentration.

a self-consistent theory and evidently $f_{C,L}(\varepsilon)$ is strongly affected by the presence of well-defined Cu bands. Loosely speaking the sharp upper edge of the Cu-like d-band has repelled the impurity resonance.

We shall now discuss the band structure corresponding to the above scattering amplitudes. For reference, in Fig. 18, we show the energy bands of pure Cu and pure Ni calculated using the same potential functions as were used to obtain the input scattering amplitudes $f_{Cu,\ell}(\varepsilon)$ and $f_{Ni,\ell}(\varepsilon)$. The calculation was for a lattice spacing appropriate to $Cu_{.87}Ni_{.13}$ (see Appendix 1). In Fig. 19 we display the Bloch spectral functions at various symmetry points for $f^{ASA}_{C,L}(\varepsilon)$ and $\bar{f}_\ell(\varepsilon)$. In each case the Bloch spectral function $\bar{A}_B(\underline{q};\varepsilon)$ was calculated by evaluating Eq. II-64 for the appropriate scattering amplitudes.

The CPA calculations give fairly well-defined Cu-like d-bands, whose positions agree well with the corresponding states in pure Cu, and a smeared out impurity band about an eV above the main d-band complex. By contrast in the ATA at the Γ point the Γ_{12} peak is lower than the Γ_{12} state in pure Cu. Also the impurity band is sharper and at lower energy than in the CPA. Clearly, the low-lying Ni resonance in Im $\bar{f}_\ell(\varepsilon)$ gave rise to a low-lying impurity band which pushed the upper Cu-like d-state (Γ_{12}) down. The same phenomena are apparent at the L and K points. [For further details, see Temmerman et al (1978)].

Before moving on describing Bloch spectral functions at the various $\underline{q}$-points for different concentrations, we wish to pause and develop an appropriate terminology. It has proved useful in the past to distinguish two broad classes of alloying behavior and talk about rigid band and split band regimes. These names usually refer to the configuration averaged density of states. The very specific meaning of the venerable "rigid band model" [Mott (1935, 1936)] notwithstanding, it is customary to describe a behavior as rigid band if the alloy density of states looks more or less like that of one of its constituents and ε_F changes with the electron per atom ratio (e/a). If the alloy density of states, or any sharp structure of it, consists of two parts each of which can be identified as due to one of the species, then one is in the split band regime. For model calculations based on simple tight-binding model Hamiltonians this language is sufficient since mostly they are aimed at calculating the density of states. However, in this paper we want to discuss the wealth of new information contained in the Bloch spectral functions. For this we need a terminology which refers to the individual states at specific $\underline{q}$-points. As it turns out the behavior of these also falls into two distinct categories : they either go in twice-one peak for each constituent - or appear as a single broad peak in the constant $\underline{q}$ Bloch spectral function. Clearly, the first case corresponds to the split band picture and

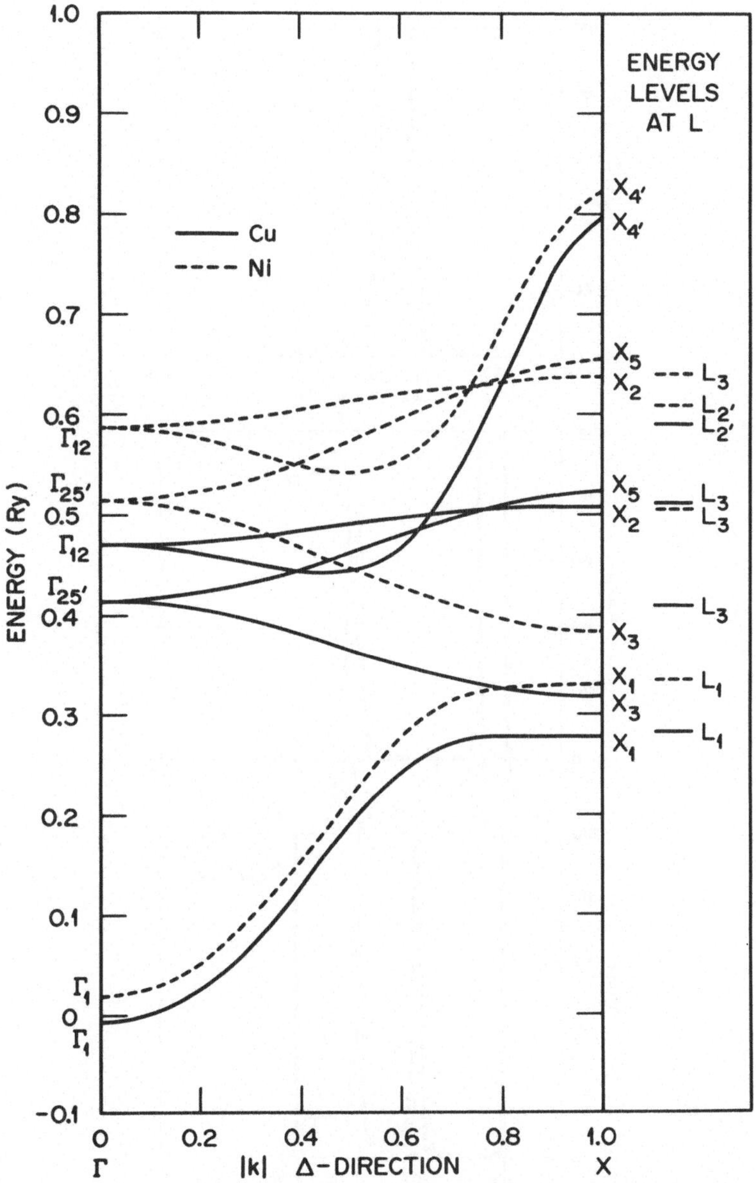

Fig. 18 : The band structure of pure Cu (full line) and pure Ni (dashed line) in the Γ-X direction of the fcc Brillouin zone. In both calculations the lattice parameter appropriate to the $Cu_{.87}Ni_{.13}$ alloy was used. The relative position of the two sets of bands was determined by lining up their V_{MTZ}.

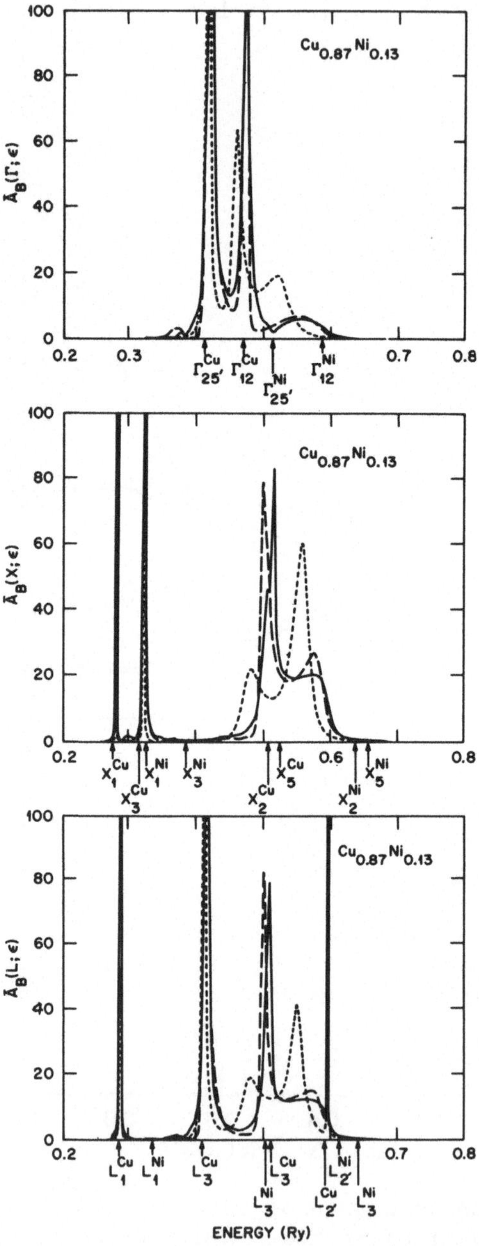

Fig. 19 : The Bloch spectral functions $\bar{A}_B(\underline{q};\varepsilon)$ at the symmetry points Γ, X and L for $Cu_{.87}Ni_{.13}$. The full line is the result of an ASA-CPA calculation. The dashed line is the prediction of the ATA.

we shall commonly refer to it as such ; we will talk about impurity states and host states. The second case is analogous to the rigid band regime and we shall call it the virtual crystal behavior [Nordheim (1931), Muto (1938)]. By this we imply that the bands are roughly where the virtual crystal approximation would put them but they are somewhat smeared out in energy. This nomenclature appears to us natural since the VCA may be regarded as the most sophisticated form of the rigid band model.

For the sake of clarity we recall that for the simplest model Hamiltonians, we get rigid band-like behavior in the CPA if the centers of mass of the pure constituents density of states are close together in energy compared with their band width. If their separation is large on that scale we get split band behavior. For a state at a given $\underline{q}$-point of the Brillouin zone it is more difficult to formulate such useful rules. Nevertheless, it is frequently the case that given the separation of two corresponding states of the pure constituents the pure metal band structures will suggest an appropriate energy scale on which this energy difference can be judged to be large or small. Often a useful energy scale is $\hbar|\Delta\varepsilon_{\underline{q}}|$. Once such a local standard of comparison is available one can make a guess as to what will happen on alloying. Though such notions cannot be backed up by rigorous calculations with simplified models they are nonetheless useful.

We now return to the discussion of the Bloch spectral functions corresponding to the scattering amplitudes displayed in Fig. 17. They are shown in Fig. 20. Evidently, the main effect of increasing Ni concentration is the development of a better defined, but still structureless impurity band and the progressively increased smearing of the high energy Cu-like d-bands. Note that there is no change in the relative positions of the Cu and Ni-related features. While the former decreases slightly in strength, the peak heights in the energy range of the latter increase somewhat. Consequently, we expect that the Fermi energy remains roughly constant. Using the scattering amplitudes of this paper, the Fermi energy and the Fermi surface were calculated by Gordon et al (1977). They found

$$\varepsilon_F^{Cu_{.87}Ni_{.13}} = .65 \text{ Ry} \; ; \; \varepsilon_F^{Cu_{.77}Ni_{.23}} = .66 \text{ Ry}$$

Such non-rigid-band behavior is consistent with the extensive photoemission data on Cu-rich Cu-Ni alloys [Seib and Spicer (1970), Yu et al (1977), Hüfner et al (1973)] and the tight-binding calculation of Stocks et al (1971).

As we have mentioned earlier the width of a peak in the constant $\underline{q}$ Bloch spectral function measures the lifetime of the state corresponding to that peak($\Delta\varepsilon_{\underline{q}} = \hbar\tau_{\underline{q}}^{-1}$). Comparing the Bloch

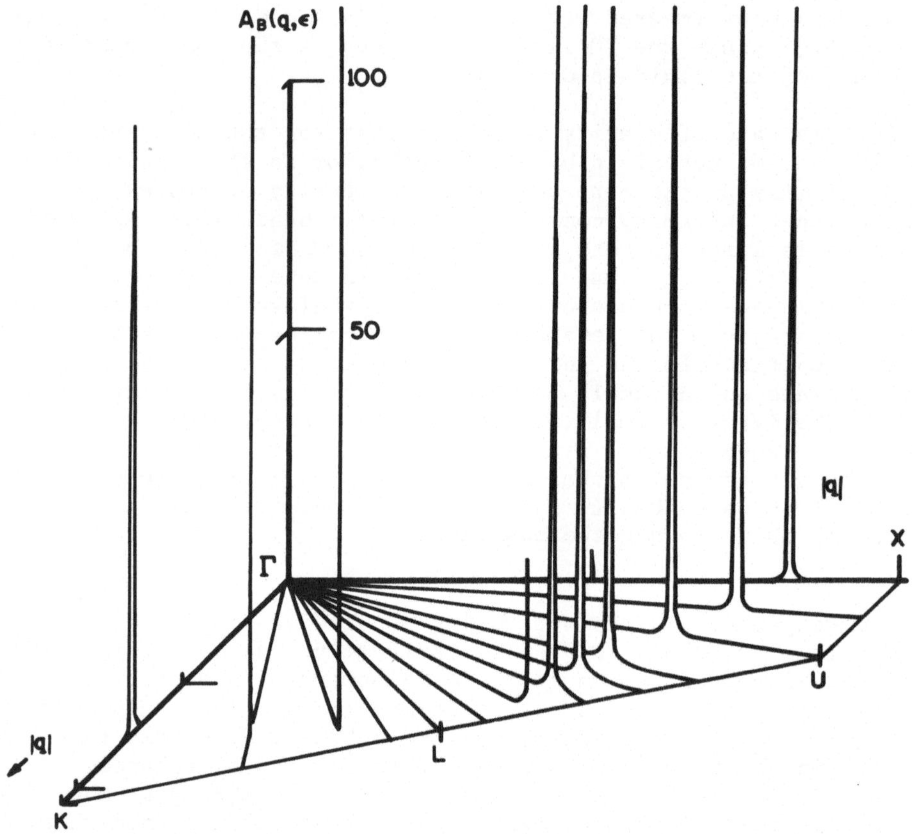

Fig. 20 : Bloch spectral density at the Fermi energy for a $Cu_{.87}Ni_{.13}$ alloy as a function of $|\underline{q}|$ for directions emanating from the Brillouin zone center, Γ, in the ΓXULKΓ plane. The sharp peaks in $\bar{A}_B(\underline{q};\varepsilon)$ define the position of the states in $\underline{q}$-space and hence the Fermi surface. The fact that for the three directions ending near the L point there is no peak indicates that the Fermi surface touches the Brillouin zone boundary. The width of the peaks gives the inverse of the coherence length.

spectral functions at the Γ and X points in Fig. 19 we see that while there is hardly any dispersion in the impurity band the impurity lifetime changes with $\underline{q}$. This effect is most pronounced in the $Cu_{.87}Ni_{.13}$ alloy from which the impurity state is much better defined at the X point than at the Γ point. This means that the self-energy is $\underline{q}$-dependent and implies that, in the language of tight-binding Hamiltonians, there is appreciable off-diagonal randomness in our description of these alloys. Evidently, the various hopping integrals connecting two Cu sites, two Ni sites, or a Cu and a Ni site are not sufficiently the same for their difference to be negligible. Experimental evidence for an anisotropic impurity band in the $Cu_{.87}Ni_{.13}$ alloy has been presented by Györffy et al (1977).

The usual methods for studying the Fermi surface require states, at the Fermi energy, extremely well defined both in Bloch vector $\underline{q}$ and energy ε [Cracknell (1971)]. In alloys the imprint of the Fermi surface disappears when the impurity broadening exceeds the effective temperature of 5°K to 10°K. This means in Cu_cNi_{1-c} alloys that for c or 1-c greater than a few hundred ppm there is no Fermi surface in the conventional sense. However, on the scale of the separation between the various energy eigenvalues at a given $\underline{q}$ at the Fermi energy ε_F can still be quite well defined at much higher concentrations. This is evident from Fig. 20.

Having determined the Fermi energy we now want to examine this Fermi surface. In Fig. 20 we show a three-dimensional plot of $\bar{A}_B(\underline{q};\varepsilon)$ for $Cu_{.87}Ni_{.13}$ [Gordon et al (1977)]. The sharp peaks along lines emanating from the Γ-point towards the Brillouin zone boundary in the ΓXULKΓ plane show the smearing of the Fermi surface in $\underline{q}$. The Fermi surface is defined by the locus of the peaks in $\bar{A}_B(\underline{q};\varepsilon)$. The width of the peaks defines the coherence length $\Delta\underline{q} = \xi^{-1}$. Clearly the Fermi surface is well defined and the "neck" around the L-point is clearly visible. In Fig. 21 we show the "neck" radius as a function of concentration. The theoretical values were read off the plots of $\bar{A}_B(\underline{q};\varepsilon)$ along the KL line. The experimental points have been determined by positron annihilation experiments [Hasegawa, Suzuki and Hirabayashi (1974)]. According to the "rigid band" model ε_F would move down towards the d-band as Ni is added to Cu and for c around .30 the Fermi surface would no longer touch the Brillouin zone boundary. Thus, the very gentle decrease of the "neck" radius is a non-rigid band effect. Clearly the KKR-CPA theory gives a good account of this behavior. We note that in our calculation ε_F moves up slightly in energy as the Ni concentration is increased. However, the s-like band at ε_F also moves up in energy slightly due to its hybridization with the emerging Ni d-band. It is the combination of these two effects which is responsible for the overall reduction in the "neck" radius. Interestingly, the belly changes in an asymmetric fashion. Its cross section in the ΓXULKΓ plane is shown in Fig. 22 for various concen-

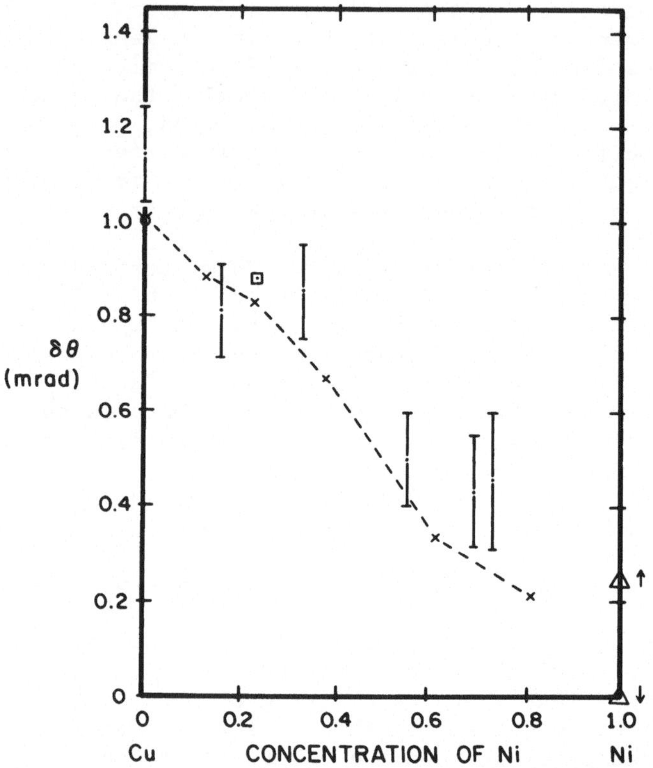

Fig. 21 : Composition dependence of the neck radius in Cu-Ni alloys. Comparison of results of KKR-CPA with data from positron annihilation experiments.

--x--	Gives the calculated results from CPA
.	Give the experimental results of Hasegawa et al. The error bars are those given by Hasegawa et al.
⊡	Experimental result of Murray and McGrevey (1970).
Δ↑ ; Δ↓	Calculated neck radius for spin up and spin down bands in ferromagnetic nickel [Snow et al (1966)]
O	Neck radius of pure Cu from de Haas-van Alphen effect [O'Sullivan and Schirber (1969)].

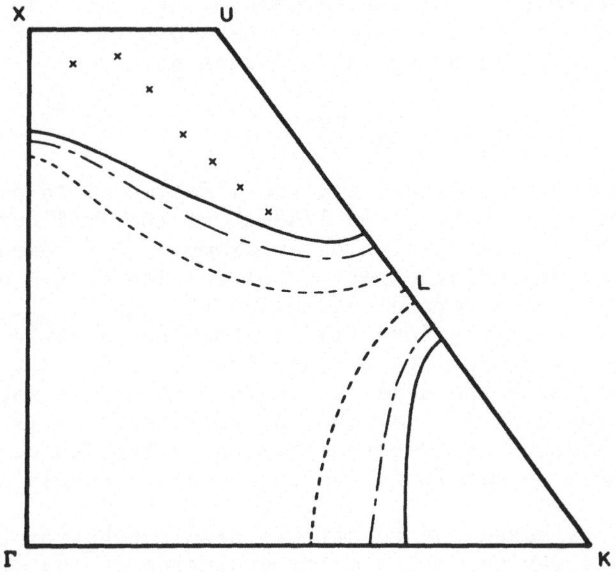

Fig. 22 : Section through the Fermi surface in ΓXULKΓ plane.

Solid line : - $Cu_{.87}Ni_{.13}$

Dot-dash line : - $Cu_{.62}Ni_{.38}$

Dotted line : - $Cu_{.19}Ni_{.81}$

For $Cu_{.19}Ni_{.81}$ there is a second 'sheet' to the Fermi surface in this plane. The peak in $\overline{A}_B(\underline{q};\varepsilon)$ is particularly sharp in the ΓX direction becoming less sharp as we go off into the ΓXULΓ plane. The locus of this peak has been indicated by the crosses. This 'sheet' is associated with the d-bands of nickel.

trations. In the ΓX direction the belly radius decreases slowly as the Ni concentration increases ; in the ΓK direction the decrease is much more rapid. A very similar effect was observed by Templeton and Coleridge (1973) at much lower Ni concentrations (50-100 ppm). The rate of change of the belly radius in various directions is in good qualitative agreement with the above calculation.

Having decided that a $\underline{q}$-point is on the Fermi surface we may ask for the lifetime, $\tau_{\underline{q}}$, of the corresponding state. In the KKR-CPA theory the lifetimes are given by the width of the Bloch spectral function $\bar{A}_B(\underline{q};\varepsilon)$ in energy at constant $\underline{q}$,

$$\Delta\varepsilon_{\underline{q}} = \frac{\hbar}{2\tau_{\underline{q}}}$$

In Fig. 23 we show the constant inverse lifetime plots per atomic percent over the Fermi surface in 1/48-th of the Brillouin zone using the $Cu_{.87}Ni_{.13}$ calculation as an example. For comparison we also give the corresponding experimental results from a de Haas-van Alphen measurement on a very dilute alloy ($Cu_{.9995}Ni_{.0005}$) by Poulsen, Randles and Springford (1974). Somewhat surprisingly, the general shape and the topology of the experimental and theoretical contours are very much the same. The agreement on the magnitude of the relative changes of $1/\tau_{\underline{q}}$ across the Fermi surface is also very good. A simple linear extrapolation to the low concentration limit yields absolute magnitudes which are also roughly correct.

As more Ni is added the relatively simple behavior of the Cu-rich alloys gives way to more complex evolution of the effective scattering amplitudes shown in Fig. 24. For reference we also show the corresponding ATA scattering amplitudes. Clearly, the simple rise of the Ni resonance and an accompanied fall in the Cu resonance as suggested by the ATA does not occur in the CPA. In the case of $Cu_{.39}Ni_{.61}$ and $Cu_{.62}Ni_{.38}$ alloys the Argand plots for $f_{C,2e_g}(\varepsilon)$ show similarly with that for $\bar{f}_2(\varepsilon)$ only in having two loops corresponding to two distinct inelastic resonances. For $Cu_{.19}Ni_{.81}$ there is only one loop which means that instead of a subsidiary Cu resonance there is only a modification of the host Ni resonance. Note that the shoulder on the low energy side of the Ni peak in the $\mathrm{Im}f_{C,2e_g}(\varepsilon)$ curve of Fig. 24 is not at the energy where $\mathrm{Im}f_{Cu,2}$ has its peak (ε_r^{Cu} = .42 Ry) but rather at the position of the pure Cu state (.48 Ry). At ε_r^{Cu} where $\bar{f}_2(\varepsilon)$ indicates a Cu impurity resonance there is no structure at all in $\mathrm{Im}f_{C,2e_g}(\varepsilon)$. Consequently, instead of the split band behavior of the impurity states in the Cu-rich alloys we expect smeared out virtual crystal bands at this end of the concentration range. Another interesting feature of $\mathrm{Im}f_{C,2e_g}(\varepsilon)$ is the variation of energy where the Ni peak occurs. For $Cu_{.62}Ni_{.38}$ it is at the same place, above ε_r^{Ni}, where it was in the other Cu-rich alloys. However, for the Ni-rich alloys it moves down to ε_r^{Ni}. The

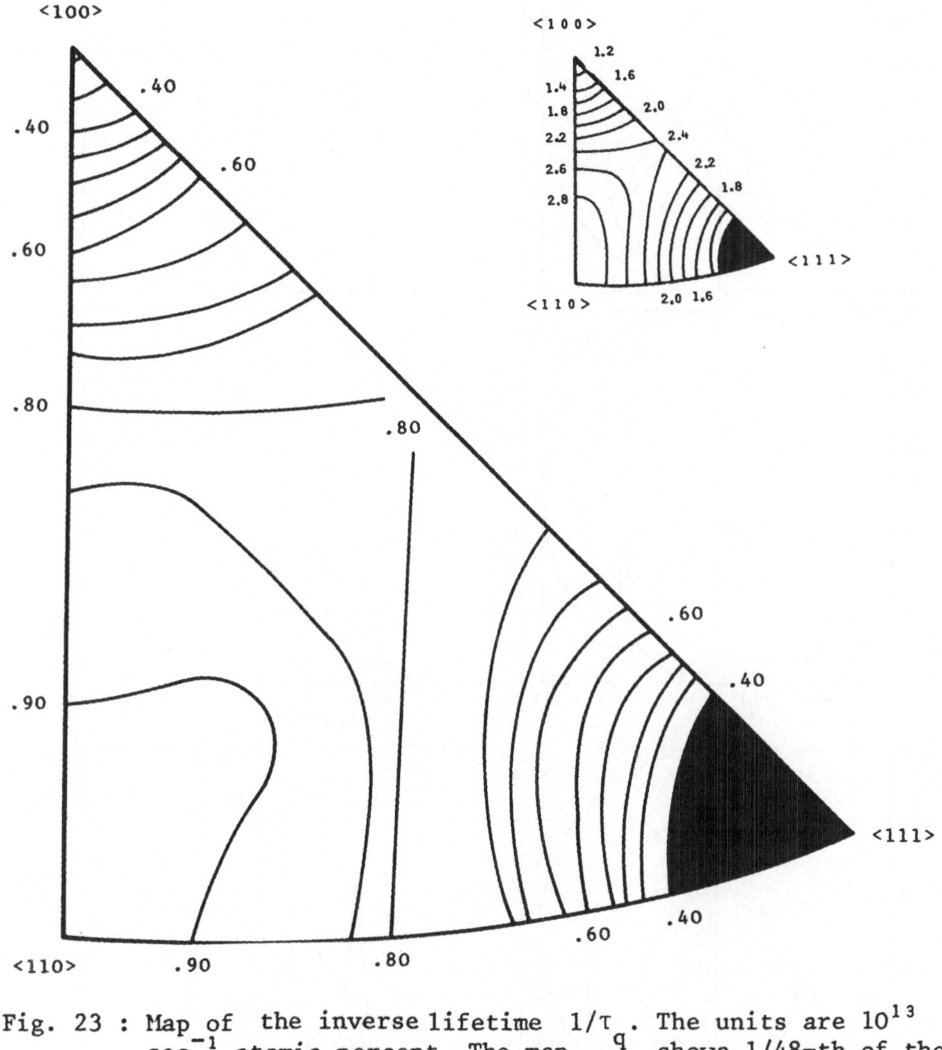

Fig. 23 : Map of the inverse lifetime $1/\tau_{\underline{q}}$. The units are 10^{13} sec^{-1} atomic percent. The map shows 1/48-th of the Brillouin zone in stereographic projection. The results are based on the KKR-CPA for a $Cu_{.87}Ni_{.13}$ alloy. Inset are the experimental scattering rates taken from Poulsen et al (1974) and were obtained from the deHvA effect using a $Cu_{.9995}Ni_{.0005}$ alloy.

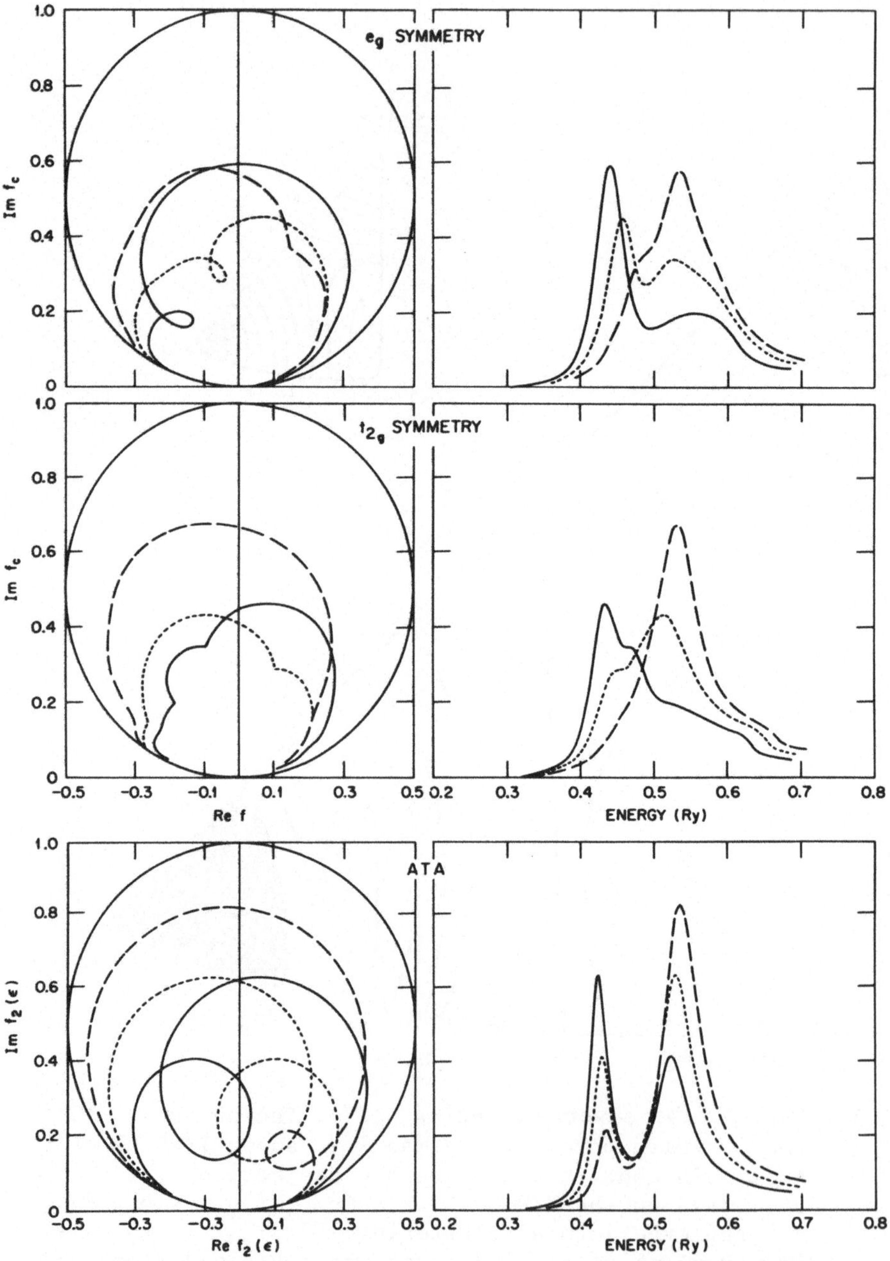

Fig. 24 : The evolution of the $\ell = 2, e_g$ and t_{2g} effective scattering amplitudes with changing concentration : $Cu_{.62}Ni_{.38}$ (full line) $Cu_{.39}Ni_{.61}$ (dotted line) and $Cu_{.19}Ni_{.81}$ (long dash line) for ASA-CPA (upper four frames) and ATA (bottom two frames).

reason for this is easy to see ; recall that it was the flat Cu related Δ_5 band which pushed the Ni resonance up in the Cu-rich alloys in spite of the fact that it is predominantly of t_{2g} symmetry. This band is now being smeared out and therefore cannot repel it.

The gradual disappearance of the second resonance is even more pronounced in the t_{2g} channel. As can be seen in Fig. 24 the Argand plots show only one loop for all three alloys. This implies that there are at most shoulders (at the cusps) to a single peak in the corresponding $\mathrm{Im} f_{C,2t_{2g}}(\varepsilon)$ versus ε curves. For $Cu_{.19}Ni_{.81}$ there is not even a shoulder which can be associated with the Cu band structure. The impurities merely broaden the Ni host d-resonance in spite of the considerable separation between ε_r^{Cu} and ε_r^{Ni}

$$(\varepsilon_r^{Ni} - \varepsilon_r^{Cu} \sim .1 \text{ Ry})$$

Surprisingly, in this channel even $Cu_{.39}Ni_{.61}$ has only one resonance with hardly any structure at ε_r^{Cu} or at the energy where the Ni resonance was in the more Cu-rich alloys. Evidently, virtual bound state-like descriptions of Ni-related features are completely inadequate at these concentrations.

Unfortunately, it is very difficult to construct a simple model which could shed some further light on the causes of this complex behavior. Nevertheless, we note that the t_{2g} and e_g lobes of the d-orbitals point towards the nearest and the next nearest neighbors respectively. Consequently, the corresponding hopping integrals of a tight-binding theory provide two different energy scales on which to judge the size of $\varepsilon_r^{Ni} - \varepsilon_r^{Cu}$ of the two; it is the e_g hopping integral which is likely to be the smaller. Some evidence for this is provided by the flatness of the Δ_2 band (from $\Gamma_{12} \rightarrow X_2$). On this smaller energy scale the difference $\varepsilon_r^{Ni} - \varepsilon_r^{Cu}$ appears larger and therefore it is plausible to expect that the bands split more easily in this channel.

The Bloch spectral functions at the Γ,X and L points for the scattering amplitudes discussed above are shown in Fig. 25. The states at the Γ point illustrate the switch from the "split-band" regime to a smeared out virtual crystal behavior. For $Cu_{.62}Ni_{.38}$ the Cu $\Gamma_{25'}$ and Γ_{12} states can be still clearly identified and the impurity state between .5 and .6 Ry is qualitatively similar to that found at lower Ni concentrations. On the other hand, for $Cu_{.19}Ni_{.81}$ there are only the Ni $\Gamma_{25'}$ and Γ_{12} states and the only impurity-related feature is the broadening of these levels. The situation is similar at the X point. The only split band feature for the $Cu_{.19}Ni_{.81}$ alloy is at the L-point where we see both the lower L_3 state of Ni at $\varepsilon = .52$ Ry and the lower L_3 Cu state which appears as a shoulder to the former at $\varepsilon = .49$ Ry.

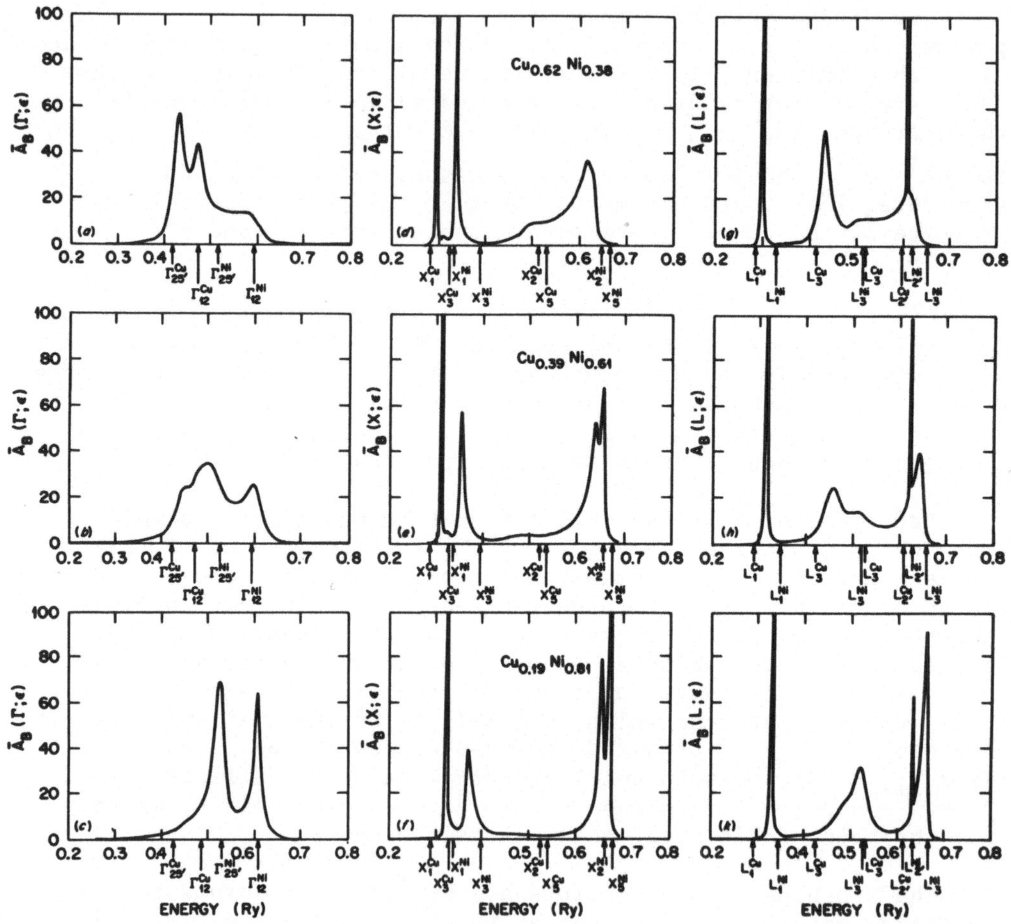

Fig. 25 : Evolution of the Bloch spectral functions at the symmetry points Γ, X, and L with concentration for the Ni rich alloys : $Cu_{.62}Ni_{.38}$, $Cu_{.39}Ni_{.61}$, $Cu_{.19}Ni_{.81}$.

Obviously these states are too widely separated to be absorbed into a single peak.

Another interesting aspect of the Bloch spectral functions in Fig. 25 is the very different rate of evolution of the Ni related peaks at the Γ and at the X points. As we have noted in the previous section the impurity peak is much better defined at the X than at the Γ point even at low Ni concentration. Evidently, a similar but more complicated effect is operating in these more concentrated alloys. For $Cu_{.62}Ni_{.38}$ at the Γ point the impurity state is a broad futureless hump with a magnitude which is hardly larger than in the more Cu-rich alloys. By contrast at the X point we no longer see an impurity state but a peak with perceptible double structure which already can be identified with the Ni X_2 and X_5 states respectively. While our calculation is for paramagnetic alloys the early development of the X_2^{Ni} and X_5^{Ni} states with increasing Ni concentration is most interesting from the point of view of ferromagnetism in these alloys. The $Cu_{.62}Ni_{.38}$ alloy for which the X_2^{Ni} - X_5^{Ni} peak is well below the Fermi energy is not ferromagnetic. However, $Cu_{.39}Ni_{.61}$ for which the well formed X_2^{Ni} and X_5^{Ni} peaks already saddle the Fermi energy, has a ferromagnetic ground state. Plausibly, if these peaks were as ill defined as the corresponding Ni related features at the Γ point they could not give rise to sufficiently large density of states at ε_F to facilitate a Stoner-type magnetic instability. The detailed picture of how the high $n(\varepsilon_F)$ occurs suggests that one should go beyond the simple Stoner model for random alloys and consider the exchange splitting $\underline{q}$-point by $\underline{q}$-point. It might be that a particular state can benefit from the exchange splitting only if the width of the state is small compared with the exchange splitting energy. Such effects could occur only indirectly in the magnetic CPA calculations based on the Hubbard model [Kanamori et al (1977)] where only the density of states functions enter explicitly. As is well known [Medina and Cable (1977)] the actual magnetic instability in the Ni-Cu systems has more to do with magnetic clustering than the simple Stoner picture we had in mind while making the above remarks. Nevertheless, our observations are suggestive of an important role played by the band theoretical aspect of the problem. However, this role might be indirect in the sense that when sharp d-states appear at ε_F the inter-site susceptibility gets enhanced by a Stoner-like mechanism and the large susceptibility allows the magnetic clustering to proceed.

We also note in passing that the large d-density of states appears at the Fermi energy not because ε_F is moving down with the electron per atom ratio as the rigid band model would have it. In fact, as in the case of Cu-rich alloys, the ε_F hardly changes :

$$\varepsilon_F^{Cu_{.62}Ni_{.38}} = .66 \text{ Ry} \quad , \quad \varepsilon_F^{Cu_{.39}Ni_{.61}} = .66 \text{ Ry}$$

and $\varepsilon_F^{Cu_{.19}Ni_{.81}} = .69$ Ry [Gordon et al (1977)]. In our calculations the increase in $n_2(\varepsilon_F)$ is due to Ni-related bands moving up as they become progressively better defined.

While the split band behavior of the Cu-rich alloys is well established both theoretically and experimentally [Yu et al (1977), Hüfner (1973), Durham et al (1978)] the understanding of the Ni-rich alloys is less well developed. Because of this we want to conclude this section by analyzing our results for $Cu_{.19}Ni_{.81}$ in more detail. Although this alloy is ferromagnetic at $T = 0°$ it is nevertheless important to have a clear picture of the theoretical paramagnetic band structure. In any case spectroscopic probes such as photoemission studies and soft X-ray spectroscopy are not very sensitive to the existence of the magnetic states.

The striking difference between the effects on the band structure due to Ni impurities in Cu and that of Cu in Ni was first noticed by Stocks et al (1971). As we have emphasized repeatedly our calculation rests on a firmer basis. We have treated the random s-d hybridization and random hopping on equal footing with diagonal randomness. Moreover, we did not have to decouple the CPA equations for the e_g and t_{2g} components of the self-energy. Nevertheless, our conclusions with regard to the density of states effects is entirely the same. Namely, the Cu impurity does not split an impurity band from the host Ni d-band. In fact our more detailed calculations show that in spite of the .1 Ry separation between the Cu and Ni resonances most of the bands are broadened virtual crystal bands.

That this is a non-trivial conclusion becomes evident from a comparison of the ASA-CPA results with the predictions of the ATA and the cluster method of solving the KKR-CPA equations. The relevant scattering amplitudes are plotted in Fig. 26. Evidently, the ATA fails completely in that it gives two resonances. In fact $f_{C,2t_{2g}}(\varepsilon)$ is more similar to the scattering amplitude $f_2^{VCA}(\varepsilon)$ calculated from the virtual crystal potential function $\bar{v}(r) = cv_{Cu}(r) + (1-c)\, v_{Ni}(r)$ and this is shown in Fig. 27. Clearly, the CPA self-consistency plays a very important role in determining the effective scattering amplitudes- if not fulfilled in detail, a second resonance begins to appear. This is what appears to happen in the CPA calculations based on clusters. In the plot of Im $f_{C,2t_{2g}}^{CCPA}(\varepsilon)$ we see a tendency towards the Ni host resonance absorbing the Cu impurity resonance. However, this absorption is not complete.

For these three scattering amplitudes we show the Bloch spectral functions at the Γ, X and L points in Fig. 28. The progression from split band to virtual crystal behavior as one goes from ATA to CCPA to ASA is clearly visible. In a recent angle resolved

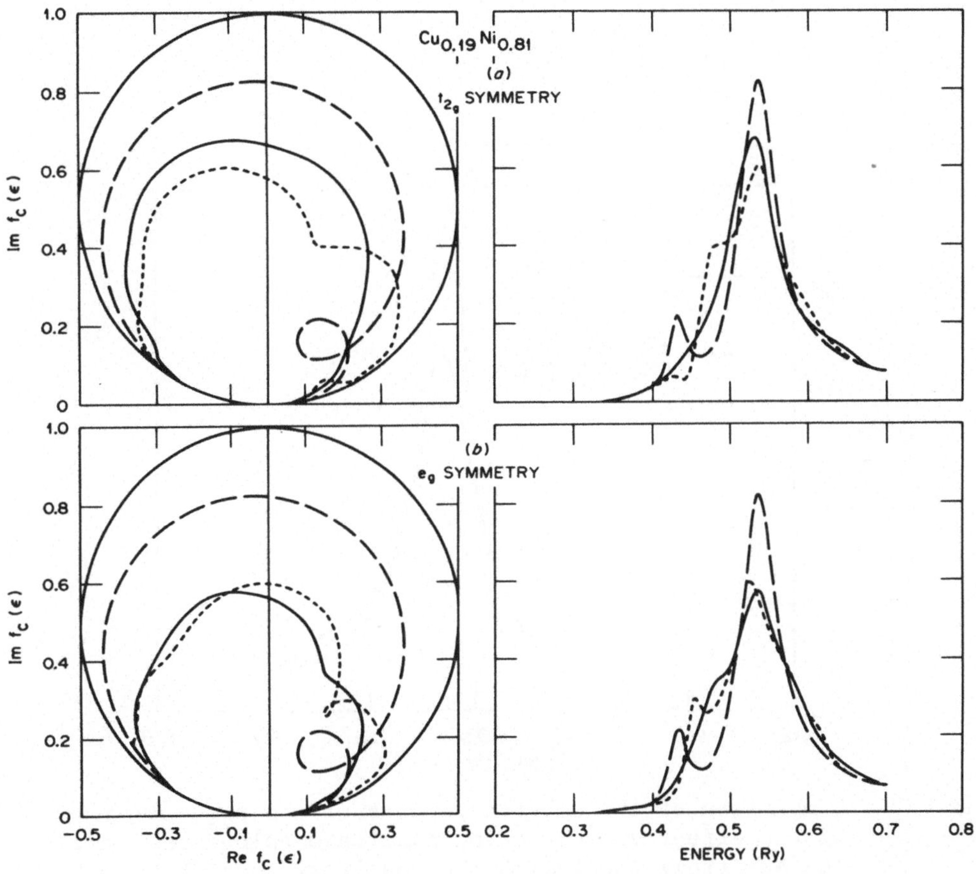

Fig. 26 : The $\ell = 2$, e_g and t_{2g} components of the effective scattering amplitude as predicted by the Atomic Sphere Approximation to the KKR-CPA theory ($f^{ASA}_{C,L}$, full line), the averaged t-matrix approximation ($\bar{f}_{C,L}$, dashed line) and the cluster method of solving the KKR-CPA equations in realspace ($f^{CCPA}_{C,L}$ dotted line) for $Cu_{.19}Ni_{.81}$. In calculating $f^{CCPA}_{C,L}$ a cluster including two nearest-neighbor shells was used.

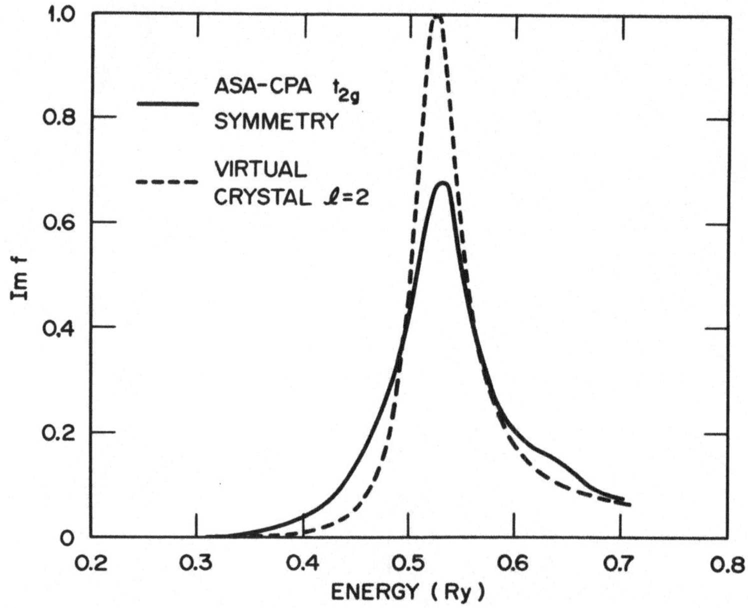

Fig. 27 : Comparison of the imaginary parts of the t_{2g} component of the effective scattering amplitude calculated using the ASA (full line) with the imaginary part of the scattering amplitude corresponding to the virtual crystal potential (dash line) for $Cu_{.19}Ni_{.81}$.

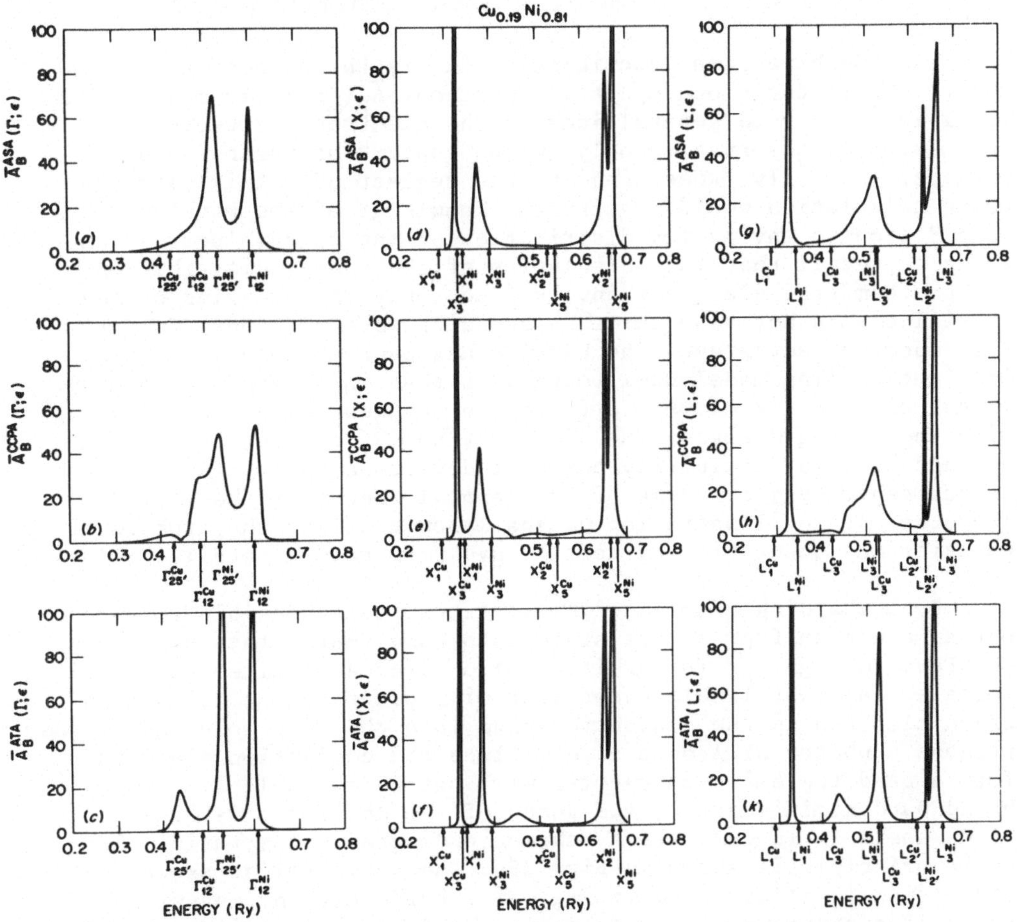

Fig. 28 : The Bloch spectral functions at the symmetry points Γ, X and L as predicted by the Atomic Sphere Approximation to the KKR-CPA theory (top frames), the averaged t-matrix approximation (bottom frames) and the cluster method of solving the KKR-CPA equations in real space (middle frames) for $Cu_{.19}Ni_{.81}$.

photoemission study of a Ni-rich Ni-Cu alloy Heimann et al (1978) have found very little Cu-related structure after the effects of surface enrichment have been carefully eliminated.

V. SOFT X-RAY SPECTROSCOPY OF RANDOM METALLIC ALLOYS

What we have been describing so far is the core of a KKR-CPA theory. While the Bloch spectral function, $\overline{A}_B(\underline{q};\varepsilon)$, is most useful in giving us a good general idea of the electronic structure, it is related to observables only in the context of further approximations. Typically, these involve the neglect of matrix elements whose calculation would require the knowledge of the wavefunctions in addition to having the distribution of the eigenvalues. As an example, recall that the calculation of the energy distribution of the photoemitted electrons involves not only the energies of the initial states but also the corresponding wavefunctions. One of the important advantages the KKR-CPA has over calculations based on tight-binding model Hamiltonians is that in this approach many, if not all, matrix element effects can be treated on equal footing with the averaged eigenvalue distribution. Of course, to take advantage of this flexibility one must develop a theory for each effect separately in terms of the general theory of the last two lectures. As an illustration we shall now consider the problem of emission and absorption of soft X-rays from random metallic alloys.

By bombarding a solid with electrons it is possible to knock out an electron from a core state into some empty state high above the Fermi energy ε_F. The excited solid, like an excited atom, will radiate. The most likely event that will occur is the transition of an electron in the conduction band into the core hole and in the process a photon will be emitted. Since the conduction electrons form a band the emission spectra will not be a single line but a broad band reminiscent of the shape of the density of states up to ε_F where it is cut off. A schematic diagram illustrating the basic principle is shown in Fig. 29a. The very similar process of absorption is illustrated in Fig. 29b. Evidently, by measuring the absorption spectrum we are learning about the unoccupied part of the conduction band. The photons involved in the above processes fall into the soft X-ray regime (100-10000 eV) and hence it is customary to refer to their study as Soft X-ray Spectroscopy (SXS).

This is one of the oldest tools for studying the electronic structure of metals [Skinner (1940), Tomboulian (1957); see also "Soft X-ray Band Spectra" Ed. Fabian (1968)] which have been somewhat eclipsed by such modern techniques as angle resolved photoemission spectroscopy. In addition to giving an example of how the KKR-CPA theory works, this lecture also aims to rekindle interest in this time honored technique by focusing attention on the very valuable information it can provide when applied to random alloys.

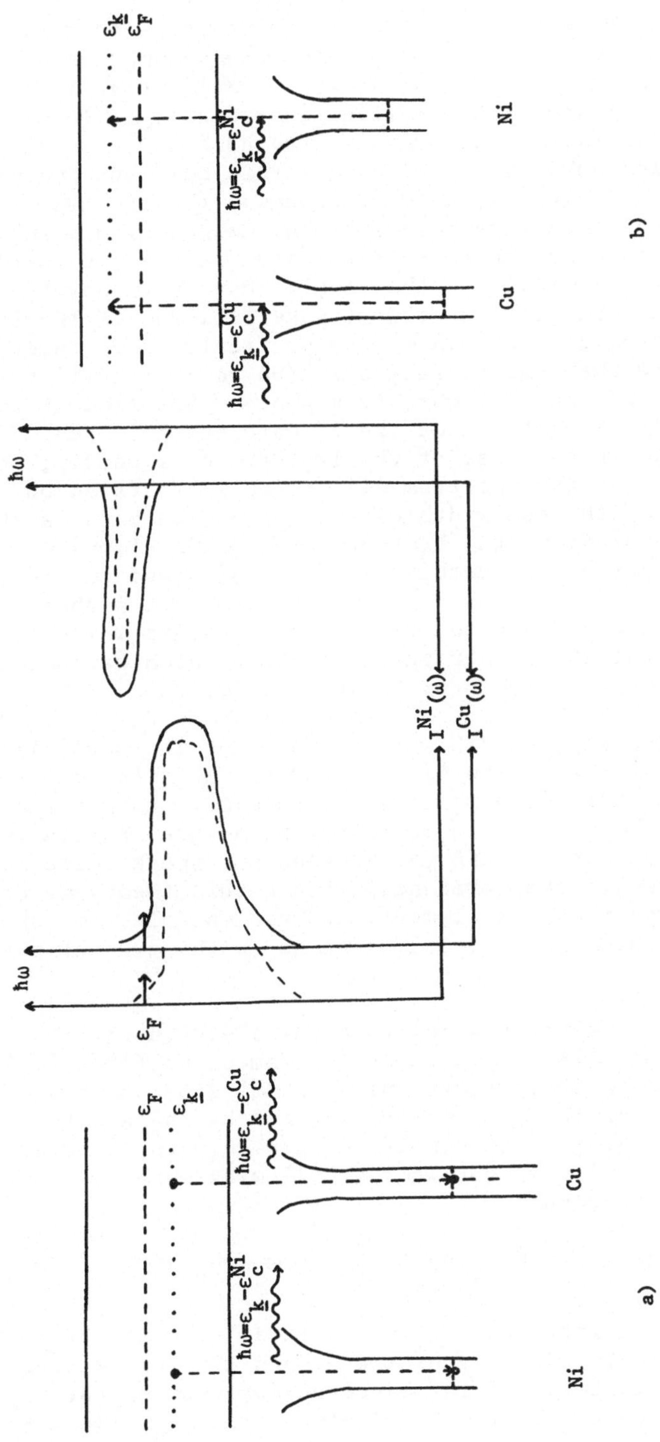

Fig. 29 : Schematic diagram of the Soft X-Ray Emission (a) and Absorption (b) processes.

The reason why SXS is a unique probe of the electronic structure is easy to see. In Fig. 30 we show the wavefunction of the 2P core states on a Cu site in a Ni-Cu alloy. Evidently, it is well localized on the scale of the muffin-tin radius r_{MT}. Thus, if a core electron was removed from this state, the hole would be also well localized and the subsequent filling of this hole and emission of a photon would be a very local event indeed. (We are assuming that the hole does not have time to hop to the neighboring site). Unfortunately, in measuring the emitted spectrum we are not identifying the source of the radiation - they could have been anywhere within the sample - and therefore we cannot make full use of the fact that the radiation process took place in a particular environment. However, if we know that the radiation is due to a transition to a 2P core hole, then we can tell whether a photon came from an event on a Cu site or on a Ni site since the 2P core state is some 70 eV higher in energy in the case of the latter. Consequently, the Ni and Cu omponents of the spectrum will be well separated on the scale of their width (the width of the conduction band is about 4 eV). Thus, in studying the L_3 spectra of Ni-Cu alloys we are probing the electronic structure on the Ni site and on the Cu sites separately. By contrast the initial state of a photoemitted electron was an extended conduction electron state which is characteristic of the whole configuration about which we know nothing and therefore must average over in the theory.

Roughly speaking, in SXS we are measuring the partially averaged density of states on the Ni site and the Cu sites separately. As we shall see, this is a most useful information from the point of view of any theory of the electronic structure. Fortunately, within the context of the KKR-CPA we need not speak quite so roughly. As we shall show presently within this theory we can calculate the relevant matrix elements and the one electron spectra are obtained without any approximation as in the case of pure metals.

We shall now sketch the relevant one electron theory [Györffy and Stott (1972) , see also Durham et al (1978)]. To simplify matters we shall treat only the emission spectra. If we denote the state of the conduction electron by $|n\rangle$ and the final core state at $\underline{R}_0$ by $|c\rangle$ and use the Golden Rule to calculate the transition probability we obtain the following general expression for the emitted spectrum

$$I_0(\omega) = \frac{2\pi}{\hbar} \rho(\omega) \sum_n |\langle c|H_{int}|n\rangle|^2 \delta(\varepsilon_c + \hbar\omega - \varepsilon_n) \qquad \text{V-1}$$

where $I_0(\omega)$ is the intensity of the emitted X-ray at the frequency ω, $\rho(\omega)$ is the photon density of states and the interaction Hamiltonian is given in terms of the electron momentum operator $\underline{p}$ and the vector potential $\underline{A}$ by the usual expression $H_{int} = (e/mc)\underline{p}.\underline{A}$

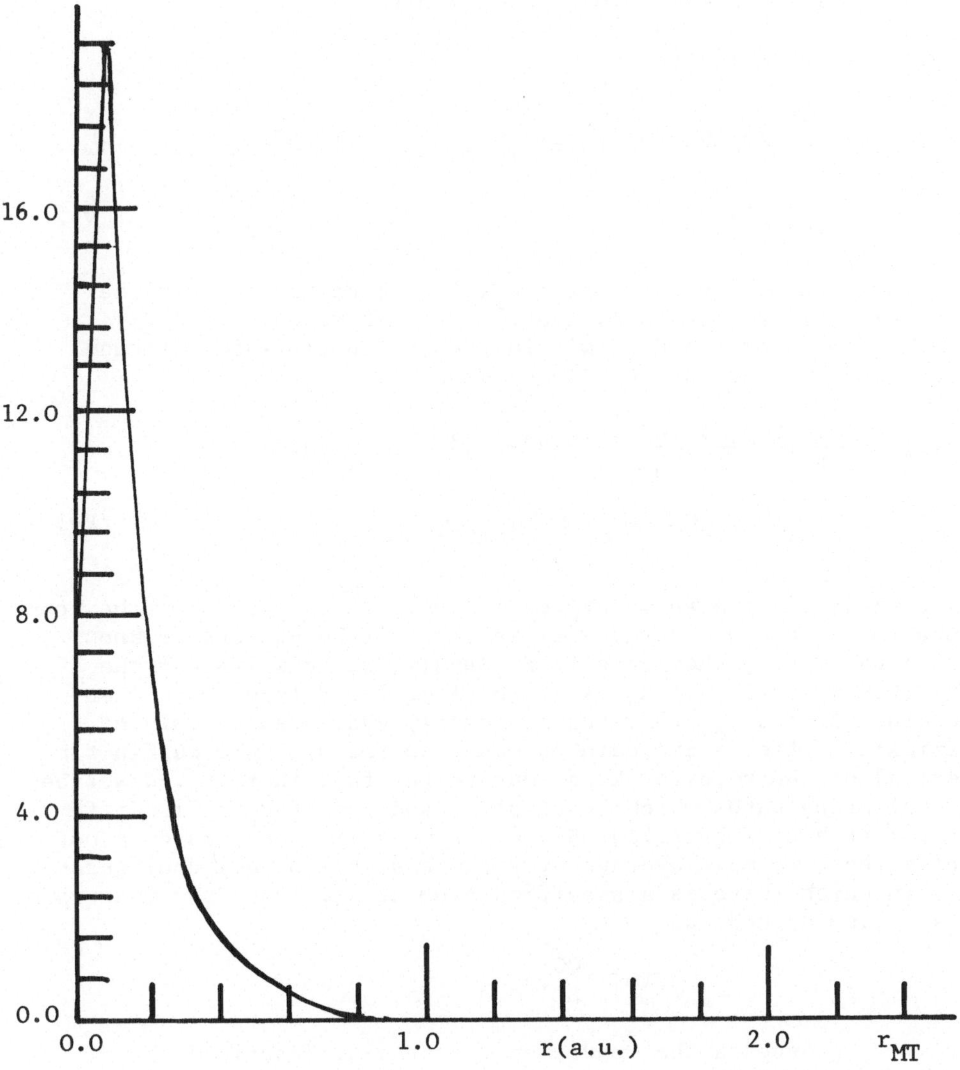

Fig. 30 : Radial part of the 2P wavefunction for Cu. The muffin-tin radius is marked r_{MT}.

(we shall work in the dipole approximation). By writing out the square and using the fact that

$$\text{Im } G^{+}(\underline{r},\underline{r}';\varepsilon)=-\pi\sum_{n}<\underline{r}|n>\delta(\varepsilon-\varepsilon_{n})<n|\underline{r}'>$$

Eq. V-1 may be rewritten as

$$I_0(\omega) = -\frac{1}{\pi}\left(\frac{2\pi\rho(\omega)}{\hbar}\right)\int d^3r\int d^3r'\ \phi_c^{\star}(\underline{r})H_{int}(\underline{r})\text{Im } G^{+}(\underline{r},\underline{r}';\varepsilon_c+\hbar\omega)\cdot$$
$$\cdot H_{int}(\underline{r}')\ \phi_c(\underline{r}') \qquad \text{V-2}$$

Since the core wavefunction $\phi_c(\underline{r})$ is localized within the muffin-tin well at $\underline{R}_0$, to evaluate Eq. V-2 we only need $\text{Im } G^{+}(\underline{r},\underline{r}';\varepsilon)$ for $\underline{r}$ and $\underline{r}'$ within the muffin-tin sphere around $\underline{R}_0$ and hence we can use Eq. II-48. Thus

$$I_0(\omega) = -\frac{1}{\pi}\left(\frac{2\pi\rho(\omega)}{\hbar}\right)\sum_{LL'}\int d^3r\int d^3r'\ \phi_c^{\star}(\underline{r})H_{int}(\underline{r})\Delta_L^{0,\star}(\underline{r})\ \cdot$$
$$\cdot\ \text{Im}\tau_{LL'}^{00}(\varepsilon_c+\hbar\omega)\Delta_{L'}^{0}(\underline{r}')H_{int}(\underline{r}')\phi_c(\underline{r}') \qquad \text{V-3}$$

Clearly, in the above expression only $\tau_{LL'}^{00}$ depends on the configuration. Since we energy analyze the emitted photons we know which kind of site they came from. Namely, we know whether the muffin-tin potential at $\underline{R}_0$ is of the A or the B type. This then determines Δ_L^0 and ϕ_c; they can be readily evaluated by solving numerically the Schrödinger equation for a single muffin-tin potential of the required kind. Due to the fact that $\tau_{LL'}^{00}$ describes all scattering paths which start at $\underline{R}_0$ and end at $\underline{R}_0$ it is different for each configuration. Since the measuring process does not specify these we must average over the ensemble of all configurations in which there is a specified atom at $\underline{R}_0$. Thus, the measurable spectrum depends on

$$\text{Im}<\tau_{LL'}^{00}>_{A,B}$$

where $<\ >_{A,B}$ denotes the conditional averaging described above. Within the CPA $<\tau_{LL'}^{00}>_{A,B}$ is given by

$$\tau_{LL'}^{A,B;00}$$

of Eq. II-56. Thus for a core state of the form

$$\phi_c(\underline{r}) = R_{L_c}^{\alpha}(r)Y_{L_c}(\hat{r})$$

$$I^{\alpha}_{L_c}(\omega) = -\frac{1}{\pi}\frac{2\pi\rho(\omega)}{\hbar}\sum_{L'}|M^{\alpha}_{L_c,L'}(\varepsilon_c + \hbar\omega)|^2 \ \text{Im}\ \tau^{\alpha;00}_{L'L'}(\varepsilon_c + \hbar\omega) \qquad \text{V-4}$$

where α = A or B as specified by the measuring process, and the matrix elements are given by

$$M^{\alpha}_{L_c;L'} = -\int d^3r Y_{L_c}(\hat{r}) R^{\alpha}_{L_c}(r;\varepsilon_c) H_{int}(\underline{r}) R^{\alpha}_{L'}(r;\varepsilon_c + \hbar\omega) Y_{L'}(\hat{r}) \frac{\sqrt{\varepsilon_c + \hbar\omega}}{\sin\delta_{L'}(\varepsilon_c + \hbar\omega)}$$

Note that $M^{\alpha}_{L_c,L}$ is non-zero only if the dipole selection rules are satisfied.
Hence for a 2P core state (L_c = 1,0) we get

$$I^{\alpha}_{L_3}(\omega) = \frac{4e^2\hbar\omega}{3m^2c^3}\left[|m^{\alpha}_{0,1}(\varepsilon_{2P} + \hbar\omega)|^2\ n^{\alpha}_0(\varepsilon_c + \hbar\omega)\right.$$

$$\left. + |m^{\alpha}_{2,1}(\varepsilon_{2P} + \hbar\omega)|^2\ n^{\alpha}_2(\varepsilon_c + \hbar\omega)\right] \qquad \text{V-6}$$

where ε_{2P} is the energy of the 2P core state, $n^{\alpha}_{\ell}(\varepsilon)$ is the partially averaged density of states per muffin-tin sphere at an α site

$$n^{\alpha}_{\ell}(\varepsilon) = \sum_{m=-\ell}^{\ell}\int_0^{r_{MT}} drr^2 \frac{|R^{\alpha}_{\ell}(r;\varepsilon)|^2}{\sin^2\delta^{\alpha}_{\ell}(\varepsilon)} \cdot \text{Im}\ \tau^{\alpha,00}_{\ell m,\ell m}(\varepsilon)$$

and the matrix elements are given by

$$m^{\alpha}_{0,1}(\varepsilon_{2P} + \hbar\omega) = \int_0^{r_{MT}} drr^2\ \overline{R}^{\alpha}_0(r;\varepsilon_{2P} + \hbar\omega)\left(\frac{d}{dr} + \frac{2}{r}\right)R^{\alpha}_{2P}(r)$$

$$\text{V-7}$$

$$m^{\alpha}_{2,1}(\varepsilon_{2P} + \hbar\omega) = \int_0^{r_{MT}} drr^2\ \overline{R}^{\alpha}_2(r;\varepsilon_{2P} + \hbar\omega)\left(\frac{d}{dr} - \frac{1}{r}\right)R^{\alpha}_{2P}(r)$$

In order to make these final expressions more transparent in Eq. V-7 we have used $\overline{R}^{\alpha}(r;\varepsilon)$ which is the same as $R^{\alpha}_L(r;\varepsilon)$ defined earlier but normalized to 1 within the muffin-tin sphere. We also took $R^{\alpha}_{2P}(r)$ to be so normalized.

Recall that the three 2P states are usually split by the spin orbit interaction into 2 $2P_{1/2}$ and four $2P_{3/2}$ states. This splitting is about 7 eV for Cu and 8 eV for Ni and it is more or less independent of whether we are talking about free atoms or atoms embedded in a solid. Thus, it is possible to study transitions into one of these multiplets separately. Above, we have calculated the spectra relevant to the transition from the conduction band into the $2P_{1/2}$ core states. In the usual nomenclature of SXS these are called the L_3 spectra. (Transitions to the $2P_{3/2}$ states give rise to the L_2 spectra).

As is clear from Eq. V-6 in measuring the L_3 spectra from both the Cu and Ni sites we are studying the s and d density of states at these sites. The important thing to note at this stage is that within the KKR-CPA we need not neglect the energy dependence of the matrix elements since the expression for $I^{\alpha}_{L_3}(\omega)$ in Eq. V-6 can be evaluated exactly. Once a KKR-CPA calculation has been performed we have $\tau^{C,00}_{LL'}$ and therefore $\tau^{A,00}_{LL'}$ and $\tau^{B,00}_{LL'}$ can be obtained by simple matrix algebra - See Eq. II-56 without much further effort. Moreover, using the same potentials as we have employed in calculating the phase-shifts $\delta^{\alpha}_{L}(\varepsilon)$ which served as inputs for our KKR-CPA calculations, the matrix elements $m^{\alpha}_{LL'}$ can be readily calculated. Thus Eq. V-6 is an example where the KKR-CPA leads to a complete theory for an observable without further simplifications beyond what is implicit in the one electron approximation.

The above discussion, and Eq. V-6 in particular, illustrates yet another important point : once a KKR-CPA calculation has been performed for a specific system the effective scattering amplitudes $f_{C,L}(\varepsilon)$ can be used to calculate many different physical observables at roughly the same expense as these can be calculated in pure systems using the phase-shifts. Thus, if many such observables are calculated, the relative effort spent on solving the KKR-CPA equations becomes modest indeed. Even more significantly, the above strategy allows us to subject our starting alloy potential to very severe tests as well as providing a language in terms of which we can correlate a diversity of electronic properties.

Let us now return to the discussion of the L_3 spectra of Ni-Cu alloys. The one electron theory is given by Eq. V-6. However, before making some predictions a number of remarks concerning many-body effects and various competing processes are in order.

Since we are dealing with a metal in its excited states, appeals to the density functional theory will not help. Under this circumstance the best that can be done, at this stage, is to rely on simplified theories and experience with interpreting other spectra. Obviously, there is no room here to go into details [see "Band Structure Spectroscopy", ed. Fabian and Watson (1972)] . We merely wish to note that there is little evidence for the celebrated edge singularity [Mahan (1974)] in the soft X-ray spectra of metals containing transition and noble metals. However, there are various broadening processes and these we attempt to take into account by folding in our one-electron intensities with a Lorentzian broadening function with a width given by

$$\Gamma(\varepsilon) = G + W\left(\frac{\varepsilon - \varepsilon_F}{\varepsilon_0 - \varepsilon_F}\right)^2$$

Here G is the core level width which can be estimated from the X-ray photoemission spectra (G = .65 eV for Cu and .7 eV for the Ni $2P_{3/2}$ states). The second term is due to the Auger decay of the conduction electron holes which have been left behind after the radiative transition [Blokhin and Sachenko (1960)] . We take W to be 1.0 eV for both the Cu and Ni spectra. Finally, in all the calculations of $I^{\alpha}_{L_3}(\omega)$ to be discussed we smeared the final theoretical result by a Lorentzian function of width .2 eV to simulate instrumental resolution.

In Fig. 31 we show the calculated local densities of states (d-components) for the 13% Ni and 81% Ni alloys. The s-components are very much smaller and broader, and the emission spectra can be expected to be dominated by the d-components.

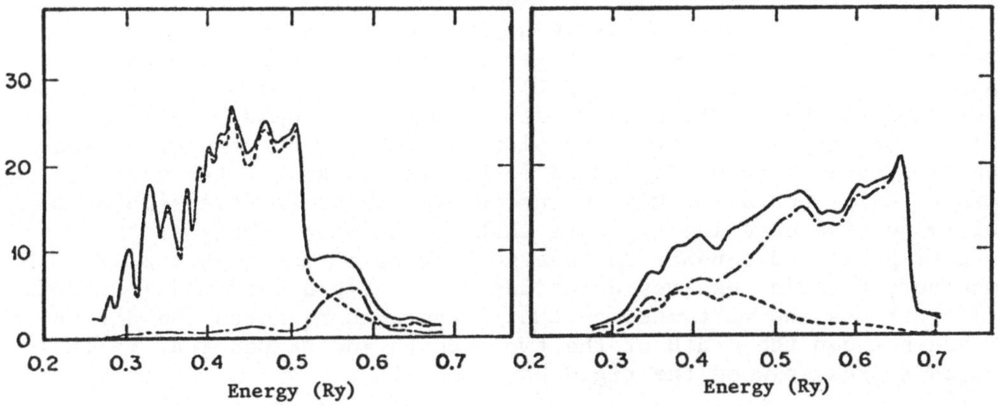

Fig. 31 : ℓ = 2 compounds of the Cu (full line) and Ni (dashed line) local densities of states for the 13% Ni (a) and 81% Ni (b) alloys.

For the Cu-rich alloy the local Cu density of states resembles that of pure Cu, while the Ni atoms introduce a rather narrow impurity state between the Cu d-bands and the Fermi level. As the Ni content increases the structure in the Cu density of states becomes blurred and the Ni states broaden out considerably, until, at 81% Ni, the local density of states on Ni sites is similar to that of pure Ni. Note that, even at the lowest Cu concentrations, the Cu states lie within the energy range spanned by the Ni d-states - a Cu impurity state is not formed, and the width of the Cu d-states hardly changes with concentration. The centers of the d-bands on Cu and Ni sites are separated by about 2 eV for all concentrations, and the gross deviations from the rigid band model are obvious. We find that the matrix elements in Eq. V-7 are smoothly and slowly varying functions of energy (varying at most 20% over the range of the d-bands). The X-ray emission spectra should therefore reflect the positions and widths of the local densities of states fairly closely, nearly all of the fine structure being smeared out by the lifetime broadening.

The calculated L_3 spectra are shown in Fig. 32, with the energy plotted relative to the Fermi level ε_F. Evidently the two spectra are well separated in energy for each concentration. This is most unlike the prediction of the rigid-band model according to which there is a common conduction band an hence the SXS spectra when measured from the Fermi energy should coincide.

In this light it is surprising that the early measurements of the M_3 spectra in Ni-Cu alloys (should be roughly the same as L_3) failed to provide as conclusive a proof of the non-rigid band behavior as was obtained a few years later by the photoemission studies of Seib and Spicer (1966). However, the reason for this is not hard to see. In Figs. 33 we show some of the data by Clift et al (1963). Since each spectrum is plotted with respect to the corresponding core state energy (this is how it appears in the measurement) their separation may be interpreted as due to the separation of respective core levels (note that the Cu peak is higher than the Ni peak) and appears to have no bearing on the question of whether the alloy behaves according to the rigid band model or not. (In fairness one must mention that these authors noted the difference between the width of the two spectra and argued that it indicated a breakdown of the rigid band model).

In a more recent experiment Durham et al (1977) measured the separation of the Ni $2P_{3/2}$ and Cu $2P_{3/2}$ core levels on the same sample on which the SXS was taken by X-ray photoemission. Consequently, they were able to shift the two experimental SXS peaks so that they had a common origin (by careful calibration of the two different experiments they were able to make this origin to be ε_F). Their results are shown in Fig. 34.

The agreement between the theory and the experiment is highly satisfactory. In the light of the fact that the calculation does not contain any adjustable parameters the very good quantitative

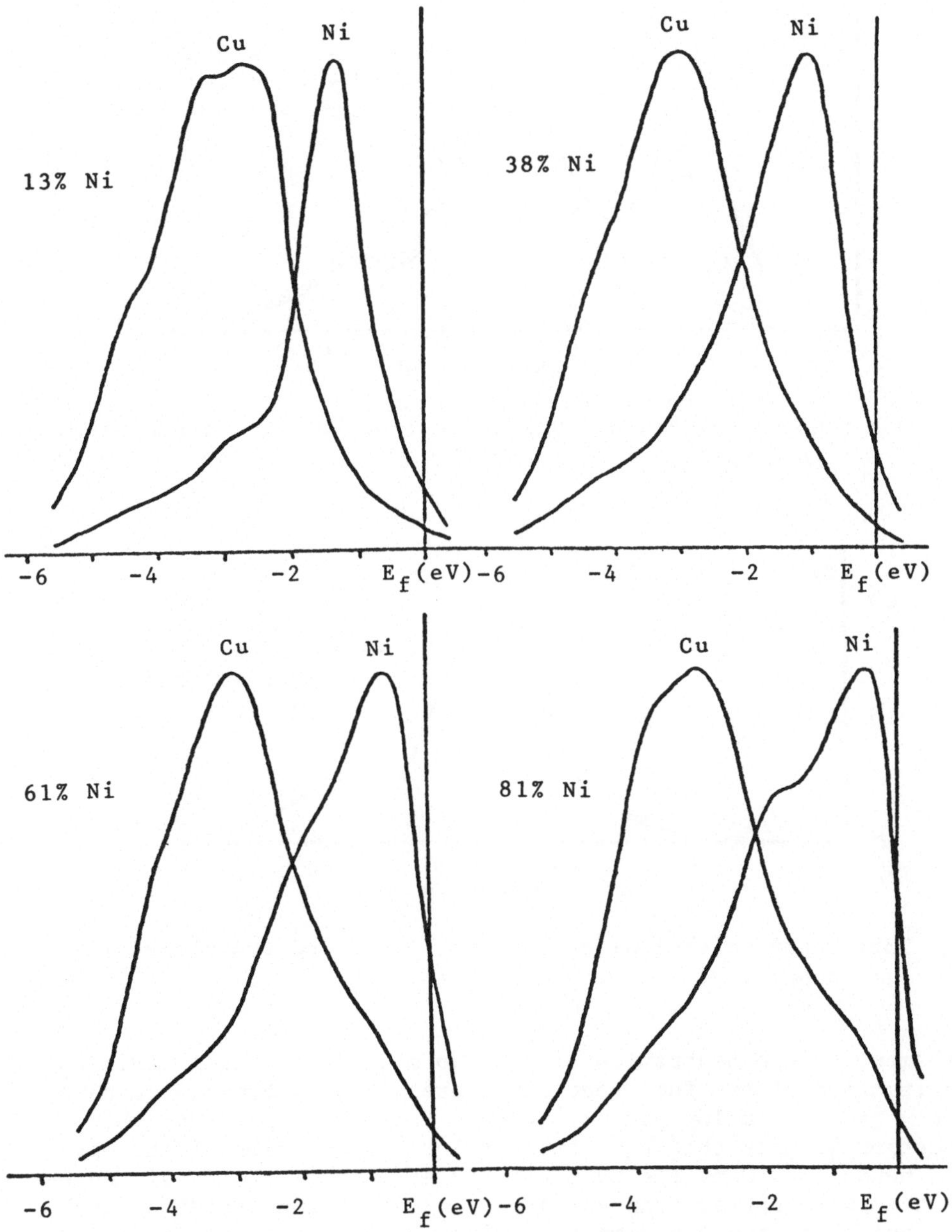

Fig. 32 : Calculated L_3 spectra (both Cu and Ni spectra are normalized to the same value).

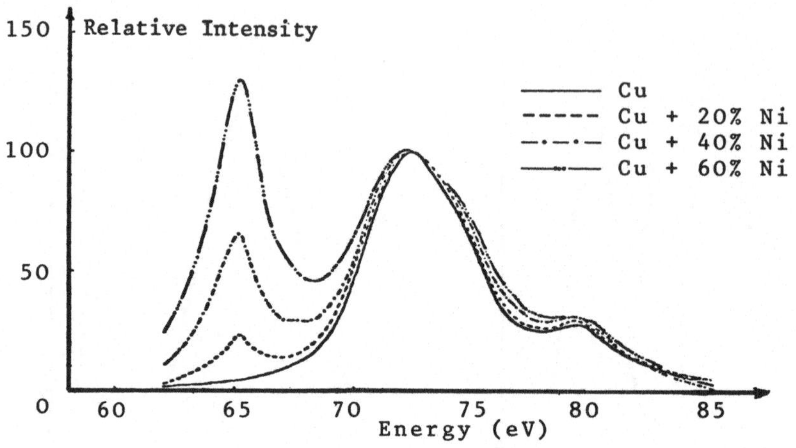

Fig. 33a : Copper-nickel alloy spectra : 0, 20, 40 and 60% nickel in copper.

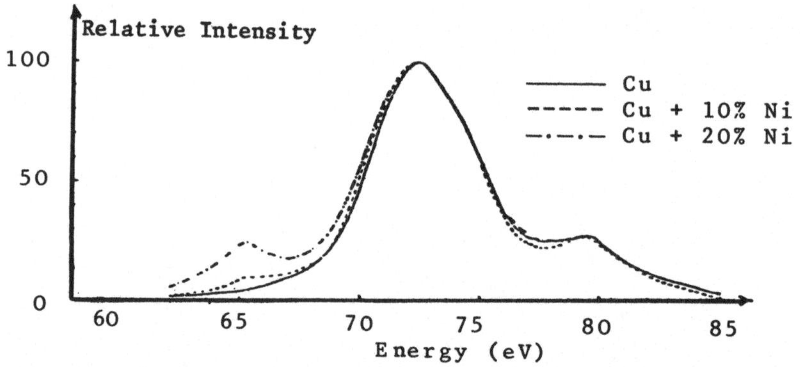

Fig. 33b: Copper-nickel alloy spectra : 0,10 and 20% nickel in copper.

agreement on the separation of the two major peaks is a striking confirmation of the fact that for these alloy systems we do have a very reliable alloy potential. As a further support for this assertion we note that even the absolute frequencies of the emission bands (∿ 930 eV for Cu L_3, ∿ 852 eV for Ni) are predicted by the theory to better than 4%. A more detailed discussion of both the theory and the experiment is given by Durham et al (1978). Here, we only wish to make one last observation.

As we have noted before, the most significant feature of our starting Cu and Ni site potential wells are the d-resonances evi-

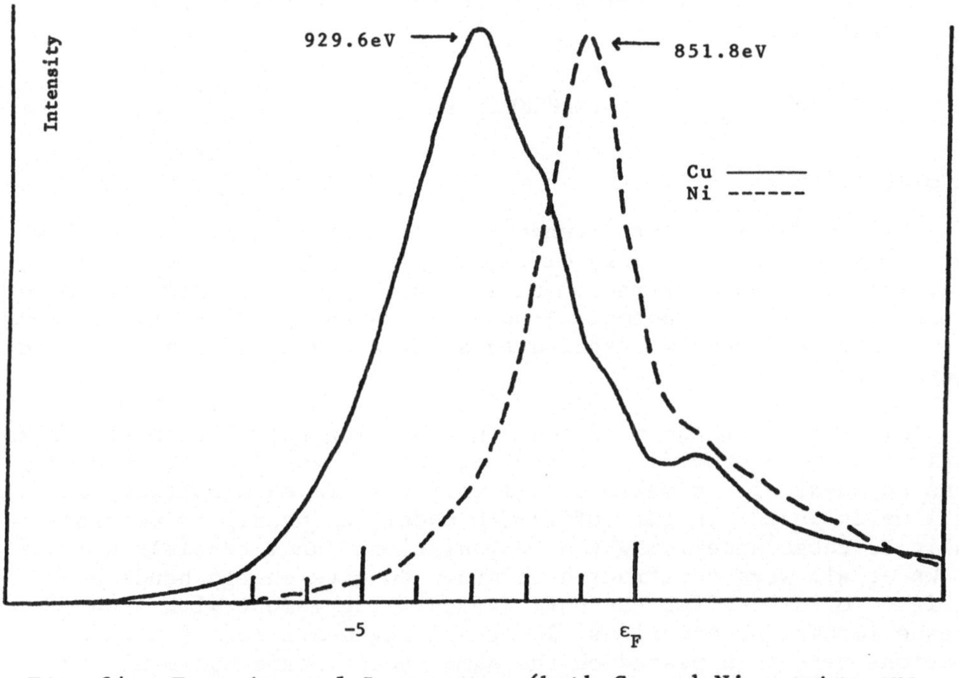

Fig. 34 : Experimental L_3 spectra (both Cu and Ni spectra are normalized to the same value).

dent in the energy variation of $\delta_2^{Cu}(\varepsilon)$ and $\delta_2^{Ni}(\varepsilon)$. From all the calculations we have been discussing it is clear that the relative position of these two resonances are the principle factors in determining the relative positions of the Ni and Cu related features. In particular it is $\varepsilon_r^{Cu} - \varepsilon_r^{Ni}$ which determines the separation of the Cu and Ni L_3 spectra discussed above. Thus, in an SXS experiment one obtains a quick and accurate check on this most important parameter. Moreover, early efforts at constructing adequate alloy potentials indicate that the relative position of the two resonances is most sensitive to charge transfer. Thus we expect that SXS will play an important role in the development of our theory of the alloy potentials for alloys where charge transfer is a significant factor.

APPENDIX 1

The Alloy Potentials

This appendix is not intended as a general description of how one should approach the very difficult problem of constructing potential functions for random alloys. Rather it is a description of how the particular potentials underlying the present calculations were constructed and is intended to allow a reader to reproduce the calculation.

The success of the modified tight-binding calculation of Stocks et al. (1972) in providing an explanation of experimental observations on Cu-Ni alloys makes it a legitimate aim to construct potentials which would, in the muffin-tin model, as nearly as possible duplicate those underlying that calculation. The potentials used by Stocks et al. were constructed in order to give energy bands and densities of states for the pure metals in agreement with a number of experimental observations. The resulting densities of states functions were then placed on the same energy scale by requiring that the energies of the bottom of the conduction bands of the pure constituents be the same. This fixed the d-band splitting parameter (δ in the notation of Stocks et al.) required in that calculation. For further details of the construction of the potentials for the pure metals the reader is referred to the original work.

In the present calculation the referencing of the pure constant potentials relative to one another is accomplished by adding constants to the pure Cu and Ni muffin-tin potentials. We choose a modified muffin-tin zero for each constituent V'^{α}_{MTZ} (α = Cu or Ni) which is coincident with the energy of the bottom of its conduction band i.e., $V'^{\alpha}_{MTZ} = V^{\alpha}_{MTZ} + E^{\alpha}(\Gamma_1)$ where V^{Cu}_{MTZ}, V^{Ni}_{MTZ}, $E^{Cu}(\Gamma_1)$ and $E^{Ni}(\Gamma_1)$ are the muffin-tin zeros and energies of the bottom of the conduction bands relative to the muffin-tin zero for Cu and

Ni of Stocks et al. δ = .134 Ry and was independent of concentration. At high Ni concentration the results of CPA calculations are not sensitive to such small differences in δ. Therefore we would expect the main structures in the densities of states calculated in this work to be comparable with those of Stocks et al. (see Fig. 18). However, as was first pointed out by Stocks et al. (1972), the extent to which the Ni impurity state is split off from the Cu host d-band in very Cu rich alloys is directly related to the magnitude of δ. Thus the fact that the Ni impurity band of the ASA-CPA calculation of Fig. 18 is not split off as much as the corresponding impurity band obtained by Stocks et al. is a reflection of the fact that δ = .108 Ry in the ASA-CPA calculation whereas δ = .134 Ry in the calculation of Stocks et al.

A summary of the data relevant in determining the present Cu and Ni potential functions is given in Table 1.

Table 1 : Summary of Data Used in Construction of Alloy Potentials.

	Lattice Spacing a_0	V_{MTZ}	$E(\Gamma_1)$	V'_{MTZ}
Cu	6.8032	− 1.1613	−.0289	− 1.1902
Ni	6.6321	− 1.1937	.0564	− 1.1373

EPILOGUE

As we have stressed before, having succeeded in implementing the KKR-CPA program is merely the first step towards a first principles theory of random metallic alloys. However, we feel that having removed the major obstacle of having to deal with a Schrödinger equation with a random potential in a realistic fashion, progress will be rapid. Clearly there are many avenues to explore.

(a) In the first place the above calculation will have to be made to some extent self-consistent in the sense of the density functional theory. We note that this cannot be done with the same generality as in the case of pure metals because the KKR-CPA only provides us with a conditionally averaged charge density $\rho_A(r)$ and $\rho_B(r)$ on the A and B sites. In principle one should at first achieve self-consistency with respect to the charge distribution and then average over all configurations. This appears to be a hopeless task. However, it might be sufficient to use $\rho_A(r)$ and $\rho_B(r)$ to construct a set of new "starting" potentials $v_A(r)$ and $v_B(r)$ and proceed to solve the KKR-CPA equations again repeating this procedure until convergence. In our opinion some such schemes will be necessary if we are to develop a reliable method for obtaining alloy potentials.

(b) A spin polarized version of the theory will have to be developed if we are to treat the fascinating variety of magnetic problems which were discussed in Dr. Edwards's lectures.

(c) Once the charge transfer problem is more or less understood in terms of such self-consistent calculation as we have mentioned, the temptation to perform total energy calculations will be great. Here the aim would be to answer the questions raised in Dr. Evans's lectures regarding the stability of alloy phases. It is not hard to guess that the relevance of these questions for alloys of transition and noble metals is no less than for those containing only simple metals.

(d) Recently, band theoretical methods have been used with considerable success for studying the electron phonon interaction in the rigid-muffin-tin approximation [Györffy (1976)] (as you might imagine at the end of this course, this approximation consists of assuming that when an atom moves, as a consequence of a lattice vibration being excited, the corresponding muffin-tin potential well moves with it without distortion). In particular, considerable progress has been made towards understanding the attractive part of the interaction between metallic electrons due to the exchange of phonons. This work led to a good quantitative understanding of the variation in the superconducting transition temperature T_c among pure metals [Butler (1977), Papaconstantopoulos et al (1977)] as well as among many interesting intermetallic compounds. [Klein et al (1974), Papaconstantopoulos et al (1975)]. Moreover, the Eliashberg equation [Fetter and Walecka (1971)], which can be regarded as the fundamental "Schrödinger equation" for the Cooper pairs has been considered for random alloys within the CPA [Lustfeld (1974), Kerker and Benneman (1974), Appel (1974)]. Thus, to study superconductors which are random alloys, using the KKR-CPA is a very real and attractive possibility. (Work along this line is in progress by Giuliano, Ruggeri and Györffy)

(e) The above theory of the electron phonon interaction can also be used to develop a first principles theory of lattice dynamics for transition metals [Pickett and Györffy (1976) ; for background, see Allen (1978)]. Though perhaps premature at this stage, [Sacchetti (1978)] the prospect of generalizing this work within the language of the KKR-CPA theory should not be viewed with misguided pessimism. This remark is particularly relevant to the line width calculations of Butler et al (1977).

This list could be continued but it is clearly time to stop. Hopefully the items included so far convey the scope of the possibilities opened up by the KKR-CPA.

ACKNOWLEDGEMENTS

We would like to thank our colleagues W. Temmerman, P. Durham and B. Gordon at the University of Bristol, J.S. Faulkner at the Oak Ridge National Laboratory, U.S.A., S. Giuliano and R. Ruggeri of the University of Messina, Italy, and A. Pindor, Warszawa, Poland, for many fruitful discussions and collaborations. One of us (B.L.G.) would like to thank the Brookhaven National Laboratory and the Department of Physics of the State University of New York at Stony Brook for the most gracious hospitality during the tenure of which this manuscript was written. We would also like to thank Dr. L. Scheire, Mr. R. Rotthier and Mr. K. Verhuyck of the University of Gent, Belgium for the painstacking editorial assistance in preparing the somewhat garbled lecture notes for publication.

REFERENCES

Allen P.B. 1978 "Dynamical Properties of Solids" Vol.3, Ed. G.K. Horten, A.A. Maradudin (North Holland).

Andersen O.K. 1973 Solid State Commun. 13, 133.

Andersen O.K. 1975 Phys. Rev. B12, 3060; also Jepsen O., Andersen O.K., Mackintosh A.R., Phys. Rev. B12, 3084.

Anderson P.W. 1958 Phys. Rev. 109, 1492.

Appel J. 1974 Solid State Commun. 15, 1043.

Bansil A. 1975 Solid State Commun. 16, 885.

Bansil A. 1978 Bull. Am. Phys. Soc. 23, 322 ; see also Bansil A., Ehrenreich H., Schwartz L. and Watson R.E. 1974 Phys. Rev. B9, 445.

Bansil A., Schwartz L. and Ehrenreich H. 1975 Phys. Rev. B12, 2893

Beeby J. 1967 Proc. Roy. Soc. (London) A302, 113.

Beeby J. 1964 Phys. Rev. 135A, 130.

Blokhin M.A. and Sachenko V.P. 1960 Izv. Akad. Nauk. 24, 397.

Bruggeman D.A.G. 1935 Ann. Phys. (Leipzig) 24, 665 and 25, 645.

Butler W.H. 1977 Phys. Rev. B15, 5267.

Butler W.H., Smith H.G. and Wakabayashi N. 1977 Phys. Rev. Lett. 39, 1004.

Callaway J. 1974 "Quantum Theory of Solids" (Academic Press).

Clift J., Curry C. and Thompson B.J. 1963 Phil. Mag. 8, 593.

Cracknell A.P. 1971 "The Fermi Surface of Metals" (Taylor and Francis).

Durham P., Györffy B.L., Hague C.F., Stocks G.M. and Temmerman W.M. 1977 "Transition Metals 1977" Ed. M.J.G. Lee, J.M. Perz and E. Fawcett (Institute of Physics Conference Series No. 39, 1978).

Ehrenreich H. and Schwartz L. 1976 "Solid State Physics" Ed. H. Ehrenreich, F. Seitz and D. Turnbull (Academic Press).

Elliott R.J., Krumhansl J.A. and Leath P.L. 1974 Rev. Mod. Phys. 46, 465.

Fabian D.J. (Ed.) 1968 "Soft X-ray Band Spectra" (Academic Press).

Fabian D.J. and Watson L.M. (Ed.) 1973 "Band Structure Spectroscopy of Metals and Alloys" (Academic Press).

Faulkner J.S. 1977 J. Phys. C. : Solid State Phys. 10, 4661.

Faulkner J.S., Davis H.L. and Joy W.H. 1967 Phys. Rev. 161, 656.

Fehlner W.R. and Vosko S.H. 1976 Can. J. Phys. 54, 2159.

Fetter A.L. and Walecka J.D. 1971 "Quantum Theory of Many-Particle Systems" (McGraw-Hill).

Friedel J. 1956 Can. J. Phys. 34, 1190.

Giuliano E.S., Ruggeri R., Györffy B.L. and Stocks G.M. 1977 "Transition Metals 1977" Ed. M.J.G. Lee, J.M. Perz and E. Fawcett (Institute of Physics Conference Series No. 39, 1978).

Gordon B., Temmerman W.M., Györffy B.L. and Stocks G.M. 1977 "Transition Metals 1977" Ed. M.J.G. Lee, J.M. Perz and E. Fawcett (Institute of Physics Conference Series No. 39, 1978).

Gunnarsson O. 1978, this volume.

Gunnarsson O. and Lundqvist B.I.1976,Phys. Rev. B13, 4274.

Györffy B.L. 1972 Phys. Rev. B5, 2382.

Györffy B.L., Jordan R., Lloyd D.R., Quinn C.M., Richardson N.V., Stocks G.M., Temmerman W.M. 1977 Solid State Commun. 23, 637.

Györffy B.L. and Stocks G.M. 1974 J. de Phys. 35, C4-75.

Györffy B.L. and Stocks G.M. 1976 "Electrons in Finite and Infinite Structures" Ed. P. Phariseau and L. Scheire (Plenum Press, NATO ASI Series Physics B24, 1977).

Györffy B.L. and Stott M.J. 1972 "Band Structure Spectroscopy of Metals and Alloys" Ed. D.J. Fabian and L.M. Watson (Academic Press).

Hasegawa M., Suzuki T. and Hirabayashi M. 1974 J. Phys. Soc. Japan 37, 85.

Heimann P., Neddermeyer H. and Pessa M. 1978 Phys. Rev. B17, 427.

Heine V. 1967 Phys. Rev. 153, 673.

Herring C. 1960 J. Appl. Phys. 31, 1939.

Hubbard J. 1967 Proc. Phys. Soc. 92, 921.

Hüfner S., Wertheim G.K. and Wernick J.H. 1973 Phys. Rev. B8, 4511.

Johnson K.H. 1966, J. Chem. Phys. 45, 3085.

Kanamori J. , Akai H., Hamada H., Miwa H. 1977 Physica 91B, 153.

Kerker G., Bennemann K.H. 1974 Solid State Commun. 14, 399 and 15, 29.

Kirkpatrick S. 1973 Rev. Mod. Phys. 45, 574.

Kittel C. 1976 "Introduction to Solid State Physics" 5th Ed. (John Wiley & Sons).

Klein B.M. and Papaconstantopoulos D.A. 1974 Phys. Rev. Lett. 32, 1193.

Kohn W., Rostoker N. 1954 Phys. Rev. 94, 1111.

Kohn W. and Sham L.J. 1965 Phys. Rev. 140A, 1133.

Korringa J. 1947 Physica 13, 392.

Korringa J. 1958 J. Phys. Chem. Solids 7, 252.

Landau L.D. and Lifshitz E.M. 1960 "Electrodynamics of Continuous Media" (Addison-Wesley).

Lax M. 1973 "Stochastic Differential Equations" SIAM-AMS Proceedings Vol. VI, 35.

Lloyd P. and Oglesby J. 1976 J. Phys. C. : Solid State Physics 9, 4393.

Lloyd P. and Smith P.V. 1972 Adv. Phys. 21, 69. This is a particularly useful reference from the point of view of these lectures since we are using all the same conventions with regard to spherical harmonics and Bessel and Neumann functions as these authors.

Loucks T.L. 1967 "The Augmented Plane Wave Method" (Benjamin) chapter 3.

Lustfeld H. 1974 Solid State Commun. 15, 301; 1975 Solid State Commun. 17, 437 and 1974 Z. Physik 271, 229.

Mahan G.D. 1974 "Solid State Physics" Ed. H. Ehrenreich, F. Seitz D. Turnbull (Academic Press) 29, 75.

Maxwell J.C. 1873 "Treatise on Electricity and Magnetism" (Dover Publications) reprint 3rd ed. (1954).

Medina R.A. and Cable J.W. 1977 Phys. Rev. B15, 1539.

Mittag-Leffler 1884 Acta. Math. A more useful reference is Barut A.O., 1967 "Theory of the Scattering Matrix"(McMillan Co.), Appendix 6.

Morruzzi V.L., Williams A.R. and Janak J.F. 1977 Phys. Rev. B15, 2854.

Mott N.F. 1935 Proc. Phys. Soc. 47,571; 1936 Proc. Roy. Soc. A153, 699 and 1936 Proc. Roy. Soc. A156, 368.

Murray B.W. and McGrevey J.D. 1970 Phys. Rev. Lett. 24, 9.

Muto T. 1938 Sci. Papers Inst. Phys. Chem. Res. (Tokyo) 34, 337.

Nickel, B.G. and Butler W.H. 1973 Phys. Rev. Lett. 30, 373.

Nordheim J.C. 1931 Ann. Phys. 9,607 and 641.

Onodera Y. and Toyozawa Y. 1968 J. Phys. Soc. Japan 24, 341.

O'Sullivan W.J. and Schirber J.E. 1969 Phys. Rev. 181, 1367.

Papaconstantopoulos D.A. and Klein B.M. 1975 Phys. Rev. Lett. 35, 110.

Papaconstantopoulos D.A., Boyer L.L., Klein B.M., Williams A.R., Moruzzi V.L., Janak J.F. 1977 Phys. Rev. B15, 4221.

Pettifor D.G. 1969 J. Phys.C. : Solid State Physics 2, 1051.

Pickett W.E., Györffy B.L. 1976 ("Superconductivity in d- and f-Band Metals"), Ed. D.H. Douglas (Plenum Press).

Poulsen R.G., Randles D.L., Springford M. 1974, J. Phys. F. : Metal Physics 4, 981.

Sacchetti F. 1978 J. Phys. F. : Metal Physics 8, 743.

Segall B. and Ham F.S. 1968 "Methods in Computational Physics" (Academic Press) Vol. 8,chapter 7.

Seib D.M. and Spicer W.E. 1970 Phys. Rev. B2, 1676.

Shevchik N.J. and Penchina C.M. 1975 Phys. Stat. Sol.(b) 70, 619.

Shiba H. 1971 Progr. Theoret. Phys. (Kyoto) 16, 77.

Shimizu M., Takahashi T. and Katsuki A. 1963 J. Phys. Soc. Japan 18, 1192.

Skinner H.W.B. 1940 Phil. Trans. A239, 95.

Snow E.C., Waber J.T. and Switendick A.C. 1966 J. Appl. Phys. 37, 1342.

Sommers C.B., Amar H. and Johnson K.H. 1966 Bull. Am. Phys. Soc. 11, 73.

Soven P. 1966 Phys. Rev. 151, 539.

Soven P. 1967 Phys. Rev. 156, 809 and 1969 Phys. Rev. 178, 1136.

Soven P. 1970 Phys. Rev. B2, 4715.

Stephen M. 1978 Phys. Rev. B17, 4444.

Stocks G.M., Györffy B.L., Giuliano E.S. and Ruggeri R. 1977 J. Phys. F. : Metal Physics 7, 1859.

Stocks G.M., Temmerman W.M. and Györffy B.L. 1978 Phys. Rev. Lett. 41, 339.

Stocks G.M., Williams R.W. and Faulkner J.S. 1971 Phys. Rev. B4, 4390.

Taylor D.W. 1968 Phys. Rev. 156, 1017.

Temmerman W.M., Györffy B.L. and Stocks G.M. 1978 J. Phys. F. : Metal Physics, to be published.

Templeton I.M. and Coleridge P.T. 1975 J. Phys. F. : Metal Physics 5, 1307 and 1317.

Thouless D.J. 1974 Physics Reports 13, 94.

Tomboulian D.H. 1957 Handb. Physik 30, 246.

Tong B.Y. 1972 Phys. Rev. B6, 1189.

Wang C.S., Callaway J. 1977 Phys. Rev. B15, 298.

Wegner F. 1976 Z. Physik B25, 327.

Yu K.Y.,Helmes C.R., Spicer W.E. and Chye P.W. 1977 Phys. Rev. B15, 1629.

Ziman J.M. 1971 "Solid State Physics" Ed. H. Ehrenreich, F. Seitz and D. Turnbull (Academic Press)

ASPECTS OF THE NUMERICAL SOLUTION OF THE KKR-CPA EQUATIONS

G.M. Stocks

Metals and Ceramics Division
Oak Ridge National Laboratory
Oak Ridge, Tenn. 37830, U.S.A.

W.M. Temmerman and B.L. Györffy

H.H. Wills Physics Laboratory
University of Bristol, Tyndall Ave.
Bristol, U.K.

ABSTRACT

In this lecture, we review some of the difficulties encountered in the solution of the KKR-CPA equations and the numerical techniques developed to overcome them.

I. INTRODUCTION

In our other lectures presented at this summer school,[1] we have reviewed the progress which has been made during the last few years in discussing electronic states in random substitution alloys (solid solutions) using the Coherent Potential Approximation (CPA) applied to the non-overlapping muffin-tin model of the crystal potential. We hope that these lectures have given you a feel not only for the immense richness of the effects of alloying on the electronic structure of metals, but also for the detailed contact which can now be made between ab initio calculations and the results of experiment, since it is through the interplay between such calculations and experiment that we will ultimately obtain a more basic understanding of the nature of alloy formation.

However, anyone paying close attention will have noticed that although the muffin-tin or KKR-CPA was formulated around 1970,[2,3] it is only recently that the equations have been solved in their full generality.[4] In this lecture, we would like to give you some feel for why this was so and what developments have led to their solution. The lecture is then of necessity fairly technical. However, we hope that we will be able to convey to you the kinds of developments which turn an intractible looking method into one of routine. Obviously, if the KKR-CPA is to be of any real value the numerical alogrithms involved must be highly efficient, (a) in order to allow calculations on many different systems and (b) ultimately, to allow iteration of the calculation to self-consistency within say the local (spin) density approximation.

We will organize the lecture as follows. In Section II, we set out the KKR-CPA and briefly review the band theory program for disordered alloys. In Section III, we discuss the nature of the difficulties encountered in the solution of the KKR-CPA equations and how these are overcome. In Section IV, we discuss calculation of the densities of states and show some illustrative results. Section V comprises some concluding remarks.

II. KKR-CPA BAND THEORY PROGRAM FOR DISORDERED ALLOYS

Since elsewhere in these summer school notes, we have set out the KKR-CPA in great detail, here we only quote the final equations. The KKR-CPA reduces, for a binary AB alloy, to solving a self-consistency condition for an effective scattering amplitude f_C of the form

$$f_C^{-1}(\varepsilon) = cf_A^{-1}(\varepsilon) + (1-c)f_B^{-1}(\varepsilon) - \sqrt{\varepsilon}[f_A^{-1}(\varepsilon) - f_B^{-1}(\varepsilon)]$$

$$\tau^C(\varepsilon)[f_B^{-1}(\varepsilon) - f_C^{-1}(\varepsilon)] \; , \tag{II.1}$$

with

$$\tau^C(\varepsilon) = -\frac{1}{\Omega_{BZ}} \int_{BZ} d^3q[\sqrt{\varepsilon}f_C^{-1}(\varepsilon) + G(\vec{q},\varepsilon)]^{-1} \; , \tag{II.2}$$

where, the matrix elements of the various matrices $f_{A(B)(C)}$, τ^C and G are given by $f_{A(B),\ell}(\varepsilon) = [\exp(2i\eta_\ell^{A(B)});- 1]/(2i)$, $f_{C,L}(\varepsilon)$; $G_{LL'}(\vec{q},\varepsilon)$ the KKR structure constants; and $\tau_{LL'}(\varepsilon)$ the on-the-energy-shell matrix elements of scattering path operator.[5] We have adopted the shorthand notation $L \equiv (\ell,m)$. Thus, given as inputs, the crystal structure and the two muffin-tin potential

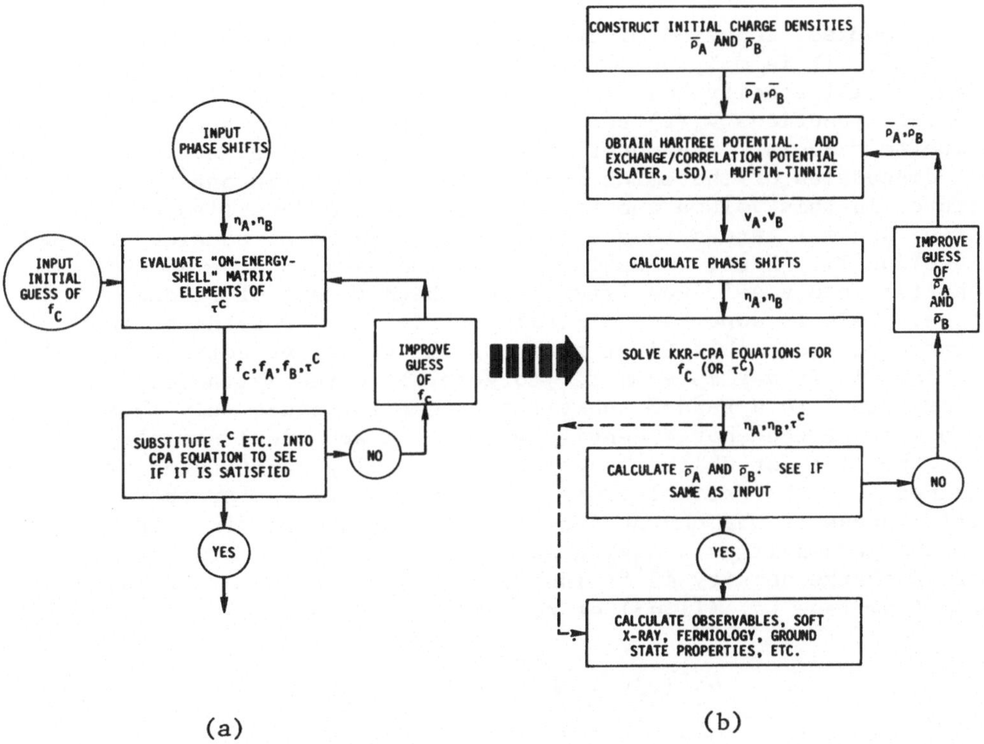

Fig. 1 : Schematic of the important elements of the KKR-CPA band theory program for random substitutional alloys. a) Steps involved in the numerical solution of the KKR-CPA equations. b) Steps involved in obtaining charge self-consistency.

functions $v_A(r)$ and $v_B(r)$, and hence, the corresponding phase shifts η_ℓ^A and η_ℓ^B, associated with the A and B sites in the alloy, we can proceed to solve the KKR-CPA equations. The iterative solution of Eqs. (II.1) and (II.2) proceeds from an initial guess of f_C through evaluation of the Brillouin zone integral of Eq. (II.2) to substitution of f_A, f_B, f_C and τ^C into Eq. (II.1) to see if the guessed f_C is indeed a solution. If it is not, an improved guess is made and the process repeated until a solution is found. This process is illustrated schematically in Fig. 1a.

Of course, solution of the KKR-CPA equation is not an end in itself, it is only one step in the overall band theory scheme. We must still specify how we are to construct the muffin-tin potential functions $v_A(r)$ and $v_B(r)$ and we must also specify how we are to calculate the observables of interest. We have given some discussion of the calculation of observables in our other lectures in this volume and for the case of the densities of states, we will return to some of the computational considerations in Section IV. Here, we would like to indicate how the KKR-CPA might fit into a self-consistent field band theory of random alloys. This is done schematically in Fig. 1b. For the purpose of making a reasonable first guess of the v_A and v_B potential functions it is a simple matter to generalize the Mattheiss prescription[6] to a random substitutional alloy since only those components of the charge density which have spherical symmetry about the site for which the potential is being constructed are retained. In the Mattheiss prescription, one takes for the initial guess of the charge density a superposition of neutral atom charge densities. Having solved the KKR-CPA equations the average charge density $\bar{\rho}^{A(B)}$ inside each muffin-tin may be calculated from [see Eq. (II.48) of ref. 1].

$$\bar{\rho}^{A(B)}(\vec{r}) = \int_0^{\varepsilon_F} \bar{\rho}^{A(B)}(\vec{r},\varepsilon)d\varepsilon \ , \qquad \text{(II.3)}$$

where

$$\rho^{A(B)}(\vec{r},\varepsilon) = \varepsilon \sum_L \frac{\mathrm{Im}\tau_{LL}^{A(B)}(\varepsilon)}{\mathrm{Sin}^2\eta_\ell} [R_\ell^{A(B)}(\vec{r},\varepsilon)Y_{\ell m}(\hat{r})]^2 \ , \qquad \text{(II.4)}$$

where $R_\ell^{A(B)}(r,\varepsilon)$ is the solution of the radial Schrodinger equation for a single A(B) muffin-tin well with the normalization $\mathrm{Cos}\eta_\ell^{A(B)}(\varepsilon) j_\ell(\sqrt{\varepsilon}\, r) - \mathrm{Sin}\eta_\ell^{A(B)}(\varepsilon) n_\ell(\sqrt{\varepsilon}\, r)$ at the muffin-tin radius r_{MT}. Knowledge of $\bar{\rho}^{A(B)}(r)$ and the total charge allows, at least in principle, iteration of the charge density to self-consistency. Clearly, solution of the KKR-CPA equations and calculation of the average charge densities is the equivalent step in our alloy theory to calculation of eigenvalues and eigenfunctions, and hence, the charge density by say the APW or KKR method in the self-consistent band theory of ordered systems.

To date, all the calculations which have been performed using the KKR-CPA have been performed without charge self consistency. The potential functions $v^{A(B)}$ were constructed in some heuristic manner, the KKR-CPA equations solved and from these solutions the relevant observables calculated as indicated by the dashed line in Fig. 1b.

Before going on to discuss the numerical aspects of the solution of the KKR-CPA equations, it is helpful if we take some advantage of the fact that the effective scattering amplitude will have the point group symmetry of the underlying lattice. Since by so doing, we can greatly simplify the structure of Eqs. (II.1) and (II.2). If in Eq. (II.1) and (II.2) we retain only terms having $\ell < 3$, as is appropriate to transition metals since the phase shifts η_ℓ for $\ell \geq 3$ are generally small, the matrices involved in these equations are 9×9. However, for fcc and bcc crystals, both f_C and τ^C are diagonal in the basis of the real spherical harmonics.[7] As a result, we can reduce Eqs. (II.1) and (II.2) to four coupled equations which have to be solved for the four effective scattering amplitudes $f_{C,\alpha}$, (or alternatively $\tau^{C,\alpha}$) corresponding to the $\alpha = a_{1g}$ $(\equiv\Gamma_1)$, $t_{1u}(\equiv\Gamma_{15})$, $e_g(\equiv\Gamma_{12})$ and $t_{2g}(\equiv\Gamma_{25'})$ irreducible representations of the cubic group. Furthermore, the calculation of the $\tau^{C,\alpha}$ from Eq. (II.2) can be greatly simplified[8] by noting that $\tau^{C,\alpha}$ can be evaluated in terms of integrals over the irreducible 1/48 of the Brillouin zone rather than over the full Brillouin zone. Specifically, $\tau^{C,a_{1g}} = 48\,\tau'_{0000}$; $\tau^{C,t_{1u}} = 16\,(\tau'_{1-11-1} + \tau'_{1010} + \tau'_{1111})$; $\tau^{C,t_{2g}} = 16\,(\tau'_{2-22-2} + \tau'_{2-12-1} + \tau'_{2121})$ and $\tau^{C,e_g} = 24(\tau'_{2020} + \tau'_{2222})$ where the $\tau'_{\ell m \ell' m'}$ are given by an expression like Eq. (II.2) except that the region of integration is confined to the irreducible 1/48 of the Brillouin zone.

III. NUMERICAL SOLUTION OF THE KKR-CPA EQUATIONS

Having sketched out the principle steps involved in the full KKR-CPA band theory program, let us now focus on how those steps indicated in Fig. 1a are actually implemented.

The KKR-CPA equations have so far been solved in three different ways, two of them approximate and one exact. As we shall emphasize momentarily, the crux of the problem lies in the calculation of the on-the-energy-shell matrix elements of the scattering path operator,[5] i.e., in the evaluation of the integral in Eq. II.2 for the values of the $f_{C,L}$'s appropriate to the current iteration. The three different methods of solving the KKR-CPA equations thus correspond to three different ways of evaluating the $\tau^{C}_{LL'}$'s.

Briefly stated these three methods are:

(1) Cluster approximation (C-CPA);[9] the τ^C occurring in Eq. (II.1) is approximated by its value at the central site of a finite cluster of effective scatterers.

(2) Atomic Sphere Approximation (ASA-CPA);[10] the limit $\varepsilon \to 0$ is taken on the CPA equations Eqs. (II.1) and (II.2) and the muffin-tin radius is replaced by the Wigner-Seitz radius everywhere it occurs. This parallels the suggestion made by Anderson[11] for simplifying the KKR method, and was called by him the Atomic Sphere Approximation.

(3) Direct evaluation of $\tau^C(\varepsilon)$ from Eq. (II.2) by performing the necessary Brillouin zone integration.[4]

In our other lectures,[1] we briefly described the first two of these methods, here we will discuss in some detail the third.

To see the nature of the difficulty involved in the evaluation of Eq. (II.2), let us consider the form of the scattering amplitude to which a complex "coherent potential" will give rise. In general, we can write

$$f_L(\varepsilon) = \frac{1}{2i}(\alpha_\ell e^{2i\eta_\ell} - 1) \ , \qquad \text{(III.1)}$$

where η_ℓ is a real phase shift and α_ℓ in a measure of the inelasticity in the "coherent potential;" $\alpha_\ell = 1$, corresponds to a real potential; $\alpha_\ell < 1$ corresponds to a complex potential which is a sink. Decomposing the inverse into real and imaginary parts as $f_L^{-1} = (f_L^{-1})_R - i(f_L^{-1})_I$ and using the fact that we can write $G_{LL'}(\vec{q},\varepsilon) = B_{LL'}(\vec{q},\varepsilon) + i\sqrt{\varepsilon}\delta_{LL'}$ Eq. (II.2) becomes

$$\tau^C_{LL'}(\varepsilon) = -\frac{1}{\Omega_{BZ}} \int_{BZ} d\vec{q}^{\,3}\{\sqrt{\varepsilon}(f^{-1})_R + B(\vec{q},\varepsilon)$$

$$- i\sqrt{\varepsilon}[(f^{-1})_I - 1]\}^{-1}|_{LL'} \ . \qquad \text{(III.2)}$$

Now if $\alpha_\ell = 1$ for all ℓ, then $(f^{-1})_I$ is the unit matrix and the integrand reduces to the inverse of the KKR matrix which is singular for $\varepsilon = \varepsilon_{\vec{q}}$ the Bloch state energies, i.e., for energies such that

Det $\| - \sqrt{\varepsilon}\ \mathrm{Cot}\eta_\ell \delta_{LL'} - B_{LL'}(\vec{q},\varepsilon) \| = 0$. If $\alpha_\ell < 1$, then $(f_L^{-1})_I \neq 1$ and the integral is intrinsically complex, the corresponding KKR determinant will have zeros only at the complex energies[12] $\varepsilon = \varepsilon_{\vec{q}}^R + i\varepsilon_{\vec{q}}^I$. Clearly, if $\varepsilon_{\vec{q}}^I$ is small, the weight of the integrand in Eq. (II.2) will be distributed on some, possibly complicated, constant energy surface in the Brillouin zone. If $\varepsilon_{\vec{q}}^I$ is large, then, the weight of the integral will be more evenly distributed throughout the Brillouin zone. In Fig. 2, we show an example of the behaviour of four of the diagonal elements of the integrand as a function of q ($\equiv|\vec{q}|$) along a line beginning at the Brillouin zone centre and ending at the zone boundary. The behavior shown corresponds to s and p scattering amplitudes which are near to the unitarity circle[14] ($\alpha_0 \approx \alpha_1 \approx 1.0$) while the d scattering amplitude is far from the unitarity circle ($\alpha_2 \ll 1$). Thus, the weight in $\mathrm{Im}\tau_{0000}$ is in a Lorentzian peak while the weight in $\mathrm{Im}\tau_{2-22-2}$ is distributed throughout the Brillouin zone (note the weight in $\mathrm{Rel}\ \tau_{LL'}$ goes roughly like the derivative with respect to $|\vec{q}|$ of

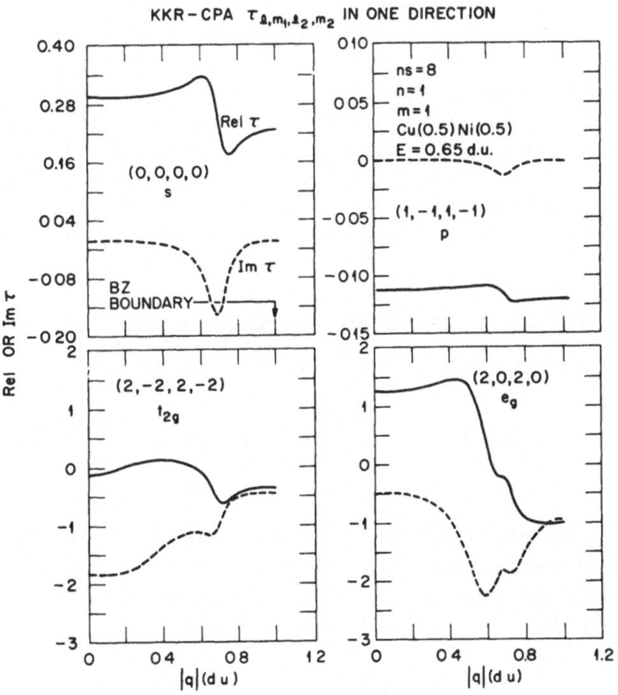

Fig. 2 : Typical behaviour of the diagonal elements of $\tau(\vec{q};\varepsilon)$ as a function of $|\vec{q}|$ along a given direction in the Brillouin zone. The solid lines gives Rel τ, the dash lines gives Im τ. The values of ns, n, and m define the direction in the BZ on the prism mesh (see text). In each frame, the numbers in parenthesis identify the particular matrix element plotted.

$\mathrm{Im}\tau_{LL'}$). Thus, the central problem in obtaining a workable method for solving the KKR-CPA equations is devising a Brillouin zone integration method which can accommodate such diverse behaviour in the distribution of the weight of the integrand (from a near δ function shell for $\alpha_\ell \approx 1$ to nearly uniformly distributed for $\alpha_\ell \ll 1$).

In fact, the problems associated with the evaluation of the integral in Eq. (II.2) are recurrent in the KKR-CPA theory. To obtain the densities of states, one needs to integrate the Bloch spectral density[1] over the Brillouin zone. To obtain the integrated density of states one needs to evaluate an integral over the Brillouin zone of $\mathrm{Im}\,\ell n \,\| t^{-1}_{c;L}\delta_{LL'} - G_{LL'}(\vec{q},\varepsilon) \|$. Thus, the evaluation of both of these functions poses problems which are essentially identical to those set in evaluating Eq. (II.2).

III.1. Brillouin Zone Integration: Directional Methods

There are obviously many ways in which one might comtemplate evaluation of Brillouin zone integrals of the form of Eq. (II.2). However, the complicated structure of the integral and the necessity for repeated evaluation place severe constraints on acceptable methods. Indeed, these considerations strongly mitigate against conventional Brillouin zone integration methods, e.g., Gilat-Raubenheimer.[15] However, the plots in Fig. 2 of the behaviour of the integrand of Eq. (II.2) are suggestive of a possible integration strategy.

Suppose we are interested in evaluating the Brillouin zone integral

$$S(\varepsilon) = \frac{1}{\Omega_{BZ}} \int_{BZ} d^3q\, s(\vec{q},\varepsilon) \;, \tag{III.3}$$

where $s(\vec{q},\varepsilon)$ is one of the functions discussed above. Let us define a set of N directions $\{\hat{q}_i;\ i = 1, \cdots N\}$ emanating from Brillouin zone centre and the line integral $S^{\hat{q}_i}$ of the function of interest, $s(\vec{q},\varepsilon)$, along the ith direction

$$S^{\hat{q}_i}(\varepsilon) = \int_0^{q_i^{BZ}} dq\,\omega(q)\, s(\vec{q},\varepsilon) \tag{III.4}$$

where $\omega(q)$ is some appropriate radial weight function and q_i^{BZ} denotes the Brillouin zone boundary along the direction $\hat{q}_i$. Even when the integrand $s(\vec{q},\varepsilon)$ is highly structured along a given direction, as is $\tau(\vec{q},\varepsilon)$, it turns out that $S^{\hat{q}_i}(\varepsilon)$ is a rather unstructured function of the radial direction $\hat{q}_i$. This allows us to approximate the required integral by a summation over the line integrals for the N directions, i.e.,

$$S(\varepsilon) \approx \sum_{i=1}^{N} w_i S^{\hat{q}_i}(\varepsilon) \quad , \qquad \text{(III.5)}$$

where w_i is again some appropriate weight function.

This then, will be our general strategy, first we will perform careful one-dimensional integrals along a number of radial directions, then we will approximate the required volume integral by a sum of such radial integrals. Again, there are obviously an infinity of ways in which one can implement the details of such a strategy, corresponding to different weight functions $\omega(q)$ and w_i and different choices for the set of directions $\hat{q}_i$ over which the one-dimensional integrals are to be performed. Below we sketch out a couple of methods which work.

III.1.1. Prism Method. The irreducible 1/48 of the fcc Brillouin zone shown in Fig. 3 is divided into a number of prisms each having apex at the zone centre (Γ-point) and base on the Brillouin zone boundary. The edges of the prisms being defined by lines joining the Brillouin zone centre with points on a square mesh defined on the (100) face of the unit cube having edges along the x, y, z axes as shown in Fig. 3. The square mesh is constructed by dividing the edges of the (100) face of the unit cube into ns intervals (Fig. 3 is drawn for ns = 8). Labeling the intervals in the y-direction m and in the z-direction n, the points on the square mesh are then labeled by the cartesian coordinates (m,n). This procedure results in the two possible elemental prisms shown at the bottom of Fig. 3 if we choose $ns = 2^k$ with $k \geqslant 2$. The irreducible part of the Brillouin zone is contained in those prisms for which $m \geqslant n$.

In the spirit of our general strategy, the directions along which we perform the one-dimensional integral in Eq. (III.4) are taken to be the centre of mass lines of the prisms. For the weight function $\omega(q)$, we make the obvious choice of the cross sectional area of the prism parallel to the Brillouin zone face which the prism intersects. The final Brillouin zone integral is then formed by summing such prism integrals according to Eq. (III.5) with the weight function w_i set equal to unity.

Convenient choices of ns are 4, 8, and 16 resulting in division of the Brillouin zone in to 11, 36, and 136 prisms, respectively. As we shall see momentarily, ns = 8 is usually adequate for our purposes.

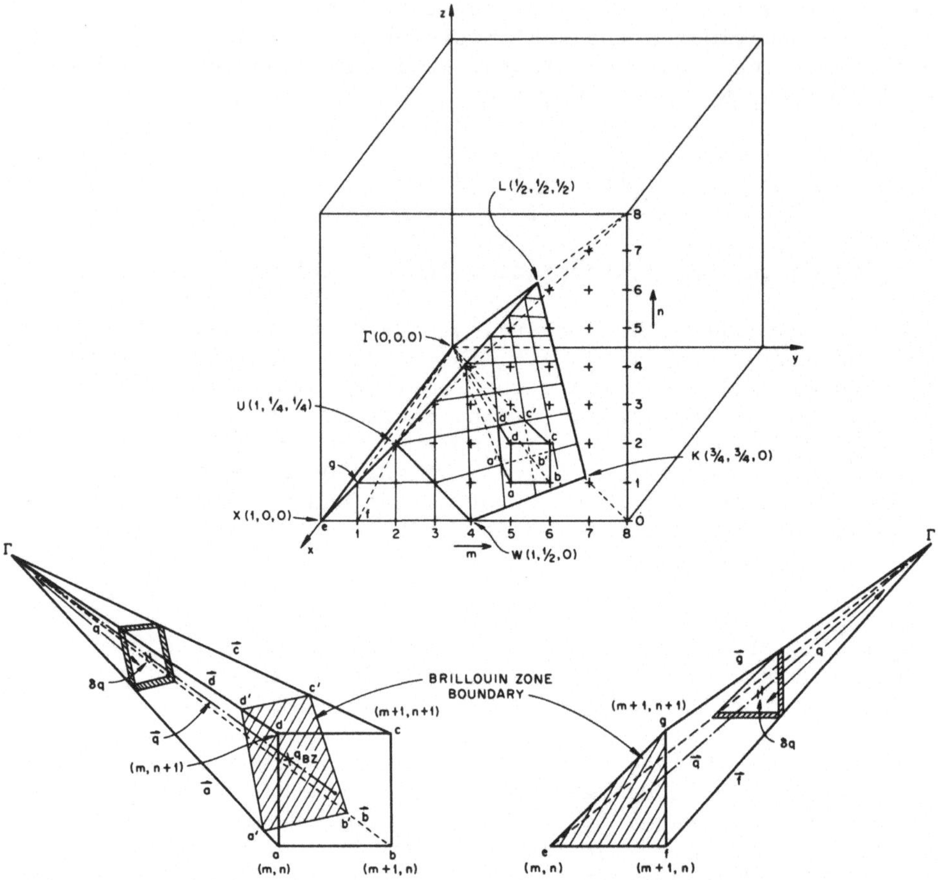

Fig. 3 : Division of the irreducible 1/48 of the fcc Brillouin zone according to the prism method (upper) and the two types of elemental prisms which result (lower).

The Brillouin zone integration method just described is a straightforward generalization of the method used by Faulkner et al.[16] in their accurate determination of the Fermi energy and density of states, in the neighborhood of the Fermi energy, in pure Cu using the KKR method.

III.1.2. Special Directions. Considerations similar to those discussed above led Bansil[17] to propose another variation on the "directional" theme. In this method, we seek to choose a set of directions such that a sum of one-dimensional integrals performed along these directions will give exactly the contribution to the volume integral from a specified number of terms in the cubic harmonic expansion of the integral. Following Bansil,[17] the argument goes as follows. We begin by defining the one-dimensional integral $\tilde{S}^{\hat{q}_i}(\varepsilon)$ of $S(\vec{q},\varepsilon)$ through Eq. (III.4) by choosing the weight function $\omega(q) = q^2$, i.e.,

$$\tilde{S}^{\hat{q}_i}(\varepsilon) = \frac{1}{\Omega_{BZ}} \int dq q^2 s(\vec{q},\varepsilon) . \tag{III.6}$$

We then make use of the fact that $\tilde{S}^{\hat{q}_i}(\varepsilon)$ is invariant under the operations of the point group of the lattice in order to expand $\tilde{S}^{\hat{q}_i}(\varepsilon)$ in cubic harmonics as

$$\tilde{S}^{\hat{q}_i}(\varepsilon) = \tilde{S}_0 + \sum_{j,\ell\neq 0} \tilde{S}^j_\ell C^j_\ell(\hat{q}) , \tag{III.7}$$

where $C^j_\ell(\hat{q})$ is the jth cubic harmonic of order ℓ. If we now use the expansion Eq. (III.7) together with the definition of $\tilde{S}^{\hat{q}_i}(\varepsilon)$, we find that the Brillouin zone integral Eq. (III.3) reduces to

$$\int_{BZ} d^3q\, s(\vec{q},\varepsilon) = \frac{1}{\Omega_{BZ}} \int \tilde{S}^{\hat{q}_i} d\Omega_q = \tilde{S}_0 . \tag{III.8}$$

Thus, the suggestion of Bansil is to choose a set of N directions such that in accordance with Eqs. (III.5) and (III.8)

$$\sum_{i=1}^{N} w_i \tilde{S}^{\hat{q}_i}(\varepsilon) \approx \tilde{S}_0 , \tag{III.9}$$

where the corrections to the expression in Eq. (III.9) involve terms having $\ell > \ell_{max}$. Clearly, the set of directions is fixed by the requirement that

$$\sum_{i=1}^{N} w_i C^j_\ell(\hat{q}_i) = 0 , \tag{III.10}$$

for all ℓ in the range $0 < \ell \leqslant \ell_{max}$.

Bansil[17] has obtained a number of sets of directions in which he chose the weight functions w_i equal to unity, increasingly larger sets of directions corresponding to Eq. (III.10) being satisfied for increasing larger values of ℓ_{max}. If then, the expansion Eq. (III.7) is convergent, performing the Brillouin zone integration according to Eq. (III.9) for larger and larger sets of directions should yield better and better approximations to the Brillouin zone integral, Eq. (III.8).

Subsequent to the work of Bansil, Fehlner and Vosko[18] produced even more powerful directional integration formulae. Using an elegant product representation of the cubic harmonics, they were able to obtain generalized Gaussian integration formulae for functions of two variables. Tables of special directions and Gaussian weights as well as a detailed discussion of the method of generating the formulae are given in the paper of Fehlner and Vosko.[18]

III.1.3. Convergence Properties. For a given direction in the Brillouin zone, it is a simple matter to ensure convergence of the one-dimensional integral, Eq. (III.4). We simply concentrate the points at which the integrand is evaluated in those regions where the integrand is large, then we perform a numerical integration using a quadratic integration formula. Depending on the nature of the structure in the function being integrated, this step will typically require evaluating the integral at between 30 and 200 points along each direction.

The question as to the accuracy of the final volume integral is simply a matter of including enough directions in our integration formula, Eq. (III.5) We have found the 13 direction method of Bansil to be adequate for most applications. Indeed, for many of the major features of KKR-CPA self-consistency, 5 directions are often adequate. In Table 1, we show some typical values for $\tau^{C,\alpha}(\varepsilon)$ using the various integration methods discussed above. The particular values of $\tau^{C,\alpha}$ shown were obtained using, Bansil's 13 direction formula, Fehlner and Vosko's 21 direction formula and the prism method with two values of ns, these being ns = 8 and ns = 16. To the accuracy to which we are normally interested, all the methods can be regarded as having given converged answers. It is reassuring that we get particularly good values of the t_{2g} and e_g components of τ^C since Imτ^C is closely related to the density of states and for the particular alloy and energy for which the table was constructed, the density of states is determined by the d-components (see Figs. 11, 12 and 13).

In practice then, we routinely solve the KKR-CPA equations using the 13 direction method of Bansil. This solution is then checked at a few energies using one of the more accurate methods. Of course, if more detailed information is required about some particular aspect of the calculation, e.g., the position of the Fermi energy or density of states at the Fermi energy, it is a simple matter to use one of the more powerful formulae over the limited energy range in question.

Table 1. Comparison of the calculated values of τ^C using various integration formulae. The values of τ^C are for the KKR-CPA scattering amplitudes for $Cu_{.5}Ni_{.5}$ at ε = 0.569 Ry. The rows labeled B13 and F31 correspond to Bansil's 13 and Fehlner and Vosko's 21 direction methods. The rows labeled P36 and P136 correspond to the prism method having ns = 8 and 16, respectively.

Integration Method	Rel τ^C			
	a_{1g}	t_{1u}	t_{2g}	e_g
B13	.2519	−.0373	.1036	.0366
F21	.2472	−.0374	.1034	.0368
P36	.2517	−.0373	.1033	.0377
P136	.2516	−.0373	.1029	.0371
		Im τ^C		
B13	−.0158	−.0012	−.3194	−.4141
F21	−.0151	−.0011	−.3158	−.4149
P36	−.0156	−.0011	−.3149	−.4146
P136	−.0156	−.0011	−.3144	−.4149

III.2 Calculation of the KKR Structure Constants

Even after we have found an efficient and reliable Brillouin zone integration method flexible enough to handle the τ^C integration, it is clear that we are still left with a formidable computational problem. We have to evaluate the inverse of the KKR matrix anywhere between 2×10^2 and 5×10^3 times per iteration of the CPA equations per energy. As a result, we must optimize each step in the setting up and subsequent inversion of the KKR matrix if we are to reduce the calculation to one of reasonable proportions. If we use conventional methods for calculating the structure constants it is this step, rather than the actual setting up or inversion of the KKR matrix, which is the time consuming element in the calculation. Since the structure constants have to be calculated at each one of the approximately 10^3 $\vec{q}$-points required to perform the Brillouin zone integration, it is here that we must look for ways to speed up the calculation. Indeed, it was the fact that in the Atomic Sphere Approximation, we can use efficient polynomial representations of the structure constants, thereby, reducing the time required for their evaluation to insignificance, which led us to propose the ASA-CPA discussed in our other lectures at this summer school. The simplifications attendant to the Atomic Sphere Approximation, which make it possible to fit the ASA structure constants with low-degree polynomials, are two fold. Firstly, the ASA structure constants are energy independent and consequently need only be fitted as functions of $\vec{q}$. Secondly, all the free electron singularities which are present at $\varepsilon = |\vec{q}_n|^2$ in the KKR structure constants condense onto $\vec{q} = (0,0,0)$ with the result that the ASA structure constants are slowly varying functions of $\vec{q}$. Once fitted, subsequent evaluation of the ASA structure constants at the particular $\vec{q}$-point required by the integration method is reduced to a few additions and multiplications. In the ASA-CPA, the price we pay for this increased numerical efficiency is that the Atomic Sphere Approximation limits the kinds of systems we are able to treat accurately, certainly to hcp and fcc structures and probably to transition metal/transition metal and transition metal/noble metal alloys.

As we shall now show, by the device of treating the regular and singular parts of the KKR structure constants separately, we are here able to fit the KKR structure constants with low-degree polynomials (in $\vec{q}$ and ε), thereby obtaining the high numerical efficiency of the ASA-CPA but without making any of the accompanying approximations.

In our other lecture notes in this volume, we give an explicit expression for the KKR structure constants. However, this is not useful for calculational purposes because of the slow convergence of the summations. Formulae more amenable to calculations are obtained by writing[19]

$$G_{LL'}(\vec{q},\varepsilon) = B_{LL'}(\vec{q},\varepsilon) - i\kappa\delta_{LL'} , \tag{III.11}$$

where $\kappa = \sqrt{\varepsilon}$ and where the $B_{LL'}(\vec{q},\varepsilon)$ are given by

$$B_{LL'}(\vec{q},\varepsilon) = 4\pi i^{\ell-\ell'} \sum_{L''} C^{L''}_{L,L'} D_{L''}(\vec{q},\varepsilon) . \tag{III.12}$$

The $C^{L''}_{L,L'}$ are the Gaunt factors,[7] the new structure constants $D_{L''}(\vec{q},\varepsilon)$ contain, apart from the trivial $i\kappa$ in Eq. (III.11), the complete $\vec{q}$ and ε dependence of the $G_{LL'}(\vec{q},\varepsilon)$. The structure constants $D_L(\vec{q},\varepsilon)$ (in the rest of this section, we will refer to the D's rather than the G's as structure constants) are now normally calculated using the Ewald procedure.[7] It is this step which is the time consuming element in the overall calculation of the $G_{LL'}(\vec{q},\varepsilon)$.

Using the Ewald method the $D_{\ell,m}(\vec{q},\varepsilon)$ naturally decompose as

$$D_{\ell,m}(\vec{q},\varepsilon) = D^1_{\ell,m}(\vec{q},\varepsilon) + D^2_{\ell,m}(\vec{q},\varepsilon) + D^3_{0,0}(\varepsilon)\delta_{\ell,0} , \tag{III.13}$$

where explicit expressions for the various terms are given in ref. 7. The quanity $D^2_{\ell,m}(\vec{q},\varepsilon)$ involves a summation over the real space lattice vectors, $D^3_{0,0}(\varepsilon)$ involves a power series in ε/η where η is the Ewald splitting parameter, both of these quantities are smoothly varying functions of $\vec{q}$ and ε and are, therefore, amenable to the same low-degree polynomial fitting procedures that proved so advantageous in the ASA-CPA. The $D^1_{\ell,m}(\vec{q},\varepsilon)$ term is not so simple; explicitly $D^1_{\ell,m}(\vec{q},\varepsilon)$ is given by

$$D^1_{\ell,m}(\vec{q},\varepsilon) = -\frac{4\pi}{\Omega}\kappa^{-\ell} e^{\varepsilon/\eta} \sum_n \frac{q_n e^{-q_n^2/\eta} Y_{\ell,m}(\hat{q}_n)}{[q_n^2 - \varepsilon]} \tag{III.14}$$

where $\vec{q}_n = \vec{q} + \vec{G}_n$; $\vec{G}_n$ is a reciprocal lattice vector and Ω is the volume of the unit cell. This term is singular at the free electron energies $\varepsilon = |\vec{q} + \vec{G}_n|^2$, and therefore, not immediately amendable to polynomial fitting. However, $D^1_{\ell,m}(\vec{q},\varepsilon)$ can be further decomposed over a limited energy range into two terms, one $D^S_{\ell,m}(\vec{q},\varepsilon)$ which is singular and one $D^{1R}_{\ell,m}(q,\varepsilon)$ which is regular at $\varepsilon = |\vec{q} + \vec{G}_n|^2$ as

$$D^1_{\ell,m}(\vec{q},\varepsilon) = D^{1,R}_{\ell,}(\vec{q},\varepsilon) + D^S_{\ell,m}(\vec{q},\varepsilon) \ , \tag{III.15}$$

where

$$D^{1,R}_{\ell,m}(\vec{q},\varepsilon) = -\frac{4\pi}{\Omega}\kappa^{-\ell}\left\{\sum_{n\in A} \frac{q_n^\ell[e^{-(q_n^2-\varepsilon)/\eta} - 1]Y_{\ell,m}(\hat{q}_n)}{[q_n^2 - \varepsilon]} + \sum_{n\notin A} \frac{q_n^\ell e^{-(q_n^2-\varepsilon)/\eta} Y_{\ell,m}(\hat{q}_n)}{[q_n^2 - \varepsilon]}\right\} , \tag{III.16}$$

and

$$D^S_{\ell,m}(\vec{q},\varepsilon) = -\frac{4\pi}{\Omega}\kappa^{-\ell}\sum_{n\in A} \frac{q_n^\ell Y_{\ell,m}(\hat{q}_n)}{[q_n^2 - \varepsilon]} \ . \tag{III.17}$$

The reciprocal lattice vectors in the set A which are included in the singular term Eq. (III.17) are those for which $\varepsilon - |\vec{q} + \vec{G}_n|^2 = 0$ for any $\vec{q}$ in the 1/48 of the Brillouin in which we are going to use the $G_{LL'}(\vec{q},\varepsilon)$ and for any energy in the energy range of interest [typically, $0 < \varepsilon < 1$ d.u.; 1 du = $(a/2\pi)^2$ Ry]. Clearly, Eq. (III.16) is regular at $\varepsilon = q_n^2$ for all $n \in A$.

Using the decomposition in Eq. (III.15), we can now write the total structure constant $D_{\ell,m}(\vec{q},\varepsilon)$ as the sum of a regular and a singular part as

$$D_{\ell,m}(\vec{q},\varepsilon) = D^R_{\ell,m}(\vec{q},\varepsilon) + D^S_{\ell,m}(\vec{q},\varepsilon) \ , \qquad \text{(III.18)}$$

where the regular part, $D^R_{\ell,m}(\vec{q},\varepsilon)$, is given by

$$D^R_{\ell,m}(\vec{q},\varepsilon) = D^{1,R}_{\ell,m}(\vec{q},\varepsilon) + D^2_{\ell,m}(\vec{q},\varepsilon) + D^3_{0,0}(\varepsilon)\delta_{\ell,0} \ , \qquad \text{(III.19)}$$

and $D^S_{\ell,m}(\vec{q},\varepsilon)$ is given by Eq. (III.17).

In Figs. 4 and 5, we have plotted $D^R_{0,0}$ as a function of ε and q along one of the Brillouin zone integration directions discussed previously. In Fig. 4, no subtraction of singular terms has been made, thus, the set A contains no members and $D^R_{00} = D_{00}$. The structure constant is singular at the free electron energies $\varepsilon = q_n^2$. In Fig. 5, the structure from the two lowest lying free electron bands, those corresponding to the reciprocal lattice vectors at (0,0,0) and (−1,−1,−1), has been removed according to Eqs. (III.18) and (III.19). Thus, by the simple expedient of removing the structure from two free electron poles, we have obtained a function which is easily fitted, both in ε and q, by low-degree polynomials. As with the ASA-CPA, these polynomial coefficients can be calculated once and for all and stored on some permanent storage device. Of course, we have not simplified the actual calculation of $D^R_{\ell,m}(\vec{q},\varepsilon)$, this still has to be done via the Ewald method, but now we need only perform the calculation once.

Our directional integration method also allows us to treat the singular term in a particularly elegant way. Representing the $\vec{q}$-points along a given direction in the form $\vec{q} = q\vec{u}$, where $\vec{u}$ is the unit vector, allows us to write the numerator in the summation in Eq. (III.17) as a polynomial of degree $\ell + 1$ in $\vec{q}$, i.e.,

$$|\vec{q} + \vec{G}_n|^{\ell} Y_{\ell,m}(\widehat{q + G_n}) = \sum_{m=0}^{\ell+1} a^n_m q^m \ , \qquad \text{(III.20)}$$

where the coefficients a^n_m can be calculated analytically in terms of the Cartesian components of $\vec{u}$ and $\vec{G}_n$ using the definitions of the spherical harmonics. Again, these polynomial coefficients can be calculated once and for all for each integration direction and for each $\vec{G}_n$ in the set A.

By making use of the polynomial expansions of D^R and D^S the calculation of the total structure constants at a particular $\vec{q}$ along one of the integration directions is reduced to a few multiplications and additions, vastly reducing the computational effort required to construct the KKR-matrix.

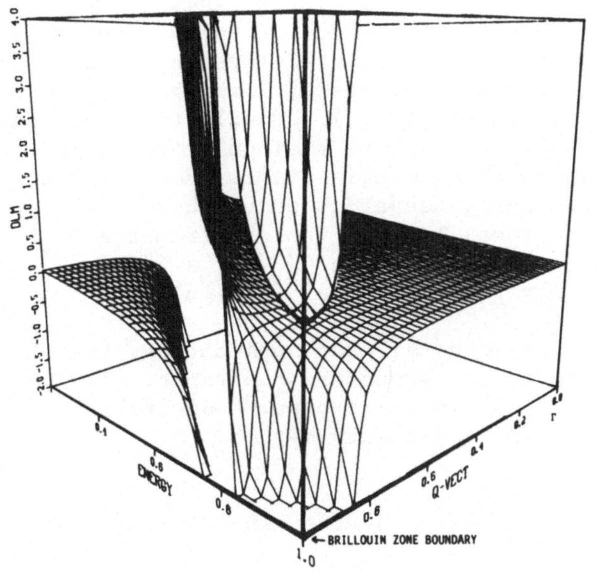

Fig. 4 : $D^R_{00}(|\vec{q}|;\ \varepsilon)$ plotted as a function of energy and as a function of $|\vec{q}|$ along a line from the BZ centre to the point (0.482, 0.350, 0.668) on the BZ boundary. No free electron poles have been removed.

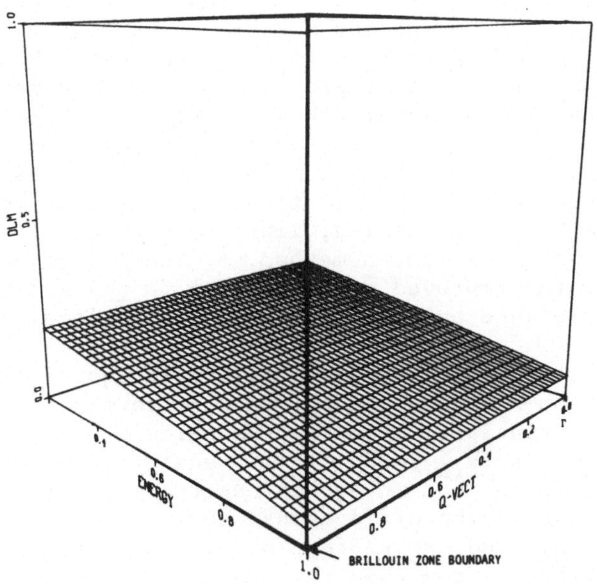

Fig. 5 : As Fig. 2, with the free electron pole contributions which result from the reciprocal lattice vectors (0, 0, 0) and (−1, −1, −1) removed.

As a practical matter for fcc, for example, we have found it possible to fit the $D^R_{\ell,m}(\vec{q},\varepsilon)$ in the energy range $-0.2 \leq \varepsilon \leq 1.0$ using Chebyshev polynomials of degree five in ε and seven in q. In order to obtain fit polynomials of low-degree , it is expedient to remove from $D^R_{\ell,m}(\vec{q},\varepsilon)$ not only the reciprocal lattice vectors which actually contribute a singularity in the fit range but also those for which $q_n^2 - \varepsilon$ is small. For the fcc crystal structure over the energy range discussed above, we found it best to remove four reciprocal lattice vectors from the $D^R_{\ell,m}$ sum. It is also expedient to multiply each $D^R_{\ell,m}$ by a factor κ^ℓ to remove the $\varepsilon = 0$ singularity present in Eq. (III.16).

Though different in detail and application, the above discussion of the fitting of the structure constants, parallels that given by Williams et al. in their paper[20] describing their methods of speeding up the KKR band-theory method.

III.3. Iteration of the KKR-CPA Equations

Having obtained efficient alogarithms for the calculation of $\tau^C(\varepsilon)$ from Eq. (II.2) for some assumed values of the effective scattering amplitudes $f_{C,L}(\varepsilon)$, we are now in position to iterate the CPA equations to self consistency.

The iteration process is initiated by assuming some reasonable values for the effective scattering amplitude. Those provided by the average t-matrix approximation are usually adequate, i.e., we begin by setting $f_C = f_{ATA} = cf_A + (1-c)f_B$, however, those provided by extrapolation to the energy currently under consideration from solutions previously obtained at other energies are usually better. The assumed values of $f_{C,L}$ are used in the evaluation of τ^C from Eq. (II.2). The result together with the assumed values of the $f_{C,L}$'s are inserted into Eq. (II.1) to see if the equation is satisfied. If it is not satisfied, improved guesses of the $f_{C,L}$'s are made via the Newton-Ralphson method.[21] The iteration procedure is repeated until the improved values of the $f_{C,L}$'s are the same to within some tolerance on two consecutive iterations.

In order to solve the coupled equations in Eqs. (II.1) and (II.2) by the Newton-Ralphson method, we require the derivatives $\delta\tau^C_{L,L}/\delta f_{C,L'}$. In principle, these can be calculated by differentiating the expression for $\tau^C_{L,L}$ in Eq. (II.2) analytically and preforming the necessary Brillouin zone integrals numerically. In the present work, we have neglected the off-diagonal derivatives $\delta\tau^C_{L,L}/\delta f_{C,L'}$ ($L' \neq L$) and approximated $\delta\tau^C_{L,L}/\delta f_{C,L}$ by values obtained by differencing the $\tau^C_{L,L}$'s from the previous two iterations. Using this approximation scheme for the derivatives does not affect the final converged values of the $f_{C,L}$'s, it merely slows down the

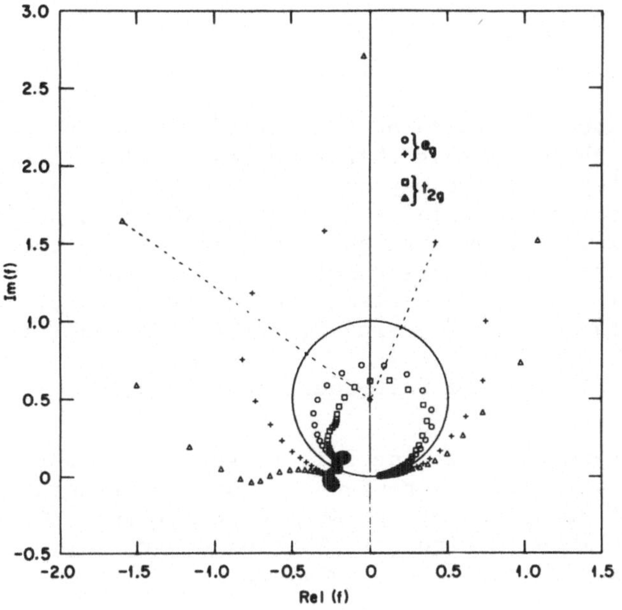

Fig. 6 : Argand diagram of the interior and exterior e_g and t_{2g} solutions of the KKR-CPA equations for $Cu_{.77}Ni_{.23}$.

approach to convergence. Using this method, we usually require some 5–8 iterations to obtain convergence. Our experience with a one-dimensional model[22] and the cluster CPA[9] leads us to believe that if we were to perform the necessary Brillouin zone integrals to evaluate all the derivatives $\delta\tau^C_{L,L}/\delta f_{C,L'}$ the resulting multi-dimensional Newton-Ralphson method would lead to convergence in 3–5 iterations.

The structure of the KKR-CPA equations is such that solutions of Eqs. (II.1) and (II.2) come in pairs. These dual solutions are illustrated, for the t_{2g} and e_g components of f_C, in Fig. 6 in the form of an argand diagram. The solution interior to the unitarity circle is obviously the required solution since it corresponds to a positive density of states. Indeed, it turns out that the solution exterior to the unitarity circle is trivially related to the interior solution. If we represent the interior solution in the standard form

$$f^{int}_{C,\ell}(\varepsilon) = \frac{1}{2i}(\alpha_\ell e^{2i\eta_\ell(\varepsilon)} - 1) \qquad \text{(III.20)}$$

where, since $f^{int}_{C,L}$ is the interior solution, $\alpha_\ell < 1$ and η_ℓ is a real "phase shift," then the exterior solution $f^{ext}_{C,L}$ is given by

$$f^{ext}_{C,\ell}(\varepsilon) = \frac{1}{2i}\left(\frac{1}{\alpha_\ell(\varepsilon)} e^{2i\eta_\ell(\varepsilon)} - 1\right) . \qquad \text{(III.21)}$$

We have speculated elsewhere on the nature of the second solution,[22] here we only wish to comment that knowledge of its existence can be put to good effect. If at any point in the iteration process, the Newton-Ralphson method projects a solution which is outside the unitarity circle, we know that this solution is not the required one and the iteration process should be redirected. A useful way of doing this is simply to reflect the incorrect solution back inside the unitarity circle using Eqs. (III.20) and (III.21).

III.4. The Self-Consistent Solutions

In the interests of making this lecture self-contained, in this subsection we will briefly review some of the features of the self-consistent solutions of the KKR-CPA equations. At the risk of walking once more on already well trodden ground, we use results on the CuNi alloy system as our examples.

In Fig. 7, we display all four components of the effective scattering amplitudes for $Cu_{.5}Ni_{.5}$. Both the s ($\equiv a_{1g}$) and p ($\equiv t_{1u}$) effective scattering amplitudes are small and unstructured. If we were to plot the s and p scattering amplitudes in the form of an argand diagram, we would also see that they fall very close to the unitarity circle, i.e., $\alpha_{0,1} \simeq 1$. That $f_{C,0}$ and $f_{C,1}$ are small results from the fact that both the Cu and the Ni muffin-tin potentials are themselves weak scatters in the s and p channels. That the scattering amplitudes fall close to the unitarity circle reflects the fact there is little disorder in these channels since the difference between the Cu and Ni scattering amplitudes in the s- and p-channels is small. The situation in the d-channel is quite different because both the Cu and Ni muffin-tin potentials have scattering resonances. This means that not only is the effective scatterer a strong scatterer but because of the separation in the position of the Cu and Ni scattering resonances that the scattering centre is highly inelastic in the d-channel, i.e., $\alpha_{t_{2g}}$ and $\alpha_{e_g} \ll 1$ in the neighbourhood of the resonances. What this means in practice is that the precise form of the effective scattering amplitude is determined by a complex interplay between the two scattering resonances of the constituent muffin-tin potentials through the CPA equations. Phrased in a slightly different way, the form of the scattering amplitude is influenced a great deal by the band-structure effects brought in through the self-consistency process. Thus, for the example shown in Fig. 7, the t_{2g} and e_g components of f_C look quite different from one another and neither one is close to the average scattering amplitude $f_{ATA,2}(\varepsilon) = cf_{A,2}(\varepsilon) + (1-c)f_{B,2}(\varepsilon)$. In Fig. 8, we show plots of the t_{2g} component of the effective scattering amplitude for three different

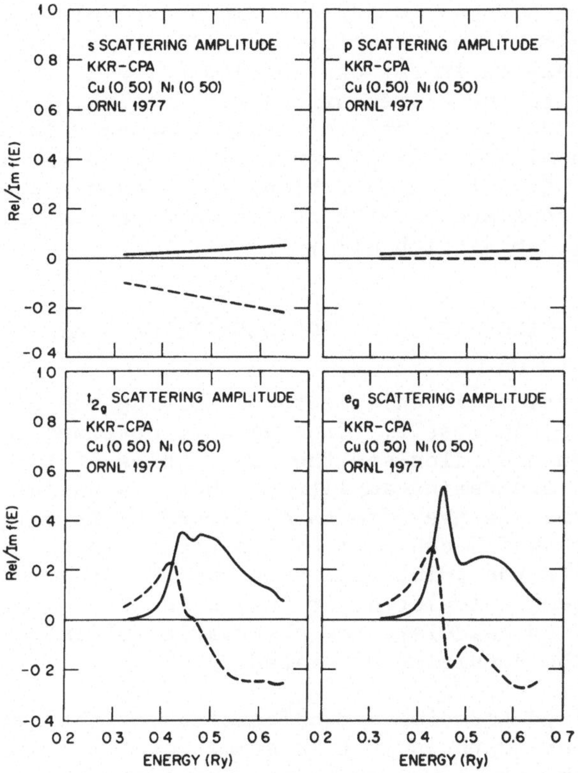

Fig. 7 : KKR-CPA s, p, t_{2g} and e_g effective scattering amplitudes as a function of energy for $Cu_{.5}Ni_{.5}$. The solid line gives Im f; the dash line gives Rel f.

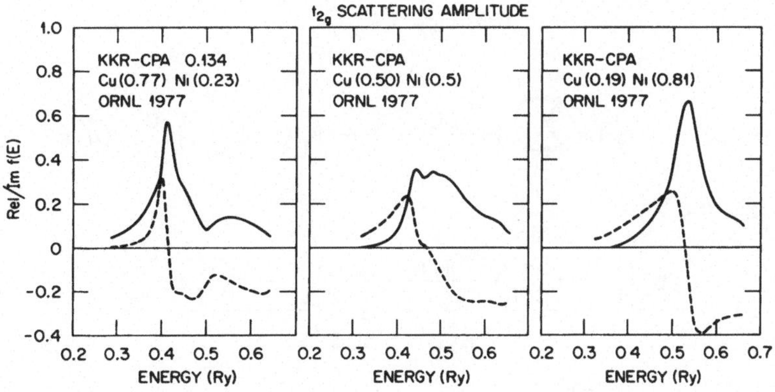

Fig. 8 : Composition dependence of $f_{t_{2g}}$ in CuNi alloys. Solid line gives Im $f_{t_{2g}}$ dash line gives Rel $f_{t_{2g}}$.

concentrations which serve to further emphasize this point. Here, we see the evolution from a scattering amplitude having a double resonance (two peaks in Im f_C; one corresponding to the Cu resonance; one corresponding to the Ni resonance) for Cu-rich alloys to a scattering amplitude having only a single resonance (one peak in Im f_C) in Ni-rich alloys. Simply averaging the Cu and Ni scattering amplitudes leads, at all concentrations, to a double peaked structure, thus, self-consistency effects are dominant in determining the form of $f_{C,t_{2g}}$ in Ni-rich alloys.

IV. CALCULATION OF THE DENSITY OF STATES

As we said earlier in this lecture, solving the KKR-CPA equations is not an end in itself. Once we have obtained the solutions, we must then go on and calculate the observables of interest. After all, the effective scattering amplitude, though a useful concept in helping us to visualize the effects of disorder, is only a construct and not a quantity which can ever be measured. In this section, in order to illustrate the point that once the effective scattering amplitudes have been calculated, it is a straight-forward, though non-trivial, step to calculate the observables of interest, we will briefly discuss the densities of states.

We have derived formulae for calculation of the density of states in our other lectures at this summer school,[1] here we just quote the final results. We can either calculate the densities of states $n(\varepsilon)$ directly from the average Bloch spectral density $\overline{A}_B(q,\varepsilon)$ via

$$n(\varepsilon) = \frac{1}{\Omega_{BZ}} \int_{BZ} d^3q \overline{A_B}(\vec{q},\varepsilon) \ , \qquad \text{(IV.1)}$$

where

$$\overline{A_B}(q,\varepsilon) = -\frac{1}{\pi}\text{Im} \sum_{LL'} \left[\frac{\partial G_{LL'}(\varepsilon)}{\partial \varepsilon} - P_L(\varepsilon)\delta_{LL'}\right]\tau^C_{L'L}(\vec{q},\varepsilon) \ , \qquad \text{(IV.2a)}$$

with $P_L(\varepsilon)$ given by

$$P_L(\varepsilon) = \frac{(t^{-1}_{A,L} - t^{-1}_{C,L})\frac{\partial t^{-1}_{B,L}}{\partial \varepsilon} - (t^{-1}_{B,L} - t^{-1}_{C,L})\frac{\partial t^{-1}_{A,L}}{\partial \varepsilon}}{(t^{-1}_{A,L} - t^{-1}_{B,L})} \ , \qquad \text{(IV.2b)}$$

and where $\tau^C(\vec{q},\varepsilon) = [t_C^{-1}(\varepsilon) - G(\vec{q},\varepsilon)]^{-1}$ is the inverse of the KKR matrix. Or, we can first calculate the integrated density of states $N(\varepsilon)$ from

$$N(\varepsilon) = N^0(\varepsilon) - \frac{1}{\pi\Omega_{BZ}} \mathrm{Im}\int_{BZ} d\vec{q}^{\,3}\ \ell n \, \| t^{-1}_{C,L}\delta_{LL'} - G_{LL'}(\vec{q},\varepsilon)\|$$

$$+ \frac{(1-c)}{\pi}\ell n \frac{\| t^{-1}_{A,L} - \langle t^{-1}_{L}\rangle \|}{\| t^{-1}_{A,L} - t^{-1}_{C,L} \|} + \frac{c}{\pi}\ell n \frac{\| t^{-1}_{B,L} - \langle t^{-1}_{L}\rangle \|}{\| t^{-1}_{B,L} - t^{-1}_{C,L} \|}\ , \qquad \text{(IV.3)}$$

where $\langle t_L^{-1}\rangle = ct_{A,L}^{-1} + (1-c)t_{B,L}^{-1}$ and $N^0(\varepsilon)$ is the integrated density of states for free electrons. From a tabulation of the integrated density of states as function of energy, we can obtain the density of states by numerical differentiation.

Provided t_C satisfies the CPA condition, it makes no difference which of the above formulae is used. This is obviously the case since we can derive Eqs. (IV.1) and (IV.2) from Eq. (IV.3) by differentiating Eq. (IV.3) and using the CPA condition, Eq. (II.1). That both sets of formulae, if used correctly, give the same results is not to say they are equally suitable for computational purposes. Both formulae involve a Brillouin zone integration, in Eq. (IV.1) we have to integrate the Bloch spectral density, in Eq. (IV.3), we have to integrate the phase of the KKR-determinant. In Fig. 9, we plot the phase of the KKR-determinant both as a function of energy at example $\vec{q}$-points in the Brillouin zone and as a function q along the (100) direction at a few energies. In Fig. 10, we show a three-dimensional plot of the Bloch spectral density along the (100) direction in the Brillouin zone. Integration of the phase of the KKR determinant involves integration of an approximate step function whilst integration of the Bloch spectral density involves the integration of a series of peaks (rather as in the integration of $\mathrm{Im}\tau^C$). Clearly, the integration of the phase of the KKR determinant poses fewer problems. Furthermore, the integrand in Eq. (IV.3) is a simpler function to calculate than the integrand in Eq. (IV.2) since we do not have to evaluate the energy derivatives of the structure constants. Thus, as a matter of routine, we evaluate the densities of states using Eq. (IV.3). We should stress, however, that the integration methods discussed in the previous section, are perfectly able to integrate the Bloch spectral density.

In Figs. 11 and 12, we show the integrated densities of states $N(\varepsilon)$ and the densities of states $n(\varepsilon) = \partial N(\varepsilon)/\partial\varepsilon$ obtained from Eq. (IV.3) for three $Cu_cNi_{(1-c)}$ alloys having c = 0.77, 0.50 and 0.19. We also show the various contributions to $N(\varepsilon)$ and $n(\varepsilon)$ according to Eq. (IV.3). The curve labelled "KKR determinant" gives the contribution from the first two terms on the right of Eq. (IV.3), the curve labelled "correction terms" gives the contribution from the remaining two terms. We have included $N^0(\varepsilon)$ with the integral of the phase of the KKR determinant since it is precisely cancelled by a contribution from the integral. This

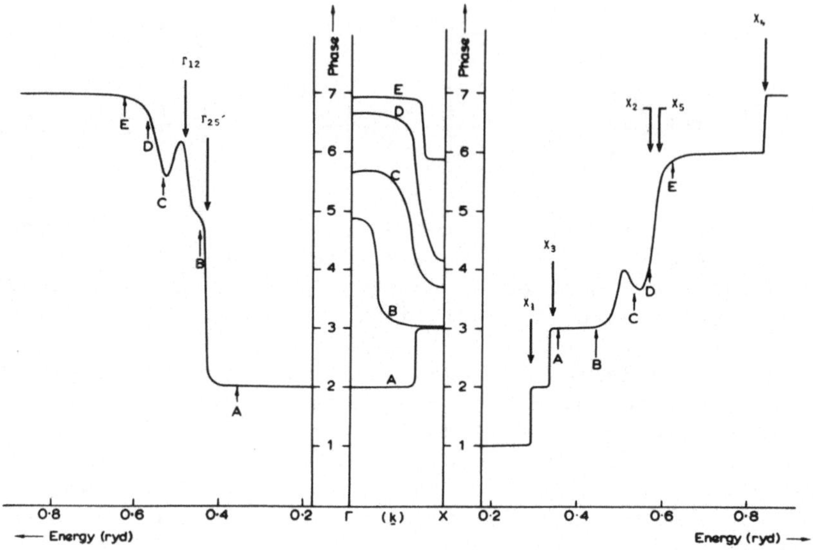

Fig. 9 : Plot of $1/\pi \ \mathrm{Im}\ \ell n \ \| t_C^{-1}\delta_{LL'} - G_{LL'}(\vec{q},\varepsilon)\|$ as a function of energy at Γ (left) and X (right) and as a function of $|\vec{q}|$ along the (001) direction (centre) at the energies indicated by the reference letters. The particular plot is for $t_C = t_{ATA}$ for $Cu_{.87}Ni_{.13}$.

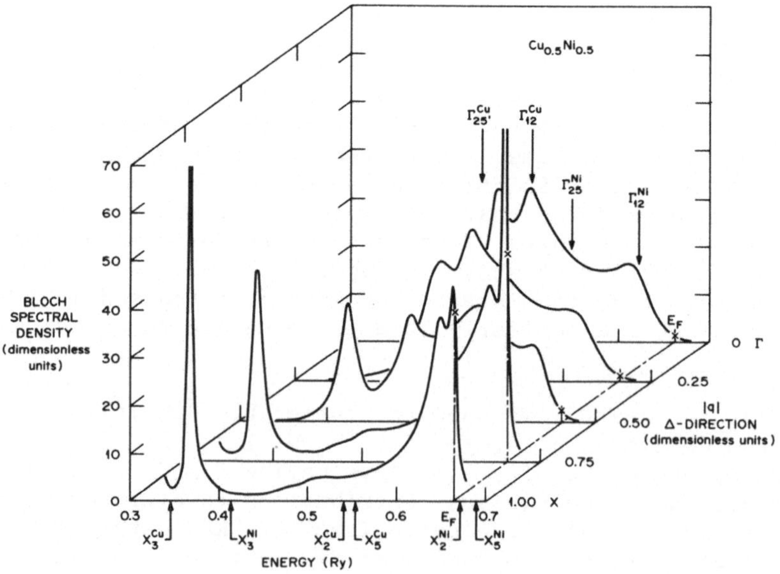

Fig. 10 : KKR-CPA Bloch spectral density as a function of energy for $Cu_{.5}Ni_{.5}$ at various $\vec{q}$-points along the Γ (background) to X (foreground) direction. Energies of states at the Γ and X points in pure Cu and pure Ni are indicated by arrows.

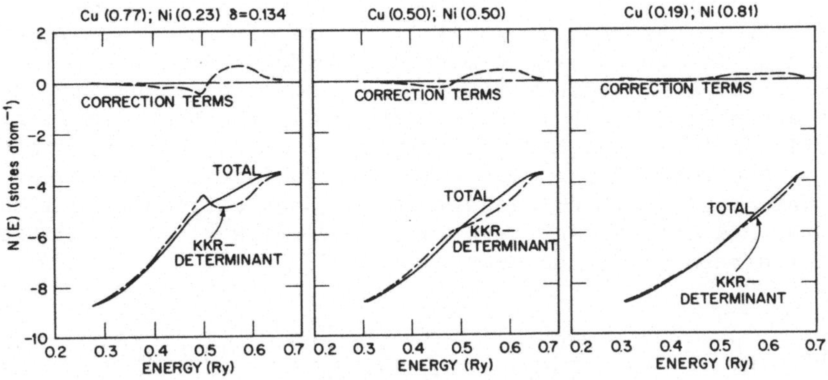

Fig. 11 : KKR-CPA integrated densities of states N(ε) for $Cu_{.77}Ni_{.23}$; $Cu_{.5}Ni_{.5}$ and $Cu_{.19}Ni_{.81}$ calculated according to Eq. (IV.1) (solid line) and contributions to N(ε) from the KKR determinant term (dash-dot) and correction terms (dash).

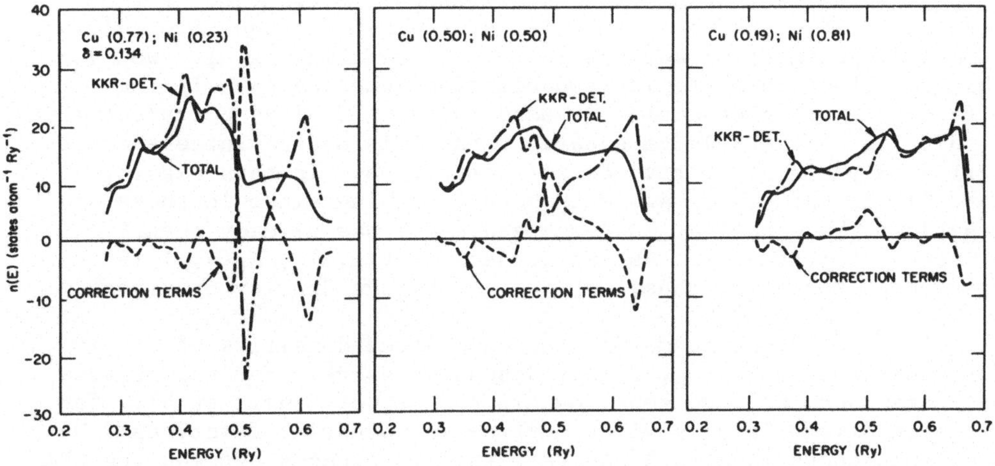

Fig. 12 : As Fig. 12 except for the density of states $n(\varepsilon) = \frac{\delta}{\delta\varepsilon} N(\varepsilon)$.

contribution arises from the changes of π in the phase of the determinant which result from the free-electron singularities present in the KKR structure constants.

The correction terms in Eq. (IV.3), terms which correct the KKR determinant term for the implicit energy dependence of the coherent potential corresponding to the effective scattering amplitude f_C, become increasingly important as we go from Ni-rich to Cu-rich alloys. If we relate this to the structure in f_C shown in Fig. 8, we see that the correction terms are increased in importance as f_C goes from a single, virtual crystal like, resonance in $Cu_{.19}Ni_{.81}$ to the average t-matrix like double resonance in $Cu_{.77}Ni_{.23}$. Indeed, in $Cu_{.77}Ni_{.23}$ in the energy range between the two resonances ($\varepsilon \sim 0.5$ Ry), it is only as a result of a delicate cancellation between the KKR determinant term and the correction terms that the total density of states is positive. The reason for the delicate numerical cancellation is not difficult to locate. Calling the KKR determinant term in Eq. (IV.3) $N^{KKR}(\varepsilon)$ and differentiating with respect to the energy we find

$$\frac{\partial N^{KKR}(\varepsilon)}{\partial\varepsilon} \equiv n^{KKR}(\varepsilon) = -\frac{1}{\pi\Omega_{BZ}}\mathrm{Im}Tr\int_{BZ} d^3q\left[\frac{\partial t_C^{-1}}{\partial\varepsilon} - \frac{\partial G(\vec{q},\varepsilon)}{\partial\varepsilon}\right]\tau^C(\vec{q},\varepsilon). \quad (IV.4)$$

It is the term $-\mathrm{Im}\,Tr\,[\partial t_C^{-1}/\partial\varepsilon]\tau^C(\varepsilon)$ which results in the large negative value of $n^{KKR}(\varepsilon)$ at $\varepsilon \sim 0.5$ Ry in the $Cu_{.77}Ni_{.23}$ alloy. However, if we differentiate the correction terms and use the CPA equations, we find a term $\mathrm{Im}Tr[\partial t_C^{-1}/\partial\varepsilon]\tau^C(\varepsilon)$ which precisely cancels the offending term in $n^{KKR}(\varepsilon)$. It is then a reflection of the accuracy of the numerical solution of the KKR-CPA equation and our numerical integration methods in general that in Fig. 12, we are able to effect this precise cancellation numerically. If the effective scattering amplitude used in Eq. (IV.3) is not precisely a solution of the KKR-CPA equations there is no guarantee that it will even yield a positive density of states. In this regard, Eqs. (IV.1) and (IV.2) are much more forgiving since in these equations the cancellation discussed above has already been effected analytically. As can be seen in Fig. 10, $\bar{A}_B(\vec{q},\varepsilon)$ is everywhere positive, thus, ensuing a positive density of states.

In Fig. 13, we summarize the calculated densities of states for the three $Cu_cNi_{(1-c)}$ alloys discussed above. For the purpose of comparison, we also show the ATA results of Bansil et al., for the same alloys using similar muffin-tin potential functions. We make this comparison to illustrate the differences between the CPA and ATA since it has been suggested that, in order to avoid the difficult task of solving the CPA equations, we use the nonself-consistent ATA. The differences between the two calculations are clearly substantial, reflecting the need for introducing the CPA self consistency.

In order to avoid any possible confusion, we should add that the ATA results shown in Fig. 13 are those of ref. 13 and are not the results of the latest version of the ATA discussed by Dr. Bansil at this summer school and in ref. 23. That there are several ATA's is a reflection of the fact that the ATA is not a self-consistent theory, and therefore, it matters which expression one uses for the calculation of the density of states. Use of Eq. (IV.3) with t_{ATA} replacing t_C gives the so-called local CPA of Györffy and Stocks.[3] Use of Eqs. (IV.2) and (IV.3) with t_{ATA} replacing t_C gives the ATA

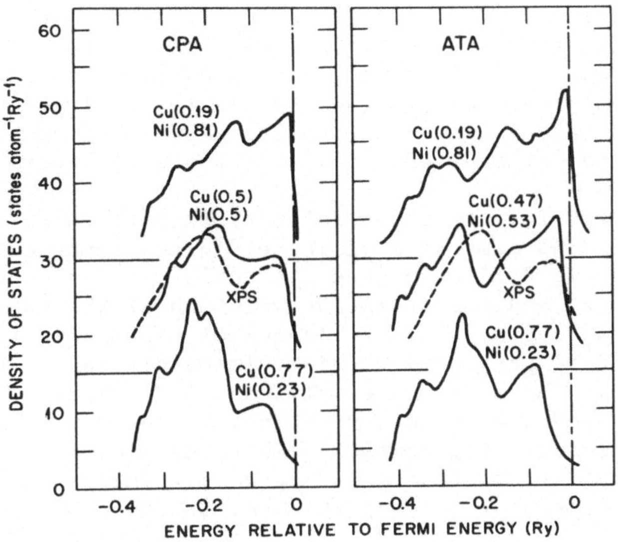

Fig. 13 : Comparison of the densities of states obtained from the KKR-CPA with those obtained in ATA by Bansil et al.[13] using similar potentials. Also shown is an XPS spectrum for $Cu_{.53}Ni_{.47}$ taken from Hüfner et al.[26]

proposed by Bansil et al. in ref. 13 and shown in Fig. 13. Use of Eq. (II.61) of our other lecture notes[1] with t_{ATA} replacing t_C gives the ATA of ref. 23. Of course, this does not exhaust the list of possible ATA's, for example, we could use t_{ATA} rather than t_C in Eq. (II.4) or we could attempt the difficult task of evaluating the density of states from the formula given by Beeby[24] in his rederivation of the ATA originally proposed by Korringa.[25]. Presumably, no two of these ATA's will yield the same density of states. The situation, then, is quite unlike the CPA where all of the proposals for calculating the density of states yield the same answer.

V. CONCLUDING REMARKS

The methods which we have described in this lecture are those which we have used in our numerical solution of the KKR-CPA equations. The devices which we have resorted to, probably represent the minimum which has to be done in order to bring the calculation down to manageable proportions. Even at this, solving the KKR-CPA equations at approximately 60 energies for a transition metal/transition metal alloy, where we include phase shifts up to $\ell = 2$ in the KKR matrix, requires in excess of one hour of CPU on an IBM 360/91 Computer. Of course, for each alloy concentration, this is a once only expenditure (unless we are involved in a self-consistent field calculation) the results of which form the input for the calculation of many quantities of physical interest.

As a footnote, it is perhaps salutary to note that, despite our endeavours in the area of devising efficient numerical procedures for evaluating Brillouin zone integrals, for calculating structure constants and for iterative solution of the KKR-CPA equation, for those of us who are blessed with IBM computers it is a simple piece of programming which ultimately makes the whole scheme tractible. The point is simply that IBM compilers handle complex multiply and divide in a very inefficient way. Writing out all manipulations involving complex numbers explicitly in terms of the real and imaginary parts, including matrix inversion, reduces the CPU time required for solving the CPA equations by a factor in excess of two.

ACKNOWLEDGEMENTS

Research sponsored by the Materials Science Division, U.S. Department of Energy under contract W-7405-eng-26 with the Union Carbide Corporation. One of us (WMT) would like to acknowledge the financial support of the United Kingdom Science Research Council. We would like to thank our colleagues in the Theory Groups in the Metals and Ceramics Division and in the H. H. Wills for their scientific input and for the stimulating atmosphere which they provide and Ms. Gail Golliher for undertaking the difficult task of typing the camera-ready manuscript.

REFERENCES

1. Györffy, B. L. and Stocks, G. M., NATO-ASI lecture notes in this volume.
2. Soven, P., Phys. Rev. B2, 4715(1970); Shiba, H., Progr. Theoret. Phys, (Kyoto) 16, 77(1971); Györffy, B. L., Phys. Rev. B5, 2382 (1972).
3. Györffy, B. L. and Stocks, G. M., J. de Phys. 5, C4–75 (1974).
4. Stocks, G. M., Györffy, B. L. and Temmerman, W. M., Phys. Rev. Lett. 41, 339 (1978).
5. Györffy, B. L. and Stott, M. J. "Band Structure Spectroscopy of Metals and Alloys," Ed. Fabian and Watson, Academic Press, 385 (1968).
6. Mattheiss, L. M. Phys. Rev. 133, A1399 (1964).

7. Davis, H. L., "Computational Methods in Band Theory," Eds. Marcus, Janak and Williams, Plenum Press, 183 (1971).
8. Butler, W. H., Olson, J. J., Faulkner, J. S. and Györffy, B. L., Phys. Rev. B14, 3823 (1976).
9. Stocks, G. M. Györffy, B. L. Giuliano, E. S. and Ruggeri, R. J. Phys. F. 7, 1859 (1977).
10. Temmerman, W. H., Györffy, B. L. and Stocks, G. M. "Transition Metals 1977" Inst. Phys. Conf. Ser. No. 39, 398 (1978) and J. Phys. F. 8, 2461 (1978).
11. Andersen, O. K., Phys. Rev. B12, 3060, (1975) and Physica, 91B, 317 (1977).
12. The complex energies have been used by Bansil et al. (ref. 13) to plot a complex energy band structure of disordered alloys using the average t-matrix approximation. They interpret $\varepsilon_{\vec{q}}^{R}$ and $\varepsilon_{\vec{q}}^{I}$ has given the position in energy and lifetimes of Bloch states in the alloy.
13. Bansil, A., Schwartz, L. and Ehrenreich, H., Phys. Rev. B12, 2893 (1975).
14. See for example the discussion of Sec. II.5 of Ref. 1.
15. See for example Janak, J. F. in "Computational Methods in Band Theory," Eds. Marcus, Janak and Williams, Plenum Press, 323 (1971) and other articles in the same volume.
16. Faulkner, J. S., Davies, H. L. and Joy, H. W., Phys. Rev. 161, 556 (1967).
17. Bansil, A., Solid State Comm. 16, 885 (1975).
18. Fehlner, W. H. and Vosko, S. H., Can. J. Phys. 54, 2159 (1976).
19. Lloyd, P. and Smith, P. V., Advances in Physics 21, 69 (1972).
20. Williams, A. R., Janak, J. F., and Moruzzi, V. L., Phys. Rev., B12, 4509 (1972).
21. Scarborough, J. B., "Numerical Mathematical Analysis," 6th edition Johns Hopkins Press, Baltimore, 215 (1966).
22. Gonis, A., Stocks, G. M., Butler, W. H. and Györffy, B. L., to be published.
23. Bansil, A., Bull. Am. Phys. Soc. 23, 260 (1978).
24. Beeby, J., Proc. R. Soc. (London), A279, 82 (1964) and Phys. Rev. 135, A130, (1964).
25. Korringa, J., J. Phys. Chem. Solids, 7, 252 (1958).
26. Hüfner, G. K., Wertheim, G. K. and Wernick, J. H., Phys. Rev. B8, 4511 (1973).

BULK ELECTRONIC STRUCTURE OF DISORDERED TRANSITION AND NOBLE METAL ALLOYS*

A. Bansil

Physics Department, Northeastern University

Boston, Massachusetts 02115, USA

I. INTRODUCTION

The recent research concerning the electronic structure of disordered transition and noble metal alloys has focussed on the properties of the muffin-tin Hamiltonian.[1] The reason is that the simple one- and two-band tight binding model Hamiltonians are not adequate for describing the detailed experimental information which is now available on specific systems.[2] In this connection, an extensive use has been made of the coherent potential (CPA) and the average t-matrix (ATA) approximations.[1] The former treats the disorder self-consistently, while the latter does not. Hence, it is generally agreed that the CPA is a better approximation from a theoretical point of view. The attractiveness of the ATA derives from its relative simplicity in application to realistic models and the fact that many aspects of the electronic spectrum of muffin-tin alloys are not sensitive to whether the disorder is treated within the framework of the CPA or the ATA.[3]

In spite of the substantial progress made with regard to the application of the CPA and the ATA to the muffin-tin Hamiltonian, basic difficulties had persisted with each of these schemes until recently. The CPA formalism has been well-developed,[4-6] but its practical implementation for the muffin-tin Hamiltonian has been reported only within the last year, due to repeated Brillouin zone integrations necessary to solve the CPA self consistency equation.[3,7,8] On the other hand, the difficulties with the ATA were formal in nature: although the commonly used ATA expression for the average density of states $\langle \rho(E) \rangle$ gives reasonable results in all instances studied so far, the corresponding expressions for the component densities $\langle \rho_{A(B)}(E) \rangle$ (i.e., the electronic charge densities associated with an A (or B) atom in the alloy) yield negative results in many cases, and are not reliable.[9] We have overcome this difficulty by developing a new version of the muffin-tin ATA, which can be expected to give reasonable results for both $\langle\varphi(E)\rangle$ and $\varphi_{A(B)}(E)\rangle$ in transition and noble metal alloys.[10]

An outline of the present paper follows. Sec. II presents the relevant formalism. Since the CPA equations are well-known, we have only reviewed the new ATA equations referred to above. The contributions to the electronic spectrum (for both the CPA and the ATA) are seen to arise from not only the usual Bloch-like states in the medium of effective atoms, but also from inpurity-like states associated with an A or B atom embedded in an otherwise perfect crystal of effective atoms. The Bloch-like complex energy bands have been discussed extensively in the literature,[9,11] but the importance of A and B impurity levels in connection with disordered metals is less familiar.

As noted in Refs. (3) and (7), the solution of the muffin-tin CPA equations was made tractable by using the special directions method of $\vec{k}$-space integration.[12-14] Sec. III reviews the technique and its application to calculations involving disordered alloys.

Sec. IV compares and contrasts the CPA and the ATA, using the illustrative example of Cu_xNi_{1-x}. The effective CPA and ATA scattering phase-shifts are seen to differ significantly in this system.[3,7] But, the physically relevant densities of states and the component densities of states in the two approximations are in good agreement.[3]

II. FORMALISM

The one electron Hamiltonian in the muffin-tin model of a disordered binary alloy A_xB_{1-x} is

$$H = p^2/2m + \sum_n v^{A(B)}(|\vec{r} - \vec{R}_n|) \quad . \tag{1}$$

The alloy potential is thus represented by a sum of spherically symmetric non-overlapping muffin-tin potentials $v^{A(B)}(|\vec{r} - \vec{R}_n|)$, centered at lattice sites $\{\vec{R}_n\}$. The potential at the site $\vec{R}_n$ is v^A or v^B depending on whether this site is occupied by an A or B atom. We will assume that the atoms occupy the lattice randomly, as is done most commonly.

The electronic density of states is discussed conveniently in terms of the Green's function $\mathcal{G}(E) \equiv (E-H)^{-1}$, or alternatively the total scattering operator $\mathcal{T}(E)$ related to $\mathcal{G}(E)$ by the equation

$$\mathcal{G}(E) = G_o + G_o \mathcal{T}(E) G_o \quad , \tag{2}$$

where $G_o \equiv (E - p^2)^{-1}$ is the free electron Green's function. For purposes of dealing with the muffin-tin Hamiltonian (1), it is useful to decompose $\mathcal{T}(E)$ into the so-called path operators $\mathcal{T}_{nn'}$ defined by[15]

$$\mathcal{T}(E) = \sum_{nn'} \mathcal{T}_{nn'}(E) \quad . \tag{3}$$

By going into the angular momentum representation, it is possible to obtain the following exact expressions for the alloy spectrum in terms of on-the-energy-shell matrix elements $T_{nn'}$ of the path operators, and $\tau_n(E)$ of the atomic t-matrices[6]

$$\langle \rho(E) \rangle = \rho_0(E) - \frac{1}{\pi} \mathrm{Im} \sum_L \left[x \langle T_{00} \rangle_{0=A} \frac{d\tau_A^{-1}}{dE} + (1-x) \langle T_{00} \rangle_{0=B} \frac{d\tau_B^{-1}}{dE} - \sum_{n'} \langle T_{0n'} \rangle \frac{dB_{n'0}}{dE} \right]_{LL} \quad , \tag{4}$$

$$\langle \rho_{A(B)}(E) \rangle = \rho_0(E) - \frac{1}{\pi} \mathrm{Im} \sum_L \left[\langle T_{00} \rangle_{0=A} \frac{d\tau_A^{-1}}{dE} - \sum_{n'} \langle T_{0n'} \rangle_{0=A(B)} \frac{dB_{n'0}}{dE} \right]_{LL} , \tag{5}$$

where $\rho_0(E) \equiv (V/4\pi^2 N)E^{\frac{1}{2}}$ is the free electron density of states. τ, B and T are supermatrices in the angular momentum ($L \equiv (\ell,m)$) and lattice site (n) space. $B_{nn'}$ is the real space representation of the usual KKR structure function, $B_{\vec{k}}(E)$.[16] The angular brackets $\langle \ldots \rangle$ in Eqs. (4) and (5) denote complete random averages and the symbol $\langle \ldots \rangle_{0=A(B)}$ denotes a one-site restricted average in the sense that the 0th site is occupied by an A(B) atom and the remaining (N-1) sites are configurationally averaged.

We now write down exact results for the path operators $T_{nn'}$ (which will be useful in connection with the new ATA equations presented below), for the problem of a single A(B) impurity placed in an otherwise perfect medium of effective atoms, t_{eff}. The relevant matrix equations (implicitly in angular momentum space) are:[1,6]

$$(T_{00})^{(imp)}_{0=A(B)} = [1 - T^{eff}_{00}(\tau^{-1}_{eff} - \tau^{-1}_{A(B)})]^{-1} T^{eff}_{00} \quad , \tag{6a}$$

$$(T_{0n})^{(imp)}_{0=A(B)} = [1 - T^{eff}_{00}(\tau^{-1}_{eff} - \tau^{-1}_{A(B)})]^{-1} T^{eff}_{0n} \quad , \tag{6b}$$

where $T^{eff}_{nn'}$ are path operators in an ordered crystal of effective atoms t_{eff} and are given by

$$T^{eff}_{nn'} \equiv N^{-1} \sum_{\vec{k}} e^{i\vec{k}\cdot(\vec{R}_n - \vec{R}_{n'})} (\tau^{-1}_{eff} - B_{\vec{k}}(E))^{-1} \quad , \qquad (7)$$

as a Brillouin zone summation.

In particular, for a single A atom placed in an otherwise perfect B crystal, the density of states $\rho_A^{(imp)}$ associated with the impurity atom is given by the relation

$$\rho_A^{(imp)}(E) = -\frac{1}{\pi} \,\mathrm{Im}\, \frac{d}{dE} \ln||1 - (\tau_B^{-1} - \tau_A^{-1})T^B_{00}|| \quad . \qquad (8)$$

The new ATA approximation to the alloy spectrum is now specified by giving the decouplings[10]

$$\langle T_{00}\rangle^{AT}_{0=A(B)} = [1 - T^{AT}_{00}(\langle\tau\rangle^{-1} - \tau^{-1}_{A(B)})]^{-1} T^{AT}_{00} \quad , \qquad (9a)$$

$$\langle T_{0n}\rangle^{AT}_{0=A} = \frac{\langle\tau\rangle^{-1} - \tau_B^{-1}}{x(\tau_A^{-1} - \tau_B^{-1})} T^{AT}_{0n} \quad , \qquad (9b)$$

for the restricted averages occurring in the exact formulae (4) and (5) in terms of the complete averages $T^{AT}_{nn'}$. [$\langle T_{0n}\rangle^{AT}_{0=B}$ is obtained from (9b) with interchanges A⟷B and x⟷y.] $T^{AT}_{nn'}$ is defined by Eq. (7) with $\tau_{eff} = \langle\tau\rangle$ where,

$$\langle\tau\rangle \equiv x\tau_A + y\tau_B \quad . \qquad (10)$$

Equations (9) when inserted into Eqs. (4) and (5) immediately yield the final expressions for the ATA spectrum

$$\langle\rho(E)\rangle^{AT} = \rho_0(E) - \frac{1}{\pi N} \,\mathrm{Im\,Tr} \sum_{\vec{k}} \left[\left(\frac{x}{1 - T^{AT}_{00}(\langle\tau\rangle^{-1} - \tau_A^{-1})} \frac{d\tau_A^{-1}}{dE} + \frac{1-x}{1 - T^{AT}_{00}(\langle\tau\rangle^{-1} - \tau_B^{-1})} \frac{d\tau_B^{-1}}{dE} - \frac{dB_{\vec{k}}}{dE}\right)\left(\frac{1}{\langle\tau\rangle^{-1} - B_{\vec{k}}}\right)\right], \qquad (11)$$

$$\langle \rho_A(E) \rangle^{AT} = \rho_0(E) - \frac{1}{\pi N} \mathrm{Im\,Tr} \sum_{\vec{k}} \left[\left(\frac{1}{1 - T_{00}^{AT}(\langle \tau \rangle^{-1} - \tau_A^{-1})} \frac{d\tau_A^{-1}}{dE} \right. \right.$$

$$\left. \left. - \frac{\langle \tau \rangle^{-1} - \tau_B^{-1}}{x\,(\tau_A^{-1} - \tau_B^{-1})} \frac{dB_{\vec{k}}}{dE} \right) \left(\frac{1}{\langle \tau \rangle^{-1} - B_{\vec{k}}} \right) \right] \quad , \qquad (12)$$

with $\langle \rho_B(E) \rangle^{AT}$ obtained from (12) with interchanges $A \longleftrightarrow B$ and $x \longleftrightarrow y$. [Here, the trace operation refers to the angular momentum space.] $\langle \rho_{A(B)}(E) \rangle^{AT}$ obviously satisfy the relation

$$x \langle \rho_A(E) \rangle^{AT} + y \langle \rho_B(E) \rangle^{AT} = \langle \rho(E) \rangle^{AT} \quad , \qquad (13)$$

which represents the constraint of charge neutrality of the average alloy.

Equations (9) for the ATA decoupling are to be compared with the corresponding CPA equations:[6]

$$\langle T_{00} \rangle_{0=A(B)}^{CP} = [1 - T_{00}^{CP}(\tau_{CP}^{-1} - \tau_{A(B)}^{-1})]^{-1} T_{00}^{CP} \quad , \qquad (14a)$$

$$\langle T_{0n} \rangle_{0=A(B)}^{CP} = [1 - T_{00}^{CP}(\tau_{CP}^{-1} - \tau_{A(B)}^{-1})]^{-1} T_{0n}^{CP} \quad , \qquad (14b)$$

in terms of $T_{nn'}^{CP}$, which are given by Eq. (7) with $\tau_{eff} = \tau_{CP}$. In contrast to the ATA equation (10), however, τ_{CP} is now determined by the solution of the CPA condition[4-6]

$$\tau_{CP}^{-1} = \langle \tau^{-1} \rangle + (\tau_{CP}^{-1} - \tau_A^{-1})\, T_{00}^{CP}\, (\tau_{CP}^{-1} - \tau_B^{-1}) \quad , \qquad (15)$$

and is much harder to compute.

A reference to the single impurity formula (6a) shows that Eq. (9a) for $\langle T_{00} \rangle_{0=A(B)}^{AT}$ is of exactly the same form as (6a) and, therefore, physically represents a single A(B) atom placed in a medium of $\langle \tau \rangle$ atoms. Equation (9b), however, is not of the same form as (6b). The reason we do not choose the single impurity form (6b) for $\langle T_{0n} \rangle_{0=A(B)}^{AT}$ in Eq. (9b) is that such a choice would lead to the violation of the charge neutrality requirement (13). The motivation for the specific form (9b) can be understood by noting that if the CPA condition (15) is used in (14b), it is easily shown that

$$\langle T_{0n} \rangle^{CP}_{0=A} = \frac{\tau^{-1}_{CP} - \tau^{-1}_{B}}{x(\tau^{-1}_{A} - \tau^{-1}_{B})} T^{CP}_{0n} \quad , \tag{16}$$

which is identical in form to Eq. (9b). It is now clear that if Eqs. (14a) and (16) are inserted into Eqs. (4) and (5), the resulting expressions for the CPA spectrum will be of identical form to the corresponding ATA expressions (11) and (12). In particular, if τ_{CP} were inserted instead of $< \tau >$ in Eqs. (11) and (12), the CPA spectrum will result. Since $< \tau >$ may be looked upon as the lowest order approximation to the CPA solution τ_{CP} of Eq. (15), the present ATA may be viewed as the lowest order approximation to the CPA. Indeed this ATA can be improved systematically by simply replacing $< \tau >$ by the successively better approximations to τ_{CP}. [We emphasize that the preceding desirable features of the new ATA emanate directly from the form of the decoupling (9). In particular the form of the muffin-tin ATA, reviewed in Ref. (1) does not possess these features.]

Equation (11) shows that the peaks in the spectral function [i.e., the quantity contained in the large square brackets in Eq. (11)] come not only from the complex zeros of the secular equation

$$|| \langle \tau \rangle^{-1} - B_{\vec{k}}(E) || = 0 \quad , \tag{17}$$

but also from those of

$$|| 1 - T^{AT}_{00} \left(\langle \tau \rangle^{-1} - \tau^{-1}_{A(B)} \right) || = 0 \quad . \tag{18}$$

Equation (17) has been discussed in the literature in the context of disordered alloys.[1,9,11] Note that for a perfect A(B) crystal $< \tau > \to \tau_{A(B)}$ and Eq. (17) reduces to the KKR equation of the band theory of perfect crystals. Therefore, this equation can be viewed as representing Bloch-like states in the alloy associated with the $< t >$ atoms.

By contrast, Eq. (18) possesses no simple analogy with a perfect crystal and its importance in connection with disordered alloys has not been realized until recently.[3,7] A reference to the single impurity formula (8) shows that in the limit of x(or y) $\to 0$ (since $< \tau > \to \tau_B$(or τ_A)), Eq. (18) will give real impurity levels for an A (or B) atom embedded in an otherwise perfect B (or A) crystal. In general, therefore, Eq. (18) may be viewed as representing the impurity levels arising when a single A (or B) atom is placed in < t > medium. We emphasize that the impurity states given by Eq. (18) are non-Bloch-like in nature, and their appearance in the alloy is a characteristic effect of disorder.

III. SPECIAL DIRECTIONS TECHNIQUE

The basic difficulty in solving the CPA equation (15) arises from the fact that to solve this equation, repeated Brillouin zone integrations must be carried out to evaluate T_{00}^{CP} via Eq. (7). As already noted, this calculation can be made tractable by using the special directions method of $\vec{k}$-space integration.[12-14] The efficiency of the calculations can be further enhanced by using the general $\vec{k}$ and E dependent numerical interpolation scheme of Williams et. al.[17] to evaluate the KKR structure functions $B_{\vec{k}}(E)$ and their derivatives $(dB_{\vec{k}}/dE)$. A brief review of the special directions method follows.

The special direction method is a Gaussian quadrature technique like the special points method introduced by Baldereschi[18] (and developed further by various authors[19,20]) for integrating periodic functions over the Brillouin zone. In the special points approach, given an integral

$$f_0 = \int_{BZ} d^3k \; f(\vec{k}) \tag{19}$$

of a function $f(\vec{k})$ periodic in the reciprocal space of the crystal, one makes use of an expansion of $f(\vec{k})$ in terms of the plane wave set $\{\Phi_n\} \equiv \{\exp(i\,\vec{k}\cdot\vec{R}_n)\}$. As usual in Gaussian integration methods,[21] a sequence of special points $\{\vec{k}_j^{(N)};\ j=1,2,\ldots N\}$ is then determined such that an average of $f(\vec{k})$ over just these points yields a reliable approximation to the value f_0 of the integral (19).[18-20] (The larger the value of N, the more accurate is the integration procedure.) While this technique can, in principle, be used with any periodic function $f(\vec{k})$, it is best suited for integrands that are slowly varying in the Brillouin zone. If (like the alloy spectral density) $f(\vec{k})$ varies rapidly, many terms in plane wave expansion are required, and the usefulness of the special points method is lost due to its slow convergence.

The central idea in the special direction method is that, rather than attempting to expand the sharp peaks of the spectral density in terms of plane waves, one integrates them exactly. Given an integral of the form (19), we define the related quantity

$$\tilde{f}(\hat{k}) \equiv \int_{BZ} dk\, k^2\, f(\vec{k}) \quad , \tag{20}$$

where the integration is now along the radial direction of the unit vector $\hat{k}$. The essential point is that $\tilde{f}(\hat{k})$ will be a relatively smoothly varying function of the angles defining $\hat{k}$, because the rapidly varying structure in $f(\vec{k})$ has now been integrated out. Because $\tilde{f}(\hat{k})$ is still invariant under the operations of the point group of the lattice (assumed to be cubic here), its angular dependence can be expanded in terms of the cubic harmonics $\{C_\ell(\hat{k})\}$:

$$\tilde{f}(\hat{k}) = f_0 + \sum_{\ell \neq 0} \tilde{f}_\ell\, C_\ell(\hat{k}) \ , \ \ell=4,6,8... \tag{21a}$$

where

$$\tilde{f}_\ell = \int_{BZ} d\Omega_{\vec{k}}\, \tilde{f}(\hat{k})\, C_\ell(\hat{k}) \quad \ell=0,4,6,8... \tag{21b}$$

The function $C_0(\hat{k})$ is taken equal to unity. In view of Eqs. (19) and (21), the quantity of principal interest is simply the coefficient

$$f_0 = \int_{BZ} d\Omega_{\vec{k}}\, \tilde{f}(\hat{k}) \equiv \int_{BZ} d^3k\, f(\vec{k}) \quad . \tag{22}$$

To illustrate the idea of special directions, suppose that a direction $\hat{k}_1$ can be found such that

$$C_4(\hat{k}_1) = 0 = C_6(\hat{k}_1) \quad . \tag{23}$$

Eqs. (21) would then imply that

$$\tilde{f}(\hat{k}_1) = f_0 + (\text{terms with } \ell \geq 8).$$

Thus, if the terms with $\ell \geq 8$ are small, the value of the function $\tilde{f}(\hat{k})$ along the single special direction $\hat{k}_1$ will give a reasonable approximation to the complete integral f_0. The evaluation of the three dimensional integral f_0 has then been reduced to the evaluation of the single one dimensional integral $\tilde{f}(\hat{k}_1)$. Despite the fact that $f(\vec{k})$ varies rapidly, computation of $\tilde{f}(\hat{k}_1)$ is quite straightforward, since a one dimensional mesh can easily be made most dense in the regions where $f(\vec{k})$ peaks.

To generalize this procedure, we solve for a set of N special directions $\{\hat{k}_j^{(N)}; j=1,\ldots N\}$ with corresponding weights $\alpha_j^{(N)}$ such that,

$$\sum_{j=1}^{N} \alpha_j^{(N)} C_\ell(\hat{k}_j^{(N)}) = 0 \qquad \text{for } \ell=4,\ldots \ell_{max} \tag{24}$$

Eqs. (21) and (22) then yield

$$f_0 \cong \sum_{j=1}^{N} \alpha_j^{(N)} \tilde{f}(\hat{k}_j^{(N)}) \tag{25}$$

where the correction terms involve only $\ell > \ell_{max}$, and the quantity f_0 is now approximated in terms of N one dimensional integrals $\tilde{f}(\hat{k}_j^{(N)})$. It should be emphasized, that once the directions $\{\hat{k}_j^{(N)}\}$ have been determined for the cubic lattice they can be utilized for the integration of any function with cubic symmetry.

Eqs. (24) are a coupled, highly non-linear set of polynomial equations for the special directions $\hat{k}_j^{(N)}$. As noted in Ref. (12), the numerical solution of these equations for $N \gtrsim 13$ becomes very difficult. In this connection Fehlner and Vosko[13] have introduced a representation for cubic harmonies in terms of the functions

$$Q(x,y,z) \equiv x^4 + y^4 + z^4 \tag{26a}$$

$$S(x,y,z) \equiv x^2 y^2 z^2 \quad . \tag{26b}$$

They show that every cubic harmonic can be expressed as an appropriate polynomial in the Q-S space. In this way, as in the corresponding Gaussian problem in one dimension, they are able to reduce the calculation of special directions to the equivalent, much more tractable problem of finding simultaneous zeros of suitable polynomials in Q-S space. Ref. (13) presents weighted sets of 16, 21, and 25 special directions. Using the techniques of Ref. (13), Prasad and Bansil[14] have recently obtained weighted sets of 28, 36,

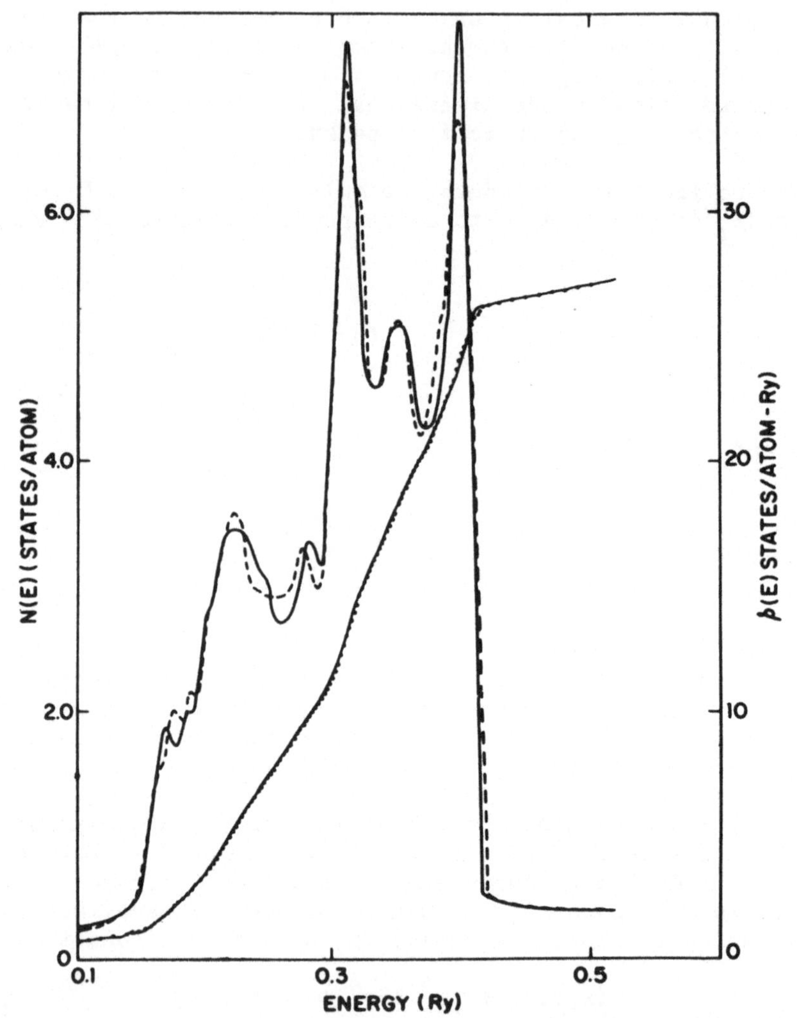

Fig. 1 : A comparison of the density of states ρ(E) (dashed) and the integrated density of states N(E) (solid dots) for fcc Cu, obtained by the use of 13 special directions with the results of a histogram (shown as smooth solid lines) using 1638 points in the irreducible 1/48th of the zone. Both calculations employ the Hodges' interpolation scheme parametrization[22] of the Cu band structure of Burdick.[23]

45, 55 and 66 special directions for cubic lattices.

Figure 1 compares the density of states, $\rho(E)$, and the integrated density of states, $N(E)$, for fcc Cu on the basis of 13 special directions with results obtained by using 1638 points in the irreducible 1/48th of the zone.[12] The agreement between the two calculations in Fig. 1 is seen to be quite good. We emphasize that for 13 special directions, the errors in the computations in disordered transition and noble metal alloys are generally expected to be smaller than those seen in Fig. 1. The reason is that the electronic states in alloys possess broadening due to disorder, and therefore, the alloy spectrum is inherently smoother than the corresponding perfect crystals. In any event, the accuracy of a given computation can be ascertained by using successively larger sets of special directions.

IV. DISCUSSION OF RESULTS

We have already noted that the ATA based on Equations (11) and (12) can be viewed as the lowest order approximation to the CPA. The following comparison of the effective scattering phase shifts and the total and component densities of states in the CPA and the ATA, therefore, reveals the effects of self consistency on these quantities.

The energy dependence of the $\ell=2$ scattering phase shifts in $Cu_{0.75}Ni_{0.25}$ is shown in Figure 2. The s and p phase shifts for Cu, Ni, and the alloy are quite similar and vary slowly in the energy regime of present interest and are not considered. Due to the cubic symmetry of the average alloy, the d phase shifts in the CPA for states of t_{2g} and e_g symmetry are seen to be different. This distinction, however, does not apply to the ATA scatterer (shown dot-dashed), which is spherically symmetric. We have plotted the function $\sin\delta_2(E)$ in Fig. 2, because the position of the d resonance in a pure A (or B) crystal (which corresponds to $\delta_2 \rightarrow \pi/2$) appears as a peak in $\sin\delta_2(E)$, making this a convenient function for representing effective scattering amplitudes. [Since the phase shifts are undetermined up to $\pm\pi$, $\sin\delta_2$ has been defined such that $\mathrm{Re}(\sin\delta_2)$ is always positive.]

Figure 2 shows that the peaks in the real as well as the imaginary parts of $\sin\delta_2$ arising from Ni impurities (marked by arrows) are shifted to higher energies by approximately 0.05 Ry in the CPA compared to the ATA.[3,7] Such movements in the positions of the impurity peaks in the alloy phase shifts are found more generally in the Cu-rich as well as the Ni-rich alloys. We emphasize, however, that this does not imply that there will be corresponding differences in the energy location of the impurity densities of states between the CPA and the ATA. This is so, because the alloy

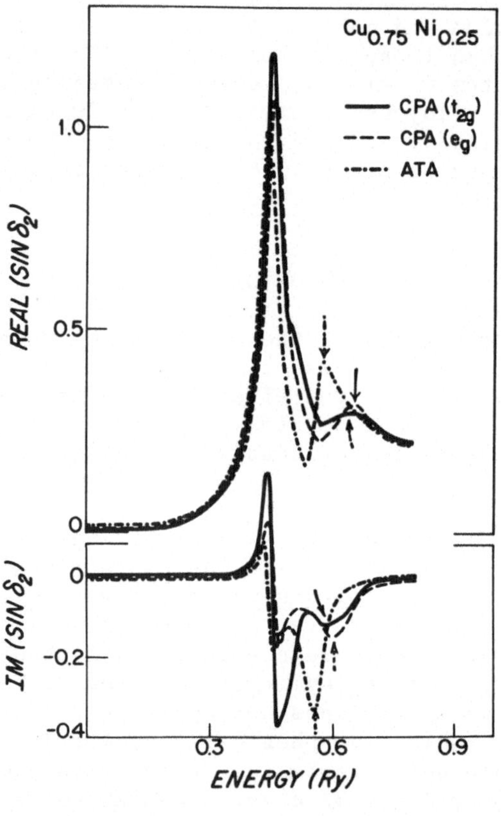

Fig. 2 : Real and imaginary parts of sin $\delta_2(E)$ in $Cu_{0.75}Ni_{0.25}$. The arrows point out the bumps in the plotted curves in the energy region of the Ni-impurity resonance in the CPA (t_{2g}, solid; e_g, dashed) and the ATA (dot-dashed).

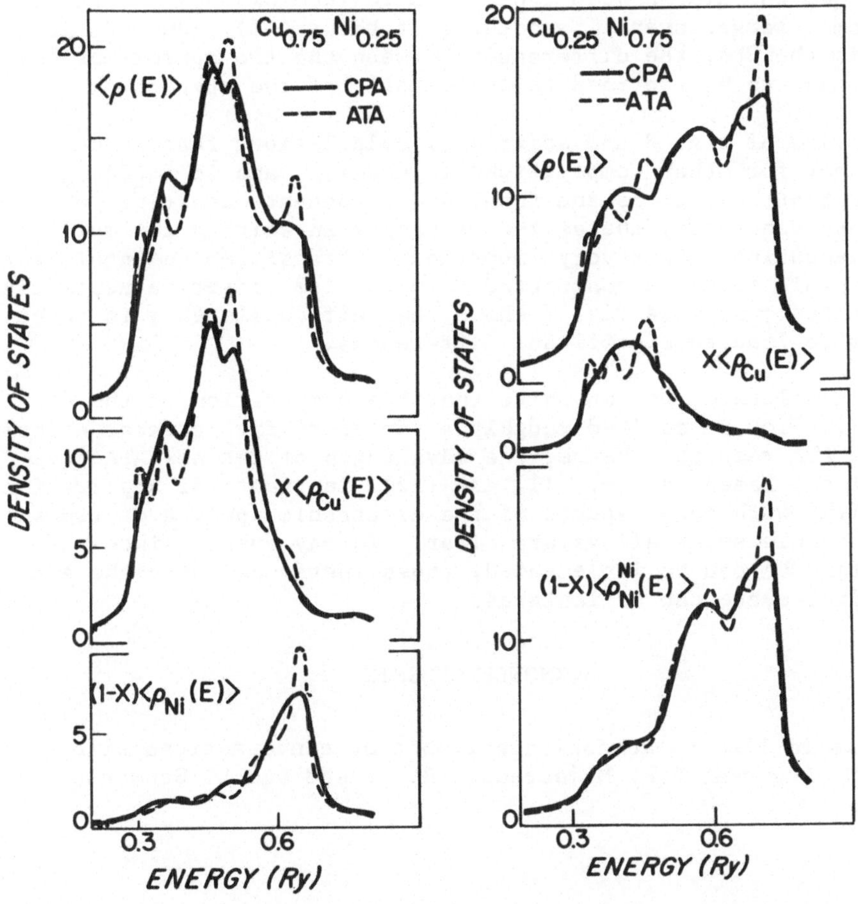

Fig. 3 : The CPA and the ATA density of states $\langle\rho(E)\rangle$ and the weighted A and B component densities of states $x\langle\rho_A(E)\rangle$ and $(1-x)\langle\rho_B(E)\rangle$ in Cu_xNi_{1-x} for x = 0.75 and x = 0.25. [Note that, $x\langle\rho_A(E)\rangle + (1-x)\langle\rho_B(E)\rangle = \langle\rho(E)\rangle$.]

spectrum is not determined solely by the properties of τ_{eff} (i.e., the Bloch-like contributions (17)). In addition, contributions from a single A or B impurity placed in an otherwise ordered medium of effective atoms τ_{eff} (i.e., from solutions of Eq. (18)) must be considered. Indeed, Fig. 3 shows that the impurity spectrum $< \rho_{Ni}(E) >$ in $Cu_{0.75}Ni_{0.25}$ and $Cu_{0.25}Ni_{0.75}$ are very similar in the CPA and the ATA in both their location and width in energy. Aside from a larger overall smoothing of the $< \rho(E) >$ and $<\rho_{A(B)}(E)>$ curves in the CPA, the differences between the two approximations would appear to be confined to the details of the spectra.

In view of Fig. 3 and additional calculations that we have carried out for other compositions in Cu_xNi_{1-x} and in Cu_xZn_{1-x},[10] it seems that insofar as the total and component densities of states are concerned, the self-consistency in solving the muffin-tin CPA equation is not very important in transition and noble metal alloys. This is to be contrasted with the $\ell=2$ effective scattering phase shifts, where as Fig. 2 shows, the self-consistency is to be expected to lead to significant differences.

In conclusion, we emphasize that the computation of the CPA spectra in Fig. 3 required roughly a factor of five greater effort than the ATA results. Hence, the advantages of the new ATA equations (i.e., based on Eqs. (11) and (12)) in practical applications for dealing with many aspects of the electronic spectra of transition and noble metal alloys are clear. In any event, since the muffin-tin CPA can be implemented, those instances where the ATA may be inadequate can be isolated.

ACKNOWLEDGEMENTS

I would like to acknowledge important conversations with Drs. P.N. Argyres, P.E. Mijnarends, R. Prasad and L. Schwartz.

---

* Work supported in part by the U.S. Department of Energy under contract No. ER-78-S-02-4720.

REFERENCES

1. For a recent theoretical review see, H. Ehrenreich and L. M. Schwartz in Solid State Physics, Vol. 31, edited by H. Ehrenreich, F. Seitz and D. Turnbull (Academic Press, New York, 1976).

2. For a recent review of the experimental literature see, D. J. Sellmeyer in Solid State Physics (1978), edited by H. Ehrenreich, F. Seitz and D. Turnbull.

3. A. Bansil, Phys. Rev. Letters (to be published).

4. P. Soven, Phys. Rev. B2, 4715 (1970).

5. B. Györffy, Phys. Rev. B5, 2382 (1972).

6. L. Schwartz and A. Bansil, Phys. Rev. B10, 3261 (1974).

7. G. M. Stocks, W. M. Temmerman and B. L. Györffy, Phys. Rev. Letters 41, 339 (1978). Approximate solutions to the CPA equation in Cu_xNi_{1-x} have also been presented by Stocks, Györffy, Giuliano and Ruggeri [J. Phys. F7, 1859 (1977)].

8. The first calculations of the CPA density of states, $< \rho(E) >$, on the basis of the muffin-tin model were reported in Cu_xNi_{1-x} by Stocks, Temmerman and Györffy [Bull. Am. Phys. Soc. 23, 216 (1978)] and Ref. (7), and by Bansil [Bull. Am. Phys. Soc. 23 360 (1978) and Ref. (3)]. In this regard the results of Refs. (7) and (3) are in agreement. The CPA component densities of states $< \rho_{Cu}(E) >$ and $< \rho_{Ni}(E) >$, however, have only been presented in Ref. (3).

9. A. Bansil, L. Schwartz and H. Ehrenreich, Phys. Rev. B12, 2893 (1975).

10. A. Bansil (to be published).

11. C. D. Gelatt, H. Ehrenreich and J. A. Weiss, Phys. Rev. B17, 1940 (1977).

12. A. Bansil, Solid State Commun. 16, 885 (1975).

13. W. R. Fehlner and S. H. Vosko, Can. J. Phys. 54, 2159 (1976).

14. R. Prasad and A. Bansil (to be published).

15. A detailed discussion of the multiple scattering theory techniques, as applied to the muffin-tin Hamiltonian is to be found in Ref. (1) and in the article by Drs. Györffy and Stocks in this volume.

16. B. Segall and F. S. Ham in Methods in Computational Physics, edited by B. Alder, S. Fernbach and M. Rotenberg (Academic, New York 1968), Vol. 8.

17. A. R. Williams, J. F. Janak and V. L. Morruzzi, Phys. Rev. B6, 4509 (1972).

18. A. Baldereschi, Phys. Rev. B7, 5212 (1973).

19. D. J. Chadi and M. L. Cohen, Phys. Rev. B8, 5787 (1973).

20. H. J. Monkhorst and J. D. Pack, Phys. Rev. B13, 5188 (1976).

21. See, for example, E. Isaacson and H. B. Keller, Analysis of Numerical Methods (Wiley, New York, 1966), p. 327.

22. L. Hodges, H. Ehrenreich and N. D. Lang, Phys. Rev. 152, 505 (1966).

23. G. A. Burdick, Phys. Rev. 129, 138 (1963).

FERMI SURFACE STUDIES IN DISORDERED ALLOYS: POSITRON ANNIHILATION EXPERIMENTS

S. Berko

Physics Department, Brandeis University

Waltham, Massachusetts 02154, U.S.A.

INTRODUCTION

In these lectures we shall outline the use of the positron (e^+) annihilation technique for the study of electrons (e^-) in metals and, in particular, in non-dilute disordered alloys. The behavior of slow e^+-s in condensed matter has been the subject of intense experimental and theoretical investigations during the last two decades, and the field has been reviewed thoroughly;[1] positron annihilation has become a useful new tool in solid state physics. By studying the various properties of annihilation quanta one obtains direct information about the electrons the positrons annihilate with. In particular the 2γ angular correlation of annihilation radiation (ACAR) reflects, by momentum conservation, the momentum density of the annihilating e^+-e^- pair. Discontinuities in the correlation curves can be used to measure the size and shape of the Fermi surface (FS) in pure metals as well as in alloys. If the behavior of the e^+ is sufficiently well understood and the effect of the e^+-e^- Coulomb interaction sufficiently well accounted for, the ACAR measurements can lead, as we shall see, to a test of the electronic wavefunctions of the system under investigation.

In most pure metals the e^+ annihilation technique cannot compete in precision with the standard methods of Fermiology, such as the measurement of magneto-resistance, the magneto-acoustic effect, and, in particular, the de Haas-van Alphen effect.[2] During the last decade, however, the central interest in metal physics has shifted from the study of pure ordered metals, to the investigation of disordered systems such as random substitutional alloys - one of the major topics of discussion in this Institute. It is well known, however, that because of scattering, the short lifetime of conduction electrons in such disordered systems limits severely the usefulness of the standard FS techniques. The dHvA measurements can be performed only in very dilute alloys, with solute concentrations of less than a few percent. As we shall see, the e^+ annihilation technique does not have these limitations, and FS measurements can also be made in high concentration random alloys. The technique has of course its own typical problems; the positron, being charged, is not an "ideal probe" compared to x rays or to neutrons. The e^+-e^- interaction can directly affect the system under investigation. We shall discuss the extent of these specific limitations later on.

Much of our knowledge about the electronic behavior of non-dilute disordered alloys has been obtained by the use of the more classical experimental techniques, such as the measurement of low temperature specific heat, resistivity, thermoelectric power and various optical properties.[3] In principle the electronic specific heat measurement can be a most valuable guide to the validity of various theoretical models, since it depends directly on the density of states at the FS. The interpretation of the experimental data can become complicated, however, by large enhancement factors due to electron-phonon and electron-electron interactions. A thorough review of specific heat measurements and of other experimental results in noble metal based alloys (Hume-Rothery phases) has recently been given by Massalski and Mizutani.[4] Far more details about the density of states in metals and alloys can be obtained from photoemission work,[5] and in particular from photoelectron spectroscopy with x rays.[6] The density of states obtained by these techniques showed most clearly the failure of the rigid band theories (in Cu-Ni for example) and played a crucial role in the early successes of the modern electronic theories of alloys.[7]

Positron annihilation is not the only technique that can, at least in principle, provide information about the FS of non-dilute disordered alloys. High precision neutron inelastic scattering reflects, via the Kohn anomaly,[8] the position of some parts of the FS in $\underline{k}$ space. Similarly, some FS dimensions have been obtained by the observation of Kohn anomaly discontinuities in x ray diffuse scattering,[9] as well as in electron diffraction.[10] Another interesting technique is the use of the polar reflection Faraday effect[11] to measure the size of parts of the FS of a class of noble metal based alloys. A technique associated closely with the e^+ annihilation measurements is the Compton profile measurement[12] using x rays or nuclear γ rays. The Compton profile reflects directly the momentum density of the electrons, and, as in the e^+ ACAR measurements, discontinuities in the momentum density indicate the position of the FS.

Before starting our detailed discussion of the e^+ experiments, we would like to describe briefly the above mentioned "non-standard" techniques, since they are not covered by the other speakers. What follows is not intended to be a systematic review of the various techniques and results, but rather a tutorial introduction to the field with some illustrative examples and with references to proper in-depth reviews.

2. THE FERMI SURFACE OF DILUTE ALLOYS; DE HAAS-VAN ALPHEN EXPERIMENTS

Before discussing the experimental techniques applicable to non-dilute alloys, we briefly outline the experimental status of FS measurements in dilute alloys. It is well known that one of the great successes of solid state physics during the last two decades has been the experimental measurement and the band theoretical computation from first principles of the Fermi surfaces of most pure metals and of some ordered alloys.[2] The experimental technique that produces the most precise FS results is the observation of the de Haas-van Alphen (dHvA) effect, i.e. the oscillatory dependence of the magnetization on the applied magnetic field. In the presence of impurities the oscillatory part of the magnetization (for a single harmonic component) is given by

$$M_{osc} = M_o(B,T)\cos\left(\frac{2\pi F}{B} + \phi\right)\exp\left[-K\,\frac{m_c}{m_o}\,\frac{T_D}{B}\right] \qquad 2.1$$

where $M_o(B,T)$ is the temperature (T) dependent amplitude in the magnetic field B,F the dHvA frequency, m_c/m_o the cyclotron mass in terms of the free electron mass, K is a constant (146.9 kG/Kelvin) and T_D is the Dingle temperature. The frequency is associated to an extremal FS area A perpendicular to B by the Onsager relation $F = \hbar cA/2\pi e$. T_D contains the effect of scattering due to impurities, $T_D = (\hbar/2\pi k_B) \langle 1/\tau(\underline{k})\rangle$; $\tau(\underline{k})$ represents the electron lifetime of the state located on the FS at vector $\underline{k}$, and the bracket is an average over the FS orbit having maximal area A.

The exponential damping term produces a drastic reduction in amplitude, and is thus responsible for the limitation of the technique; at a 1% at. concentration the amplitude can drop by several orders of magnitude from its value in the pure metal. It is at these low amplitudes that the experimentalist has to observe the small changes in the dHvA frequencies upon alloying. Nevertheless, because of great improvements in precision over the last decade, many experiments have been performed both on FS changes and Dingle temperatures in various dilute alloys. Early work in alloys has been reviewed by Heine;[13] more recently Coleridge[14] has given summaries of experimental results as well as of the theoretical analyses of the data.

Much theoretical work has been published on the existence, definition, and blurring of the FS in alloys[15] - Dr. Györffy will no doubt refer to some of these in his lectures. Experimentally, as long as the dHvA amplitudes permit, one can observe the change ΔF upon alloying corresponding to a change ΔA in area of a characteristic FS feature of the host metal. For example, in early experiments on Cu based alloys, Coleridge and Templeton[16] have studied the effect of alloying on the <111> neck cross sections of various heterovalent and transition metal solutes. They have also observed the variation of the Dingle temperature with concentration, by studying the magnetization amplitude at a fixed T as a function of B (Eq. 2.1). They compare the results of the FS change to predictions of the rigid band (RB) model.

The earliest versions of the RB model assumed that the band structure of the host metal remains unchanged and the solute electrons simply raise the Fermi energy. Various corrections can be made, at least for dilute alloys, to take into account the change of lattice parameters with alloying. In his classic paper on impurity scattering Friedel[17] has shown that in the dilute limit the Fermi energy stays fixed and the bands shift due to the impurity. In an analysis of optical data of disordered Cu(Zn) (α phase) alloys Lettington[18] found indeed that he could fit the experimental results with the aid of copper-like energy bands where the Fermi energy did not move appreciably, but the conduction band was displaced downward with respect to the other bands with increasing Zn concentration. A similar "sinking conduction band" model was used more recently by Rea and De Reggi[19] to explain their optical measurements in Cu(Al) alloys. It was found in this model that although the energy bands do not remain "rigid" the FS dimensions change along the RB predictions. The initial slope S of the relative change ΔA of the FS areas with the change in electron concentration Δn given by the RB model can be expressed purely in experimentally observable quantities:[16]

$$S = \frac{(\Delta A/A)}{(\Delta n/n)} = \left(\frac{16\pi^4}{3}\right)^{1/3} \left(\frac{\hbar c}{e}\right) \frac{1}{a^2 F} \left(\frac{m_c}{m_{th}}\right) \qquad 2.2$$

where a is the lattice parameter, A and n the FS area and electron concentration of the host metal, F the dHvA frequency, m_c the cyclotron mass and m_{th} the thermal mass obtained from specific heat measurements. The value of S for copper based alloys is 6.2 for the neck area and 0.69 for the belly compared to the free electron value of 2/3. Coleridge and Templeton find that while Cu(Zn) and Cu(Al) dilute alloys follow well the RB predictions, Cu(Si) and Cu(Ge) do not. The transition metal impurities showed no simple correlation with valence. The Dingle temperature results indicate a much larger lifetime anisotropy with transition metal solutes than with heterovalent solutes. T_D varies linearly with impurity concentration, the variation ranging from 26 K/at% for the neck to 50 K/at% for the belly orbit in Cu(Ni) alloys.

With the increasing precision in experimental technique it has been possible more recently to obtain sufficient dHvA data in dilute alloys to actually invert the data and obtain parametrized FS-s, as has been done earlier in pure metals. The inversion is obtained either by a purely empirical parametrization of the FS or by the use of specific band model parameters. Similarly, one can invert the Dingle temperature data and obtain a map of lifetimes on the FS in $\underline{k}$ space. Thus, Coleridge, Holzwarth and Lee[20] have developed a general framework for noble metal based alloys for the self-consistent phase shift analysis of the dHvA data. Their method uses the Korringa-Kohn-Rostoker (KKR) phase shift parametrization of the FS and the description of the impurity scattering by the introduction of a set of effective ("Friedel") scattering phase shifts (s, p and d) using the experimentally obtained Dingle temperatures. In applying their technique to the high precision Dingle-temperature measurements of Poulsen, Randles and Springford[21] on the Cu(Ni), Cu(Ge) and Cu(Au) systems they find that the complex Dingle temperature anisotropies can indeed be well parametrized by the use of only three scattering phase shifts.

An analysis of new high precision dHvA data on Cu(Al) and Cu(Ni) by Coleridge and Templeton[22] indicates that the FS changes in Cu(Al) are approximately rigid band like, but deviate for Cu(Ni); as expected, scattering is dominated by the d like character of the solute in Cu(Ni).

In alloys where the host metal has nearly-free-electron (NFE) like bands, the impurity scattering is formulated within the pseudopotential framework. The change in the FS is described using pseudopotential model parameters rather than scattering phase shifts. An example of such an alloy system treatment is the work by Hornbeck, Fung and Gordon[23] on Mg(Li) and Mg(In) alloys. Using a nonlocal pseudopotential model they find that a modified RB approach considering both Fermi energy and local band-edge changes gives an overall satisfactory description of the observed dHvA frequency shifts.

As we have indicated, the study of electron lifetimes has become the focus of much experimental and theoretical activity during the last decade. The experiments are not confined to dHvA measurements alone. Results with many other techniques as well as applications to other solid state problems can be found in papers presented at the 1974 Oregon meeting on electron lifetimes.[24]

3. KOHN ANOMALY EXPERIMENTS

As mentioned in the introduction, some features of the FS can be studied by inelastic neutron scattering or diffuse x ray or electron scattering. In 1959 Kohn[8] has pointed out that because of the conduction electron response to the ionic motions, images of the FS are expected to appear in the phonon spectra of metals. Specifically, one expects to observe a discontinuity in the phonon dispersion curves at a reduced phonon wave vector $\underline{q}$ satisfying the relation $\underline{q}+\underline{K} = \underline{k}_{2f}-\underline{k}_{1f}$, where $\underline{K}$ is a reciprocal lattice vector, and $\underline{k}_{1f}$ and $\underline{k}_{2f}$ define two points on the FS with parallel tangent planes. (For spherical FS-s this reduces to a discontinuity at $|\underline{q}+\underline{K}| = 2k_f$, k_f being the Fermi radius.) This discontinuity stems from the well known logarithmic infinity of the dielectric function at $2k_f$.[25] The theory of the Kohn effect was further developed in several papers,[26] in order to extend it to realistic FS topologies. The first observation of the effect was made in 1961 by Brockhouse, Rao and Woods[27] using neutron scattering, and Paskin and Weiss[28] using x ray diffuse scattering. The effect is extremely small and high precision measurements are required; in neutron scattering it appears as a discontinuity in the derivative of the dispersion curves with respect to $\underline{q}$. Since these first results many detailed experiments have been performed on some pure metals as well as on a few alloy systems. Since the effect does not require long electronic lifetimes (the experiments can be performed at elevated temperatures, for example), the technique would be well suited to study non-dilute alloy FS features. In pure metals, after the early measurements in Pb[27] and in Al,[29] Kohn anomalies were studied in other metals, and were related to the FS in these materials.[30] We would like to point out the more recent high precision measurement[31] in Cu. The Kohn anomalies were found to be surprisingly weak, even along the <110> direction where one would expect[26] the nearly parallel FS perpendicular to <110> to contibute a strong anomaly. About twenty discontinuities were studied and were related to the FS of Cu in the (100) and (110) planes. Good agreement (a few percent) was found with the FS obtained from dHvA measurements. It would be quite useful to study some copper based disordered alloys with the same precision, in order to compare the results with the e^+ annihilation data.

Two of the few alloy systems studied in detail by the neutron technique are the Bi-Pb-Tl alloys[32] and the Nb-Mo alloys.[30] The Kohn anomalies observed in the Nb-Mo system correlated reasonably well with some of the FS features predicted by a RB estimate. We should note, however, that the nature of the anomalies appearing in neutron scattering cannot always be attributed to the Kohn effect; many anomalies have been observed that do not correlate with FS calipers. Because of the great interest in superconductivity and the role of the electron-phonon coupling in the transition temperature T_c, many experiments have been performed correlating phonon anomalies with T_c. A review of such experiments and of other phonon anomaly studies was given recently by Smith, Wakabayashi and Mostoller.[33]

We should of course mention at this stage the explosive growth during the last few years in the field of neutron (as well as x ray) scattering experiments[34] designed to elucidate the physics of one-dimensional and two-dimensional metallic systems, exhibiting charge density waves, soft phonon modes, structural phase transformations, etc. TTF-TCNQ[35] and the transition-metal dichalcogenides[36] are good examples of such 1D and 2D systems.

We now proceed to a brief discussion of the use of x ray and electron diffuse scattering experiments from disordered alloys. As analysed some time ago by Clapp and Moss,[37] the intensity of diffuse scattering (scattering from concentration fluctuations in disordered alloys) depends on the pair interaction of the alloy constituents resulting mainly from the Coulomb interaction of the ions screened by the conduction electrons. Thus the singularity in the dielectric constant has a direct effect via the pair interaction on the diffuse scattering amplitude, similar to the Kohn anomaly in neutron scattering, as discussed theoretically by Krivoglaz.[38] The theory and the experimental situation was reviewed in 1974 by Moss and Walker.[39] Among the several studies perhaps the most complete is on the Cu(Al) α-phase by Scattergood, Moss and Bever.[40] They observe strong anomalies in the x ray intensity which they identify as being due to the nearly parallel FS flats perpendicular to the <110> direction. Their data fits well a RB-like scaling of the FS of copper; the data is also in agreement with FS results obtained by e^+ annihilation.[41]

As an example of electron diffuse scattering we mention the study of the FS of Cu(Pt) and Cu(Pd) by Ohshima and Watanabe;[42] their data on Cu(Pd) will be compared later in these lectures with e^+ experiments. on Cu(Pd). As pointed out by Moss and Walker[39] the detailed connection between these x ray and electron FS images and the neutron Kohn anomalies have not been completely explained theoretically: the neutron data in Cu does not show the strongly enhanced <110> contributions that seem to be so pronounced in the x ray and electron diffuse scattering case in Cu-based alloys.

Just as neutron scattering, x ray scattering exhibits unusually large anomalies in the case of one- and two-dimensional metals.[34] As an outstanding example of this we can mention the detailed mapping of the FS in $Ta_{0.9}Hf_{0.1}S_2$ by Yamada et al.[43] Another example, perhaps of greater interest to the theory of random alloys is the work on non-stoichiometric alkali-tungsten bronzes; Kamitakahara et al.[44] report changes in the FS of Na_xWO_3 with the Na concentration x ranging from 0.56 to 0.83, changes that seem to follow RB predictions.

4. POLAR REFLECTION FARADAY EFFECT EXPERIMENTS

The use of the polar reflection Faraday effect (PRFE) in the study of FS measurements of random alloys bas been developed by Stern[45] and co-workers. Stern[11] reviewed the theory and the existing PRFE data in 1974. We outline briefly the essentials of the technique and a few relevant results. When linearly polarized light is reflected from a metal with a magnetic field normal to the surface, the polarization becomes slightly elliptical and the axis of the ellipse is rotated from the initial linear polarization direction. The experimentally measured rotation angle and ellipticity are directly connected with off-diagonal elements of the complex conductivity tensor σ_1.

Stern has shown that for a photon energy well below the interband energies of the system $\sigma_1 = (e^3\ H\ I_{FS})(4\pi^3\ \hbar^4 c\ \omega^2)^{-1}$, where H is the applied field, ω is the light frequency, and I_{FS} is an integral over the FS of a complex expression involving ω, the electron lifetime $\tau(\underline{k})$ and the shape and size of the FS. For the condition $\omega\tau >> 1$, I_{FS} simplifies to

$$I_{FS} = \frac{h^2}{6} \int_{FS} v^2 \left(\frac{1}{\rho_1} + \frac{1}{\rho_2}\right) ds \qquad 4.1$$

where v is the Fermi velocity and ρ_1 and ρ_2 are the principal radii of curvature at a point on the FS, and the integration is over the FS. Stern then shows how Eq. 4.1 can be combined with other optical data to obtain a quantity

$$R = \text{Const.} \frac{I_{FS}}{\left(\int_{FS} \underline{v}\cdot d\underline{s}\right)^2} \qquad 4.2$$

R depends only on the size and shape of the FS and on the anisotropy of the Fermi velocity v. Finally a simple parametrization of the FS of the noble metals allows the theoretical computation of R as a function of the <111> neck radius. The experiments are performed with a laser beam scattered from thin polycrystalline alloy foils. The rotation angles are extremely small (of the order of 10^{-6} minute/Gauss) and thus a lock-in technique has to be used with a slowly varying magnetic field. The experimentally obtained values of R are in good agreement with theory for the pure metals Au and Ag; R vs solute concentration is used to deduce the change in the (111) contact area with alloying.

To date experiments have been performed on Ag-Au,[45] Ag(Mg),[46] Ag(Cd)[46,47] as well as on Ag(Sn)[48] alloys. In the Ag-Au study the results indicate that the FS contact area and shape vary linearly from pure Ag to pure Au. Ag(Mg) and Ag(Cd) were found to behave similar to each other up to the α-β phase transition of Ag(Cd). In Ref. 46 various model computations of R are discussed, indicating the sensitivity of R to FS topology as well as to the velocity anisotropy on the FS.

In Fig. 4.1 we reproduce the results of Ref. 46 on the α-phase of Ag(Mg) and Ag(Cd) alloys, in terms of the ratio of the neck radius r_n to the radius of the free electron FS sphere r_0 as a function of the electron-per-atom ratio, compared to a RB model computation based on the parametrized Ag FS of Halse.[49] In a more recent experiment on Ag(Cd),[47] the precision of the experimental results has been improved, resulting, however, in essentially similar trends as indicated in Fig. 4.1. The FS necks were found to grow in radius at about one half the rate predicted by the RB model. In Ref. 47 the data is compared with RB predictions based on the Ag band computations of Christensen,[50]

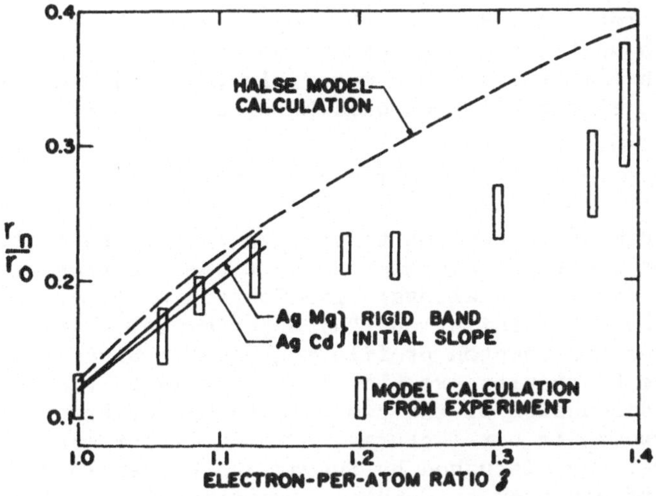

Fig. 4.1. : The ratio of neck radius r_n to the free-electron radius r_0 in Ag(Mg) and Ag(Cd) as a function of electron per atom ratio.

with essentially similar conclusions. The new data of Ref. 47 indicates that the FS of Ag(Cd) does not make contact with the (100) BZ in the α-phase, thus ruling out the simple explanation[51] of the Hume-Rothery rules for this system. The slow growth of the necks is attributed by Vail and Stern[47] to a decrease in the hybridization between the d and the s-p states when the deeper lying d levels of Cd replace those of Ag in the alloy. Such a "modified rigid band" has been previously suggested by an analysis of optical constant measurements.[52] The study of PRFE has been extended into the β phase (BCC) for Ag(Cd) alloys;[47] touching of the FS with the BZ in the <110> directions is observed with a rapid growth of the necks with alloying.

In a study of Ag(Sn) alloys, Kumar and Stern[48] extend their analysis to the case where lifetime effects on R are not negligible. In Ag(Sn) they observe an anomalous behavior at a concentration larger than 2 at% Sn, and interpret this as a very large smearing of the FS in the neck region. A comparison with optical data leads them to even question the validity of the Boltzman equation for high concentration Sn alloys (>4% at% Sn). It would be very important to test these conclusions with e^+ annihilation experiments.

The PRFE technique has of course its limitations: instead of a direct observation of the FS anisotropies in oriented single crystals, a single number (R) is obtained at each alloy concentration, thus requiring a theoretical model for the topology of the FS. Within this limitation its usefulness has been demonstrated by Stern and co-workers, and further experiments on materials measured with other techniques such as e^+ annihilation will be of great interest.

5. COMPTON PROFILES

We now turn our attention to the two techniques that can yield perhaps the most direct information about the electron wavefunctions and the FS of disordered alloys: The Compton profile (CP) measurements[12] and the e^+ annihilation (PA) experiments.[1] In this section we briefly outline the Compton profile technique. Since theoretically the CP and the PA results are closely interrelated via the momentum density, the discussion of the two effects and of experimental results will be interwoven in our lectures. The CP measurements have a long history[53] carefully discussed by Suewar and Cooper in an introductory article of a recent book on Compton scattering.[12] During the last decade the field has undergone a true renaissance due chiefly to the early efforts by Weiss in the US and Cooper in England. Briefly, the Compton profile is the energy distribution of inelastically scattered x rays or γ rays at a fixed scattering angle; the Compton scattered photons are Doppler broadened about the Compton wavelength due to the initial momentum distribution of the electrons producing the scattering. The pertinent geometry of the CP experiments is shown in Fig. 5.1, with $\underline{k}_i$ and $\underline{k}_f$ being the incident and scattered photon momenta, $\underline{p}_i$ and $\underline{p}_f$ the corresponding electron momenta; $\underline{k} = \underline{k}_i - \underline{k}_f$ is the scattering vector defining the z axis of a coordinate system. The well known Compton wavelength shift, corresponding to $\underline{p}_i = 0$, is changed, in the simple case of a non-relativistic free electron of finite $\underline{p}_i$, by the Doppler term depending on the p_z component of $\underline{p}_i$ in the $\underline{k}$ direction,

$$\lambda'-\lambda = \frac{2h}{mc}\sin^2\frac{\theta}{2} - \frac{2\lambda\sin\frac{\theta}{2}}{mc}p_z \qquad 5.1$$

Since in a typical experiment one does not detect the recoiling electron, all momenta with component p_z will contribute for a given $\underline{k}$. For high energy scattering with $|k_f|$ much larger than the typical atomic momenta, the probability of measuring $\underline{k}_f$ should depend, to a good approximation, on the probability $I(p_z)dp_z$ of having a momentum component between p_z and $p_z + dp_z$, given a momentum probability (density) of $\rho(\underline{p})$:

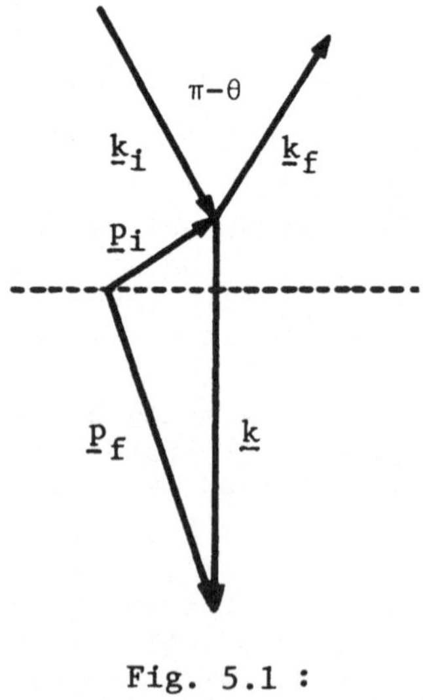

Fig. 5.1 : The geometry of CP experiments

$$I(p_z)dp_z = \left(\iint \rho(p)dp_x dy_y\right) dp_z \qquad 5.2$$

The integration is over a plane ⊥ to $\underline{k}$, as indicated by the dotted line in 5.1. This argument is, of course strictly classical, based only on energy and momentum conservation.

Quantum mechanically, the cross section for Compton scattering from a single electron with initial wavefunction ψ_i is given, in the non-relativistic limit, by

$$\frac{d^2\sigma}{d\Omega\ d\omega} = \left(\frac{d\sigma}{d\Omega}\right)_T \frac{\omega_f}{\omega_i} \sum_f |\langle\psi_f|e^{i\underline{k}\cdot\underline{r}}|\psi_i\rangle|^2 \delta(E_f-E_i-\omega) \qquad 5.3$$

where ω_i, E_i and ω_f, E_f are the initial and final photon and electron energies, $\omega = \omega_i - \omega_f$, ψ_f is the final electron wavefunction and $(d\sigma/d\Omega)_T$ is the Thompson cross-section.

For the case where final electron energy is large with respect to its binding energy prior to scattering, one can use the "impulse approximation", which then reduces Eq. 5.3 to the desired form:

$$\frac{d^2\sigma}{d\Omega\ d\omega} = \left(\frac{d\sigma}{d\Omega}\right)_T f(\omega_i\omega_f,k)I(p_z) \qquad 5.4$$

In this approximation the cross section is indeed proportional to $I(p_z)$, i.e. to the projection of the three-dimensional momentum density onto one dimension. Note that the final electron wavefunction ψ_f has dropped out in Eq. 5.2; the essential feature of the impulse approximation is the assumption that the interaction time between the photon and the electron is so short that the potential the electron is moving in is essentially constant during the interaction, and the final state of the electron is a plane wave. Platzmann[54] and co-workers and others have investigated the applicability of the impulse approximation, the effects of relativistic corrections as well as extensions to the many-electron system. For nearly backscattering and for photon energies appreciably larger than the electron binding energies the impulse approximation is quite reliable; deviations are expected to be on the one percent level or less.

In experiments one uses either x rays or γ rays for CP measurements. Most x ray experiments use Mo K_α or Ag K_α radiation, Soller slits, and a single crystal spectrometer for energy analysis (usually LiF), leading to a typical effective momentum resolution of 0.1-0.25 a.u. (one atomic unit equals mc/137). Unfortunately most x ray photons suffer too much absorption in high Z elements and x ray CP measurements are limited to medium and low Z solids. Various nuclear

γ ray sources have been used recently for CP measurements: ^{241}Am (59.57 keV), ^{123m}Te (159 keV) and ^{198}Au (412 keV) for example provide high energy monochromatic γ lines. The energy analysis is performed with a Ge(Li) or intrinsic Ge spectrometer that has the advantage over the step-by-step scanning of the x ray spectrometer of analysing simultaneously the full CP with a digital multi-channel analyser. On the other hand the energy resolution of these detectors is fundamentally limited, corresponding to a momentum resolution of 0.3-0.6 a.u. Since most momentum profiles are a few a.u. wide, these γ ray spectrometers are indeed low resolution machines.[55] Yet many of the recent CP experiments are being performed with γ spectrometers in view of the Z limitation of low energy x rays, with a great deal of care being taken about various resolution unfolding techniques.

In the independent particle model (IPM), the momentum density $\rho(\underline{p})$ is the sum of the momentum densities $\rho_j(\underline{p})$ of the independent j electrons of the system,

$$\rho(\underline{p}) = \sum_j \rho_j(\underline{p}) = \sum_j \left| \int e^{-i\underline{p}\cdot\underline{r}} \psi_j(\underline{r}) d^3\underline{r} \right|^2 \qquad 5.5$$

The application of Eq. 5.5 to periodic systems, and in particular to metals will be discussed after the description of the PA technique.

We note that a measurement of $\rho(\underline{p})$ gives independent information about the wavefunctions from the information gained in a typical diffraction experiment measuring the charge density in real space; $\rho(\underline{p})$ is not to be confused with form factors that depend on the Fourier transform of the square of the wavefunction. Because tightly bound electrons contribute a broad distribution to $\rho(\underline{p})$, the CP at low momenta is most sensitive to valence electrons in general and to conduction electrons in metals. CP measurements have been performed in a variety of materials, including metals, but little work has been done so far in disordered alloys. We shall discuss a few of these experiment later.

Closely related to the photon inelastic experiments are the medium energy electron inelastic scattering experiments (10 keV-200 keV) as well as the newly developed (e,2e) coincidence experiments. For light atoms the momentum resolution of these electron experiments is superior to the x ray technique, and many high precision experiments of the CP have been obtained on atoms and molecules. To date only a few measurements have been performed on metallic foils and we shall not discuss any further this rapidly growing field.[56]

6. THE POSITRON ANNIHILATION TECHNIQUE

As mentioned in the introduction, one can obtain information about the electrons of a system by studying the various properties of annihilation quanta from e^+ annihilation experiments. Low energy positron physics is a very rich field, involving problems ranging from pure tests[57] of QED on positronium (Ps), the bound state of an electron with a positron, to the atomic physics of e^+ collisions with atoms and molecules as well as Ps formation,[58] Ps chemistry,[59] quantum liquid studies, applications to metallurgy[60] and even to medical physics.

Due to QED, the e^+-e^- annihilation probability decreases rapidly with increasing number of annihilation photons. One-photon annihilation with weakly bound electrons is ruled out by momentum conservation; conservation rules require low energy e^+-e^--s to annihilate by 2γ-s for spin-singlet states and by 3γ-s for spin-triplet states. In a spin unaligned system the yield for 3γ to 2γ annihilations is 1/372, if no Ps is formed. In the case of Ps formation the large 2γ to 3γ cross section ratio exhibits itself in the ratio's of lifetimes: the $n = 1$, S singlet Ps state has a lifetime of 1.25×10^{-10} sec against 2γ annihilation, compared to the $n = 1$, S triplet state 3γ lifetime of 1.4×10^{-7} sec. In metals it has been found that no Ps is formed in the bulk, leading to a $2\gamma/3\gamma$ yield of 372 [this is only true for non-ferromagnetic systems where the spins are not aligned with respect to the e^+ spin].

There are four categories of experiments in e^+ annihilation physics: 1) 2γ angular correlation (ACAR) experiments; 2) the energy measurement of the annihilation γ-s; 3) e^+ lifetime experiments, and 4) $3\gamma/2\gamma$ yield measurements. Briefly, the ACAR curves depend on the momentum distribution of the e^+-e^- pair. The energy measurement of one of the γ-s from the 2γ annihilation mode also provides information about the momentum distribution of the annihilating pair. Using an internal nuclear γ ray following promptly the emission of e^+ in ^{22}Na, a nuclear β^+ emitter, one can measure accurately the lifetime of the positrons; the lifetime (of the order of 10^{-10} sec in metals) depends on the local density of electrons at the positron site, and is thus a sensitive test of the electronic densities sampled by the e^+, and of the importance of e^+-e^- interactions. Finally, the $3\gamma/2\gamma$ yield is important when one studies systems where the e^- and e^+ spins are relevant, such as in the case of Ps formation, or in ferromagnetic studies using spin aligned e^+-s. Because our interest is in FS studies we shall now focus our attention on ACAR measurements.

The usual e^+ experiments proceed by injecting fast e^+-s from a long-lived radioactive source, such as ^{22}Na, ^{58}Co or ^{64}Cu, into the samples to be studied. Usually these samples are kept at low temperatures ($\approx$ 70K); a magnet can be used to focus the e^+ onto the sample. Since slow e^+-s are known to be captured in low density defects such as vacancies and dislocations,[60] well annealed samples are required. The e^+-s slow down to thermal equilibrium with the sample prior to annihilation.[1] In 2γ annihilation the photons are non-collinear in order to conserve the momentum of the e^+-e^- pair. The deviation from

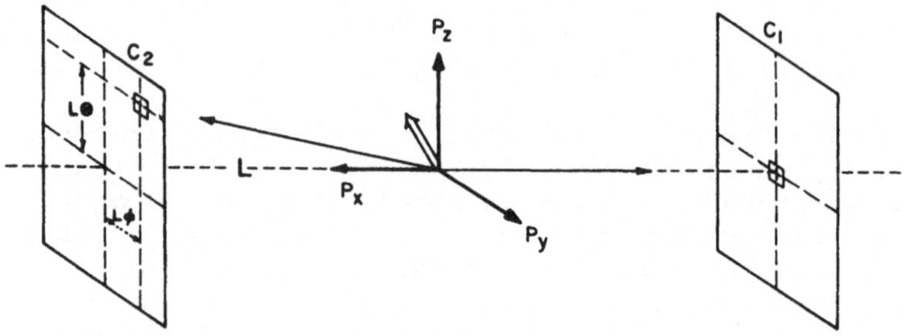

Fig. 6.1. : Geometry of 2γ ACAR experiments

collinearity can be measured by counting coincidences between two detectors. In Fig. 1 we have drawn (not to scale) the geometry of the typical ACAR measurement. Given a typical electronic momentum $\underline{p}$, and the large momenta of the annihilation photons ($\approx$ mc), the angles θ and ϕ measuring the non-collinearity between the 2γ-s detected by detectors c_1 and c_2 are only a few milliradians. Thus, to a high precision θ and ϕ are directly connected to the p_z and p_y components of $\underline{p}$, by $p_z = \theta mc$ and $p_y = \phi mc$. One can therefore express momenta in ACAR measurements in "milliradians", one mrad corresponding to a momentum of $mc \times 10^{-3} = 0.137$ a.u. The p_x component of $\underline{p}$ produces a Doppler shift in the photon energies; one finds, for $|\underline{p}| \ll mc$, that the photon energies are $E_{2,1} = mc^2 - E_B/2 \pm (cp_x)/2$. We can neglect the binding energy E_B (few eV) of the e^+ and e^- to the system being studied with respect to mc (511 keV); the Doppler term $(cp_x)/2$ produces an energy shift of a few keV for typical electron momenta.

In "the ideal experiment" one would use small detectors c_1 and c_2 in coincidence, simultaneously measure the energy of both photons with ultrahigh resolution, and obtain the full "spectral density" $\rho(\underline{p},E_B)$ of the system. Because of obvious counting rate difficulties as well as limitations on energy resolution, one usually does not measure the energy, thus integrating over E_B as well as p_x via the Doppler term. In order to maximize the count rate, most ACAR experiments were performed with the so-called "long-slit geometry" apparatus, using two long counters behind collimating slits, as sketched in Fig. 6.2 - thereby integrating experimentally over one of the momentum components. Thus for a given momentum density $\rho(\underline{p})$, such a setup measures, for perfect resolution, a coincidence rate proportional to

$$I(p_z) = \iint \rho(\underline{p})dp_x dp_y \quad , \qquad 6.1$$

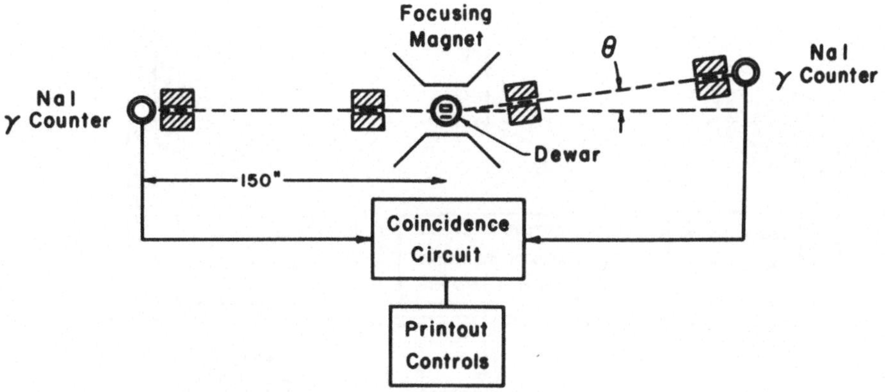

Fig. 6.2 : Sketch of typical long-slit ACAR apparatus

where $p_z = \theta mc$ is connected to the angle scanned by the moving NaI γ counter. The one-dimensional ACAR curve $I(p_z)$ is of course convoluted by the finite momentum resolution of the apparatus, typically 0.2-1.0 mrad = 0.03-0.14 a.u., an order of magnitude better than the available γ ray Compton profile resolution. It is for this reason that one prefers to perform ACAR rather than CP experiments for FS studies. On the other hand, the details of the ACAR results depend, as we shall see, in a subtle way on the behavior of the e^+-s, a problem not encountered in CP experiments.

Clearly one loses details about the fine structures of $\rho(\underline{p})$ by the planar integration corresponding to long-slit geometry. To compensate for counting rate losses, crossed-slit geometry experiments were performed by Fujiwara and Sueoka[61] on Cu and Cu-based alloys using high e^+ activities by producing in situ ^{64}Cu e^+ sources in the samples to be studied. Recently several groups have designed two-dimensional (2D) "point-slit geometry" systems, with small detectors, the counting rate problem being compensated by multiple-counter techniques. Thus at Brandeis University we have built two moveable sets of 32 NaI detectors each,[62] measuring simultaneously 32 x 32 coincidences, with an adjustable resolution presently set at 0.5 x 1.5 $mrad^2$. Fig. 6.3 shows the schematic diagram of our setup. The distance between each counter plane and the annihilation site (source and sample in cryostat, external focusing magnet) is 10 meters. A PDP-8/E on-line minicomputer stores the data, as well as positions one of the detector assemblies to measure the required p_z and p_y correlations. Other laboratories have built or are building 2D machines based on proportional wire chambers,[63] on NaI ANGER cameras,[64] or on multiple Ge detectors.[65] The ultimate resolution of such detectors will be limited by source strength and by the thermal motion of the positron - 0.2 x 0.2 $mrad^2$ resolutions seem feasible in the near future. The 2D

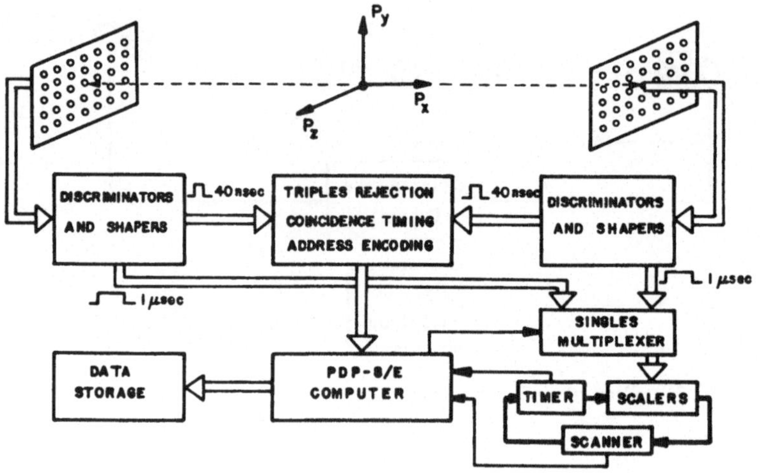

Fig. 6.3. : Schematic diagram of the Brandeis 2D ACAR apparatus

setups measure the 2D projection of $\rho(\underline{p})$, i.e.

$$N(p_z,p_y) = \int \rho(\underline{p})dp_x \qquad 6.2$$

We require of course a theoretical proof that the ACAR distributions are related to $\rho(\underline{p})$ and its integrals. The cross section for 2γ annihilation of a slow free positron with a free low energy electron is given by $\sigma_{2\gamma} = \pi r_0{}^2c/v$, where v is the relative e^+-e^- velocity ($v \ll c$) and $r_0 = e^2/mc^2$. Thus the annihilation rate $\lambda_{2\gamma}$ (inverse lifetime) of a slow e^+ in a non-interacting electron gas is $\lambda_{2\gamma} = \sigma_{2\gamma}nv = \pi r_0{}^2cn$, n being the density of electrons. These formulae do not include the Coulomb interaction. In a 1956 paper,[66] Ferrell presented the second quantization formulation of the fully interacting theory. The probability $P(\underline{p})$ for a 2γ annihilation event with momentum $\underline{p}$, from initial state $|i\rangle$ to a specific final state $|f\rangle$ is given by

$$P_f(\underline{p})d^3\underline{p} = \text{const}\left|\langle f|\int d^3\underline{x}\; e^{-i\underline{p}\cdot\underline{x}}\; \psi_-(\underline{x})\psi_+(\underline{x})\,|i\rangle\right|^2 d^3\underline{p} \qquad 6.3$$

where $\psi_+(\underline{x})$ and $\psi_-(\underline{x})$ are the annihilation operators for the e^+ and e^- respectively; $|i\rangle$ stands for the initial state of the one positron-n electron system that includes all Coulomb interactions, and $|f\rangle$ stands for the final (n-1) electron state and the two annihilation quanta after annihilation. Conservation of energy should, of course,

momentum distribution appreciably. Using a Green's function formalism, Kahana and Carbotte have studied extensively this many-body effect.[70] The importance of the e^+-e^- attraction shows up most clearly in the lifetime results: the annihilation rates are higher[71] than those predicted by the IPM model. Kahana and Carbotte compute $\rho(\underline{p})$ as a function of the electron density. Their result can be expressed as a momentum dependent 'enhancement factor' $E(\underline{p}) = \rho(\underline{p})/\rho^0(\underline{p})$ where $\rho^0(\underline{p})$ stands for the non-interacting electron gas momentum density. They find that as one approaches the Fermi momentum, the annihilation probability increases; above the Fermi momentum, due to a surprisingly effective cancellation of terms, the momentum density is effectively zero, compared to the well known electron-electron correlation high momentum tail of the interacting electron gas. We also plot in Fig. 7.1 their result for the e^+-e^- system and $\rho(\underline{p})$ for the non-interacting electron gas ("e_{free}"). For easy comparison we have renormalized the $\rho(\underline{p})$-s to equal volume in $\underline{p}$ space.

The most important feature of Fig. 7.1 is that the discontinuity is not shifted from the IPM FS due to the many-body interactions[72] as shown for electrons by Luttinger - and for the e^+-e^- interactions by Majumdar.

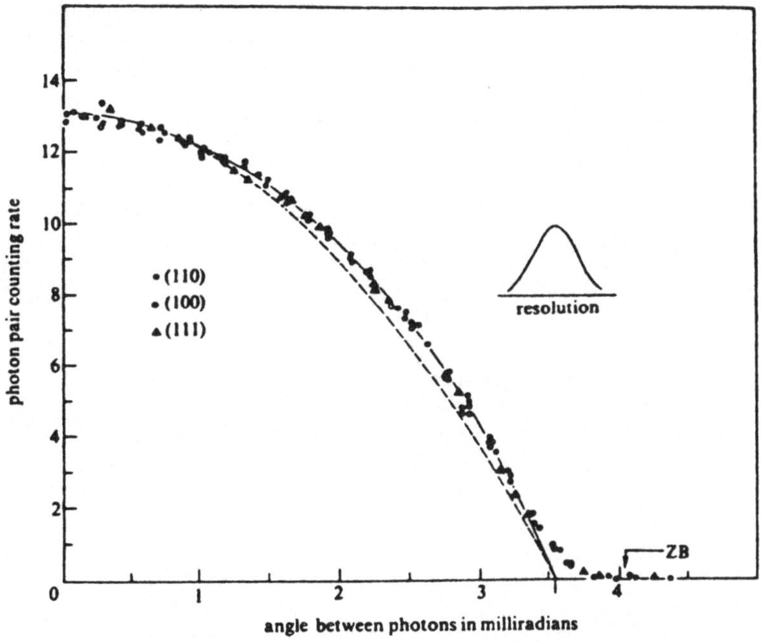

Fig. 7.2.: Long-slit ACAR curve in Na, by Donaghy and Stewart. The experiment is compared to the theoretical many-body prediction (——) and to a parabola (---). Data along different orientations indicate isotropy in Na.

The effect of the e^+-e^- many-body interaction on the predicted long-slit ACAR curve is to produce a slight "bulge" from the IPM parabolic distribution. This effect has been observed in Na by Donaghy and Stewart;[73] their highly isotropic results (as expected from Na band computations) are plotted in Fig. 7.2; a low intensity broad component has been subtracted from the raw data, corresponding to the annihilation with core electrons. Notice that the cutoff momentum (corrected for resolution) can be obtained without any guidance from theory - one of the important features of the ACAR technique.

The effect of e^--e^- interactions (without the e^+'s presence) has been observed in Na by Compton profile measurements with x rays.[74] In a real metal, however these many-body effects become more complex to handle theoretically, because of the presence of the periodic potentials of the lattice.[74,75]

8. MOMENTUM DENSITIES IN REAL METALS; THE DIFFERENCE BETWEEN k AND p SPACE

We now focus our attention on the effect of the periodic lattice on the momentum density; we neglect many-body effects and discuss the extension of the IPM $\rho(p)$, Eq. 6.6, to a metal with a periodic potential. The IPM theory of ACAR experiments in crystalline solids was given in the pioneering work of De Benedetti, et al,[76] and was applied for the first time in detail to the oriented single crystal metal problem by Berko and Plaskett.[77]

For a periodic potential, $\rho(\underline{p})$ at T=0 takes the form

$$\rho(\underline{p}) = \sum_{\underline{k},n} \rho_{\underline{k},n}(\underline{p}) = \text{Const} \sum_{\underline{k},n} \left| \int d^3\underline{r}\ \exp(-i\underline{p}\cdot\underline{r}) u_0^{*+}(\underline{r}) u_{\underline{k},n}(\underline{r}) \right|^2 \quad 8.1$$

where $u_0^+(\underline{r})$ is the ground state ($k^+=0$) positron Bloch wavefunction, (no Pauli exclusion between e^+-s and e^--s), $\psi_{\underline{k},n} = u_n(\underline{r})\exp(i\underline{k}\cdot\underline{r})$ is the electron wavefunction with crystal momentum $\underline{k}$ (reduced zone) and band index n, $\hbar \equiv 1$, and the summation is over all occupied stated. One obtains

$$\rho_{\underline{k},n}(\underline{p}) = \text{Const} \sum_{\underline{G}} |A_{\underline{G}}(\underline{k},n)|^2 \delta(\underline{p}-\underline{k}-\underline{G}) \quad 8.2$$

where the $\underline{G}$-s are the reciprocal lattice vectors and $A_{\underline{G}}(\underline{k},n)$ is given by $u_0^{*+}(\underline{r})u_{\underline{k},n}(\underline{r}) = \sum_{\underline{G}} A_{\underline{G}}(\underline{k},n)\exp(-iG\cdot\underline{r})$. Equation 8.2 shows the essential difference between the quantum number $\underline{k}$ and the momentum $\underline{p}$:[77] an electron with quantum number $\underline{k}$ will contribute to the momentum density not only at $\underline{p}=\underline{k}$, but at all $\underline{p}=\underline{k}+\underline{G}$, with amplitude $|A_G(\underline{k},n)|^2$. Once the summation over occupied states is performed, the following

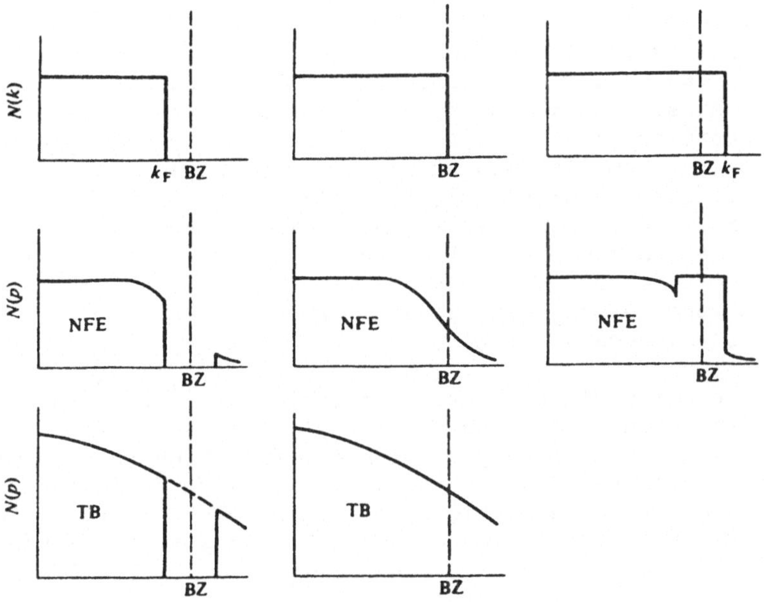

Fig. 8.1. : $\underline{k}$ vs. $\underline{p}$ space densities for simple bands as discussed in the text.

characteristics of $\rho(\underline{p})$ are noted: 1) Filled bands contribute a continuous distribution in $\underline{p}$ space; 2) partially filled bands appear as distributions in the repeated zone scheme, bounded by discontinuities at the FS and modulated by the Fourier amplitudes.

Figure 8.1 exhibits the $\underline{k}$ vs $\underline{p}$ space distributions in one direction of $\underline{p}$ for three different cases: a) FS in first zone (left); b) no FS (middle), and c) FS in second band (plotted for $\underline{k}$ space in the extended zone). These cases are plotted for a nearly-free-electron (NFE) band, and for a tight binding band (TB) (from Berko and Plaskett, Ref. 77). Notice that $\rho(\underline{p})$ becomes continuous just as the $\underline{k}$ distribution touches the Brillouin zone (BZ). Also, it is not necessary for the largest break to be in the first zone - thus in the NFE case the Fermi break for case c) is largest in the second zone. In the case of TB we have similar trends, but with appreciably larger high momentum components (HMC). In the literature the $\underline{G}\neq\underline{0}$ contributions are called "Umklapp annihilations".

Since the $\underline{k}$ vs. $\underline{p}$ analysis only depends on the $A_{\underline{G}}(\underline{k},n)$-s, similar trends hold for the momentum densities measured by Compton experiments ($u_0^+ \equiv$ const). The measurement of $\rho(\underline{p})$ via ACAR or CP measurements contain two kinds of information: a) Discontinuities in $\rho(\underline{p})$ reflect the topology and size of the FS, whereas b) the actual shape of $\rho(\underline{p})$ depends on the wavefunctions. One problem is to separate these two effects; this can be achieved for single NFE bands, but for complicated FS topologies, such as in transition metals, theory is needed to guide the experimental interpretation. The $\rho(\underline{p})$ in ACAR measurements has of course the complication of the e^+ wavefunction. We note that in the IPM formula one single $u_0^+(\underline{r})$ is needed for any one material (for T=0). For finite temperature one modifies the master equation (Eq. 6.6), to contain the proper $T \neq 0$ statistical weights. It is interesting to note that the kT energy smearing for finite temperature results in a much larger corresponding momentum smearing by the e^+ than by the e^--s, because of $\underline{k}^+ = 0$.

Since the early Wigner-Seitz computation of $u^+(\underline{r})$,[77] various techniques have been used to obtain a realistic, anisotripic e^+ Bloch wave, by using the sign-inverted potential from band theory, without an exchange term.[1] Besides a plane wave expansion technique used by some, KKR and APW e^+ wavefunctions have also been computed. An interesting approximation was introduced by Kubica and Stott,[78] based on writing the e^+ wavefunction as a product $u_0^+(\underline{r})=u_{ws}(\underline{r}-\underline{R})\psi^{ps}(\underline{r})$, where $\underline{r}$ lies within an atomic cell centered on $\underline{R}$, and u_{ws} is a Wigner-Seitz function. The residual "pseudo wavefunction" $\psi^{ps}(\underline{r})$ obeys an effective Schroedinger equation containing a weak e^+ pseudopotential responsible for the interstitial anisotropy of the e^+ wavefunction. This technique combines the good behavior of the Wigner-Seitz function in the ion core region with the correct behavior of the plane-wave expansion in the interstitial region. All techniques lead to very similar e^+ behavior.

Mijnarends[79] has discussed the consequences of symmetry in Eq. 6.6 using group theory. The contribution to $\rho(\underline{p})$ of a band not belonging to the totally symmetric representation possesses nodes in certain symmetry directions. In a recent paper Harthoorn and Mijnarends[80] have tabulated the results of the group theoretical selection rules for simple cubic, fcc and bcc structures; these rules can lead to momentum densities that are very different from those obtained by the naive assumption that ACAR measurements correspond to integrals through the FS in k space. To compare the simple curves of Fig. 8.1 with a theoretical $\rho(\underline{p})$ (including the e^+ wavefunction) for a real metal, we have reproduced in Fig. 8.2 radial momentum densities for the conduction bands of Cu calculated by Mijnarends[79] Although the bands are quite complex, the net momentum distribution is quite similar to the NFE model of Fig. 8.1. Notice that along the <111> directions, where we have no FS (due to the <111> necks) the distribution is indeed continuous.

The experimental ACAR curves can be analyzed two ways: If a band theoretical computation exists, one can compute the 1D and 2D ACAR curves, fold in the experimental resolution, and compare theory directly with experiments. If sufficient data is available with many crys-

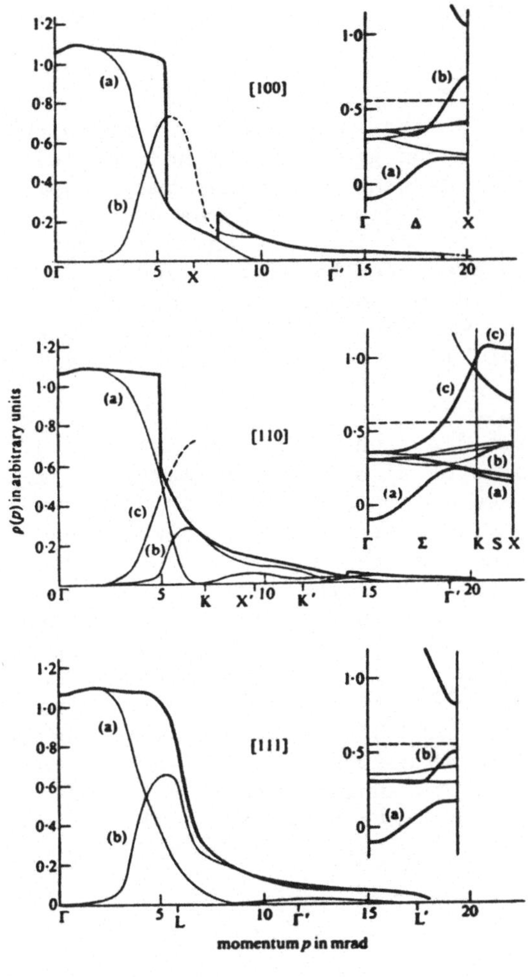

Fig. 8.2.: Radial momentum densities in copper along the three main directions as computed by Mijnarends, and the relevant band structure with bands that cannot contribute for symmetry reasons marked with thin lines.

tal orientations, one can also "reconstruct" the 3 dimensional $\rho(\underline{p})$ from its projections;[81] the reconstruction is not without its dangers - it introduces complex statistical correlations and an "effective resolution" which is hard to evaluate.

9. EXPERIMENTAL ACAR MEASUREMENTS IN PURE METALS

A great wealth of data has been published on various ACAR measurements in metals. Most of the experiments to date have been performed with the "long-slit" geometry; the results are thoroughly discussed in the review papers on e^+ annihilation.[1] In the following we shall only give a few illustrative examples for pure metals and then we shall discuss the experiments on disordered alloys more carefully; wherever possible we shall illustrate our examples with new results from recent 2D ACAR experiments.

Alkali metals. We have discussed the experiments on Na in connection with the many-body effects. Although conventional techniques such as the de Haas-van Alphen effect can give much higher resolution FS values, in Li the low temperature martensitic transformation precludes complete dHvA measurements.[82] Long slit ACAR measurements in single crystals of Li above the transition indicate an anisotropy (∿3-4%)[73,83] which is smaller than predicted by simple band theory. Lithium has also been studied by Compton profile measurements.[84] With the new 2D ACAR instruments it will be possible to obtain a much more accurate FS shape and size for Li as well as the other alkali metals.

Aluminum, having a nearly-free-electron (NFE) conduction band, has been used to study many features of e^+ behavior in metals. Fig. 9.1 shows the recent 2D ACAR surface obtained at Brandeis University with the NaI multicounter system. The distribution is nearly spherical in the extended zone, reflecting the NFE feature of the conduction band. The influence of the BZ on the FS and the wavefunctions appears as small deviations from sphericity, particularly around the rectangular BZ face.[62] The detailed shape of the ACAR surface agrees quite well with an OPW computation, as indicated in Fig. 9.2, for a single curve ($p_y \equiv 0$) of the full 2D ACAR surface.[85] At high momenta the rather flat distribution of Fig. 9.1 exhibits marked anisotropies when magnified ten-fold, due to the high momentum components (HMC) of the conduction electrons (the so-called Umklapp annihilations). A study of these HMC reveals that one can differentiate between a real OPW and a

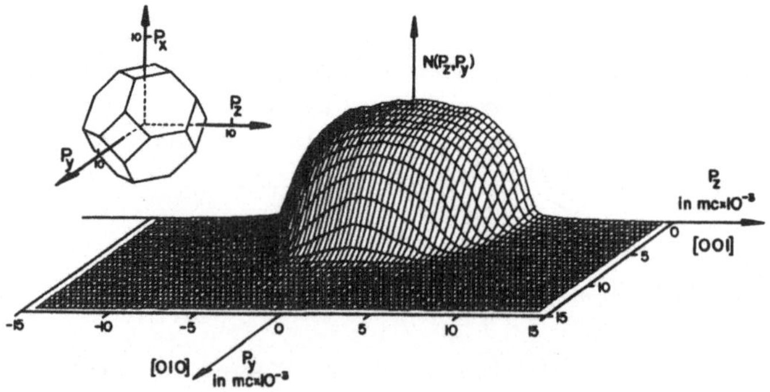

Fig. 9.1.: 2D ACAR surface from Al, Ref. 62. The orientation of the crystal is illustrated in terms of the Brillouin zone (BZ) in the insert. Each crossing of lines is an independent measurement. Sample at 100 K.

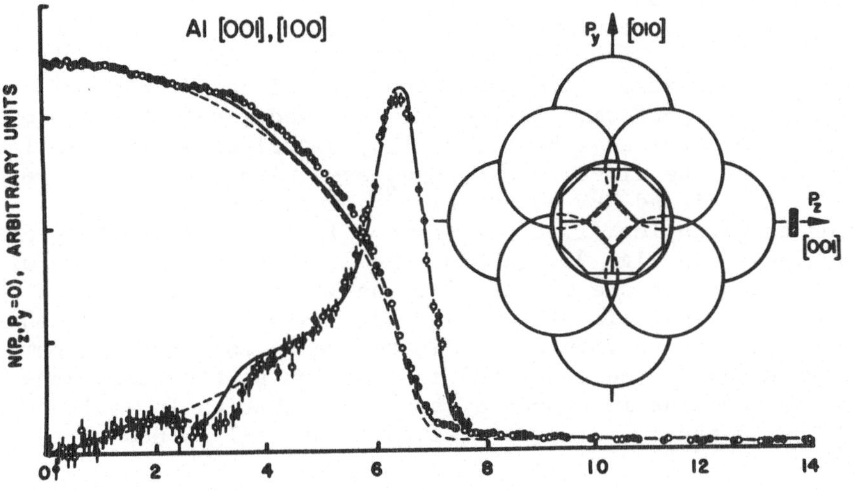

Fig. 9.2.: Al ACAR curve with p_y=0, and the absolute magnitude of the first derivative (second nearest neighbors). Open circles correspond to p_z < 0 folded about p_z=0 to exhibit the inherent symmetry of the data. Full curve is the theoretical OPW prediction; dashed curve corresponds to a simple sphere model. The inset indicates the projection of the BZ and of the (111) and (200) spheres used in the simple model.

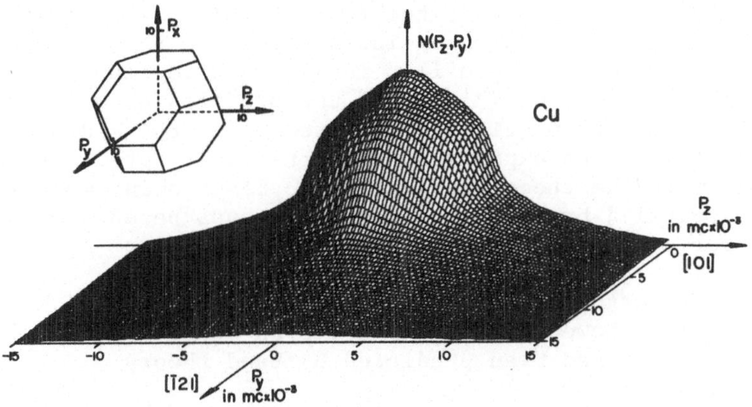

Fig. 9.3. : 2D ACAR surface from Cu, ref. 62.

pseudo-wavefunction, the experiment preferring the full OPW. Recently an ultra-high precision long-slit measurement with 0.1-0.2 mrad resolution was performed in Al,[86] indicating that one can measure maximal FS orbits with a precision of 0.25% in k_F.

Copper. Because of the well known FS of copper many e^+ experiments have been performed to test various model calculations of $\rho(\underset{\sim}{p})$, as well as to serve as a base for alloy experiments. The well known <111> necks can be directly observed, even with long-slit experiments.[87] A modified APW computation[88] predicted surprisingly well the observed anisotropies (i.e. differences in ACAR curves along different orientation). In Fig. 9.3 we plot the 2D ACAR surface in Cu with an orientation showing one of the <111> necks at $p_y = p_z = 0$, as obtained in Ref.62. An analysis of this surface, as well as of $\rho(\underset{\sim}{p})$ obtained from long-slit measurements[1] indicates reasonably good agreement with the Halse Cu FS.[49] We note that in addition to the first zone contribution, one obtains in copper a large contribution in $\rho(p)$ corresponding to the 3d bands, the core electrons, as well as HMC of the unfilled bands.

Transition metals and rare earths. Most transition metal work to date has used the long-slit geometry. Because of their complex FS structure, $\rho(\underset{\sim}{p})$ requires band computations for guidance in interpretation. The rare earths also exhibit a rich anisotropic structure, since their 5d-6s hybridized bands are similar to those of transition metals. From the many transition metal e^+ experiments we single out the recent work on V, Cr, Mo and Nb by Shiotani et al.[89] and the theoretical APW computations by Wakoh et al.[90] The theoretical anisotropies are in good agreement with experiments, in particular for Nb - see Fig. 9.4 for V and Nb. The actual individual ACAR shapes are somewhat less well predicted; the theoretical curves (based on the APW method) predict a high momentum intensity which is larger than found experimentally. Compton profiles in V have also been performed[91] using the 60 keV γ rays of ^{241}Am, and the profiles as well as the anisotropies have been theoretically computed using the KKR method by Wakoh and Yamashita.[92] The experiments and the thoery are in reasonable agreement, with the theoretical anisotropies showing more fine structure than revealed by the experiment - perhaps because of the intrinsically poor resolution of γ spectrometers.

The rare earths have been studied some time ago with long-slit geometry,[93] and showed a complex FS vis an anisotropic $\rho(\underset{\sim}{p})$, the main features of which had been predicted by band theory.

We note in connection with transition metals and rare earths the development of the polarized e^+ annihilation technique for the study of spin aligned momentum densities in ferromagnetic substances. The theory, as developed by Berko[95] uses the spin polarization of e^+ by parity non-conservation in beta decay and the selection rule forbidding 2γ decay from spin-triplet collisions. Experiments have been

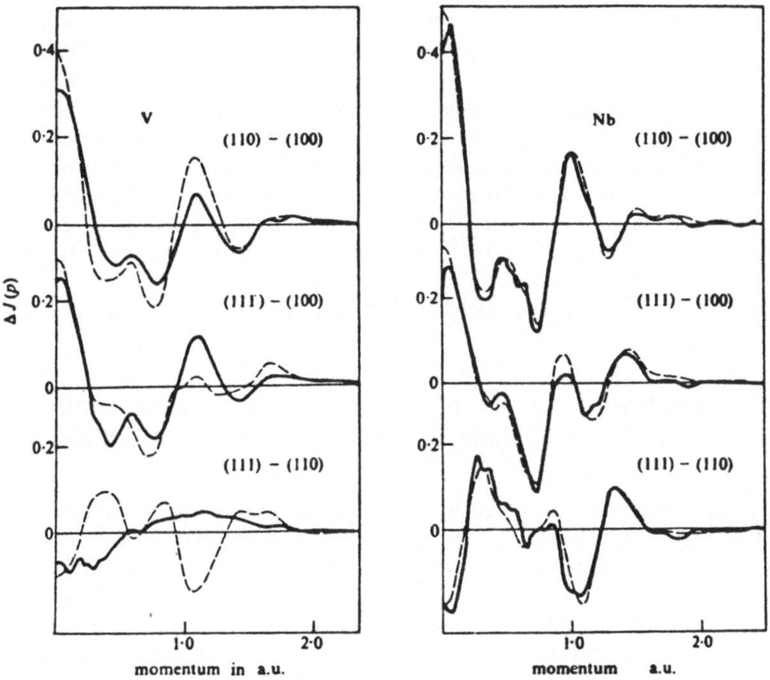

Fig. 9.4. : The momentum anisotropies (differences between long-slit ACAR curves along the directions indicated) from V and Nb. The experimental anisotropies (——, Ref. 89) are compared with the theoretical APW predictions (---, Ref. 90).

performed in Fe,[96] in Ni[97] and in Gd.[98] Mijnarends compared the experiments in Fe with spin polarized band computations of $\rho_{\uparrow}(\underline{p}) - \rho_{\downarrow}(\underline{p})$.[96] This quantity is the "spin density" in momentum space (as sampled by the positron), as compared to the "spin density" in real space, obtainable by diffraction experiments using polarized low energy neutrons. The analog of the polarized e^+ experiment is the measurement of the Compton profile of spin-aligned ferromagnetic metals using circularly polarized γ rays; for circularly polarized γ rays the Compton cross section depends on the spin orientation of the scattering electron. Such a CP experiment has been performed recently by Sakai and Ono,[99] in iron and has been theoretically analyzed by Wakoh and Kubo,[100] who also compare their theory with the polarized ACAR experiments.

The above experiments are but a few examples from the long list of ACAR studies that have been conducted in pure metals, as well as other solids. We have indicated that one can measure the sharp breaks in $\rho(\underline{p})$ to obtain the position of the FS, as well as study the nature of the electron wavefunctions by a detailed comparison between the experimental and theoretical $\rho(\underline{p})$ shapes. For FS studies the ACAR measurements are by far the more precise, in view of the inherent resolution problem of the CP experiments. On the other hand the exact shape of $\rho(\underline{p})$ as measured by the ACAR technique deviates somewhat from the predictions of the IPM computations, because of the many-body enhancement due to the e^+-e^- interaction. Although theoretical progress has been made to extend the prediction of the "enhancement" factors based on electron gas computations[70] to more realistic electron bands,[101] further theoretical work will be necessary if the finer details of $\rho(\underline{p})$ as measured by ACAR experiments are to be fully explained.

10. MOMENTUM DENSITY AND FERMI SURFACE MEASUREMENTS IN NON-DILUTE DISORDERED ALLOYS; THEORETICAL REMARKS

Since in disordered alloys the Bloch theorem does not hold rigorously, and thus $\underline{k}$ is not a good quantum number, even the existence of a Fermi surface can be questioned;[15] as the concentration of solute atoms increases one expects to see an increasing smearing of the Fermi surface, due to the finite lifetime of electrons caused by the disorder, as we have seen already in the case of dilute alloys.

During the last decade the theory of electronic structure of concentrated alloys has undergone a major development, with the coherent potential approximation (CPA) playing a central role. Realistic calculations of energy bands, density of states, as well as lifetimes have been performed in transition and noble-metal alloys, using the average t-matrix approximation (ATA),[102] and more recently the full CPA treatment.[103] One essential conclusion of these computations is that in the alloys treated the Fermi surface can still be well defined, since the smearing at the FS, although appreciably larger than the effect of the temperature, is still a small fraction of the FS dimensions.

Central to the theories of disordered alloys based on the multiple scattering formalism is the configuration averaged single particle Green's function and the associated quantity $A(\underline{k},E)$, the spectral density function ($\underline{k}$ is in the reduced zone). For a perfect crystal the spectral density is simply a set of δ functions $\delta(E-E^0_{\underline{k}})$; the principal effect of the disorder is to broaden these δ functions - to Lorentzian distributions in the case of weak scattering, for example. The width of these distributions at the Fermi energy gives the lifetimes at the FS, as discussed in Ref. 7. Sometimes it is convenient - but not necessary - to characterize the alloy spectrum in terms of complex energy bands, with the imaginary part of the energy connected to the lifetimes.

Although the computation of the spectral density is sufficient to determine the FS and its smearing, it does not permit the computation of the momentum density $\rho(\underline{p})$ in a disordered alloy. Mijnarends and Bansil[104] have discussed the theory of momentum densities for random alloys, corresponding to Compton profile measurements (i.e. no e^+ present). One needs to compute the momentum representation of the Green's function

$$G(\underline{p},\underline{p};E^+) \equiv G(\underline{p},E^+) = \sum_\alpha \frac{\psi_\alpha(\underline{p})\psi_\alpha(\underline{p})}{E-E_\alpha+i0^+} \quad 10.1$$

where the $\psi_\alpha(\underline{p})$-s denote the electronic wavefunction for a given configuration, and E_α the corresponding eigenvalues. Then the formula for $\rho(\underline{p})$ (Eq. 5.5) becomes

$$\rho(\underline{p}) = -\frac{1}{\pi}\int_{-\infty}^{E_F} dE \; \mathrm{Im}\, G(\underline{p},E^+) \quad 10.2$$

For the random alloy one then computes the configurational average of $\rho(\underline{p})$

$$\langle\rho(\underline{p})\rangle = -\frac{1}{\pi}\int_{-\infty}^{E_F} dE \; \mathrm{Im}\langle G(\underline{p},E^+)\rangle \quad 10.3$$

and obtains $\langle G(\underline{p},E^+)\rangle$ by one of the various approximations developed for random alloys. Mijnarends and Bansil derive a closed form for $\langle\rho(\underline{p})\rangle$ via Eq. 10.3 in terms of the ATA model, and show that for the pure metal the expression reduces to the $\rho(\underline{p})$ expression in the KKR band model formalism. One could formally define a "spectral density function in momentum space" $A'(\underline{p},E) = -\frac{1}{\pi}\mathrm{Im}\langle G(\underline{p},E)\rangle$ which, for a pure metal is simply connected to the regular spectral density function by $A(\underline{k},E) = \sum_{K_n} A'(\underline{k}+\underline{K}_n,E)$, where $\underline{K}_n$-s are the reciprocal lattice vectors, $\underline{k}\varepsilon$BZ. Unfortunately the actual computation of $\langle\rho(\underline{p})\rangle$ is quite tedious, since $A^1(\underline{p},E)$ involves the evaluation of the off-the-energy shell scattering matrix elements, as compared to only the on-the-energy shell elements required for the evaluation of $A(\underline{k},E)$. We understand that a realistic computation of $\langle\rho(\underline{p})\rangle$ for the Ag_xPd_{1-x} system will be soon completed.[105]

The theory of ACAR measurements in random substitutional binary alloys, i.e. the extension of the master equation 6.6 describing the momentum distribution $\rho(\underline{p})$ with a e^+ included to random alloys has been recently discussed within the CPA model by Hong and Carbotte[106] for the idealized case of plane waves, a spherically averaged BZ, and a constant potential difference between the two constituents over the unit cell. They start their description in terms of an e^- and a e^+ single particle Green's function formulation of the master formula equation 6.6:

$$\rho(\underline{p}) = \mathrm{const}\int d^3\underline{x}\int d^3\underline{y}\, e^{-i\underline{p}\cdot(\underline{x}-\underline{y})}\int\frac{dE}{\pi} f(E)\int\frac{dE^1}{\pi} f_+(E^1)\,\mathrm{Im}G_-(\underline{x}\underline{y},E)\,\mathrm{Im}G_+(\underline{x}\underline{y},E^1)$$

where f_+ and f_- are the Fermi distributions. Great care must be taken in the configuration averaging to obtain $\langle\rho(\underline{p})\rangle$, because of the product of two Green's functions, leading to so called "vertex corrections"; also the exact procedure to be taken to go to the low density limit of the "e^+ band" can lead to problems. Hong and Carbotte conclude, that for their idealized model, the positron contributes significantly to the smearing at the FS, as observed by ACAR measurements. It will be most important to extend this computation to a more realistic alloy system, in order to interpret correctly the high precision ACAR alloy data that will be forthcoming with the new generation of 2D ACAR machines.

Another problem associated with positron annihilation experiments is the possibility of the e^+-s preferential annihilation with electrons in the neighborhood of one of the constituents. It is known that the positrons do localize in low density point defects such as vacancies and dislocations.[60] The question of e^+ localization around one of the constituent atoms or cluster of atoms in an alloy is a more delicate question. Stott and Kubica[107] have addressed themselves to this theoretical question, by using the e^+ pseudowavefunction technique[78] discussed in Chapter 8. They compute relative e^+ affinities in alloys by obtaining the pseudopotential differences between the constituents of a binary alloy. The computation is delicate since it involves an estimate of the contribution of the e^+-valence electron correlation energies at the two sites, and an estimate of the importance of charge transfer. They conclude that indeed the positron can sample preferentially one of the constituents, but that actual localization is unlikely. In a series of 50 at.% alloys they find a preference for the following underligned atoms: Al-Zn, Mg-Al, Mg-Zn, Li-Mg, Mg-Cd, In-Cd and Ag-Au. Stott and Kubica also propose to use a combination of e^+ lifetime and ACAR measurements to obtain experimentally a measure of e^+ affinities; a similar analysis was suggested by Lock and West,[108] and applied to Pb-In, Mg-In, Cu-Al and Cu-Ni alloys, with an indication that clusters do provide an attractive site to the e^+. More detailed experiments along these lines will be necessary using single crystal specimens, in order to understand in detail the momentum densities $\rho(\underline{p})$ obtained by ACAR measurements in alloys. Breaks in the density corresponding to FS positions are less sensitive to the e^+ affinity discussed above.

11. MOMENTUM DENSITY AND FERMI SURFACE MEASUREMENTS IN NON-DILUTE DISORDERED ALLOYS: EXPERIMENTAL ACAR RESULTS

We finally turn to a brief description of a few ACAR experiments on non-dilute random alloys. Most experiments to date have been performed on substitutional binary alloys. These experiments have been reviewed in detail up to 1974 by Berko and Mader[41] and the list was updated recently by Mijnarends.[1c] We shall not discuss every recent experiment in detail, but try to emphasize the general trends observed so far. Most experiments to date have been performed on copper-based alloys, a few alloys of the nearly-free-electron (NFE) metals, and some on metal hydrides.

Li-Mg. The first ACAR measurements on a random alloy were made by Stewart on the Li-Mg system,[109] with a long-slit apparatus. This system is almost an ideal NFE case; it remains bcc up to 70 at.% Mg concentration, and is hcp above 85 at.% Mg, with a mixed phase in between. The early long-slit experiments showed a RB-like growth of the spherically averaged FS (the experiments were performed on polycrystalline samples), with a noticeable smearing in the alloy phase. More recently Kubica et al.[110] have re-measured long-slit ACAR curves in Li-Mg, and conclude again a RB behaviour, with the smearing however only apparent for the Mg-4% Li and Mg-11% Li samples. They interpret this smearing as positron localization in Li clusters of ≃20 Å diam., i.e., in higher than average Li rich regions. The e^+ affinity computation using the e^+ pseudowavefunction approach[107] indicates indeed higher affinity to Li than to Mg, but too weak for localization at a single Li site. A similar conclusion is reached by Koening,[111] who studied the spatial behaviour of e^+-s in binary alloys using a CPA approach on the e^+ pseudowavefunction. By analyzing the shape of the ACAR curves and measuring the annihilation rates, Kubica et al.[110] obtain the partial "core annihilation" rates as a function of concentration; their result provides an independent evidence from the FS smearing result of a preferential annihilation with Li clusters. The authors do not mention the possibility that their result could be perhaps attributed to the bcc to hcp phase transition in the low concentration Li region. In a recent theoretical analysis Tsuchiya and Tamaki[112] attribute the e^+ annihilation results in Li-Mg alloys to the effect of a charge transfer from Mg to Li sites rather than to Li clustering. More work would be required to further elucidate these questions. 2D ACAR measurements would be also useful to ascertain directly the indirect result from optical data[113] and Hall measurements[114] that the FS contact takes place with the [110] BZ faces at only 30 at.% Mg. Existing x ray CP measurements[115] on polycrystalline Li-Mg alloys are too poor in resolution to measure the smearing at the FS; some evidence has been obtained about the FS touching of the BZ at ≃30-40 at.% Mg.

Copper-based alloys. The study of the various Cu-based alloy phases played a fundamental role in the development of alloy physics; these phases still form the classic testing ground of various theoretical computations. As we have discussed earlier, pure Cu has been thoroughly studied by ACAR measurements; these show directly the well known FS topology of Cu. Thus the Cu alloys are also well suited for ACAR studies. Several long-slit as well as crossed-slit measurements have been performed to date with various degrees of precision. Most experiments use by now well annealed oriented single crystals, and measure the change of the FS upon alloying. Usually the neck radii k_N and the Fermi dimensions k_{100} and k_{110} are obtained from the ACAR curves, by various fitting techniques or by differentiation of the data to establish the discontinuities in $\rho(\underline{p})$ at the FS.

Several experiments also use the so-called "rotating specimen method",[116] where one sets the γ detectors to the colinear position (ϕ = 0 and θ = 0 in Fig. 6.1), and measures coincidences at various angular positions of the sample rotated about a crystalographic axis. It appears that the values of k_N obtained by this method are consistently lower compared to those obtained by the full ACAR measurement.[41]

In the following figures we have assembled most of the data obtained on the Cu-based alloys, but have excluded some early measurements as well as the results obtained by the "rotating specimen method". Table 11.1 contains the list of references for the data used in these figures. We first discuss the α phases of Cu with the polyvalent metals Cu(Zn), Cu(Al), Cu(Ga), Cu(Si) and Cu(Ge). Figures 11.1, 11.3 and 11.4 contain the measured <111> neck radii k_N in units of mrad $\equiv$ mc$\times10^{-3}$. The Cu(Zn) data includes values measured recently with the 2D multi-counter apparatus at Brandeis University ("CuZn 1"); we shall return to these new measurements at the end of this chapter.

In Fig. 11.1 the Cu(Zn) data is compared with four theoretical curves: The rigid band prediction (RB1) and two curves marked SMT and CR based on ATA computations by Bansil et al.,[130] and the curve marked SB corresponding to a "sinking band" model computation by Rea and

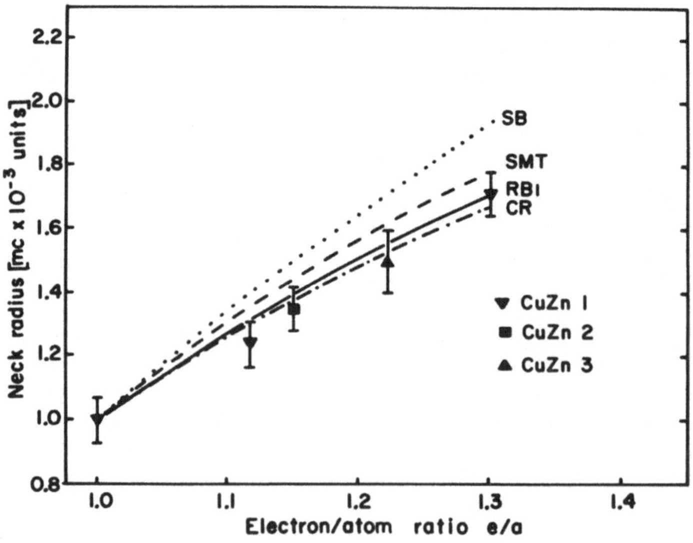

Fig. 11.1.: Neck radius vs. e/a in Cu(Zn) alloys. The theoretical curves are discussed in the text.

Table 11.1

Sources of data used in Figs. 11.1, 11.3, 11.4, 11.5 and 11.6

Symbol	Notation	Source	Ref. #	Method
▼	CuZn 1	Berko et al; Haghgooie et al	117 118	point slit point slit
■	CuZn 2	Berko and Mader	119	point slit
▲	CuZn 3	Triftshäuser and Stewart	120	long slit
▲	CuAl 1	Murray and McGervey	121	long slit
◆	CuAl 2	Thompson et al	122	long slit
⊘	CuAl 3	Fujiwara and Sueoka	123	crossed slit
▩	CuAl 4	Mader	124	point slit
▼	CuAl 5	Akahane et al	125	long slit
+	CuGa	McLarnon and Williams	126	point slit
△	CuGe 1	McLarnon and Williams	127	point slit
□	CuGe 2	Hasegawa et al	128	crossed slit
○	CuGe 3	Suzuki et al	129	crossed slit
×	CuSi	Suzuki et al	129	crossed slit

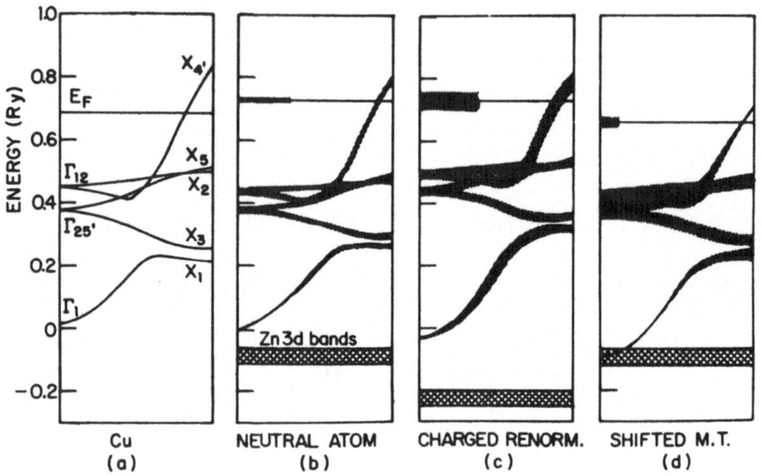

Fig. 11.2.: Calculated energy bands for pure Cu and α-$Cu_{0.7}Zn_{0.3}$ from $\Gamma \rightarrow X$: (a) pure Cu, (b)–(d) α-$Cu_{0.7}Zn_{0.3}$ for different charge-transfer models. The shading of the bands corresponds to four times the imaginary part of the complex energies. The shading around the Fermi energy corresponds to four times the average damping on the Fermi surface. The Zn 3d bands are shown as a hatched band. The energy zero is taken to be -0.8341 Ry. From Bansil et al., Ref. 130.

DeReggi,[19] used by them to explain their optical data for Cu(Al) alloys (this curve is plotted for reference, since it is used in Fig. 11.3 and 11.4). The theoretical curves have been adjusted vertically slightly to predict the correct k_N for Cu. The RB1 curve as computed by Bansil et al.[130] includes a correction for the lattice expansion with increasing Zn concentration. The two ATA results correspond to two different models used in Ref. 130 to account for charge transfer between the Cu and Zn atoms: The "shifted-muffin-tin" (SMT) model involves lowering the sp-muffin-tin constant of the alloy by an adjustable amount, while the charge-renormalized (CR) model has charge transferred and spread evenly over the whole of the muffin-tin sphere; both models use an extension of the renormalized-atom approach[131] to construct the ℓ-dependent Cu and Zn potentials. Complex energy bands are computed and the smearing at the FS is predicted to be a few percent of the FS dimension. We note that the k_N data seems to prefer the CR (and RB1) model over the SMT model, but more precise measurements will be needed to clearly select from the various ATA model computations. It is interesting that the smearing at the FS does change substantially in the two charge-transfer models; higher precision ACAR data will be able to measure the smearing as a function of orientation in $\underline{k}$ space. The changes in the band structure obtained in the various ATA models for Cu-Zn are shown in Fig. 11.2.

Figures 11.3 and 11.4 contain the data on k_N in the Cu(Al),Cu(Ga) and Cu(Si),Cu(Ge) alloys. We notice that the trivalent alloys in Fig. 11.3 follow reasonably the SB curve, which was indeed developed for Cu(Al).[19] The several sets of Cu(Ge) and the Cu(Si) data have been obtained in different laboratories. We note that they could be brought in much better agreement with each other by a small parallel displacement of the various sets; there is a slight indication that most of the data is consistently above the SB (and therefore well above the RB1) curve. It would be of great importance to perform Cu(Zn) and Cu(Ga) measurements in the same laboratory to see if the difference between Figs. 11.1 and 11.4 persist; such a difference would mean a clear deviation from RB behaviour, since the simple RB predicts a universal k_N vs. e/a curve.

We have plotted all existing measurements of k_{110} and k_{100} in Figs. 11.5 and 11.6. We notice a much better agreement between different measurements for k_{110} than for the k_{100} values. This might be due to the sharper break along the 110 ACAR curves due to the flat FS regions perpendicular to 110, or might reveal a deviation from RB behaviour. Clearly better measurements are required before this can be settled. In Fig. 11.6 the data is compared with two rigid band curves RB1 and RB1', corresponding to the Cu(Zn) lattice parameter corrected (RB1), vs the no lattice parameter corrected (RB1') curve, in order to exhibit the magnitude of this correction. The important feature of the data for the k_{100} measurement is that even for the highest e/a ratios the FS is far from touching the BZ (indicated for Cu(Ge) and Cu(Ga) in the figure). We now discuss briefly the Cu-transition metal alloys. From a theoretical point of view these alloys are of greater challenge, since they contain the d bands of the transition metal solutes close to the FS (leading to the well known d resonances for the dilute alloys). Much fewer ACAR experiments have been performed to date on these important alloys than on the Cu-polyvalent systems. The reason for this is two-fold: a) it is not easy to grow good homogeneous single crystals of these materials, and b) because the <111> neck shrinks, rather than expands upon alloying, it is harder to measure its value with reliable accuracy.

There are several experiments on Cu-Ni, with somewhat contradictory results.[1,41] Perhaps the most reliable experiment to date on Cu-Ni is that of Hasegawa et al.,[132] using crossed-slit geometry. In Fig. 11.7 we plot their results on the decreasing neck radius with increasing Ni concentration, which is statistically consistent with a linear e/a dependence. Since the Cu-Ni system is a single phase (α) alloy it is ideally suited for theoretical studies. We have also sketched in Fig. 11.7 the recent theoretical results of Gordon et al.[103] using a full CPA band treatment, showing good agreement with the data. Using the experimental slope of k_N vs at.% Ni, and calculations based on a simple RB model, Hasegawa et al. assign an effective valence of ≈0.8 to Ni in Cu. In Fig. 11.7 we also plot the theoretical estimates of k_N for pure Ni by Snow et al.[133] for two configurations, and the

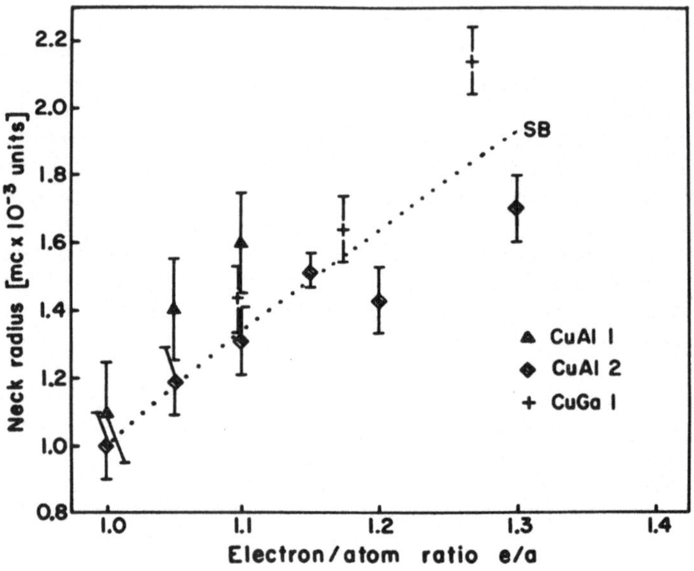

Fig. 11.3. : Neck radius vs. e/a for Cu(Al) and Cu(Ga). Theoretical curve SB discussed in text.

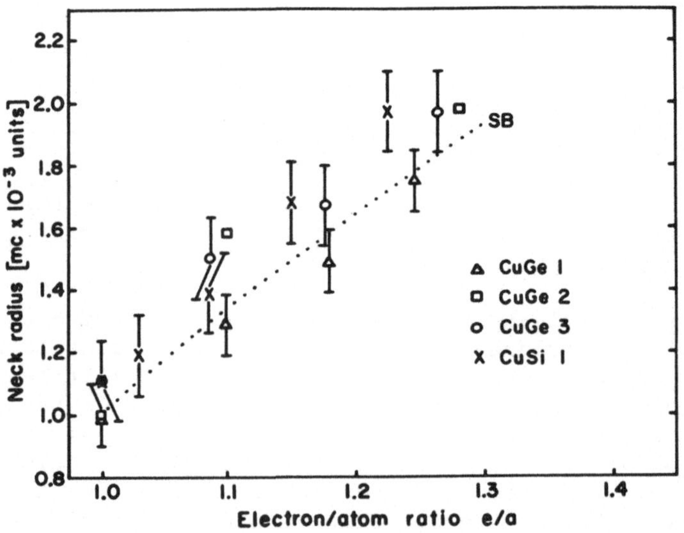

Fig. 11.4. : Neck radius vs. e/a for Cu(Ge) and Cu(Si). Theoretical curve SB discussed in text.

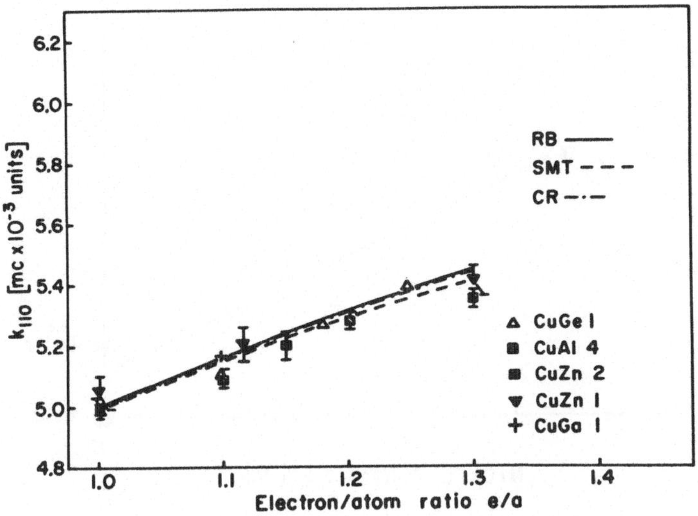

Fig. 11.5. : The FS dimension k_{110} for polyvalent metals.

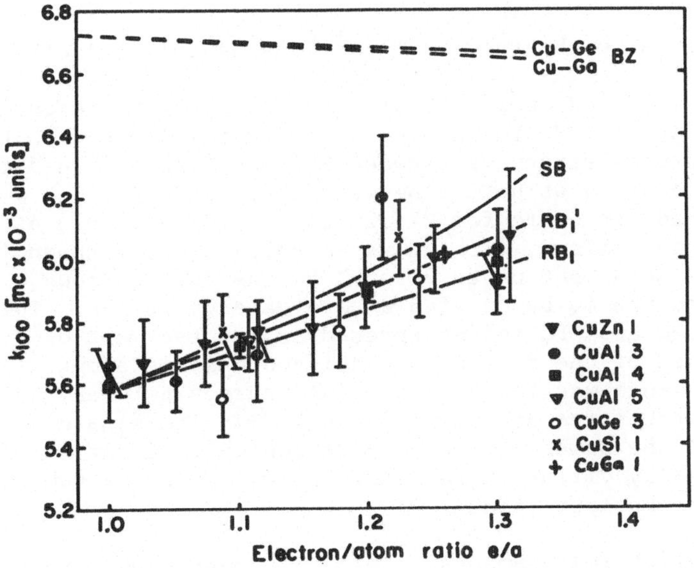

Fig. 11.6. : The FS dimension k_{100} for polyvalent metals.

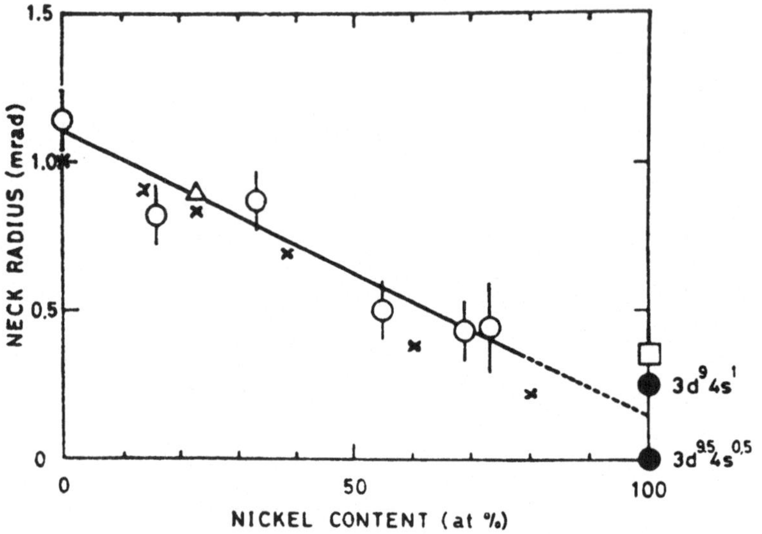

Fig. 11.7.: Neck radius vs e/a for Cu(Ni). ɸ Hasegawa et al. Ref. 132; △ Murray and McGervey Ref. 121; ● Snow et al. Ref. 133; □ Tsui Ref. 134; ✖ Theoretical CPA results by Gordon et al. Ref. 103.

experimental value observed by Tsui[134] using the dHvA effect.

The FS of Cu(Pd) below 63 at.% Pd content was recently studied by Hasegawa et al.[135] The phases are more complicated in this system with a Cu_3Au type structure between 8-18 at.% Pd, a one-dimensional long period ordered structure between 18-27 at.% Pd, and a two-dimensional ordered structure for 27-32 at.% Pd. The values of k_{110}, k_{100} and k_N were determined, with k_N vs Pd content being shown in Fig. 11.8. It is seen that up to 20 at.% Pd the data follows a RB prediction based on the Cu bands with a zero valency for Pd. The constancy of k_N above 20 at.% Pd is interpreted as a detachment of the FS from the (111) face of the BZ, but a continued bulging along <111>. No detectable change was found in the ACAR curves between the ordered and disordered structures at 15 and 19 at.% Pd. The k_{100} and k_{110} data follows well the same RB model prediction up to 30 at.% Pd; the k_{110} data obtained by diffuse x ray scattering[42] show the same trend, but with a lower absolute value.

Other alloy measurements. Among the many other experiments performed on various alloy systems we should mention the study of the ordered phase of Cu(Ge) by Suzuki et al.[136] designed to test the suggestions of Massalski and Cockayne[137] regarding the nature of the hcp phases in copper based alloys. Cu(Ge) goes from the α phase to a stable ζ hcp phase above 9.5 at.% Ge; Suzuki et al., having studied

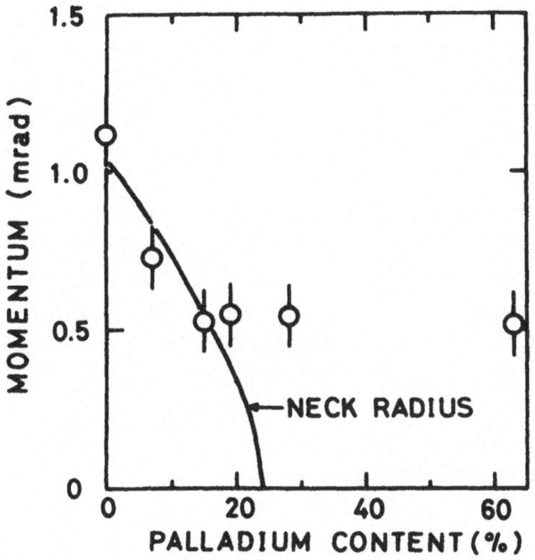

Fig. 11.8. : Neck ("bulge") radius vs. e/a for Cu(Pd), Hasegawa et al., Ref. 135. For larger than 20 at.% Pd, the authors conclude that the FS does not touch the BZ, leading from a "neck" to a "bulge" in the <111> FS direction.

12.6 and 15.0 at.% Ge, suggest a very anisotropic hcp FS with large bulges toward the zone faces perpendicular to the <10.1> directions, and necks in the <10.0> directions, leading to a density of states with three peaks which is used to explain the stability of the Hume-Rothery phases.

Recently Harthoorn and Mijnarends[138] have finished a detailed study of a Ag(Pd) crystal with 32 at.% Pd, showing that the FS detached itself from the (111) BZ face, in agreement with the Cu(Pd) findings of Hasegawa et al.[135]

Finally we mention experiments performed on metal hydrides.[139] The experiments to date have been performed with long-slit geometry on polycrystalline materials. It is found that in all cases studied the ACAR curves broaden, suggesting that the FS swells with increasing hydrogen concentration. Much more study will be needed using 2D geometry and oriented single crystals before a consistent picture can be drawn of the electronic structure of non-stoichiometric metal hydrides; ATA computations have been performed[140] on the Pd-H system, and it will be interesting to check forthcoming high precision results with these computations. At the present dHvA data[141] on Pd-H indicate a much slower change in the FS with H concentration than predicted either by RB, or the ATA computation.[140]

We conclude our discussion of the experimental ACAR results by describing the recent 2D ACAR surface measurements[117,118] in Cu(Zn). Using the new 2D apparatus described earlier, full ACAR surfaces were obtained on Cu(Zn) with 11.6 and 30 at.% Zn concentrations. In order to show the inherent high precision obtainable with such multicounter systems, we plot in Fig. 11.9 the contour maps of $N(p_y,p_z)$ for two orientations of the Cu-11.6 at.% Zn sample. A quantitative analysis of such contour maps leads to a measure of the FS dimensions along various orientations; one obtains the position of the sharpest slopes and corrects these for the influence of the resolution function. In Fig. 11.10 and 11.11 we show these results for pure Cu, Cu-11.6 at.% Zn, and Cu-30 at.% Zn, on different relevant planes in $\underline{k}$ space. (See also Figs. 11.1, 11.5, 11.6.) For pure Cu we also plot the FS dimensions from Halse's analysis of the dHvA data.[49] The results of the FS in the "belly" plane (Fig. 11.11) (plane $\perp$ to <111> direction, going through the center of the zone) corresponds to only an estimate of the FS position in the angular regions where the <111> necks line up above or below the plane; the good fit of the Cu data to the Halse values in these regions is fortuitous. We see from Figs. 11.10 and 11.11 that the FS is growing essentially isotropically with Zn concentration, following well the theoretical predictions of Bansil et al.[130]

An important conclusion in this study is that the smearing of the FS is small, even in the Cu-30 at.% Zn alloy, i.e., it is not observable with the present apparatus. It is estimated that along the <110> directions the FS smearing is less than 0.15 mrad., i.e., less than 3% of k_{110}. It will be most interesting to study the FS smearing and its anisotropy in the Cu-Ni system, where one expects a larger effect due to disorder.

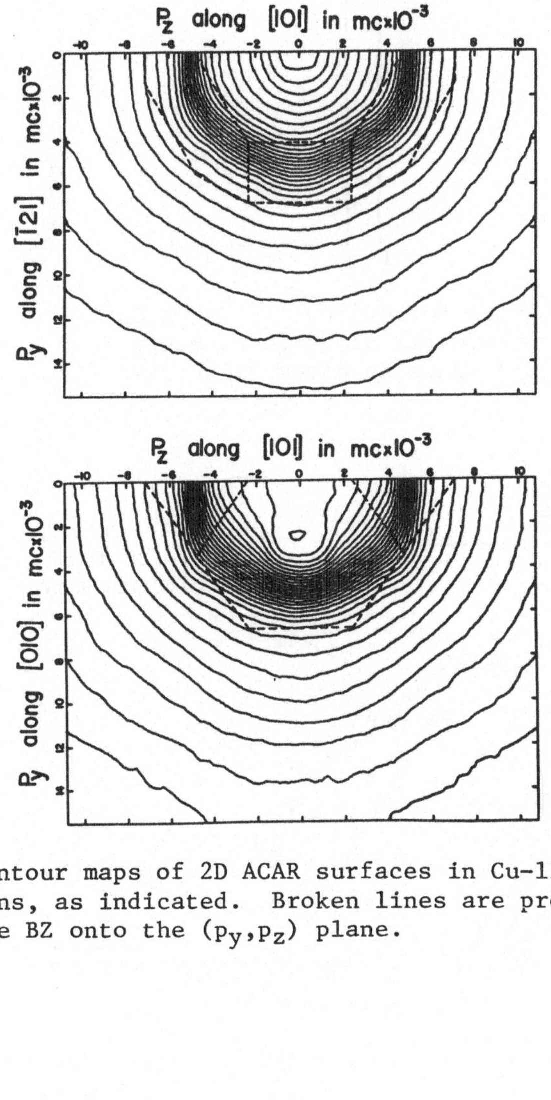

Fig. 11.9. : Contour maps of 2D ACAR surfaces in Cu-11.6 at.% Zn for two orientations, as indicated. Broken lines are projections of the top half of the BZ onto the (p_y, p_z) plane.

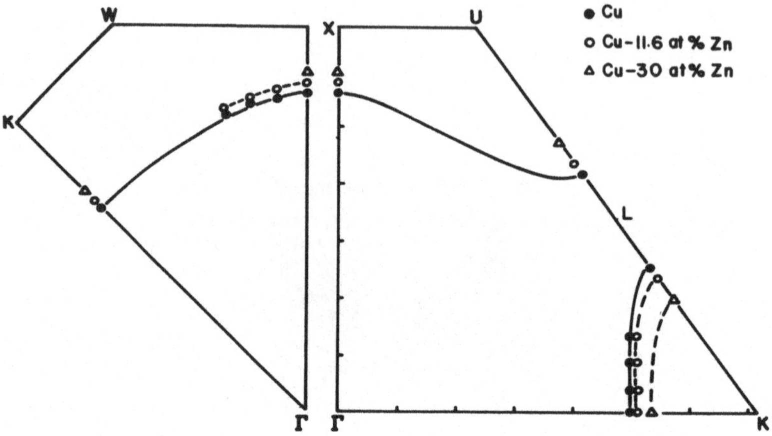

Fig. 11.10. : FS results for Cu, Cu-11.6 at.% Zn and Cu-30 at.% Zn alloys, Ref. 118. The marks along ΓK correspond to 1 mrad = mc x 10^{-3}. Full line is the FS of pure Cu after Halse, Ref. 49.

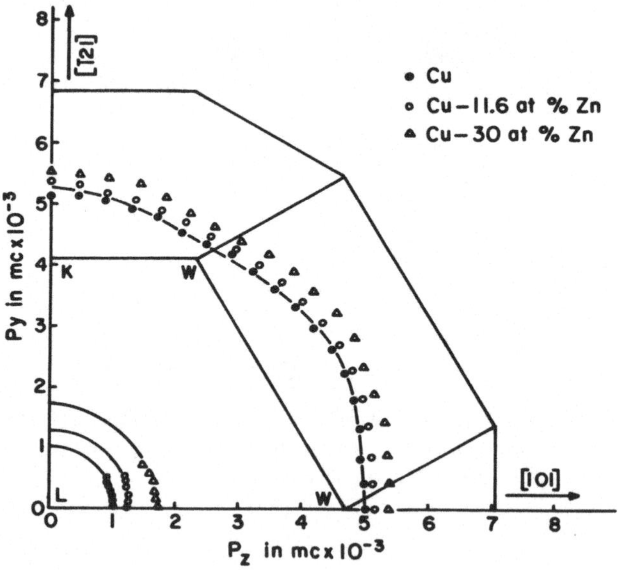

Fig. 11.11. : FS results on the hexagonal BZ face (neck results) and the belly plane for Cu and Cu(Zn) alloys. The projection of the BZ onto the (p_y,p_z) plane is also indicated.

12. CONCLUSION

In the first part of these lectures we have discussed briefly several techniques that provide experimental information about the existence, the shape, as well as the sharpness of the FS in disordered alloys. In the second part we reviewed the theory of positron annihilation experiments in solids and described some of the experiments designed to measure the electronic momentum density and the FS of pure metals as well as substitutional binary alloys. We have tried to show that the positron annihilation technique, although not without its specific set of problems, can yield reasonably detailed maps of the FS even in non-dilute disordered alloys. With the improved resolution of the new generation of 2D ACAR setups we shall see, no doubt, a re-measurement of various alloy systems, providing more detailed direct information about the FS, as well as the anisotropic smearing at the FS due to disorder.

We conclude that most experiments performed to date with various techniques agree that the FS concept remains valid even in high concentration substitutional random alloys, and that the CPA and associated approximations such as ATA provide a reasonable theoretical description of most of the available data. Much more theoretical as well as experimental work is of course necessary, particularly if one is to understand the detailed shapes, rather than just the position of the FS breaks of the $\rho(\underline{p})$ distributions obtained by Compton profile and positron annihilation experiments.

In finishing we would like to make a very brief detour to the other topic of this Institute, namely, the physics of metallic surfaces. As you have no doubt heard from the other lecturers a great deal of experimental data has been accumulated on clean metal surfaces using standard techniques such as optical experiments, LEED, etc. We have indicated that in the standard ACAR experiments the fast positrons are injected from a radioactive source, and slow down in the bulk prior to annihilation. During the last several years a new technique has been developed that permits the production of slow (few eV and up) positron beams.[141] These beams can be used to study the interaction of slow positrons with solid surfaces. Even before these beams have been developed annihilation experiments in powders, as well as on internal surfaces such as microvoids, revealed a complex interaction between e^+-s and e^--s, leading to positronium (Ps) formation and possible e^+-surface state formation.[142] Recently it was discovered that when a slow e^+ beam interacts with a solid surface in vacuum, Ps is formed with a high probability[143]; this discovery has been verified even for very pure metal surfaces in an ultra-high vacuum system.[144] One also observes that slow positrons injected into the bulk are ejected from some surfaces with a positive energy[145] (negative work function). Thus the interaction of slow positrons with surfaces appears to be a field rich in new phenomena which will be studied in much greater detail in the future. Such studies, including scattering and diffraction of slow positrons, could lead to the development of a new technique to probe the electronic structure of solid surfaces.

ACKNOWLEDGEMENTS

I am grateful to Professors A. Bansil and L. M. Schwartz for many illuminating discussions regarding the theory of disordered alloys, to Professors T. B. Massalski and D. G. Sellmyer for sending me their respective review articles on alloys, and to Dr. P. E. Mijnarends for several preprints of his recent work. I would also like to thank G. M. Beardsley and M. Haghgooie for reading the manuscript, and M. Moppett and M. Fricks for their patience in typing the manuscript. My sincere thanks to Professor B. L. Györffy and Professor P. Phariseau for the opportunity to present these lectures at the 1978 NATO Advanced Study Institute in Gent.

This work was supported in part by the US National Science Foundation.

REFERENCES

1. a) R.N. West, Adv. Phys. 22, 263 (1973);
b) S. Berko in "Compton Scattering" edited by B. Williams, McGraw-Hill, London, 1977, Chapter 9.
c) P.E. Mijnarends in "Positrons in Solids", edited by P. Hautojarvi, Springer Verlag, 1978, to be published.

2. See for example A.P. Cracknell and K.C. Wong, "The Fermi Surface", Clarendon Press, Oxford, 1973.

3. See for example the review by L. Muldawer in "Charge Transfer-Electronic Structure of Alloys", edited by L.H. Bennett and R.H. Willens, Metallurgical Society of AIME, 1974, p. 291.

4. T.B. Massalski and V. Mizutani in "Progress in Materials Science", edited by B. Chalmers, J.W. Christian and T.B. Massalski, Pergamon Press, Oxford, 1978, to be published.

5. See for example W.E. Spicer in "Band Structure Spectroscopy of Metals and Alloys", edited by D. Fabian, Academic Press, NY, 1972.

6. G.K. Wertheim and S. Hufner in "Charge Transfer-Electronic Structure of Alloys", edited by L.H. Bennett and R.H. Willens, Metallurgical Society of AIME, 1974, p. 69.

7. See for example the extensive review by H. Ehrenreich and L.M. Schwartz in "Solid State Physics", edited by H. Ehrenreich, F. Seitz and D. Turnbull, Academic Press, NY, 1976, Vol. 31, p. 149.

8. W. Kohn, Phys. Rev. Letters 2, 393 (1959).

9. See review by S.C. Moss and R.H. Walker, J. Appl. Cryst. 8, 96 (1974).

10. See for example K. Ohshima and D. Watanabe, Acta Cryst. A29, 520 (1973).

11. See the review by E.A. Stern in "Charge Transfer-Electronic Structure of Alloys", edited by L.H. Bennett and R.H. Willens, Metallurgical Society of AIME, 1974, p. 197.

12. "Compton Scattering" edited by B. Williams, McGraw-Hill, London, 1977.

13. V. Heine, Proc. Phys. Soc. A69, 505 (1956).

14. P.T. Coleridge in "Low Temperature Physics-LT13", edited by K.D. Timmerhaus, W.J. O'Sullivan and E.F. Hammel, Plenum Press, NY,

1974, Vol. 4, p. 9; P.T. Coleridge, to be published 1979.

15. See for example the papers by E.A. Stern, Phys. Rev. 157, 544 (1967); 168, 730 (1968); 188, 1163 (1969); B5, 366 (1972); 7B, 1303 (1973) and references therein; see also R.S. Sorbello, Phys. Rev. B15, 3045 (1977), as well as ref. 7.

16. P.T. Coleridge and I.M. Templeton, Can. Journal Phys. 49, 2449 (1971).

17. J. Friedel, Adv. Phys. 3, 446 (1954).

18. A.H. Lettington, Phil. Mag. 11, 863 (1965).

19. R.S. Rea and A.S. DeReggi, Phys. Rev. B9, 3285 (1974).

20. P.T. Coleridge, N.A.W. Holzwarth and M.J.G. Lee, Phys. Rev. B10, 1213 (1974); Phys. Rev. B13, 3249 (1976).

21. R.G. Poulsen, D.L. Randles and M. Springford, J. Phys. F4, 981 (1974).

22. I.M. Templeton and P.T. Coleridge, J. Phys. F5, 1307 (1975); P.T. Coleridge, J. Phys. F5, 1317 (1975).

23. L.J. Hornbeck, W.K. Fung and W.L. Gordon, Phys. Rev. B15, 4750 (1977); W.K. Fung and W.L. Gordon, Phys. Rev. B15, 4762 (1977).

24. "Electron Lifetimes in Metals", edited by D.H. Lowndes and F.M. Mueller, Physics of Condensed Matter, Vol. 19 (1975).

25. See for example J.M. Ziman, "Principles of the Theory of Solids", second edition, University Press, Cambridge, 1972, p. 155.

26. E.J. Woll, Jr. and W. Kohn, Phys. Rev. 126, 1693 (1962); P.L. Taylor, Phys. Rev. 131, 1995 (1963); S.H. Vosko, R. Taylor and G.H. Keech, Can. J. Phys. 43, 1187 (1965); M.I. Kaganov and A.I. Semenenko, Soviet Phys. JETP 23, 419 (1966); L.M. Roth, H.J. Zeiger and T.A. Kaplan, Phys. Rev. 149, 519 (1966).

27. B.N. Brockhouse, K.R. Rao and A.D.B. Woods, Phys. Rev. Lett. 7, 93 (1961).

28. A. Paskin and R.J. Weiss, Phys. Rev. Lett. 9, 199 (1962).

29. J.W. Weymouth and R. Stedman, Phys. Rev. B2, 4743 (1970).

30. A.D.B. Woods and B.M. Powell, Phys. Rev. Lett. 15, 778 (1965); B.M. Powell, P. Martel and A.D.B. Woods, Phys. Rev. 171, 272 (1968).

31. G. Nilsson and S. Rolandson, Phys. Rev. B9, 3278 (1974).

32. S.C. Ng and B.N. Brockhouse, S.S. Comm. 5, 79 (1967).

33. H.G. Smith, N. Wakabayashi and M. Mostoller in "Superconductivity in d- and f-band Metals", edited by D.H. Douglass, Plenum Press, NY, 1976, p. 223.

34. See for example the review by R. Comes and G. Shirane in "Highly Conducting One-Dimensional Solids", edited by J.T. DeVreese, Plenum Press, NY, 1978, to be published.

35. G. Shirane, S.M. Shapiro, R. Comes, A.F. Garito and A.J. Heeger, Phys. Rev. B14, 2325 (1976).

36. D.E. Moncton, J.D. Axe and F.J. DiSalvo, Phys. Rev. B16, 801 (1977).

37. P.C. Clapp and S.C. Moss, Phys. Rev. 142, 418 (1966); Phys. Rev. 171, 754 (1968); Phys. Rev. 171, 764 (1968).

38. M.A. Krivoglaz, "Theory of x ray and Thermal Neutron Scattering by Real Crystals", Plenum Press, NY, 1969.

39. S.C. Moss and R.H. Walker, J. Appl. Cryst. 8, 96 (1975).

40. R.O. Scattergood, S.C. Moss and M.B. Bever, Acta Met. 18, 1087 (1970).

41. See Fig. 13 in S. Berko and J. Mader, Appl. Phys. 5, 287 (1975).

42. K. Ohshima and D. Watanabe, Acta Cryst. A29, 520 (1973).

43. Y. Yamada, J.C. Tsang, G.V. Subba-Rao, Phys. Rev. Letters 34, 1389 (1975).

44. W.A. Kamitakahara, B.N. Harmon, J.G. Taylor, L. Kopp, H.R. Shanks and J. Rath, Phys. Rev. Lett. 36, 1393 (1976).

45. J.C. McGroddy, A.J. McAlister and E.A. Stern, Phys. Rev. 139, A1844 (1965); Phys. Rev. 140, A2105 (1965).

46. J.M. Tracy and E.A. Stern, Phys. Rev. B8, 582 (1973).

47. W.B. Vail, III, and E.A. Stern, to be published.

48. C.S. Kumar and E.A. Stern, to be published.

49. M.R. Halse, Phil. Trans. R. Soc. London A265, 507 (1969).

50. N.E. Christensen, Phys. Stat. Sol. B54, 551 (1972).

51. H. Jones, Proc. Phys. Soc. London 49, 250 (1937); W. Hume-Rothery and D.J. Roaf, Phil. Mag. 6, 55 (1961).

52. C.J. Flaten and E.A. Stern, Phys. Rev. B11, 638 (1975); see also G.P. Pells and H. Montgomery, J. Phys. C3, Suppl., S330 (1970).

53. J.W.M. DuMond, Rev. Mod. Phys. 5, 1 (1933).

54. See review paper by P. Platzman and N. Tzoar in "Compton Scattering" edited by B. Williams, McGraw-Hill, London, 1977, p. 28.

55. For experimental details see discussion by R.J. Wein, W.A. Reed and P. Pattison in "Compton Scattering" edited by B. Williams, McGraw-Hill, London, 1977, p. 43; see also R.M. Singru, Phys. Stat. Sol. (a) 30, 11 (1975).

56. For a review see R.A. Bonham and H.F. Wellenstein in "Compton Scattering", edited by B. Williams, McGraw-Hill, London, 1977, p. 234.

57. See for example the review by A.P. Mills, Jr., S. Berko, K.F. Canter in "Atomic Physics", 5, edited by R. Marrus, M. Prior and H. Shugart, Plenum Press, NY, 1977, p. 103.

58. See the review by H.S.W. Massey, E.H.S. Burhop and H.B. Gilbody in "Electronic and Ionic Impact Phenomena", V, University Press, Oxford, 1975.

59. See for example V.I. Goldanskii and V.P. Shantarovich, Appl. Phys. 3, 335 (1974).

60. A. Seeger, Appl. Phys. 4, 183 (1974).

61. K. Fujiwara and O. Sueoka, J. Phys. Soc., Japan 21, 1947 (1966).

62. S. Berko, M. Haghgooie and J.J. Mader, Phys. Letters 63A, 335 (1977)

63. A.P. Jeavons, G. Charpak, and R.J. Stubbs, Nucl. Inst. Methods 124, 491 (1975); A.A. Manuel, G.H. Bongi, Ø. Fischer, and M. Peter, Helv. Phys. Acta 50, 166 (1977); R.J. Douglas and A.T. Stewart in H14 IV Int. Conf. on Positron Annihilation, Helsingør (1976).

64. R.N. West, private communication.

65. W. Triftshauser in IInd Int. Conf. on Positron Annihilation, Kingstron, 4.77-86 (1971).

66. R.A. Ferrell, Rev. Mod. Phys. 28, 308 (1956).

67. Chang Lee, J. Exptl. Theor. Phys. (USSR) 33, 365 (1957).

68. See the review by L. Hedin and S. Lundqvist in "Solid State Physics" 23, edited by F. Seitz, D. Turnbull and H. Ehrenreich, Academic Press, NY, 1969, p. 1.

69. D.J.W. Geldart, A. Houghton, and S.H. Vosko, Can. J. Phys. 42, 1939 (1964).

70. S. Kahana, Phys. Rev. 117, 123 (1960); Phys. Rev. 129, 1622 (1963); J.P. Carbotte and S. Kahana, Phys. Rev. 139, A213 (1965). See also other references quoted in e^+ reviews (Ref. 1).

71. H. Weisberg and S. Berko, Phys. Rev. 154, 249 (1967).

72. C.K. Majumdar, Phys. Rev. 140, A227 (1965).

73. J.J. Donaghy and A.T. Stewart, Phys. Rev. 164, 396 (1967).

74. W.C. Phillips and R.J. Weiss, Phys. Rev. 171, 790 (1968); P. Eisenberger, L. Lam, P.M. Platzman and P. Schmidt, Phys. Rev. B6, 3671 (1972); see also the discussion of K.F. Berggren, et al. in "Compton Scattering" edited by B. Williams, McGraw-Hill, London, 1977, p. 138.

75. K. Fujiwara, T. Hyodo and J. Ohyama, J. Phys. Soc. Japan 33, 1047 (1972).

76. S. De Benedetti, C.E. Cowan, W.R. Konneker and H. Primakoff, Phys. Rev. 77, 205 (1950).

77. S. Berko and J.S. Plaskett, Phys. Rev. 112, 1877 (1958).

78. P. Kubica and M.J. Stott, J. Phys. F4, 1969 (1974).

79. P.E. Mijnarends, Physica 63, 235 (1973).

80. R. Harthoorn and P.E. Mijnarends, J. Phys. F., to be published.

81. P.E. Mijnarends, Phys. Rev. 160, 512 (1967); Phys. Rev. 178, 622 (1969).

82. M.J.G. Lee, CRC Crit. Rev. in Solid State Sci. 2, 85 (1971).

83. J.J. Paciga and D.L. Williams, Can. J. Phys. 49, 3227 (1971).

84. See the reconstruction of $\rho(p)$ for Li from CP measurements by W. Schulke, Phys. Stat. Sol.b80, K67 (1977), and references therein.

85. J.J. Mader, S. Berko, H. Krakauer, and A. Bansil, Phys. Rev. Letters 37, 1232 (1976).

86. A.T. Stewart, private communication.

87. S. Cushner, J.C. Erskine and S. Berko, Phys. Rev. B1, 2852 (1970).

88. H. Bross and H. Stohr, Appl. Phys. 3, 307 (1974).

89. N. Shiotani, T. Okada, T. Mizoguchi and H. Sekizawa, J. Phys. Soc. Japan 38, 423 (1975).

90. S. Wakoh, Y. Kubo and J. Yamashita, J. Phys. Soc. Japan 38, 416 (1975).

91. O. Terasaki, T. Fukamachi, S. Hosoya and D. Watanabe, Phys. Letters 43A, 123 (1973); T. Paakkari, S. Manninen, O. Jukinen and E. Liokkonen, Phys. Rev. B6, 361 (1972); W.C. Phillips, Phys. Rev. B7, 1047 (1973).

92. S. Wakoh and J. Yamashita, J. Phys. Soc. Japan, 35, 1406 (1973).

93. R.W. Williams and A.R. MacKintosh, Phys. Rev. 168, 679 (1968).

94. R.P. Gupta and T.L. Loucks, Phys. Rev. 176, 848 (1968).

95. S. Berko, Positron Annihilation, eds. A.T. Stewart and L.O. Roellig (Academic Press, NY, 1967) pp. 61-79.

96. a) P.E. Mijnarends, Physica 63, 248 (1973); b) S. Berko and A.P. Mills, Jr., J. Physique 32, C1-287 (1971).

97. N. Shiotani, T. Okada, H. Sekizawa, T. Mizoguchi and T. Karasawa, J. Phys. Soc. Japan 35, 456 (1973); see also Ref. 96b.

98. C. Hohenemser, J.M. Weingart and S. Berko, Phys. Lett. 28A, 41 (1968).

99. N. Sakai and K. Ono, J. Phys. Soc. Japan 42, 770 (1977).

100. S. Wakoh and Y. Kubo, J. Magnetism Magn. Mat. 5, 202 (1977).

101. J.P. Carbotte and A. Salvadori, Phys. Rev. 162, 290 (1967); see also T. Chiba, G.B. Dürr and W. Brandt, Phys. Stat. Sol. (b) 81, 609 (1977).

102. A. Bansil, H. Ehrenreich, L. Schwartz and R.E. Watson, Phys. Rev. B9, 445 (1974); L. Schwartz and A. Bansil, Phys. Rev. B10, 3261 (1974); A. Bansil, L. Schwartz and H. Ehrenreich, Phys. Rev. B12, 2893 (1975).

103. G.M. Stocks, B.L. Györffy, E.S. Giuliano and R. Ruggeri, J. Phys. F., 7, 1859 (1977); B. Gordon, W.M. Temmerman, B.L. Györffy and

G.M. Stocks, Inst. Phys. Conf. Ser. 39, 402 (1978), G.M. Stocks, W.M. Temmerman and B.L. Györffy, Phys. Rev. Letters 41, 339 (1978).

104. P.E. Mijnarends and A. Bansil, Phys. Rev. B13, 2381 (1976).

105. P.E. Mijnarends, private communication.

106. K.M. Hong and J.P. Carbotte, Can. J. Phys. 55, 1335 (1977).

107. M.J. Stott and P. Kubica, Phys. Rev. B11, 1 (1975).

108. D.G. Lock and R.N. West, J. Phys. F4, 2179 (1974).

109. A.T. Stewart, Phys. Rev. 133, A1651 (1964).

110. P. Kubica, B.T.A. McKee, A.T. Stewart and M.J. Stott, Phys. Rev. B11, 11 (1975).

111. C. Koening in "IVth Int. Conf. on Positron Annihilation," Helsingor, 1976, Paper D6, also, to be published.

112. Y. Tsuchiya and S. Tamaki, J. Phys. F8, L29 (1978).

113. A.G. Mathewson and H.P. Myers, J. Phys. 3, 623 (1973).

114. M. Ide, J. Phys. Soc. Japan 30, 1352 (1971).

115. K. Berndt and O. Brümmer, Phys. Stat. Sol. (b) 78, 659 (1976) and references therein.

116. O. Sueoka, J. Phys. Soc. Japan 23, 1246 (1967); D.Ll. Williams, E. H. Becker, P. Petijevick and G. Jones, Phys. Rev. Lett. 20, 448 (1968).

117. S. Berko, M. Haghgooie and J. J. Mader, Inst. Phys. Conf. 39, 94 (1978).

118. M. Haghgooie, S. Berko and U. Mizutani, to be published.

119. S. Berko and J.J. Mader, Phys. Cond. Matter 19, 405 (1975).

120. W. Triftshäuser and A.T. Stewart, Phys. Chem. Solids 32, 2717 (1971).

121. B.W. Murray and J.D. McGervey, Phys. Rev. Lett. 24, 9 (1970).

122. A. Thompson, B.W. Murray and S. Berko, Phys. Lett. 37A, 461 (1971).

123. K. Fujiwara, O. Sueoka and T. Imura, J. Phys. Soc. Japan 24, 467 (1968).

124. J.J. Mader, PhD Thesis, Brandeis University, 1975.

125. T. Akahane, O. Sueoka, H. Morinara, K. Fujiwara, J. Phys. Soc. Japan 36, 135 (1974).

126. J.G. McLarnon and D.Ll. Williams, Sol. St. Commun. 13, 1469 (1973).

127. J.G. McLarnon and D. Ll. Williams, J. Phys. Soc. Japan 43, 1244 (1977).

128. M. Hasegawa, T. Suzuki, M. Hirabayashi and S. Yajima, Acta Cryst. A28, S102 (1972).

129. T. Suzuki, M. Hasegawa and M. Hirabayashi, Appl. Phys. 5, 269 (1974).

130. A. Bansil, H. Ehrenreich, L. Schwartz and R.E. Watson, Phys. Rev. B9, 445 (1974).

131. L. Hodges, R.E. Watson, and H. Ehrenreich, Phys. Rev. B5, 3953 (1972).

132. M. Hasegawa, T. Suzuki and M. Hirabayashi, J. Phys. Soc. Japan 37, 85 (1974).

133. E.C. Snow, J.T. Waber and A.C. Switendick, J. Appl. Phys. 37, 1342 (1966).

134. D.C. Tsui, Phys. Rev. 164, 669 (1967).

135. M. Hasegawa, T. Suzuki and M. Hirabayashi, J. Phys. Soc. Japan 43, 89 (1977).

136. T. Suzuki, M. Hasegawa and M. Hirabayashi, J. Phys. F6, 779 (1976).

137. T.B. Massalski and B. Cockayne, Acta Mat. 7, 762 (1959).

138. R. Harthoorn and P.E. Mijnarends, to be published.

139. See list of references in Ref. 1c and Ref. 41.

140. A. Bansil, S. Bessendorf and L. Schwartz, Inst. Phys. Conf. 39, 493 (1978).

141. R. Griessen, W.J. Venema, J.K. Jacobs and F.D. Manchester, Inst. Phys. Conf. 39, 490 (1978).

142. See for example the reviews by H.S.W. Massey in Physics Today 29, 42-51 (1976); H.S.W. Massey, E.H.S. Burhop, and H.B. Gilbody in "Electronic and Ionic Impact Phenomena," Vol. 5, Oxford U. Press (1975).

143. See the review by W. Brandt in "Advances in Chemistry Series 158," edited by M. Kaminsky, Am. Chem. Soc., 1976, and references therein.

144. K.F. Canter, A.P. Mills, Jr., and S. Berko, Phys. Rev. Lett. 33, 7 (1974).

145. A.P. Mills, Jr., to be published, Phys. Rev. Lett. 1978.

146. S. Pendayala, D. Bartell, F.E. Girouard and J.W. McGowan, Phys. Rev. Lett. 33, 1031 (1974); A.P. Mills, Jr., to be published 1978.

OPTICAL ABSORPTION AND PHOTOEMISSION FROM RANDOM ALLOYS

H. Neddermeyer

Ruhr-Universität Bochum

4630 Bochum, F.R.G.

A. INTRODUCTION

The purpose of these lectures is to show what kind of information can be obtained from studies of the optical properties and of the photoemission spectra. The discussion of photoemission will play a major role because of the selected interest and experience of the author. As main examples results from Cu-Ni alloys will be given from both optical and photoemission measurements. To support the interpretation basic mechanisms of photoemission will be illustrated by results of pure metal single-crystals.

The copper-nickel system is a typical representative of a binary substitutional alloy. At high temperatures it forms a metallurgical simple system with fcc lattice throughout the entire composition range. Because of this simplicity and some other interesting properties Cu-Ni alloys are often used as model substances in calculations of the electronic structure of random binary alloys. A number of density-of-states calculations by means of the coherent-potential approximation are now available (1-3) as well as calculations of the complex alloy band structure along main symmetry directions of the Brillouin zone based on the coherent-potential (4,5) and the average-t-matrix approximation (6).

As experimental methods for the study of the electronic structure of binary alloys the measurement of optical properties and of photoemitted electrons are currently used. It has recently been demonstrated that by measuring directional photoemission normal to clean single-crystals (e.g. tungsten, see Ref. 7) very direct information on the electronic structure along the corresponding symmetry lines in the Brillouin zone can be obtained. Following these ideas it seemed promising to also study directional photoemission at high energy and angle resolution for random alloy single-crystals, which has recently been reported by Heimann, Neddermeyer, and Pessa (8).

The lectures are organized as follows. Firstly the theories for the analysis of photoemission, optical properties and for the electronic structure of random alloys will briefly be discussed. Then experimental procedures will be summarized and photoemission data from metal single-crystals, surface states of metals and random alloys are described. What follows is a discussion of photoemission from ordered surface layers, which might be present at surfaces of random alloy single-crystals. Temperature effects may also be responsible for the occurrence of disorder, some results of temperature dependencies in photoemission spectra are given. Finally, results from optical measurements of Cu-Ni alloys are explained and some concluding remarks are made.

B. THEORY

1) Photoemission

If we consider electron states near surfaces of ordered systems then these states may be characterized by a wavevector

$$\vec{k} = \vec{k}_{\parallel} + k_{\perp} .$$

Here $\vec{k}_{\parallel}$ is the wavevector component parallel and $k_{\perp}$ perpendicular to the surface. For bulk states $\vec{k}_{\parallel}$ and $k_{\perp}$ is real, but for surface states $k_{\perp}$ is complex

$$k_{\perp} = k'_{\perp} + i \cdot k''_{\perp}$$

and $k''_{\perp}$ describes the damping of the wavefunction ϕ. This follows from the transformation properties of the wavefunction ϕ by translation through a lattice vector $\vec{R}_1$.

$$\phi(\vec{r} + \vec{R}_1) = \exp(i \cdot (\vec{k}_{\parallel} + k_{\perp})\vec{R}_1)\ \phi(\vec{r}) .$$

The energy distribution of photoemitted electrons I(E) in the one-electron approximation is usually written as (Fermi's golden rule)

$$I(E) \propto \sum_{i,f} |M(\vec{k}_{\parallel}, k_{\perp}, E_i, E_f)|^2 \cdot \delta(E_f - E_i - \hbar\omega) \cdot$$

$$\cdot (E - E_f).$$

Here M is the optical matrix element (dipole approximation) connecting the initial state E_i and the final state E_f, $\hbar\omega$ the energy of the incident photons, $\delta(E_f - E_i - \hbar\omega)$ describes the optical energy surface in $\vec{k}$ space, $\delta(E - E_f)$ the principle of the measurement, where only electrons of energy E are counted. The sum runs over all possible pairs of initial and final states i and f, respectively. Because of the translational symmetry parallel to the surface $\vec{k}_{\parallel}$ is conserved modulo a two-dimensional reciprocal lattice vector of the surface in the photoemission process, this remains true also during the transmission of the photoelectron through the surface.

What can be said about the conservation of $k_{\perp}$? For the first case when the imaginary part of $k_{\perp} = 0$, then it follows for the matrix element

$$M \propto \delta(k_{\perp}^{i\,\prime} - k_{\perp}^{f\,\prime}).$$

This means that during the optical absorption $k_{\perp}$ is also conserved and we call such transitions as direct. This formulation is valid in the reduced zone scheme. Actually conservation of $\vec{k}$ occurs modulo threedimensional reciprocal lattice vectors and the photoemission process has to be discussed in the repeated or extended zone scheme. Direct transitions are easily recognized in the photoelectron energy distribution curves (EDC's) of single-crystals, where they may be identified by their characteristic dependencies on photon energy. For the second case the imaginary part of $k_{\perp} \neq 0$, then

$$M \propto ((k_{\perp}^{i\,\prime} - k_{\perp}^{f\,\prime})^2 + (\kappa + k_{\perp}^{i\,\prime\prime} + k_{\perp}^{f\,\prime\prime})^2)^{-1},$$

where κ is the absorption coefficient of the vector-potential of the light.

These transitions have the shape of a Lorentzian and $k'_{\perp}$ must not be conserved strictly dependent on the imaginary parts $k_{\perp}^{i\,\prime\prime}$ and $k_{\perp}^{f\,\prime\prime}$. In other words, non-direct transitions become possible in this case.

For more details of the photoemission theory see, e.g., papers by Mahan (9), Feibelman and Eastman (10) and Bross (11). See also the recently published monograph entitled "Photoemission and the Electronic Properties of Surfaces", edited by Feuerbacher, Fitton and Willis (12).

2) Optical Properties

Essentially is the determination of the complex dielectric function

$$\varepsilon(\omega) = \varepsilon_1(\omega) + i\cdot\varepsilon_2(\omega).$$

Here ε_2 is related to interband transitions by a formula similar to the expression for I(E) of the photoemission process

$$\varepsilon_2(\omega) \propto \sum_{i,f} |M(\vec{k}, E_i, E_f)|^2 \cdot \delta(E_f - E_i - \hbar\omega).$$

However, the dielectric function is not directly accessible experimentally from optical measurements: directly accessible are the reflectance power $R(\omega)$, the refractive index $n(\omega)$ and the extinction coefficient $K(\omega)$.

Between the experimentally observable quantities and the imaginary part of the dielectric function the following relation exists

$$\varepsilon_2(\omega) = 2\,n\,K.$$

For the determination of the optical properties the Kramers-Kronig relations between the real and imaginary parts of the dielectric function play an important role, see also "Introduction to Solid State Physics" by C. Kittel (13).

The main differences to photoemission are the following. (i) Initial and final state energies cannot be measured directly, since only energy differences are observable. The position of energy bands can therefore not be determined in a straightforward way. (ii) In general it is not possible to identify single transitions in $\vec{k}$ space since states throughout the whole Brillouin zone may contribute to optical properties. Sometimes critical point behavior is used to assign observed features to transitions near certain symmetry points or directions in $\vec{k}$ space. (iii) Optical measurements are not as surface sensitive as photoemission.

3) Theory of Random Alloys

The reader is referred to the lectures of Gyorffy and Stocks on "First principles band theory for random metallic alloys" in this volume for an adequate description of the theory of electron states in random alloys. Since mainly experimental results on Cu-Ni alloys are discussed here, we further refer to calculations of the electronic properties of this system, which were performed by means of the coherent-potential and average-t-matrix approximation (see Refs. 1-6). Theoretical results are given there on the partial and total electronic density of states and on the complex alloy band structure. In the complex alloy band structure the electron states are characterized by a complex wavevector, where the imaginary part describes the lifetimes of the states. In the discussion of the experimental results it will be attempted to relate experimental data to these calculated quantities.

In the theoretical description of photoemission and optical transitions it has to be considered that disorder effects in alloys weaken the $\vec{k}$ conservation rules. Compositional and other disorder effects on the photoemission process have recently been discussed by Shevchik (14). If $\vec{k}$ conservation breaks down we expect that the EDC's reflect properties of the electronic density of states.

C. EXPERIMENTAL PROCEDURES

In a photoemission experiment monochromatized light in the uv or the soft x-ray range is falling onto a sample and the photoemitted electrons are analysed by their kinetic energy and sometimes also by their direction. In the latter case the so called angle-resolved photoelectron energy distribution curves (AREDC's) are determined, which are especially useful for the elucidation of electron states of oriented surfaces and single-crystals. In Fig. 1 a typical experimental arrangement for photoemission measurements is reproduced. As incident radiation the resonance lines of noble gases are used. They are generated in a dc gas discharge (above). As energy analyser of the photoelectrons an electrostatic spherical condenser is chosen, which has high energy and angle resolution. The whole system is build into a UHV chamber. Since the mean free path of the photoelectrons in the energy range of 10 to 30 eV above the Fermi level is only a few atomic layers the

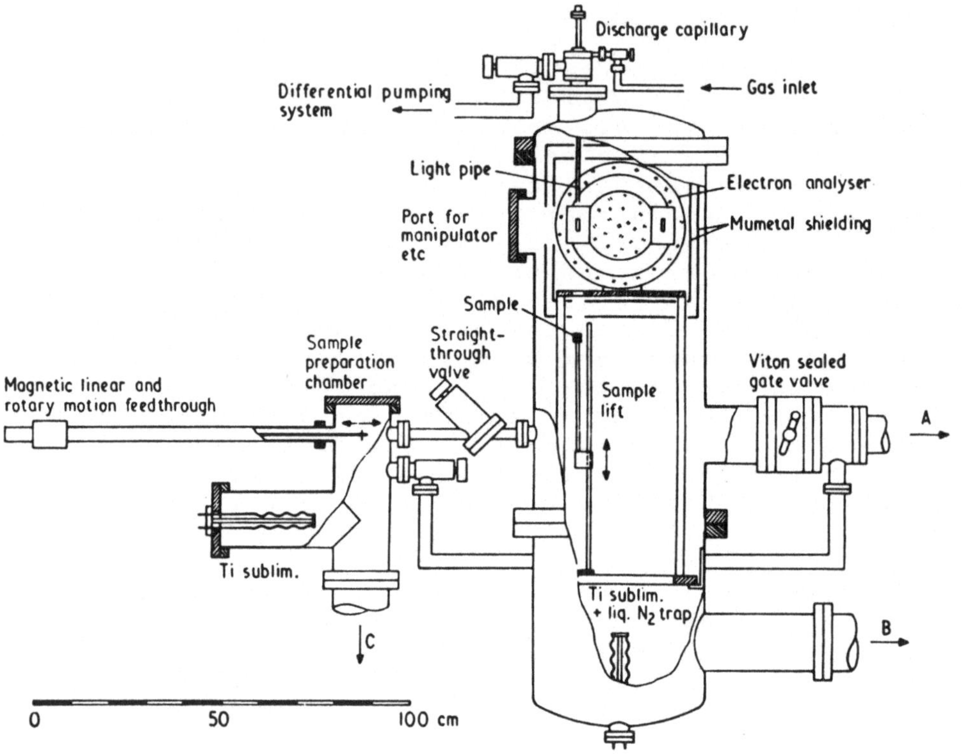

Fig. 1 : uv-Photoelectron spectrometer with separate sample preparation chamber (Ref. 15)

method is very sensitive to the quality and characteristics of the surface.

The preparation of the surfaces of the samples and their characterisation is therefore of great importance. In the experimental set-up of Ref. 15 a separate sample preparation chamber allows cleaning of the surfaces by argon ion bombardement and annealing. The structural order of the surface may be checked by LEED (not shown in Fig. 1) and the chemical composition by spectroscopy of the Auger electrons. In most cases discussed here the samples are bulk single-crystals. It is also possible to obtain clean surfaces by evaporating or sputtering of the materials which, in general, results in polycrystalline films.

For the measurements of optical properties one needs a tunable light source in the uv range and facilities to determine, e.g., the reflectivity as function of the photon energy. The optical constants are commonly evaluated by Kramers-Kronig analysis of the experimental data. Sample preparation is similar to that of the photoemission technique, although the cleanliness of the surfaces is less critical.

D. PHOTOEMISSION FROM METAL SINGLE-CRYSTALS: BULK EFFECTS

In the following paragraphs the photoemission process will be illustrated by a few representative results. In particular, the role of the wavevector $\vec{k}$ and its conservation rule will be demonstrated for ordered systems and the appearance of surface states will qualitatively be described. Differences obtained for disordered systems will later be discussed.

In an ordered material the optical transitions occur between occupied and unoccupied energy bands under conservation of the wavevector $\vec{k}$ (direct transitions). It is actually possible to observe single direct transitions in the AREDC's of single-crystals, see for example Fig. 2, where peaks b', c', d', e', and f' of the experimental spectrum of Au (111) correspond to direct transitions between occupied d bands and a free-electron -like final state band.

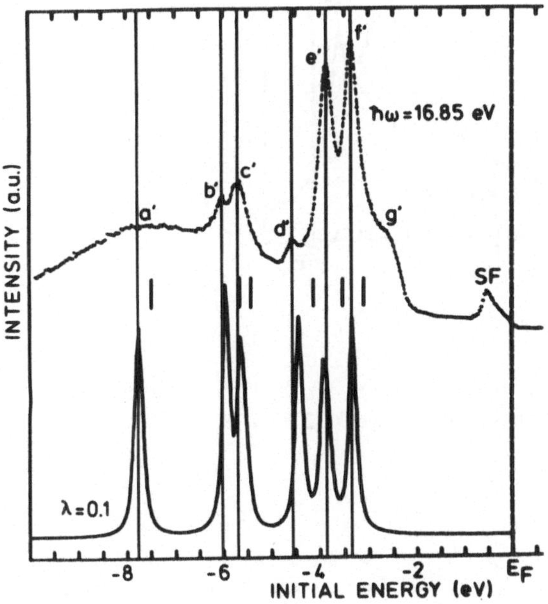

Fig. 2 :

Angle resolved photoemission normal to Au(111) (19) Comparison between measured and calculated EDC's neglecting matrix element effects. For details see Ref. 19. The experimental EDC's are as usually referred to the Fermi level $E_F = 0$.

Characteristic for direct transitions is their dependence from photon energy: the associated peaks show shifts in their initial energy position, which is dependent on the dispersion of the initial and final state energy bands. An example is shown in Fig. 3, where a small shift of peaks a (→a'→a") and b (→b'→b") towards lower energies is observed for decreasing photon energy.

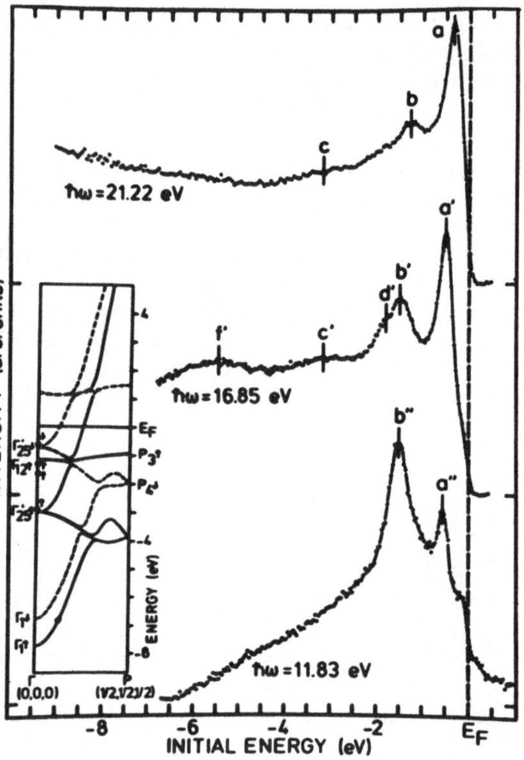

Fig. 3 :

Angle resolved photoemission normal to Fe(111) (20) and band structure of Fe along the ΓP line (21).

The interpretation of results like those displayed in Figs. 2 and 3 is facilitated by the fact that only those photoelectrons are measured, which are emitted in a narrow cone normal to the surface of a low-index face.

In this case $\vec{k}_{\parallel}$ is approximately zero, which means that the observed transitions can be directly related to the corresponding symmetry directions in $\vec{k}$ space (7).

From photoemission data the position of bands or of special points in $\vec{k}$ space can approximately deduced. However, only the initial and final state energies of a transition can be measured directly. Very recently an attempt was made by Heimann et al. (16) to also measure $\vec{k}$ vectors by means of a coincidence technique. The idea was already proposed by Kane (17). A first attempt to evaluate $\vec{k}$ vectors was reported by Turtle and Callcott (18). In Ref. 18 only general points of the Brillouin zone were accessible, whereas in Ref. 16 $\vec{k}$ vectors along main symmetry directions could be obtained.

The basic idea to determine $\vec{k}$ vectors of states contributing to a certain peak is to identify the same transition in AREDC's of different single-crystal faces. Consider, for example, normal photoemission from Au(111) (Fig. 4, above).

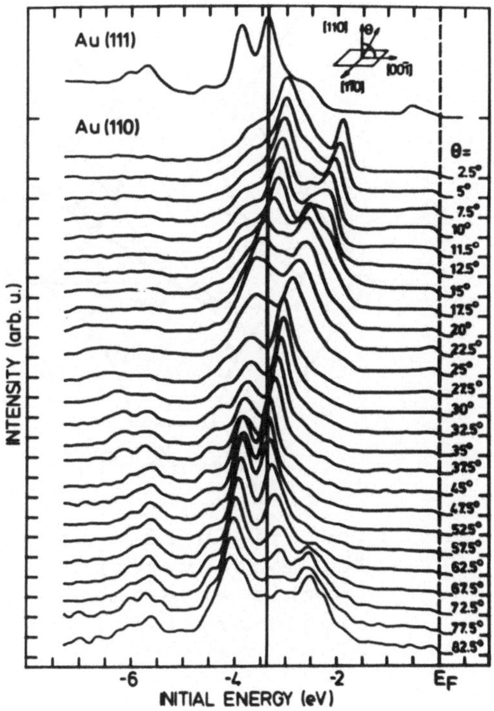

Fig. 4 :

Normal take-off EDC of Au(111) (see Fig. 2) and angle resolved EDC's of Au(110) at different polar angles θ according to Ref. 22. The photon energy is 16.85 eV.

We know that the wavevectors of the participating states are directed along the ΓL line (Fig. 5) of the threedimensional Brillouin zone, but not the actual value of the wavevectors. We now compare AREDC's obtained at different polar angles of, e.g., Au(110) with the results of Au(111) and examine, whether the features of Au(111) can also be recognized in the AREDC's of Au(110). This is in fact the case. At $\theta \simeq 47.5°$ the EDC of Au(110) is very similar to that of Au(111). From the known polar angle of the transition (e.g. $E_i = -3.87$ eV) and its vacuum energy $E_v = \hbar^2 \cdot k_{v,110}^2/2m$ the wavevector component $\vec{k}_{\parallel,110}$ can be calculated by $\vec{k}_{\parallel,110} = \vec{k}_{v,110} \cdot \sin\theta_{110}$. Since $\vec{k}_\parallel$ is conserved during the transmission of the photoelectron through the surface we know $\vec{k}_\parallel$ inside the crystal and may project it on the ΓL direction. We thus obtain the internal wavevector $\vec{k}_{b,111}$ of the final state. It is therefore possible to measure at the same time $\vec{k}$ vector and the initial and final energies of a transition and are able, in principle, to experimentally map bulk energy bands.

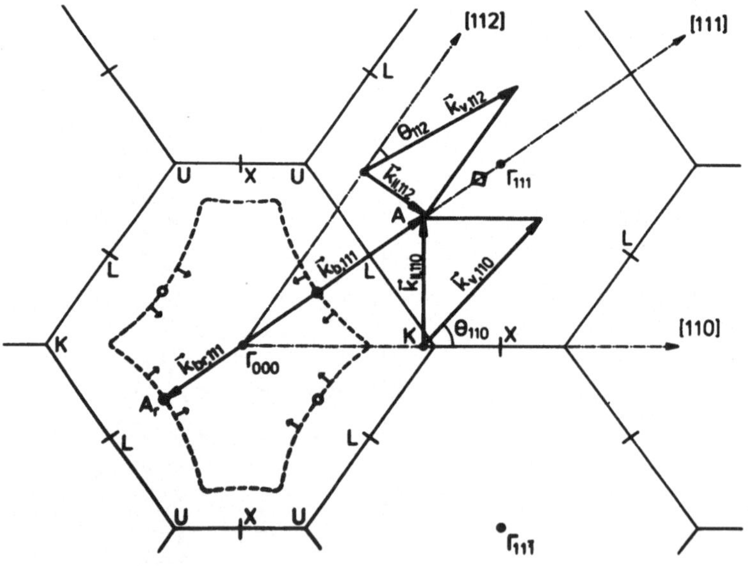

Fig. 5 : The origin of a transition from Au(111)(16). The conservation rule of $\vec{k}_\parallel$ is also established in this experiment, since a different pair of single-crystal faces (Au(112) and Au(111)) gives the same origin.

E. PHOTOEMISSION FROM SURFACE STATES OF METALS

In the gaps of the bulk band structure of ordered materials surface states may exist. The most simple case is the occurrence of occupied surface states in energy gaps of nearly-free electron-like bands. Such surface states were first abserved by Gartland et al. (23) for Cu. Their results were later confirmed by Heimann et al. (24), who found surface states at Ag and Au, in addition. Emission from surface states is seen in Fig. 2 by the small peak S_F immediately below the Fermi level E_F. Surface states are characterized by $\vec{k}_{\parallel}$ as quantum number. The dispersion of surface state bands in the twodimensional reciprocal lattice can immediately be derived from the experiment. It is only necessary to measure the emission of surface states for different polar angles θ and then to calculate $\vec{k}_{\parallel}$ as has been described for the bulk states in chapter D. As typical result the dispersion of occupied surface states is shown in Fig. 6.

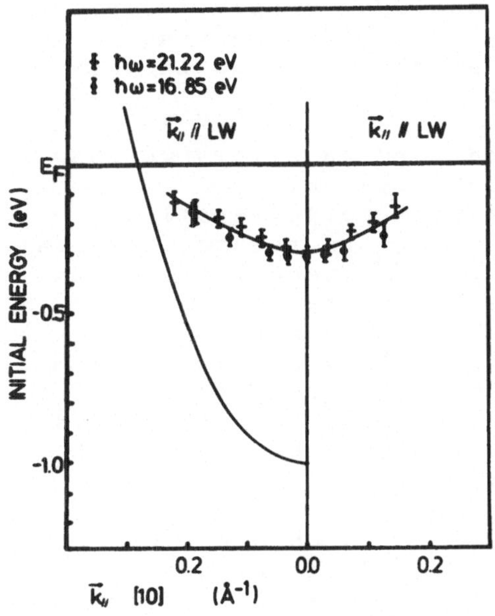

Fig. 6 :

Experimental dispersion of an occupied surface state of Au(111) (25). The full line without experimental data points (left side of the Fig.) corresponds to a bulk band. $\vec{k}_{\parallel} = 0$ corresponds to electron states described by $\vec{k}$ vectors along the ΓL line.

Some important characteristics of surface state emission are the following. (i) Sometimes a remarkable sensitivity to chemisorption or contamimation is observed, which fact is often used to support their identification.

(ii) Due to the break-down of the direct-transition model for surface state emission the spectral features are representative for the density of surface states. It is therefore clear that in normal photoemission the position of surface state peaks is independent from the photon energy. (iii) The intensity of emission from surface states generally increases with decreasing photon energy.

Surface state emission from random alloys has not yet been observed.

F. PHOTOEMISSION FROM RANDOM ALLOYS

Since the mean free path of the photoelectrons in metals is usually only a few atomic layers the properties of the surface region are of special importance for photoemission work on alloy systems. A main difficulty is an estimation of the concentration of the alloy constituents, which may be different from the bulk and dependent on the preparation of the surface. Auger electron spectroscopy or a careful analysis of x-ray or ultraviolet photoemission (XPS and UPS, respectively) results is therefore neccessary to allow a meaningful interpretation. Some of the observed phenomena in this respect will later be discussed.

Representative for photoemission work on alloys we show here mainly results from the Cu-Ni system.
The first results were obtained from polycrystalline samples. It is therefore not possible to relate the data with certain symmetry directions in $\vec{k}$ space but one can compare the measured spectra with density of states calculations. Such comparisons have proven to be very useful because the general applicability and validity of theoretical models could be tested directly. One of the first consequences was, for example, that the rigid band model could not be used to explain the density of states behavior of Cu-Ni alloys.

As examples XPS valence band spectra as obtained by Hüfner et al. (26) are shown in Fig. 7. If we use x-rays for the excitation of photoelectrons it is commonly assumed that the measurements reflect density of states character. From Fig. 7 it is clearly seen that the spectra of the alloy samples consist of two peaks which are associated with the Cu and Ni d-electrons. Due to the comparably low resolution of XPS experiments the spectra do not show much fine structure details.

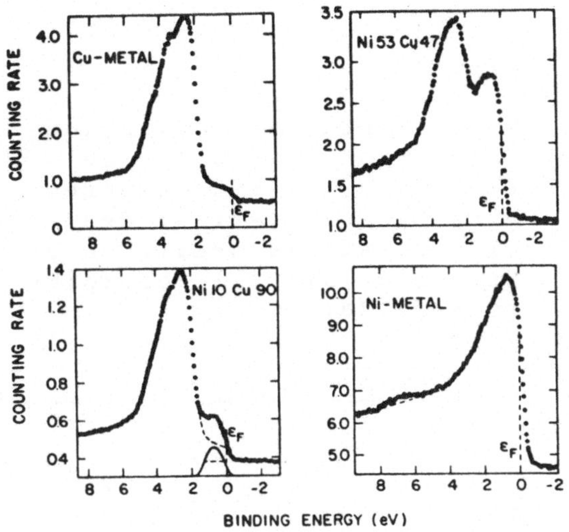

Fig. 7 :

XPS valence band spectra of vacuum-evaporated Cu, Ni ana Cu-Ni alloys as measured by Hüfner et al. (26).

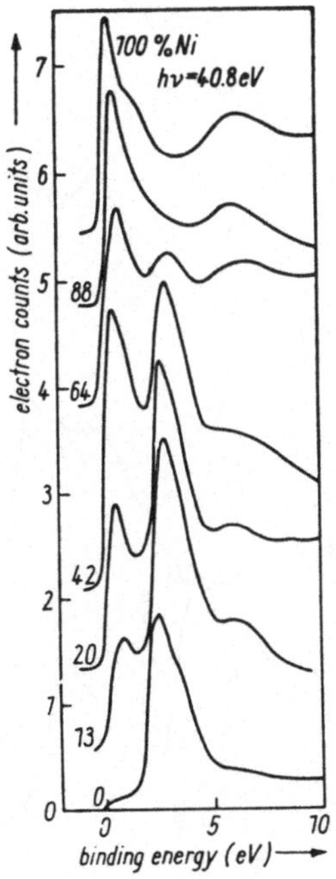

Fig. 8a :

Ultraviolet photoelectron emission spectra from Cu-Ni alloys using a photon energy of 40.8 eV (Shevchik and Penchina (28)).

UPS investigations of Cu-Ni alloys were first reported by Spicer and coworkers (see, e.g. Ref. 27) and later by Shevchik and Penchina (28). Some of the results of the latter authors are displayed in Fig. 8a, where spectra of a series of polycrystalline Cu-Ni samples are plotted.

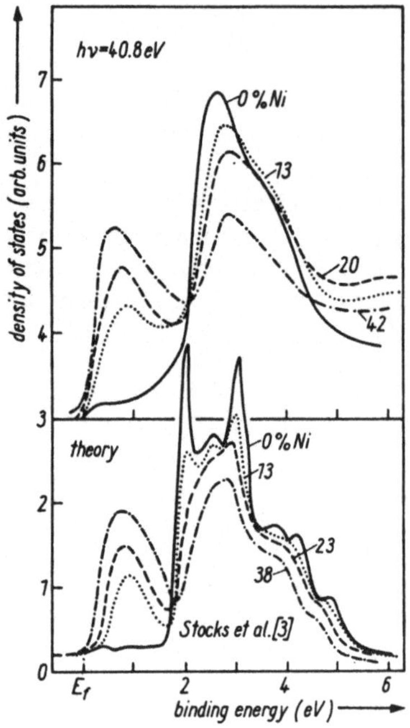

Fig. 8b:

Experimental photoemission aensity of states for Cu rich Cu-Ni alloys comparea with the theoretical density of states preaictea by Stocks et al. (2) accoraing to Ref. 28.

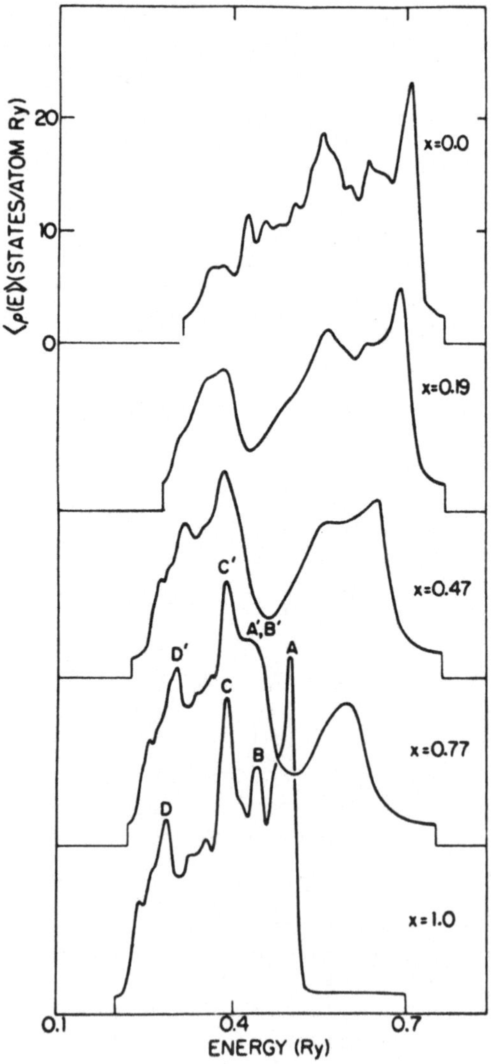

Fig. 9 :

Density of states in the alloy Cu_xNi_{1-x} by Bansil et al.(6)

In Fig. 8b a comparison between theory and experiment is made. The energy range of the Ni d electrons between E_F and 2eV binding energy is generally characterized by a structureless peak, whereas the Cu region between roughly 2eV and 5eV should theoretically reveal fine structure features which are actually not observed.

In Fig. 9 density of states of Cu-Ni alloys are reproduced. It shall be demonstrated that always a systematic behavior with concentration of the alloy constituents is to be expected. With decreasing concentration of one constituent features of the density of states attributable to this constituent should increasingly be smeared out. In the dilute limit an impurity level should appear. In the experimental data the effects are less pronounced possibly due to additional broadening effects in the photoemission process. However, the general properties of the experimental data are certainly explained by the recent theories although it might be difficult to favour one or the other theoretical method on the basis of experimental spectra like those shown above.

Angle-resolved photoemission results from single-crystalline Cu-Ni samples have recently been reported by Gyorffy et al. (4) and Heimann et al. (8). Since only the normal take-off photoelectrons were measured in this case, the AREDC's should reflect the electronic states of the corresponding symmetry directions in $\vec{k}$ space. Some of the results of Ref. 8 are reproduced in Figs. 10 - 12. The spectra were taken at photon energies of 11.83 eV, 16.85 eV and 21.22 eV, so that direct $\vec{k}$ conserving transitions should be visible by characteristic shifts with photon energy, whereas peaks at constant initial energies could be associated with maxima in the one-dimensional density of states.

Since the Cu-Ni single-crystals have to be cleaned in situ by repeated anneal-sputter cycles the surface concentration need not be the same as in the bulk. The most dramatic effect can be observed after heating, when Cu atoms have segregated to the surface and form an ordered Cu monolayer for bulk Cu concentrations around 20 at.% . This effect has first been established by Helms (29) and is also demonstrated in Figs. 10 - 12. The full lines corresponds to the spectra obtained after heating, when the Cu enriched surface layer is present. The intensity of the Cu d electrons between -2 eV and -5 eV is then comparable to that of the Ni d electrons between E_F and -2 eV especially for higher photon energies. Due to an increased mean free path of the photoelectrons at lower photon energies the Cu related peaks are less intensive at a photon energy of 11.83 eV. The Cu layer can be removed by shortly bombarding the surface with argon ions. The intensity of the Cu d electrons is then markedly reduced and the spectra may essentially be regarded as representative for the bulk. For more details of this effect the reader is referred to Ref. 8.

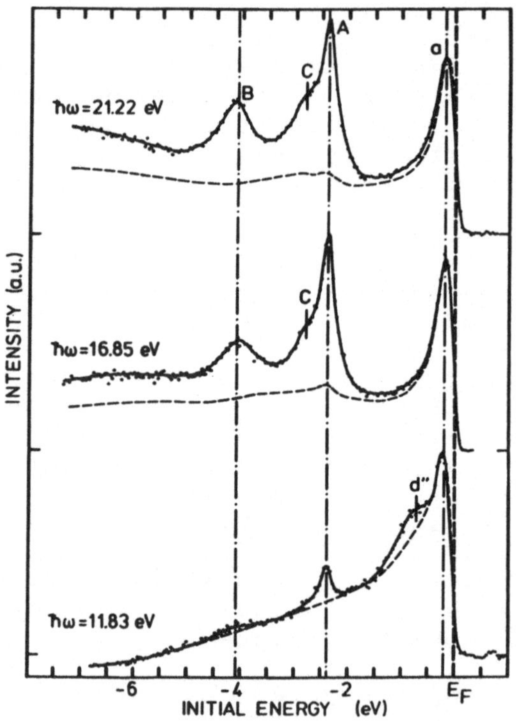

Fig. 10 :

EDC's emitted normal to the (110) face of annealed (full lines) and sputtered (dashed lines) $Cu_{0.24}Ni_{0.76}$ (8).

Compared to the results of polycrystalline Cu-Ni samples the AREDC's shown in Figs. 10-12 are characterized by more and better resolved fine structure details. This is especially true for the Ni region of the spectra, where the peaks of the (111) faces show small shifts with photon energy indicating their origin in direct transitions. This is what one would expect on the basis of the alloy band structure (6) in the dilute Cu limit, because the Ni bands are then not influenced too much by alloying. The Cu region of the spectra from the bombarded surfaces is not well pronounced due to the relatively low Cu concentration. However, the fact that the observed details are practically independent from photon energy can be realized and leads to the conclusion that these structures may be related to density of Cu d states in the alloy sample.

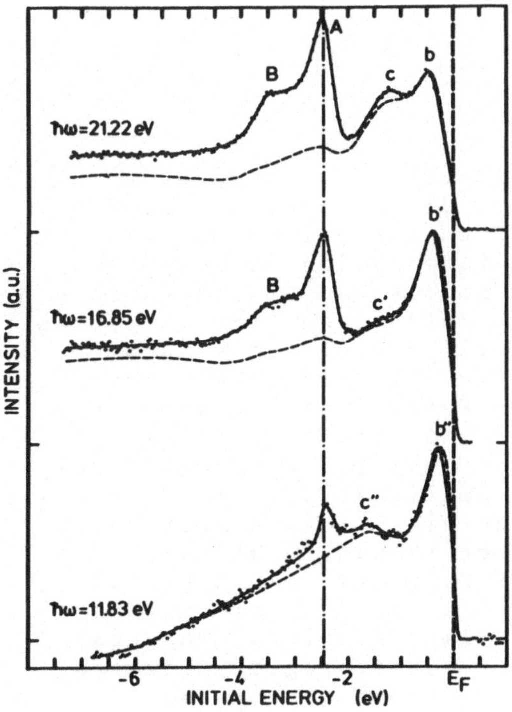

Fig. 11 :

EDC's emitted normal to the (111) face of annealed (full lines) and sputtered (dashed lines) $Cu_{0.24}Ni_{0.76}$ (8).

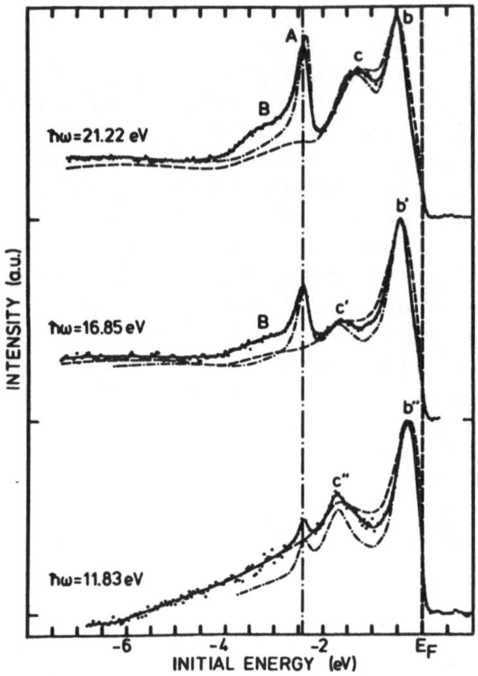

Fig. 12 :
EDC's emitted normal to the (111) face of annealed (full lines) and sputtered (dashed lines) $Cu_{0.16}Ni_{0.84}$ (8). Emission from a 2 Å thick Cu overlayer on Ni(111) is also shown (dashed-dotted lines).

The few results on Cu-Ni single-crystals shown above illustrate the possibilities of directional photoemission for alloy work. As in the case of clean metal single-crystals more detailed information is obtained and the available theoretical models can be tested more confidentially. The next step would be the investigation of angular distributions from alloy single-crystals. One example is reproduced in Fig. 13. Here the angle-resolved spectra from a random Fe-Ni single-crystal are shown for different polar angles of electron emission. Compared to the AREDC's of a clean metal single-crystal (Fig. 4) the effects are very small. A lack of fine-structure details is also surprising. To qualitatively understand the observed behavior it has to be considered that the energy range of Fe and Ni d electrons overlaps and that we may assume a statistical arrangement of Fe and Ni atoms on the lattice sites. Therefore the real part of the wavevector describing the electronic states is no longer a good quantum number and effects due to conservation of $\vec{k}$ are not expected. We essentially see characteristics of the density of states which is also supported by independency from photon energy. It is interestingly to note that striking similarities to results of polycrystalline Fe (Fig. 14)

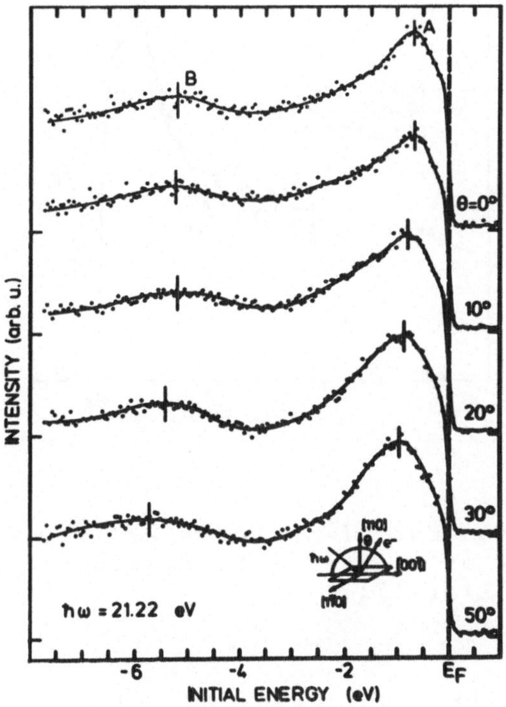

Fig. 13 :

EDC's from $Fe_{0.65}Ni_{0.35}(110)$ (30).

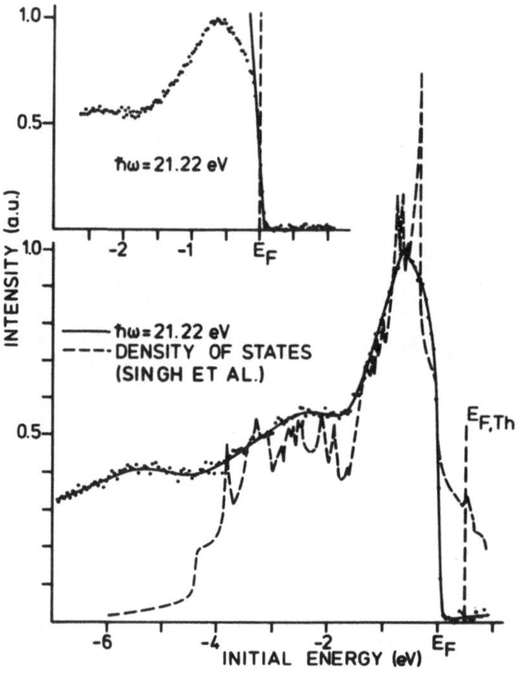

Fig. 14 :

Comparison between EDC of polycrystalline (evaporated) Fe and a density of states of ferromagnetic Fe as calculated in Ref. 21 (31).

are present. On the other hand, emission from single-crystalline Fe (111) (Fig. 3) is very different.

G. PHOTOEMISSSION FROM ORDERED OVERLAYERS

From Figs. 10 - 12 we have concluded on the existence of a monoatomic Cu overlayer on the alloy single-crystal surface after heating. It therefore seems to be interesting to investigate the properties of such overlayers more thoroughly. It has to be mentioned that a Cu overlayer with similar characteristics can be produced by evaporating Cu onto a clean Ni (111) face. The result for a 2 Å thick Cu layer is included in Fig. 12 (dashed-dotted line) and for a series of films with increasing thickness in Fig. 15. The latter spectra clearly demonstrate the development of structural features related to the Cu d electrons with increasing Cu over-

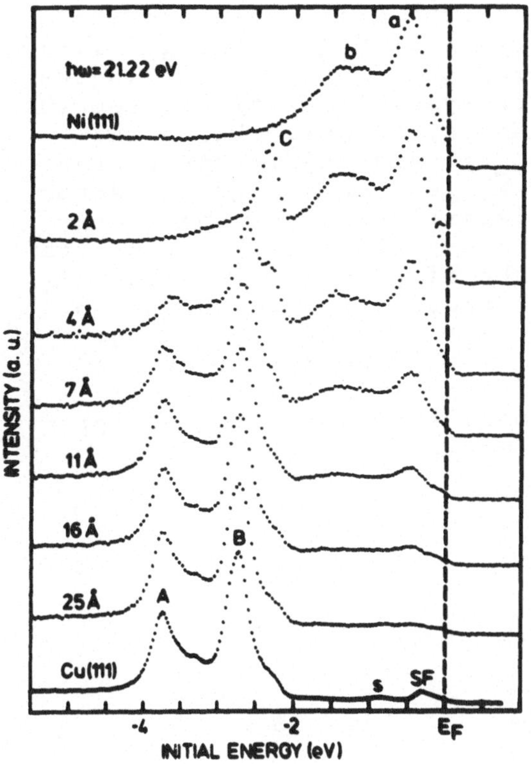

Fig. 15 :

Photoemission at normal take-off from ordered overlayers on Ni (111) (32). Emission from bulk Ni (111) and Cu (111) is also shown (above and below). SF corresponds to emission from surface states of bulk Cu (111).

layer thickness. Already a few atomic layers show peaks A and B, which are characteristic for bulk Cu (111).

These peaks may be explained by bulk $\vec{k}$ conserving transitions. The good agreement between the Cu part of the emission from the evaporated films and that of the bulk Cu single-crystal allows the conclusion that the films grow in an ordered structure (LEED facilities to control the surface geometry directly were not available during this experiment). The only difference is the observation that emission from surface states near E_F are not seen for the thin Cu films. It cannot be decided at present, whether this is an artifact due to experimental difficulties in preparation of the films or whether it can be explained by influences of the Ni substrate. Ordering in a plane parallel to the surface should give rise to a twodimensional band structure.

In fact, evidence for the existence of two dimensional

energy bands has been derived from angle-resolved photoemission of the annealed surface of $Cu_{0.16}Ni_{0.84}$ (111) (16). In Fig. 16 AREDC's for polar angles θ between 0° and 85° are displayed. Transitions from the Cu d bands below − 1.7 eV show systematic shifts and changes in intensity with θ. By using different photon energies we could show that $\vec{k}_{\parallel}$ is indeed a good quantum number for describing the dispersion of the two dimensional energy bands of the Cu electrons (16). It has to be remarked, however, that the dispersion is relatively small amounting to values of about a few tenths of an eV. The more dramatic appearance of different energy levels separated by amounts in the order of 1 eV has possibly to be explained by crystal field splittings through the substrate and could eventually be interpreted by means of symmetry arguments.

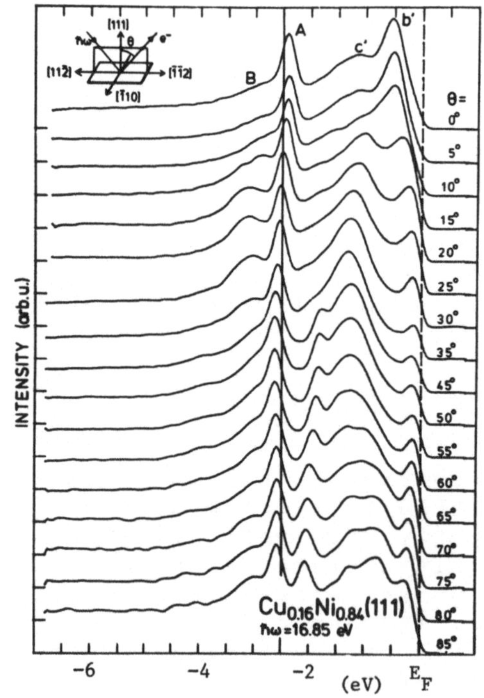

Fig. 16 :

Angle resolved photoemission of an annealed Cu-Ni single-crystal, where an ordered Cu layer is present on the surface (16). The Ni derived features (above an initial energy of -1.5 eV) are similar to those of pure Ni (22).

H. TEMPERATURE EFFECTS IN PHOTOEMISSION

Recently, effects of thermal disorder on photoemission results of metal single-crystals have been reported by Williams et al. for Cu (33) and by Heimann et al. for Au. According to a theoretical paper of Shevchik (14) essentially two contributions are responsible for the emission of photoelectrons. One may be described by direct and the other by phonon-assisted non-direct transitions, the latter one becoming more important with increasing temperature. In Ref. 33 the intensity of a pronounced direct transition was found to decrease exponentially with temperature according to the temperature dependent Dedye-Waller factor. In Ref. 34 another effect was observed: direct transitions showed a broadening, which increased with temperature. This behavior is illustrated in Fig. 17.

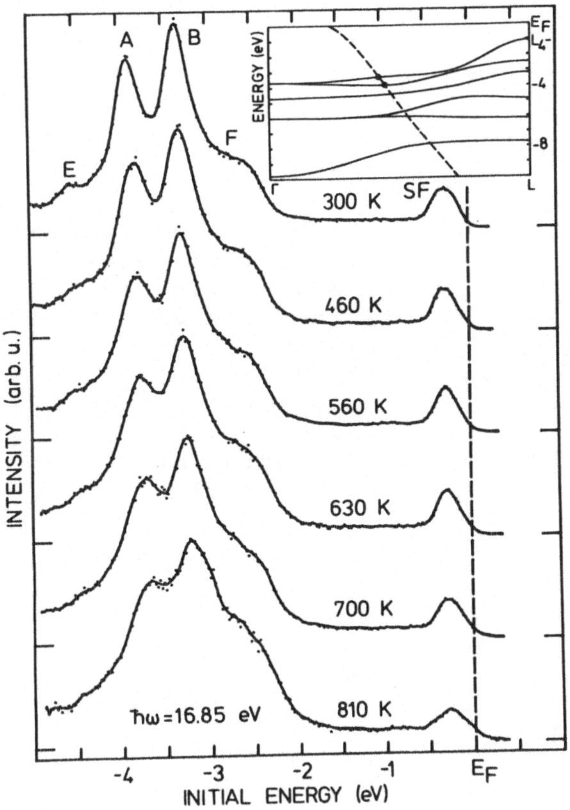

Fig. 17 :

EDC's from Au (111) obtained at different temperatures. For more details see Ref. 34. Peaks A and B are explained by direct transitions.

In Fig. 18a the broadening effect of the AREDC's shown in Fig. 17 has been determined by fitting the observed direct transitions with Lorentzians. A quantitative explanation of the broadening is not yet possible. It is believed, however, that the effect is due to an increasing weakening of the $\vec{k}$ selection rule with increasing temperature. Consequently, the broadening should be dependent on the dispersion of the energy bands participating in the transitions in question. In some cases this has been observed. Other cases, for example the sp band transition analysed by Williams et al., does not reveal such broadening, which is not understood at present.

In addition to thermal influences on the shape of the spectra the position of experimental peaks show shifts towards the Fermi level E_F with increasing temperature (Fig. 18b). This behavior can qualitatively be understood by thermal expansion of the lattice.

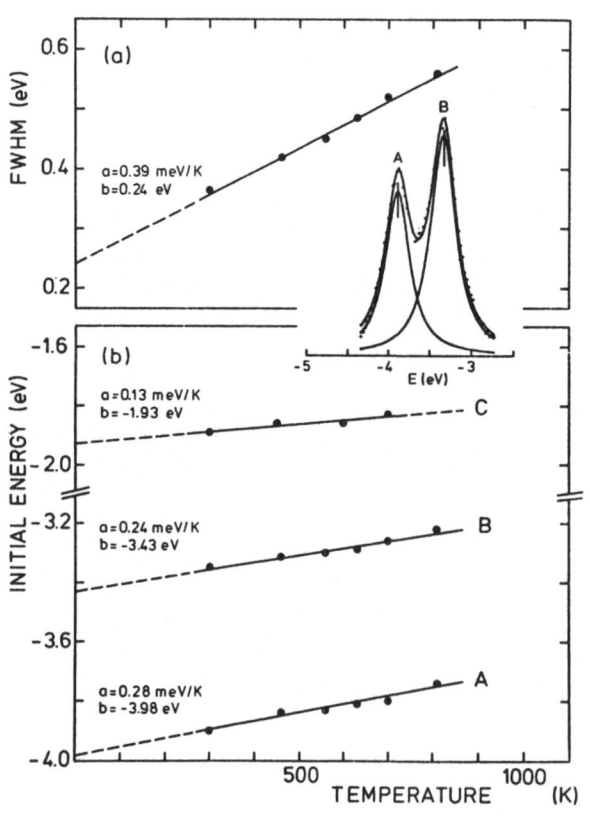

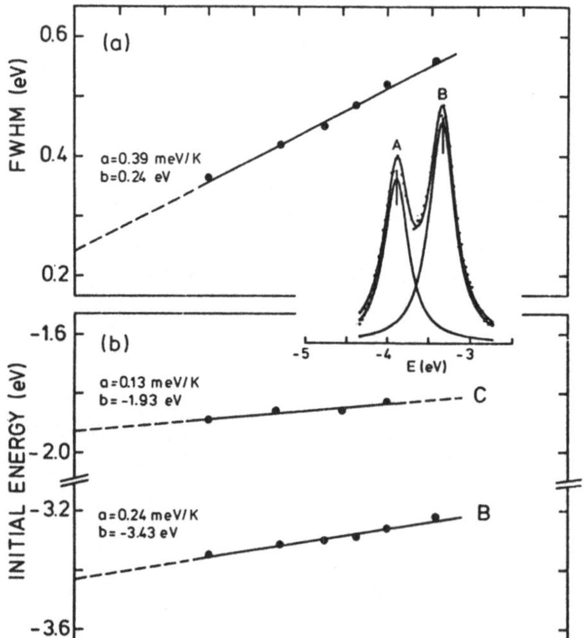

Fig. 18 :

a) FWHM of a Lorentzian fit of peaks A and B of Fig. 17 as a function of temperature.
b) Energy positions of peaks A, B and C (not shown here) as a function of temperature (34).

Another effect was investigated for a Fe-Ni alloy (Fig. 19), which is ferromagnetic at room temperature (29). Heating the sample above the Curie temperature causes a small shift of the maximum of the AREDC. However, this shift is much smaller than is expected on the basis of the itinerant-electron theory of ferromagnetism (8, 30).

I. OPTICAL ABSORPTION FROM RANDOM ALLOYS

It will not be attempted here to give a full survey on optical studies of random alloys. Only one example of measurements obtained from dilute Cu-Ni alloys is briefly discussed in order to demonstrate the capabilities of such investigations. For more details the reader is referred to Refs. 35 - 38, where recent results of optical measurements of Cu-Ni alloys are reported.

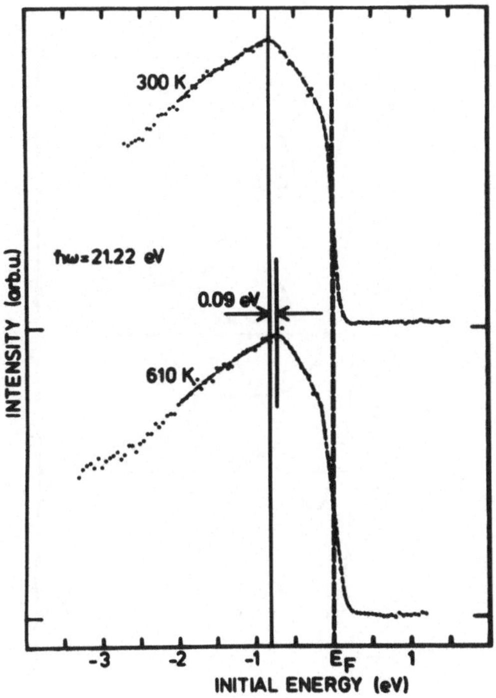

Fig. 19 :

Photoemission from $Fe_{0.65}Ni_{0.35}$ (110) below and above the Curie temperature (22).

In Figs. 20 and 21 results of Tokumoto et al. (38) are reproduced. The optical conductivity, which is given in these Figs., is related to ε_2 by the expression

$$\sigma_1(\omega) = \omega \cdot \varepsilon_2(\omega) / 4\pi .$$

Fig. 20 shows the optical conductivity of pure Ni. The main structure is a peak at a photon energy between 4 eV and 5 eV. As has already been mentioned the entire Brillouin zone contributes to ε_2, which explains the fact that the spectrum does not show much fine structure details.

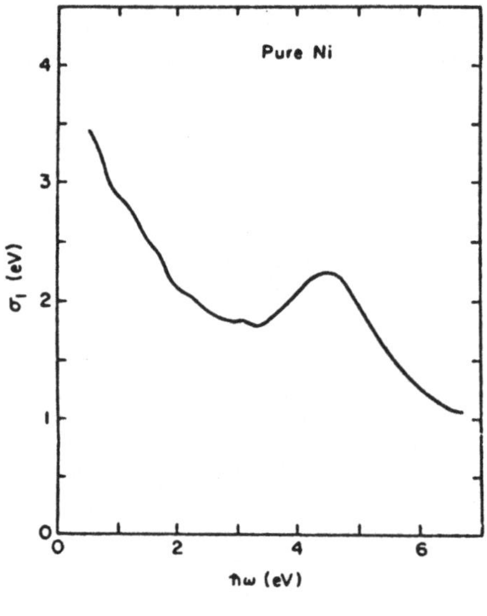

Fig. 20 :

Optical conductivity of pure Ni as abtained from a Kramers-Kronig analysis of reflectivity data (Fig. from Ref. 38).

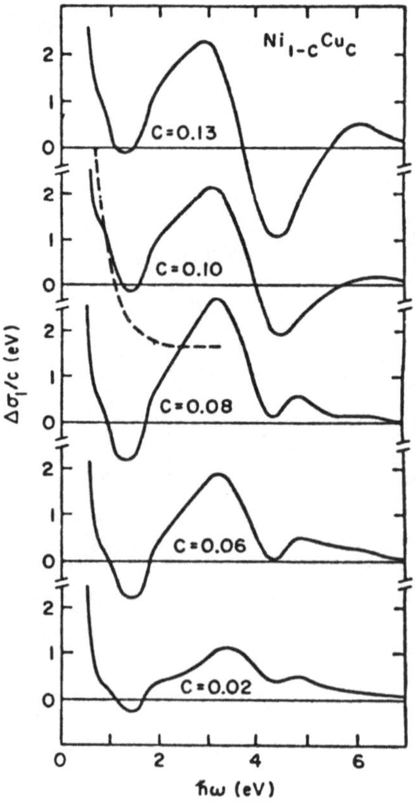

Fig. 21 :

Measurements of the differential optical conductivity per atomic percent for a series of Cu-Ni alloys (Fig. from Ref. 38)

Although the information obtained from the optical data of Ni does not seem to be very detailed when compared with photoemission work on metal single-crystals, in many cases important parameters of the band structure have been extracted from optical measurements. Another advantage is the possibility to investigate binary alloys in the dilute limit of one constituent by means of sensitive differential techniques. Fig. 21 shows results for alloys of Cu diluted in Ni. In the discussion of these data Tokumoto et al. concentrated on features like the shoulder at 1.9 eV or the peak around 3 eV. It was possible to interpret the measured optical properties of the Cu-Ni alloys in terms of the current theoretical ideas on the Cu impurity density of states. To the opinion of the present author, however, it seems to be difficult to really discriminate between the results of the different recent theories on the basis of such measurements alone.

J. CONCLUSION

It was the purpose of these lectures to give typical examples of photoemission work on metal single-crystals and alloys. It was not attempted to review photoemission results on alloys and to discuss the data in detail. Instead, processes and important effects contributing to the experimental electron energy distributions were summarized. Some examples of optical measurements were also given. It was shown, that the access to the electronic states in metals and alloys is especially direct, when the measurements are performed on single-crystalline materials. In this case, from photoemission results information on wavevectors can also be obtained. Another advantage of photoemission is the fact that initial and final energies are measured directly, whereas in the optical properties only differences of these energies are observable. Due to the short mean free path of the photoelectrons the photoemission technique is very sensitive to effects of the surface. Optical properties probe more the bulk electronic structure. Information on dilute alloys can be obtained from optical measurements by means of sensitive differential techniques. Due to these characteristic differences between photoemission and optical measurements it seems to be valuable to consider both methods in order to obtain a full experimental description of the electron states in random alloys.

K. REFERENCES

1) S. Kirkpatrick, B. Velický, and H. Ehrenreich, Phys. Rev. B 1, 3250 (1970).

2) G.M. Stocks, R.W. Williams, and J.S. Faulkner, Phys. Rev. B 4, 4390 (1971).

3) V. Srivastava and S.K. Joshi, Phys. Rev. B 12, 2871 (1975); F. Sacchetti, Nuovo Cimento 32B, 285 (1976).

4) G,M. Stocks, B.L. Györffy, E.S. Giuliano, and R. Ruggeri, J. Phys. F: Metal Phys. 7, 1859 (1977); B.L. Györffy, G.M. Stocks, W.M. Temmerman , R. Jordan, D.R. Lloyd, C.M. Quinn and N.V. Richardson, Solid State Commun. 23, 637 (1977).

5) G.M. Stocks, W.M. Temmerman , and B.L. Györffy, Phys. Rev. Lett. 41, 339 (1978).

6) A. Bansil, L. Schwartz, and H. Ehrenreich, Phys. Rev. B 12, 2893 (1975).

7) B. Feuerbacher and N.E. Christensen, Phys. Rev. B 10, 2373 (1974).

8) P. Heimann, H. Neddermeyer, and M. Pessa, Phys. Rev. B 17, 427 (1978).

9) G.D. Mahan, Phys. Rev. B 2, 4334 (1970).

10) P.J. Feibelman and D.E. Eastman, Phys. Rev. B 10, 4932 (1974).

11) H. Bross, Z. Physik B 28, 173 (1977).

12) B. Feuerbacher, B. Fitton and R.F. Willis (edts.), Photoemission and the Electronic Structure of Surfaces, John Wiley and Sons, New York (1978).

13) C. Kittel, Introduction to Solid State Physics, John Wiley and Sons, New York (1976), 5th edition.

14) N.J. Shevchik, Phys. Rev. B 16, 3428 (1977).

15) H. Neddermeyer, P. Heimann, and H.F. Roloff, J. Phys. E: Sci. Instr. 9, 756 (1976).

16) P. Heimann, H. Miosga and H. Neddermeyer (unpublished).

17) E.O. Kane, Phys. Rev. Lett. 12, 97 (1964).

18) R.R. Turtle and T.A. Callcott, Phys. Rev.Lett. 34, 86 (1975).

19) P. Heimann and H. Neddermeyer, J. Phys. F: Metal Phys. 7, L37 (1977).

20) P. Heimann and H. Neddermeyer, Phys. Rev. B (to be published).

21) M. Singh, C.S. Wang, and J. Callaway, Phys. Rev. B 11, 287 (1975).

22) P. Heimann and H. Neddermeyer (unpublished).

23) P.O. Gartland, S. Berge, and B.J. Slagsvold, Phys. Rev. Lett. 30, 916 (1973).

24) P. Heimann, H. Neddermeyer and H.F. Roloff, J.Phys. C: Solid State Phys. 10, L17 (1977).

25) P. Heimann, thesis (University of Munich, 1977) unpublished.

26) S. Hüfner, G.K. Wertheim, and J.H. Wernick, Phys. Rev. B 8, 4511 (1973).

27) D.H. Seib and W.E. Spicer, Phys. Rev. B 2, 1694 (1970).

28) N.J. Shevchik and C.M. Penchina, Phys. Stat. Sol. B 70, 619 (1975).

29) C.R. Helms, J. Catalysis 36, 114 (1975); C.R. Helms and K.Y. Yu, J. Vac. Sci. Technol. 12, 276 (1975).

30) P. Heimann, H. Neddermeyer and M. Pessa, J. Magn. Magn. Mater. 7, 107 (1978).

31) M. Pessa, P. Heimann, and H. Neddermeyer, Phys. Rev. B 14, 3488 (1976).

32) P. Heimann, H. Neddermeyer and H.F. Roloff, Proc. 7th Intern. Vac. Congr. and 3rd Intern. Conf. Solid Surfaces(Vienna 1977) p. 2145

33) R.S. Williams, P.S. Wehner, J. Stöhr, and D.A. Shirley, Phys. Rev. Lett. 39, 302 (1977).

34) P. Heimann and H. Neddermeyer, Solid State Commun. 26, 279 (1978).

35) B.Y. Lao, R.E. Doezema, and H.D. Drew, Solid State Commun. 15, 1253 (1974).

36) H.D. Drew and R.E. Doezema, Phys. Rev. Lett. 28, 1581 (1972).

37) D. Beaglehole, Phys. Rev. B 14, 341 (1976).

38) M. Tokumoto, H. D. Drew, and A. Bagchi, Phys. Rev. B 16, 3497 (1977).

ACKNOWLEDGEMENTS

This work has been supported financially by the Deutsche Forschungsgemeinschaft through special funds of Sonderforschungsbereich 128.

POLARIZED ELECTRONS FROM METALLIC SYSTEMS

M. Campagna, S.F. Alvarado and E. Kisker

Institut für Festkörperforschung, KFA, Postfach 1913

D-5170 Jülich, F.R.G.

ABSTRACT

In these notes we briefly introduce and review the present status of spin polarized electron emission studies from metallic systems. We confine ourselves to the field emission and photoemission technique. We then discuss two examples : 1) field emission from clean 3d metal surfaces and the spin filter effect of EuS on tungsten and 2) photoemission from clean Ni(100) and Fe(111) surfaces ; temperature and photon energy dependent spin polarization measurements on magnetite are used to illustrate the potentiality of this method to investigate the temperature dependence of the magnetism of surfaces of bulk solids.

INTRODUCTION

Spin polarized photo- and field emission have made significant progress in the last years and we can say today that both techniques are, from an experimental point of view, basically understood /1/. The final theoretical understanding of the data relay on progress to be made especially in the field of the electronic structure of transition metal surfaces and photo- and field emission calculations.

The major aims followed by the spin polarized electron emission techniques, i.e. by the measurement of the spin polarization $\vec{P} = <(\sigma_x,\sigma_y,\sigma_z)>$ where σ_i are the Pauli spin matrices, of the emitted electrons are :

1) determination of the bulk and surface electronic structure (electronic excitations) of :

a) clean and adsorbate covered magnetic materials. Examples are Fe, Co, Ni, alloys, transition and Rare Earth compounds.
b) Nonmagnetic materials, e.g. W,GaAs,... . While in case 1a) polarized electrons are present in the ground state because they are the origin of the magnetism and therefore unpolarized light (or an external electric field for field emission) is sufficient to obtain polarized electron emission, for the case b) circularly polarized light is needed so as to make use of matrix element effects to obtain a different spin "up" and spin "down" population of the excited electrons in the final state. This last method first used with minor success on polycrystalline Cs films /2/ has been recently extended to single crystals of clean and cesiated tungsten. Preliminary results are quite encouraging /3/. But since these studies are just beginning we shall not comment on them.

2) As by-product one expects from these studies a better understanding of both the photo- and field emission process, especially from correlated metals. And furthermore one or both methods will be soon used as a source for polarized electrons to be used in other experiments /4/.

BASIC IDEA BEHIND THE POLARIZED ELECTRON EMISSION EXPERIMENT

In Fig. 1 we show a scheme of the polarized electron emission experiment.

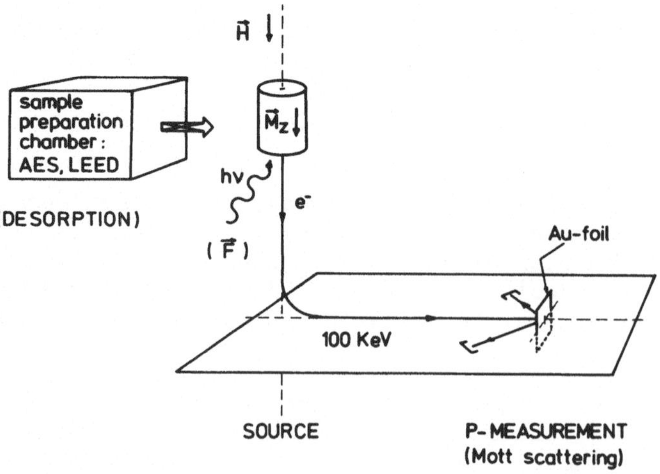

Fig. 1 : Scheme of the basic idea of the spin polarized photo- and field emission experiment.

As mentioned before the quantity of interest in this experiment is the vector $\vec{P} = \langle(\sigma_x,\sigma_y,\sigma_z)\rangle = (P_1,P_2,P_3)$, or depending on the reference system, one of its components P_i. For simplicity we confine ourselves here to the description of the case involving magnetic materials. A sample (usually a single crystal) is treated in a preparation chamber according to conventional surface preparation techniques. A clean surface is being obtained by sputtering and annealing (involving the usual Auger and Leed checks) or else obtained by cleaving under UHV conditions. In the case of field emission controlled UHV or possibly hydrogen promoted field desorption in the main chamber can be used to remove from the apex of the tip the various oxides and impurities /5/. A channeltron is in this case used to detect the desorbed ions. The field emission image is then a good check of the success of the cleaning procedure. Once the cathode, i.e. the surface of the crystal under investigation is characterized, then it is placed on the axis of an electron optical system and the main experiment can be performed. It involves emission of the polarized electrons (either by using photons $h\nu$ or by means of an electric field F) and detection of their degree of spin polarization by Mott scattering. Mott scattering, an elastic scattering experiment at 100 kV, has been often described in the literature /6/. It makes use, in addition to the Coulomb part, also of the spin-orbit contribution to the scattering cross section. In the praxis its sensitivity S to transverse polarized electrons (i.e. electrons with the components of the spin angular momentum perpendicular to their linear momentum different from zero) is the order of 30 % for 100 % polarized electrons and the counting efficiency E around 10^{-4}. This should be compared to the detection of polarized light, for example using a Nicol prism with both S and E in favourable cases of the order of 1. Mott scattering is therefore a reliable but somewhat inefficient method of detection.

EXPERIMENTAL TECHNIQUE

As an example of the experimental realization of the ideas schematically described by Fig. 1, we show in Fig. 2 a scheme of the field emission apparatus /1/. It is basically a modified field emission microscope. It makes use of the so called "probe hole" method (a 2 mm diameter hole in the 10 cm diameter field emission screen) to select well defined emission directions from the field emission pattern. Those electrons passing the probe hole are then spin analyzed. The movable superconducting coil acts as an image steering element. The effective acceptance angle of the probe hole varies with the applied external magnetic field via image compression. It is of the order of a few degrees (from about 2 to 8^0). The consequence is that for a high symmetry plane all the electrons having $\underline{k}_{\parallel} \approx 0$ ($\underline{k}_{\parallel}$ is the component of the crystal momentum parallel to the surface) emitted only from an essentially atomically flat plane are selected for P measurements. An example is shown in Fig.3

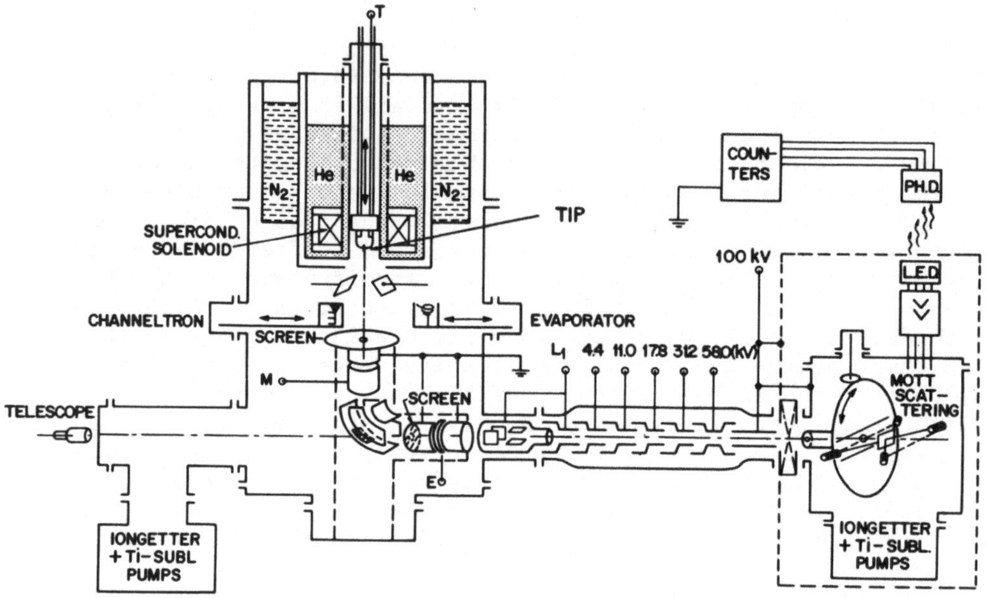

Fig. 2 : Schematic diagram of the apparatus : T = tip holder, movable vertically along the axis of the dewar ; M = main lens, it is used to form an image of the tip through the first probe hole on the second screen located at the entrance of the Einzel-lens, E ; L.E.D. and PH.D. : light-emitting diodes and photo-diodes for optical decoupling of the signal from 100 kV to ground potential.

for the case of Ni(100) in a field of 0.9 kOe. The image is observed on a mirror at an angle of about 40°. The situation is rather different for emission along high step density regions (bright regions on the field emission patterns). Surface scattering effects open up the effective acceptance angle of the probe hole so that also electrons with $k_{\parallel}$ different from 0 may contribute to the field emission current passing the probe hole.

In the photoemission experiment where in contrast to field emission also electron states far below the Fermi energy E_F can be tested, the effect of the external field on the electron trajectories cannot be followed visually but it is expected that for photon energies at few eV larger than photo threshold all the emission directions within 2π are being measured. Recent numerical calculations /8/ demonstrate that also in this case it will be possible in the near future to identify exactly which emission angle contributes at a given photon energy for a given set of electron optical parameters. It is to be mentioned that so far energy selection

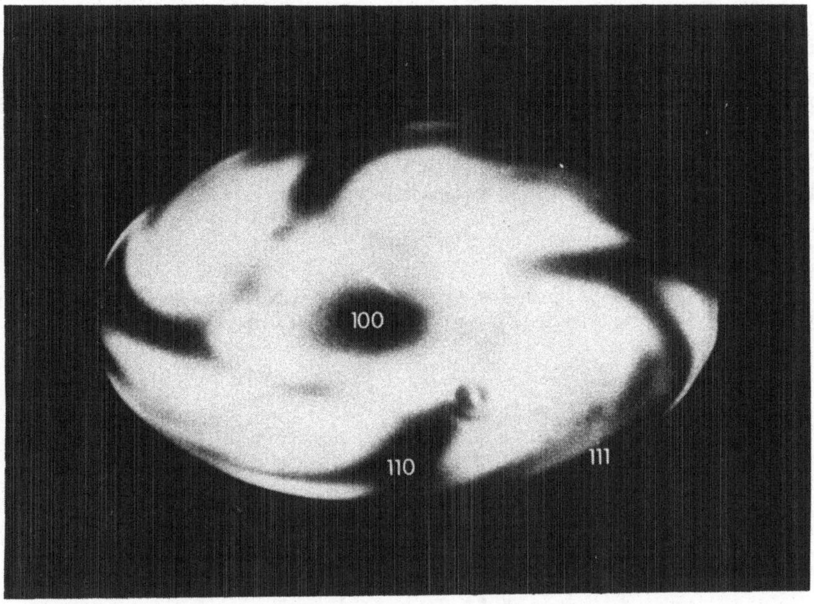

Fig. 3 : Field emission pattern of a (100) oriented Ni tip during polarization measurement ; external field H = 0.9 kOe and P = $-3\pm1\%$.

(energy distribution curves) has not been achieved for photoemission experiments with polarized electrons and will therefore represent a major improvement if realized in the future.

FIELD EMISSION RESULTS

A. 3d transition metals Fe and Ni

In the following table we have summarized the published results for single crystalline Fe and Ni samples (data from Ref. 9).

Ni (hkl)	P(%)
100	-3 ± 1
110	$+6\pm1$
111	-6 ± 1
321	$+7.3\pm0.5$

In each case the external magnetic field $\vec{H}$ is used both as an image steering device and as axial vector along the measured (hkl) direction. So the measured P is the component of $\vec{P} = (0,0,P_z)$ parallel to $\vec{H}$.

Fe (hkl)	P(%)
110	-5 ± 10
100	$+26\pm5$
111	$+20\pm5$

The complete understanding of the above data, representing the polarization of electrons emitted with an exponentially weighting factor from about 200 meV below E_F is yet beyond the present theoretical capabilities. This is not due to the status of the field emission theory but to that of the calculation of magnetic properties of 3d transition metal surfaces /10/. These results represent therefore a possible test for such self consistent surface calculations and future field emission calculations based on them. It is clear that in addition high resolution field emission energy distribution curves from both Ni and Fe surfaces will be extremely helpful within this framework.

B. The spin filter EuS-W

In Fig. 4 we show a scheme of the interface between a nonmagnetic metal (W) and the well known ferromagnetic insulator EuS.

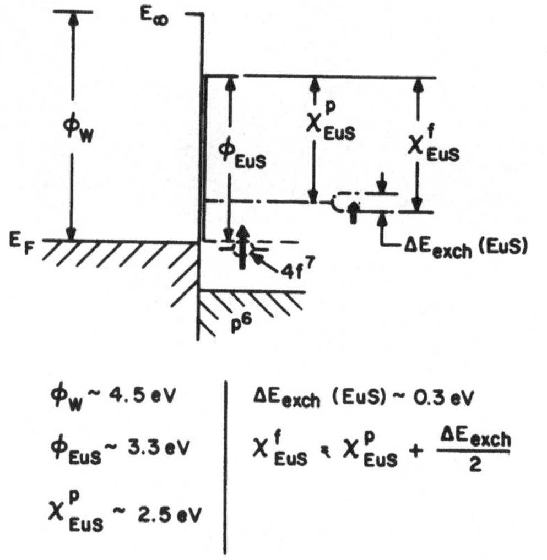

Fig. 4 : Schematic diagram of the W-EuS interface. For simplicity we have not indicated the band bending expected to occur at the interface.

The most interesting feature of EuS, first reported by Busch and Wachter /11/, is the magnetically driven exchange splitting of the empty conduction band states for temperatures T below $T_c \simeq 16°K$ and due to the f-d coupling. This means that for $T \to 0$ below T_c majority-spin conduction band states will be energetically favoured with respect to minority ones. The magnetic exchange splitting of the empty conduction band is of the order of 300 meV. The field emission current for metal-insulator interfaces, where the external electric field $\vec{F}$ penetrates the semiconductor, although reduced by a factor determined by the static dielectric constant ε of the insulator, is known to be proportional to $\exp{-a\Phi_i(\uparrow)^{3/2}}, (\downarrow)$ with $\Phi_i(\uparrow)$, $(\downarrow)$ being in this case the internal barrier for electrons with spin parallel and spin antiparallel to the 4f-spins, i.e. the distance from the Fermi energy E_F to the bottom of the majority (minority) conduction band states. Tunneling occurs from the otherwise unpolarized tungsten conduction band states. Since the exchange splitting is proportional to the reduced spontaneous magnetization $m(T) = M(T)/M(T=0)$, for a given emission potential applied to the tip the field emission current of majority-spins will therefore increase exponentially below T_c. The minority-spin contribution can then be neglected. It is expected that for $T \to 0$ not only $P \to 1$ but also that, for an ideally stoichiometric EuS film, the field emission energy distribution curve will be characteristic of that of a metal. Both expectations are surprisingly rather well fulfilled /12/. This can be inferred from the data presented in Fig. 5. And all this despite the simplicity of the discussed model and the complicated EuS-W interface problem. EuS on tungsten acts therefore as a spin filter for the tungsten conduction band electrons. Because of technical reasons (the field emission current near T_c is rather small) a precise measurement of the spin polarization and of the field emission current near the Curie point T_c is still not available. It should give interesting information on the relation between magnetism and probing range of field emission.

PHOTOEMISSION RESULTS : Ni(100), Fe (111)

Photoemission from 3d levels in metals is not yet well understood, the basic unresolved question being the extent to which the hole left behind is being localized and not a simple band state. This might be rather important for hole states a few hundred meV away from E_F. Here we would like to mention that so far only Ni(100) and Fe(111) have been investigated, using photon energies of up to only 11 eV and in an energy integrated measurement /13/. Since the Ni electronic structure is being discussed by other authors in these lecture notes we shall just comment on the spin polarized photoemission result with respect to field emission data. We reproduce the Ni(100) data of Ref. /13/ in Fig. 6. In the photon energy range used for the data represented in Fig. 6 the escape depth is such that predominantly bulk properties of Ni are being tested. The negative spin polarization at photothreshold confirms the picture of a Stoner-Wohlfarth-Slater band theory at these energies near the Fermi energy. However, the rapid change in sign at photon energies exceeding threshold by about 75 meV is difficult to reconcile with most of the band structure calculations available for Ni. Furthermore far above threshold the spin polarization curve seems to be rather structureless, indicative, in opinion of the authors, of the importance of hole lifetime effects (i.e. importance of electron correlation effects), first pointed out by Anderson /14/. Such effects seem to be even more dramatic for the Fe(111) data, where the polarization curve /13/ is even more smooth and falling from a value of about + 60% at threshold to about 16 % for

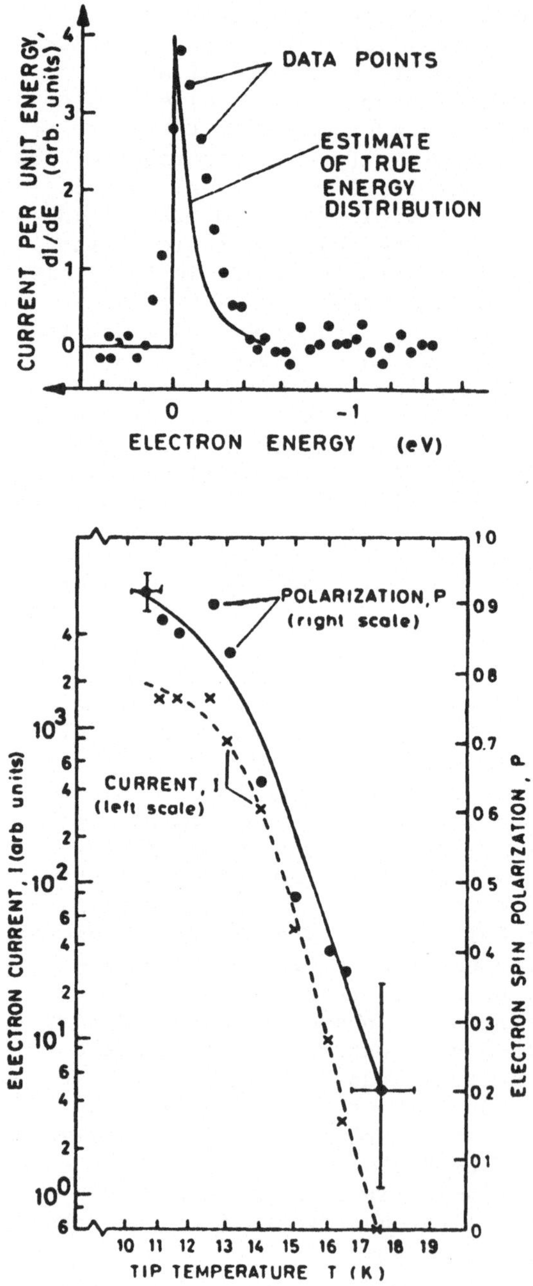

Fig. 5 :

Upper part : Energy distribution of the EuS-W interface measured after annealing of the EuS layer at a temperature of 840 K. The energy resolution of the retarding field analyzer was determined to be 150 meV. The distribution function given above as a solid curve has a width of 80 meV. This function folded with a Gaussion of 150 meV half width gives a good approximation to the data points.

Lower part : Electron current going through the probe hole and electron spin polarization versus tip temperature (data from Ref. 12).

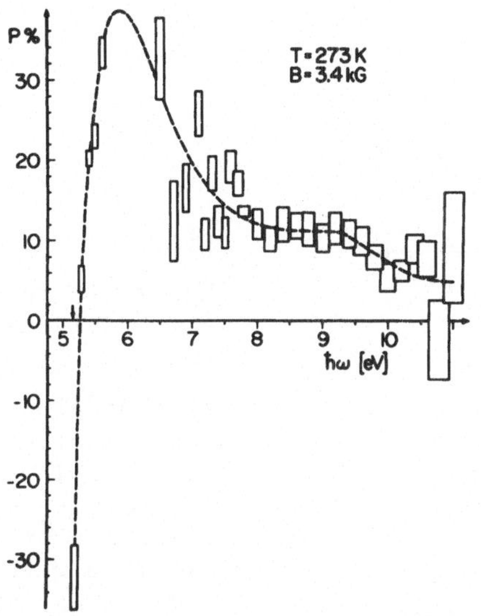

Fig. 6 : Dependence of the spin polarization in photoemission from Ni (100) on photon energy. The threshold is indicated by the arrow.

photon energy larger than 7.5 eV. A puzzle is still the fact that $P_{bulk}(Fe) = n_B/n = 2.2/8 \cong 28$ % whereas the observed spin polarization is only 16%. This may indicate that the magnetic moment of the probed region is smaller than the one in the bulk, namely smaller than 2.2 Bohr magnetons. Clearly much more work both experimentally and theoretically is needed in this direction. Synchroton radiation should be extremely helpful in the future in such investigations. This can be easily extrapolated from the available results obtained using a hydrogen continuum as a light source. The agreement in sign between photo- and field emission data for the Ni(100) surface indicates that magnetic properties of the bulk and of the Ni surface must be similar, especially of the same sign magnetically. This because field emission probes only the surface while photoemission probes the bulk at the photon energies used in the present experiments. The different magnitude of spin polarization observed in the two experiments (about -30% in photo- and - 3% in field emission) is related to the different role played by the various electron states probed in the photo- and in the field emission experiment, respectively (d and sp). Before closing this paragraph we note that the exact interpretation of angle resolved photoemission data for single crystal Ni surfaces has been matter of controversy /15/. This point, not yet fully clarified, is being treated in other lectures.

TEMPERATURE DEPENDENCE OF SURFACE MAGNETISM

Using Fig. 7 we discuss the possibility of using spin polarized photoemission to obtain information on temperature dependence pro-

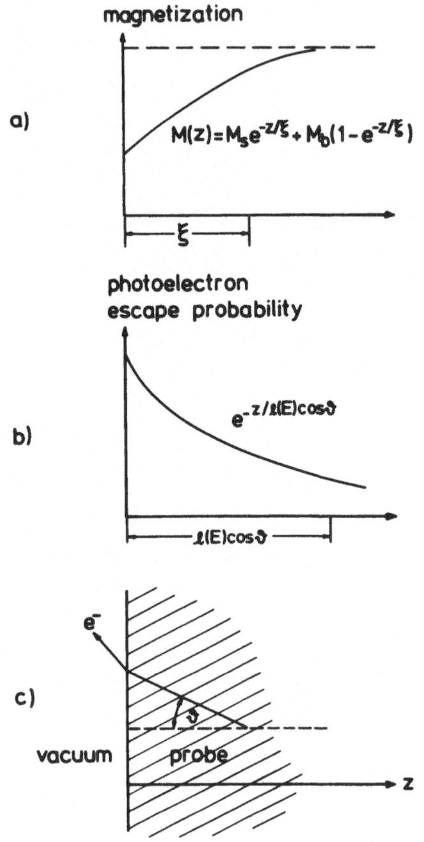

Fig. 7 :

Schematic representation of the model described in the text

a) Spatial dependence of the magnetization $M(z)$. ξ is related to the bulk magnetic coherence length.
b) Escape probability for an electron with kinetic energy E above E_F propagating in a direction at an angle θ from the surface normal and excited at z below the solid vacuum interface.

perties of the surface magnetization of bulk solids. Fig. 7a) represents the spacial dependence of the magnetization $M(z)$ of a hypothetical ferromagnet, where z is the direction perpendicular to the surface. M_s is the surface magnetization, which at $T = 0$, is in general expected to be smaller than the bulk magnetization M_b (this might not necessarily always be the case). In the following we make the assumption that at $T = 0$; $M_s = M_b$. This is surely resonable for systems with localized moments. Since the spin polarization is proportional to the magnetization of the probe region $P(H,T) \propto M(H,T)/M_0$ with M_0 the saturation magnetization, we obtain the "Ansatz" :

$$P(z) = P_s \exp(-z/\xi) + P_b[1-\exp(-z/\xi)]$$

Within the three-step model of photoemission /16/ the number $n_s(n_b)$ of photoelectrons carrying surface (bulk) information can be determined. We note first that : $n_s + n_b = 1$ and for the measured spin polarization $P = n_s p_s + n_b p_b$. Let :

$$T(E,z) = 1/2 \int_0^{\theta_0} d\theta \sin \exp(- z/\cos\theta)$$

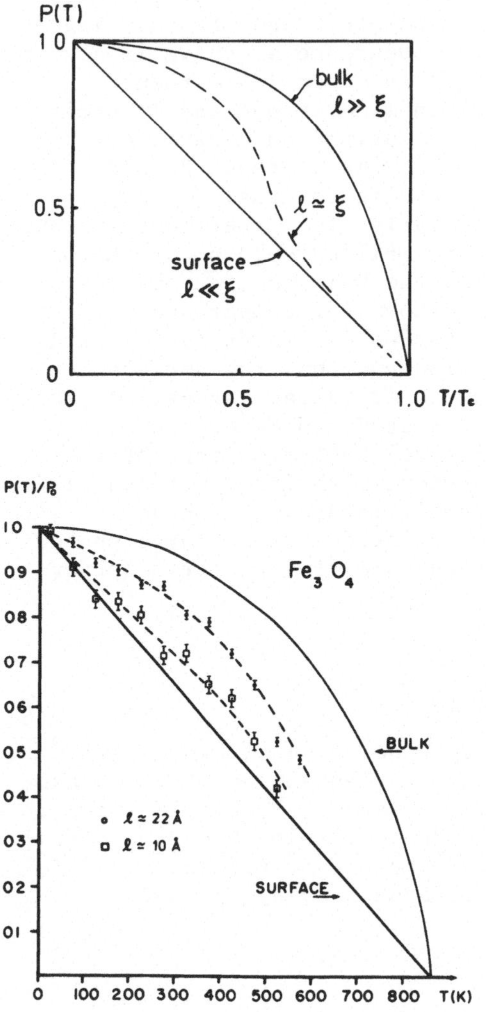

Fig. 8 :

Upper part : Temperature dependence of the degree of spin polarization of photoelectrons emitted from a hypothetical ferromagnet. See text for description.

Lower part : P(T) data collected on a single crystal of magnetite at ℓ = 22Å (ħω = 7.8 eV) and ℓ = 10 Å (ħω = 10.3 eV). The bulk (experimental) and surface magnetization (assumed to be linear) curves are also shown for comparison. A field H = 8.44 kOe was applied to magnetize the sample.

be the semiclassical photoelectron escape function with the usual notations. Then one obtains /17/ :

$$n_s = \frac{2\xi}{\ell}\left[\frac{1}{1+\cos\theta_0} - \frac{\xi}{\ell(1-\cos^2\theta_0)}\ \ln\left(\frac{\ell+\xi}{\ell\cos\theta_0+\xi}\right)\right]$$

For small excitation energies E above E_F, $\cos\theta_0 \simeq 1$ so $n_s = \xi/(\xi+\ell)$. Thus, within this model, one obtains the intuitive result that when the escape depth of the photoelectrons ℓ is much larger than the magnetic coherence length only bulk effects will be observed ($n_s \simeq 0$) and if the contrary is true surface effects will predominate ($n_s \simeq 1$). In general it is expected that the bulk and the surface magnetization have different temperature dependence. Let us consider the case of a ferromagnet such that its surface magnetization decreases linearly with temperature and the bulk magnetization has a different temperature dependence as shown in Fig. 8, upper part. If $\ell \ll \xi$ then $P(T) \propto M_s(T)$ and when $\xi \ll \ell$ then $P(T) \propto M_b(T)$. With this simple model in mind we can easily understand the temperature dependent spin polarization measurement performed on magnetite, Fe_3O_4, at escape depths of about 22 and 10 Å /17/. This corresponds to measurements done using photon energies $\hbar\omega$ = 7.8 and 10.2 eV respectively. The experimental results are given in Fig. 8, lower part. The normalized spin polarization of the photoelectrons $P(T)/P(T=0)$ and the bulk spontaneous magnetization curve of magnitite /18/ are shown for comparison. It can be clearly seen that the spin polarization decreases faster with temperature than the bulk magnetization and that the surface contribution, responsible for this effect, is indeed largest at the shortest escape depth. This pioneering study on Fe_3O_4 confirms qualitatively the findings of theoretical investigations of the critical properties of Heisenberg systems with free surfaces /20/. Again in the future it will be extremely interesting in doing these measurements using synchroton radiation over a large photon energy range so as to vary the surface sensitivity over a larger range as it has been done so far.

REFERENCES

/1/ For recent brief review of the spin polarized field emission and photoemission techniques see : M. Landolt and M. Campagna Surf. Sci. 70, 197, 1978 and S.F. Alvarado, W. Eib, F. Meyer, H.C. Siegmann and P. Zürcher, in Photoemission Electronic Properties of Surfaces, Eds. B. Feuerbacher, B. Fitton, and R.F. Willis (John Wiley and Sons Ltd., London 1978), p. 437.

/2/ K. Koyama and H. Merz, Z. Physik B20, 131, 1975.

/3/ P. Zürcher and F. Meyer, private communication.

/4/ In the paper by D.T. Pierce and F. Meyer (Phys. Rev. B13, 5484, 1976) some of the proposed experiments using polarized electron beams are being discussed.

/5/ For a comprehensive review of the field desorption techniques see : E.W. Müller and T.T. Tsong in Progress in Surface Science, Eds. S.G. Davison, Pergamon Press Ltd., Oxford 1973.

/6/ The Mott scattering technique is described for example in : M. Campagna et al, Adv. Electr. and Electron Physics 41, 113, 1976.

/7/ J.-N. Chazalviel, Surf. Sci. 62, 720, 1977.

/8/ E. Kisker and M. Campagna, Verhandlungen der DPB 1, 222, 1978 and to be published.

/9/ M. Landolt and M. Campagna, Phys. Rev. Lett. 39, 568, 1977, and M. Landolt and Y. Yafet, Phys. Rev. Lett. 40, 1401, 1978.

/10/ J.-N. Chazalviel and Y. Yafet have given a tentative theoretical interpretation of the Ni data, but neglecting influence of the surface on the electronic structure (Phys. Rev. B15, 1062, 1977).

/11/ For a review of exchange effects in magnetic semiconductors as detected by optical methods see : P. Wachter, C R C, critical reviews in Sol. Stat. Sci. 3, 189, 1972.

/12/ E. Kisker, G. Baum, A.H. Mahan, W. Raith and B. Reihl, Phys. Rev. B (in print).

/13/ W. Eib and S.F. Alvarado, Phys. Rev. Lett. 37, 444, 1976 and W. Eib and B. Reihl, Phys. Rev. Lett. 40, 1674, 1978.

/14/ P.W. Anderson, Phil. Mag. 24, 203, 1971.

/15/ P. Heimann and H. Neddermeyer, J. Phys. F 6, L 257, 1976. D. Dietz, U. Gerhardt and C.J. Maetz, Phys. Rev. Lett. 40, 892, 1978. D.E. Eastman, F.J. Himpsel and J.A. Knapp, Phys. Rev. Lett. 40, 1514, 1978.

/16/ For a review of the three-step model see : D.E. Eastman in "Techniques of Metals Research VI", ed. by E. Passaglia (Interscience, New York 1972).

/17/ S.F. Alvarado to be published.

/18/ S.F. Alvarado and W. Eib, J. Mag. and Mag. Materials 7, 16, 1978.

/19/ R. Pauthenet, Ann. Phys. Paris 7, 710, 1952.

/20/ K. Binder and P.C. Hohenberg, I.E.E.E. Transactions on Magnetics Vol. Mag.-12, 2, 66, 1976.

TRANSITION FROM THE ORDERED SOLID TO THE DISORDERED LIQUID STATE VIEWED BY PHOTOEMISSION

P. Oelhafen, U. Gubler

Institut für Physik, Universität Basel

Klingelbergstrasse 82, CH-4056 Basel, Switzerland

F. Greuter

Laboratorium für Festkörperphysik ETH Zürich

CH-8093 Zürich, Switzerland

ABSTRACT

For several metals like Hg, In, Ga and Tℓ, we show that the solid-liquid phase transition can generate large changes in the photoemission properties. The temperature or phase dependence of the yield Y(T), workfunction Φ(T), vector-ratio α(T) and the energy distribution curves (EDC) are of special interest here. Thallium shows an additional solid-solid phase transition and offers the opportunity of studying a transition between two different ordered states. Possible reasons for the observed changes at a phase transition are discussed.

The structure sensitive photoemission parameters, in combination with their established surface sensitivity makes it possible to investigate experimentally phase transitions in the first few atomic layers of a surface.

1. INTRODUCTION

The periodic potential has been a very successful concept for the understanding of the physical properties of crystalline materials. However, by destroying this strict order, the same classification of metals, semiconductors and insulators can still be found as in the ordered state. This problem is a challenge for both theorist and experimentalist and for about twenty years encouraging progress has been achieved in the understanding of disordered systems. Different kinds of disorder may exist; these include thermal vibrations, chemical disorder as in substitutional alloys, defects, the introduction of a surface or grain boundaries or positional disorder as in the amorphous and liquid states.

In this article, we want to discuss the influence of the melting transition on the electronic structure and photoemission properties of some 2B/3A group metals. The electronic and structural properties of liquid metals have been studied in a large number of experiments and are reviewed in several papers (see e.g. Busch and Güntherodt 1974, Faber 1972 and references therein).

The information on the electronic structure obtained in most of these experiments is, however, restricted to a very narrow energy range close to the Fermi energy. The deeper-lying electron states can be tested by photoemission and a few measurements on simple and noble liquid metals and alloys have already been reported (Norris and Wotherspoon 1977, Baer and Myers 1977, Oelhafen 1976, Koyama and Spicer 1971, Eastman 1970 and references therein). In general, it was observed that the photoemission spectra are not much affected by the loss of periodicity. This behavior will be discussed in 3.1 with a special emphasis on the s- and p-like conduction electrons, which mainly determine the transport properties in these 2B and 3A elements. In 3.2 we briefly discuss some experiments concerning the photoemission excitation mechanism in liquid metals and present a parameter, which turned out to be very surface structure sensitive and may be quite useful for the study of surface phase transitions. Finally in 3.3 we present the first experiments involving a direct observation of photoemission properties during the solid-liquid and solid-solid phase transitions.

2. EXPERIMENTAL

The preparation of an atomically clean surface of a liquid needs special care and techniques. At the melting transition contaminations may either segregate to the surface or dissolve into the bulk. To exclude such surface contaminations, we used ultra high purity starting materials, which were further cleaned by either distillation or degassing under ultra high vacuum (UHV) conditions. Mercury can easily be distilled, whereas the evaporation rates of In and Ga are so low, that we carefully degassed these metals at 600-700°C for at least two days to eliminate gas inclusions that may exist even in highest purity materials. During this degassing procedure, a thin oxide layer normally appears at the surface which can be eliminated by draining the metal through a small pipe. The nominal purities of the metals we studied here are given by: Hg: 99,99999 %, In: 99,9999 %, Ga: 99,9999 %, and Tℓ: 99,9999 %. All the preparations and measurements were done in Pyrex or quartz systems at a typical base pressure of $2 \cdot 10^{-10}$ torr (Oelhafen 1976).

The samples of liquid thallium were prepared by distillation as well as by degassing. The photoemission results for these different preparation methods turned out to be identical within the measuring accuracy, which was better than 10 meV for the workfunction measurements.

After this cleaning procedure, highly reproducible surfaces (reproducibility of workfunction better than 3-10 meV) were obtained by pouring the metal (under UHV) into a stainless steel or quartz emitter. For photon energies up to 6.6 eV the energy distribution curves (EDC) of the photoemitted electrons were measured with a hemispherical 2-grid retarding field analyzer with a best resolution of 0.06 eV.

3. RESULTS AND DISCUSSION

Photoemission is by no means a simple process. Different parameters are involved in this process and give rise to observable changes in the order-disorder phase transition at the melting point.

The photoemission process itself may be described by the volume or the surface effects or a combination of both (Feuerbacher and Willis 1976 and references therein), depending on whether it is the interaction with the periodical or the surface potential, that leads to the simultaneous conservation of energy and momentum in the optical excitation process. In general, we have to expect that both excitation processes are effective, although there is, at least to the authors' knowledge, no straight-forward experimental proof for the surface effect up to now. At the melting point, changes in the surface ion- and electron-densities may influence the local electronic structure and the surface potential, which may result in a variation of the excitation probability for the surface effect. In the present photon energy range a powerful description of the volume effect is given by the three step model (Berglund and Spicer 1964), where the optical excitation, the transport of the excited electron to the surface and its escape over the surface barrier are considered as independent processes. The optical excitation process is essentially influenced by the loss of long range order at the solid-liquid phase transition. In the crystalline state conservation of the electron momentum k is a strong selection rule for the optical transition, whereas in the liquid k is not a well-defined quantum number and hence its conservation is not important (Enderby 1972). This change in the character of the optical transition will, of course,influence both the optical and photoemission properties.

As an illustration, most of the 3A elements show remarkable changes in the optical conductivity between the solid phase, where parallel-band absorption dominates the spectra, and the Drude-like liquid or amorphous state (e.g. Hunderi and Ryberg 1974, Ament and de Vroomen 1977 and references therein). Such variations in the optical constants influence the reflectivity and penetration depth of the exciting light and hence will affect the photoemission properties. Thus from the optical excitation process alone we might expect changes in the number as well as in the angular- and energy distribution of photoelectrons.

The transport of a photoelectron from the region of excitation to the surface is governed by the mean free paths for electron-electron and electron-phonon scattering. These problems have recently been reviewed by Fitting et al. 1978, with a special emphasis on the interaction of hot electrons with phonons. We might expect, that in the liquid state, the electron-phonon scattering is much stronger than in the solid and occasionally may become the leading term. The electron-phonon interaction is characterized by a small energy exchange and large changes in the electron propagation direction.

The escape of the photoelectrons over the surface barrier is mainly determined by the electron energy, its k-vector and the workfunction. Usually, the free electron approximation and a step potential are used to describe the refraction of the electrons at the metal-vacuum interface. Only for electrons with very low kinetic energies does the actual form of the surface potential become important, due to quantum mechanical reflection effects (Read and Jennings 1978). Thus changes of the escape probability at the solid-liquid phase transition are essentially due to workfunction changes and occasionally to changes in the electron dispersion relation.

3.1 Energy Distribution Curves for Solid and Liquid Hg,In,Ga and Tℓ

By comparing photoemission spectra of the ordered polycrystalline phase with the disordered liquid state we can get an idea of the importance of the periodical potential for the electronic structure of a metal.

Measurements with low photon energies belong to the so-called "band structure regime of photoemission", where the initial as well as the final state determine the measured spectrum and hence an enhanced sensitivity for any changes in the electronic structure might be expected. In figure 1-4 we compare the measured energy distribution curves (ECD) for the solid and liquid phases of Hg, In, Ga and Tℓ. The solid samples were obtained by cooling the liquid specimen to below its melting point, where they crystallized in the polycrystalline form. The only exception is thallium, which always crystallized in large grains of some millimeters in diameter, so that the measurements were actually done on single crystal surfaces.

The spectra for the ordered and disordered phases of Hg (Cotti et al. 1973) and In are nearly identical, whereas in Ga and Tℓ the loss of longrange order obviously strongly affects the photoemission properties.

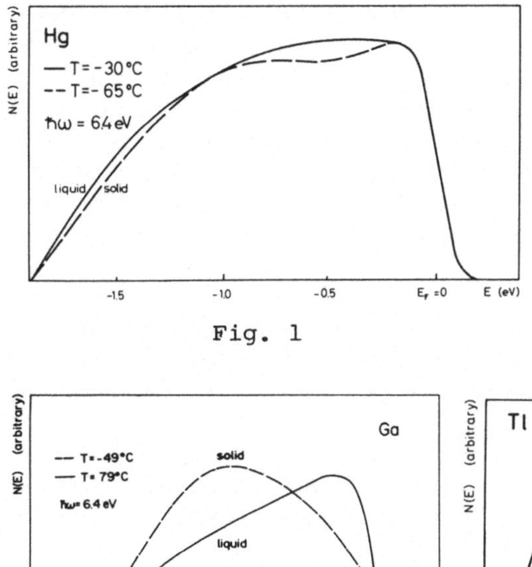

Fig. 1

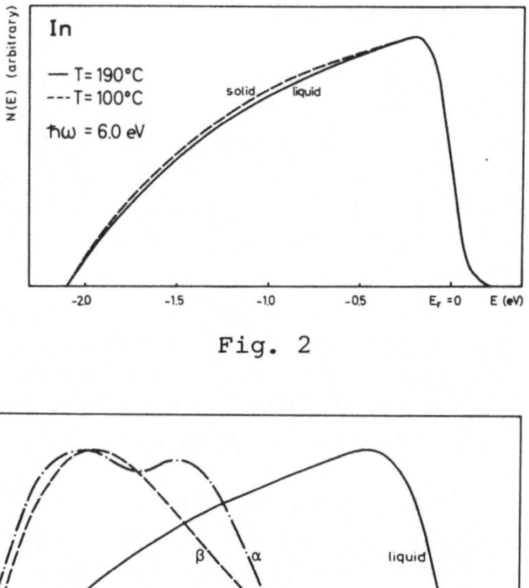

Fig. 2

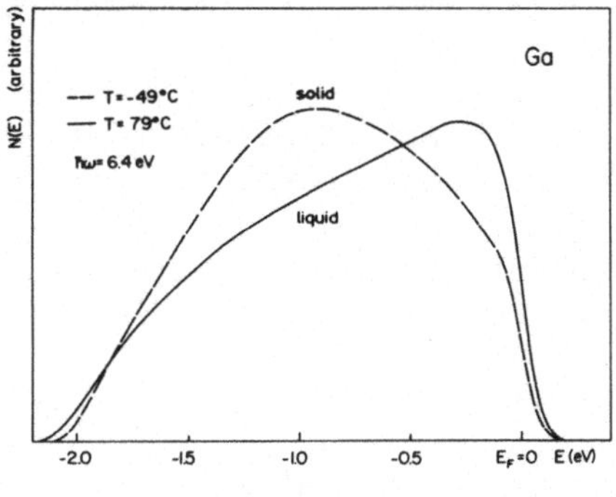

Fig. 3

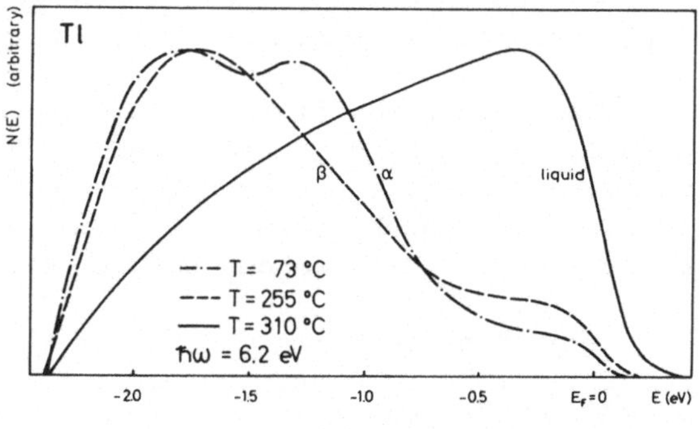

Fig. 4

Fig. 1-4 : Energy distribution curves for the liquid and solid state of some 2B/3A metals.

The behavior of the spectra of Hg and In are in agreement with most photoemission experiments on simple and noble metals (Norris 1977, Eastman 1971, Koyama and Spicer 1971), which show the persistence of the general features of a solid spectrum into the liquid.

All these metals show a nearly close packed structure in the solid state, which is only slightly distorted to a "loosened close packed" structure at the melting transition. For the interpretation of photoemission results it is therefore generally assumed, that the short range order persists into the liquid state and that the local atomic arrangement alone determines the gross electronic features seen in photoemission. This idea is supported by our measurements on Ga, where the open, low symmetry structure of the α-phase collapses to a nearly close packed arrangement in the liquid state, resulting in a large change in both the short range order and the electronic structure near the Fermi level (Greuter and Oelhafen, to be published). The spectra for liquid Ga and In are well explained on the basis of the three step model by a free electron approximation for the uppermost few electron volts of the conduction band, whereas liquid Hg shows the characteristics of the pseudogap proposed by Mott (Cotti et al. 1973). The results for polycrystalline Ga and the single crystalline α-phase of Tℓ are quantitatively understood on the basis of bandstructure calculations (Hunderi and Ryberg 1974, Ament and de Vroomen 1977). Up till now the comparison of liquid Tℓ with its solid α- and β-phases has, at least to the authors' knowledge, not been studied by photoemission.

The results for the hexagonal close packed α-phase and the (probably) body centered cubic high temperature β-phase are, in fact, angle resolved measurements on single crystals due to the narrow escape cone for the low-energy electrons used in our experiment, and the large grains of the sample.

The comparison of the α- and β-phase spectra indicates, that despite the extremely small changes in short range order, changes in the band structure may occur. However, it is to be expected that measurements which average over the whole Brillouin zone would not show essential differences between the two phases.

The pronounced minimum in the number of photoelectrons near the Fermi level must be attributed to the energy gap that appears along most of the symmetry lines of the 6p-band due to a weak spin orbit interaction and which gives rise to a well-developed minimum in the density of states at the Fermi energy (Soven 1965, Ament and de Vroomen 1977). This spin orbit interaction must also exist in the β-phase, but is destroyed at the melting point resulting in a nearly free electron-like behavior for the top of the probably very narrow conduction band. Our measurements may indicate, that the 6p-band is almost separated from the 6s-band even in the liquid state, quite similar to the situation of Bi and Pb (Baer and Myers 1977, Norris 1977), but further investigations with higher photon energies are necessary in order to clarify this situation.

With the exception of mercury, the common feature of all the studied liquid metals is the free electron behavior of the s- and p-like conduction electrons near the Fermi level. This is in agreement with the fact, that the weak scattering formalism of Ziman is a useful description for the transport properties of these metals. Obviously this model is restricted to the top of the conduction band and is not able to describe correctly the deeper lying electron states, especially in the heavy elements like Bi, Pb and Tℓ.

3.2 Excitation with Polarized Light

The use of non-normally incident, polarized light is increasingly recognized as an important tool for the study of photoemission properties. In this context, the 50 year-old problem of the relative strengths of volume- and surface-excitations turns out to be relevant (Endriz 1973, Sass 1975, Kliewer 1977 and references therein).

The excitation probability of the surface effect is given by the scalar product of the electric field vector and the surface normal. Thus the surface effect can be excited only by p-polarized light, where the electric field vector is parallel to the plane of incidence, whereas s-polarized light will never be effective. The experimental evidence for the surface effect is complicated by the problem of producing a well-defined, smooth surface (Flodström et al. 1975, Endriz 1973) and the difficulty of separating it, even in polycrystalline samples, from anisotropic bulk effects (Sass 1975). These problems might be eliminated by applying polarized light studies to liquids, with their well-defined surface geometry and isotropic bulk properties.

The definition of the vector ratio $\alpha = Y_p/Y_s$, where Y_p and Y_s are the yields per incident photon for parallel and perpendicular polarized light respectively, and the optical absorption ratio $\beta = (1-R_p)/(1-R_s)$, determined from reflectivity measurements, are useful parameters for further discussions. For pure isotropic volume excitation we would expect $\alpha = \beta$ and for normal incidence $\alpha = \beta = 1$.

The measured vector- and absorption-ratios for liquid Hg, In and Ge are shown in figure 5 for an angle of light incidence of 50 degrees. The corresponding results for the crystalline samples are also shown.

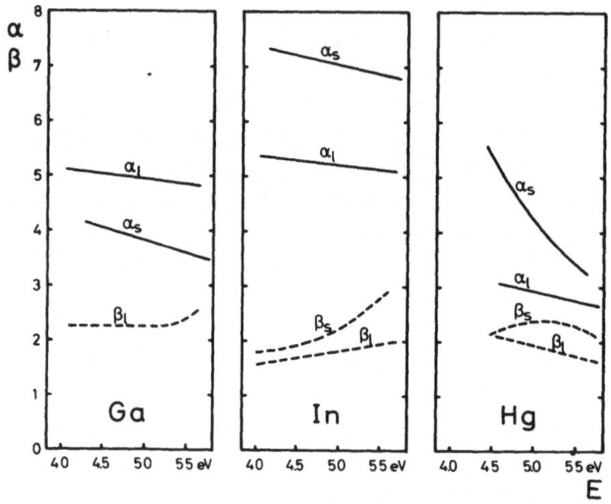

Fig. 5 : Vector ratio α and optical absorption ratio β for Ga, In and Hg in the solid (s) and liquid (ℓ) state.

It is interesting to notice, that in the present photon energy range, the absorption ratio β is almost identical for all these liquid and solid metals with a typical value of about 2. The vector ratio α , on the other hand, is quite different for the different liquids and, in general, shows large variations at solidification. A detailed discussion of these results is beyond the scope of this article and will be given elsewhere (Greuter and Oelhafen, to be published). The close agreement of the α- with the β-values for liquid Hg is what one would expect for an isotropic volume excitation process, whereas the high α/β-values ranging from 2-3 for liquid Ga and In might indicate a surface- and isotropic volume-excitations of approximately equal strengths (Gisler et al. 1977).

From theoretical predictions (Endriz 1973, Kliewer 1977), it is known that the surface effect is strongest near threshold and that the EDC of electrons emitted by this excitation process show a nearly triangular shape. Thus the presence of surface- and volume-excitation of equal strengths should result in remarkably different shapes for s- and p-light excited EDCs. However, our measurements shown in figure 6 on liquid Ga showed, just as in mercury (fig. 7), no polarization dependence of the spectra, although the expected differences should be larger than our measuring uncertainty (Gubler et al. 1978).

To account for the high α-values we therefore have to conclude, that for the studied liquid metals the excitation process is dominated by an anisotropic volume effect, where the electron final momenta are preferentially parallel to the electric field vector. First estimations based on the theory of Schaich and Ashcroft (1970) showed similar anisotropies as proposed by Sass (1975) for polycrystalline materials.

The low vector ratio for liquid Hg must be attributed to the strong electron-phonon scattering in this metal, which will smear out most of the initial excitation anisotropy before the electrons reach the surface.

In liquid Ga and In, on the other hand, the electron-phonon interaction must be much weaker, as is already indicated by the mean free paths for thermal electrons, resulting in only a small disturbance of the excitation anisotropy by the transport process and thus explaining the observed high α-values for these metals.

At the phase transition, changes in the electron-phonon scattering will, in a similar way, contribute to the observed large variations of the vector ratio, together with probably smaller contributions from altered electron-electron scattering and directional effects in the excitation process. Our preliminary results show that the vector ratio is, in addition to its significance for the character of the excitation process, a very sensitive parameter for the study of surface phase transitions. It is of special interest due to its non-destructive, strongly structure dependent and extremely surface sensitive properties. Experiments on crystalline and liquid surfaces, which were slightly contaminated by residual gases, always showed a remarkable increase of the vector ratio even before changes in the workfunction could be detected. Thus the determination of the vector ratio might also be promising for adsorbate studies.

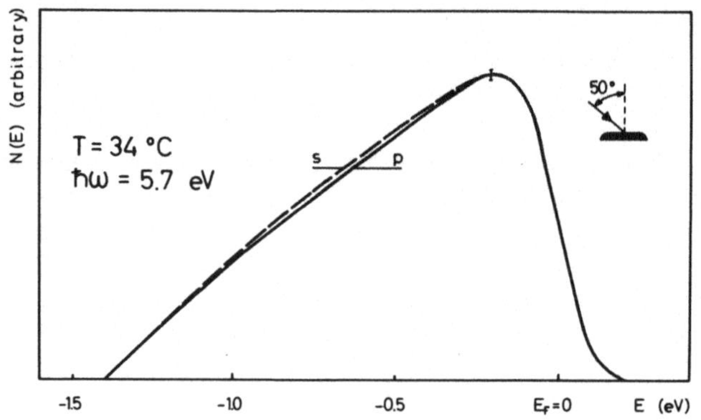

Fig. 6 : EDCs for liquid Ga measured with s- and p-polarized light.

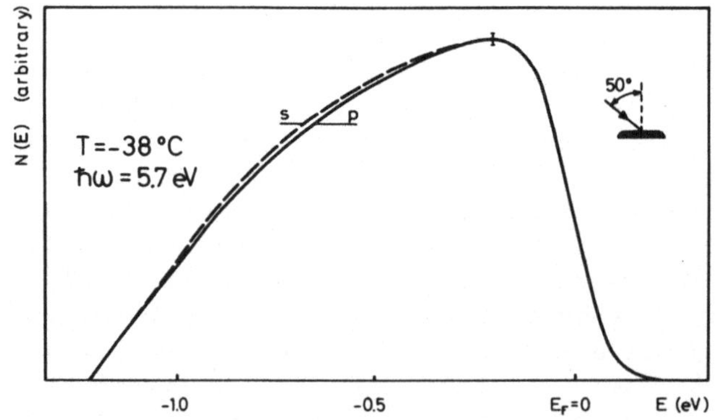

Fig. 7 : EDCs for liquid Hg measured with s- and p-polarized light.

3.3 Yield Measurements at the Solid-Liquid Phase Transition

The photoelectric yield $Y(\hbar\omega)$ is given by the total number of emitted electrons per incident photon of energy $\hbar\omega$. Fig. 8a shows the yield $Y(T)$ measured for Hg as a function of temperature with a photon energy of 5 eV and near normal light incidence. Fig. 8b shows the yield and the temperature as a function of time, measured at 4.55 eV, which is very close to the photoelectric threshold.

A first essential feature of the yield curves is the distinct jump at solidification. The height of this jump depends on different experimental parameters which determine the freezing conditions of the surface, such as the cooling rate, the Hg vapour pressure and the temperature of the surroundings. The time dependence of yield and temperature show a constant yield during solidification of the specimen which was cooled from the bottom. The yield curve shows a decrease just before the large jump related to the freezing of the sample surface.

In order to explain these variations we have to consider the different effects which may alter the yield at a solid-liquid phase transition. Applying the three step model of photoemission, we expect that the largest effects are due to the optical excitation and the emission step. The former is influenced by the change of the optical constants at the phase transition and the increase of the mean angle of light incidence measured relative to the local surface normal, an effect which is related to the surface roughness. For the emission step we have to consider changes of the workfunction, especially for measurements close to the photoelectric threshold.

From optical measurements of Choyke et al. (1971) on solid and liquid Hg we get a ratio for the absorbed light intensity at a photon energy of 5 eV

$$\frac{1 - R_c}{1 - R_\ell} = 1.13$$

where R_c and R_ℓ are the reflectivities for the crystalline and liquid state, respectively. For the number of absorbed photons in a surface layer from which the photoelectrons originate we have also to take into account the penetration depths of the light. From the optical constants we get a penetration depth of 121 Å for solid Hg (W_c) and 91 Å for the liquid state (W_ℓ). The solid to liquid ratio of the number of absorbed photons in the surface layer from which the photoelectrons originate is given by

$$\frac{1-R_c}{1-R_\ell} \; \frac{1-\exp(-\lambda_{e-e}/W_c)}{1-\exp(-\lambda_{e-e}/W_\ell)} = 0.90 \, .$$

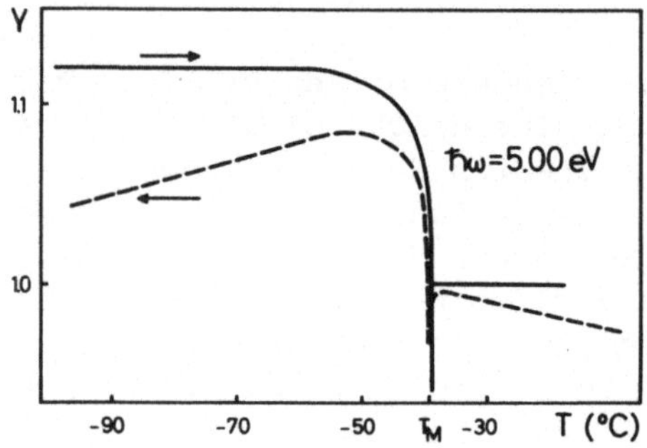

Fig. 8a : Yield of Hg at the solid-liquid phase transition. The cooling and heating rates were 3.7 and 3.3 °C/min respectively.

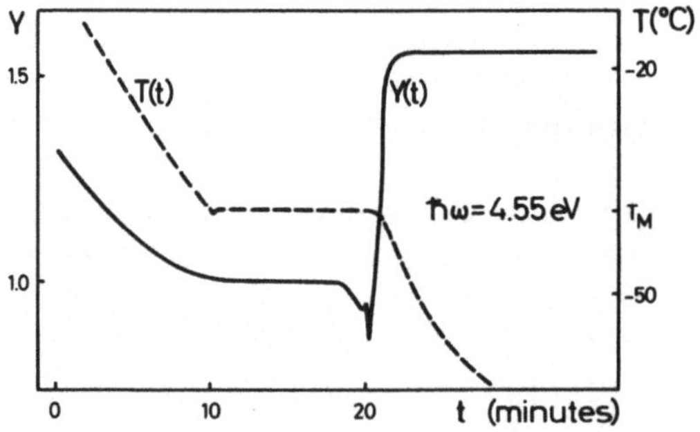

Fig. 8b : Yield and temperature as a function of time.

This ratio is not sensitive to the choice of the escape depth λ_{e-e} and we may assume it to be the same (50 Å for the present energy range, Brundle 1974) for the crystalline and liquid states. The value of 0.9 shows that the change in the optical absorption would rather <u>reduce</u> the yield in the solid state and therefore cannot be responsible for the large increase of the yield at solidification.

Table I : Workfunctions for Hg, In, Ga and Tℓ in the solid and liquid state deduced from Fowler plots.

	Φ liquid (eV)	T (°C)	Φ solid *) (eV)	T (°C)
Hg	4.48 ± 0.01	-25	4.46 ± 0.05	-100
In	4.06 ± 0.01	174	4.09 ± 0.01	130
Ga	4.30 ± 0.01	50	4.35 ± 0.04	0
Tℓ	3.84 ± 0.02	303	3.81 ± 0.04 α-phase	25
			3.78 ± 0.04 β-phase	260

*The measurements were performed on polycrystalline samples. Since the measured workfunction depends on the actual distribution of the different crystal orientations present on the surface, the values for different samples show relative large variations compared to the highly reproducible values for the liquid state.

As can be seen from table I, changes of the workfunction at the solid-liquid phase transition are generally small. In the case of Hg (Fig.9) the change at the melting point is less than 0.01 eV. Assuming a maximum change of $\Delta\Phi = 0.01$ eV for the workfunction, we calculate a relative yield variation of

$$\left|\frac{\Delta Y}{Y}\right| = 2\,\frac{\Delta\Phi}{\hbar\omega-\Phi} = 0.04$$

using the relation $Y \sim (\hbar\omega-\Phi)^2$ for photon energies $\hbar\omega$ near threshold This small value shows, that variations of the workfunction cannot generate the observed jump in the yield at the phase transition which is of the order of 20 % . The most reliable explanation for the great increase of the yield at solidification is the development of surface roughness just below the melting point. This is followed by an increase of the mean angle of the incident light measured to the local surface normal and therefore results in an increased yield. From a simple estimate of the angular dependence of the yield in the volume excitation model it can be shown that an increase of less than 50 degrees of the angle of the incident light can explain a 50 % increase of the yield at solidification. The variations of the increase in the yield at solidification for different cooling cycles can be explained by a variable formation of the surface roughness which depends strongly on the experimental parameters.

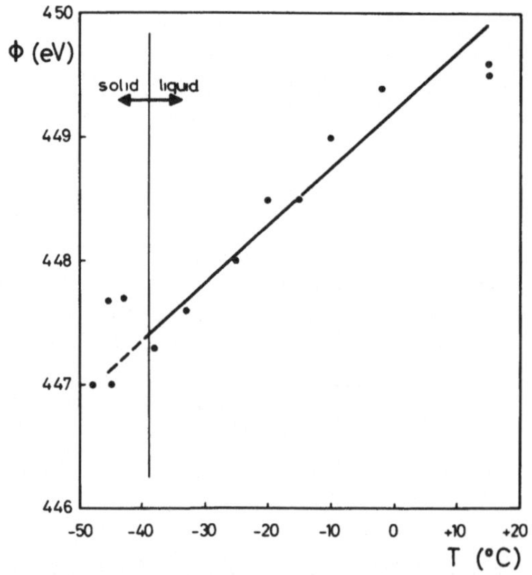

Fig. 9 : Workfunction of Hg at the solid-liquid phase transition.

The decrease of the yield near the melting point just before the surface roughness becomes effective at solidification is probably due to the optical effects discussed above, which tend to decrease the yield by a factor of 0.9 . The yield minimum represents the photoemission from the solid bulk materials where its surface is similar to the surface of a liquid sample: the roughness is not yet developed and it is possible that, at this time, the first few atomic layers are still in the liquid state. Analogous effects can be observed by heating solid Hg: first the surface roughness disappears, melting begins at the surface due to the heat radiation from the surroundings and again we get a decrease of the yield which is typical for the solid bulk with a "liquid like" surface (fig.8a).

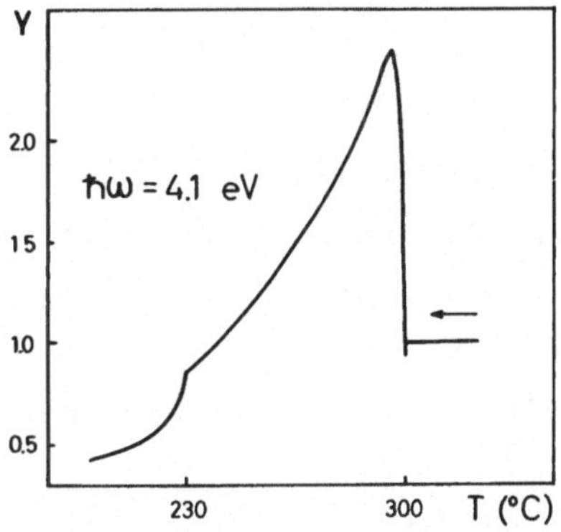

Fig. 10 : Yield of Tℓ as a function of temperature.

Fig. 10 shows the yield for thallium from room temperature up to 325 °C . The measurement was performed with a photon energy of 4.1 eV , which is close to the threshold (see table I). Beside the large change in the yield at the melting point, the phase transition from the α- to the β-phase can be seen clearly. This measurement demonstrates that the temperature dependence of the yield is even increased when photon energies close to the threshold are used. In this case the yield measurements at the phase transitions reflects first of all the workfunction changes: the workfunction is decreased at the transitions from the α- to the β-phase and from the liquid- to the β-phase (see table I) and consequently the yield is increased at the corresponding transitions. Details of these measurements will be published elsewhere.

We would like to thank Dr. K. Agyeman for careful reading of the manuscript. Financial support of the Swiss National Science Foundation is gratefully acknowledged.

REFERENCES

Ament M.A.E.A. and de Vroomen A.R., J.Phys. F 7, 97, 1977

Baer Y. and Myers H.P., Solid State Commun. 21, 833, 1977

Berglund C.N. and Spicer W.E., Phys.Rev. 6A,. 1030, 1964

Brundle C.R., J.Vac.Sci.Technol. 11, 212, 1974

Busch G. and Güntherodt H.-J., Solid State Phys. 29, 235, 1974

Choyke W.J., Vosko S.H. and O'Keeffe T.W., Solid State Commun. 9, 361, 1971

Cotti P., Güntherodt H.-J., Munz P., Oelhafen P and Wullschleger J., Solid State Commun. 12, 635, 1973

Eastman D.E., Phys. Rev. Letters 26, 1108, 1971

Enderby J.E., Liquid Metals 1972, edited S.Z. Beer, Dekker New York

Endriz J.G., Phys.Rev. B7, 3464, 1973

Faber T.E., "An Introduction to the Theory of Liquid Metals", Cambridge University Press 1972

Feuerbacher B. and Willis R.F., J.Phys. C9, 169, 1976

Fitting H.J., Glaefeke H and Wild W., Surface Sci. 75, 267, 1978

Flodström S.A., Hansson G.V., Hagström S.B.M. and Endriz J.G., Surface Sci. 53, 156, 1975

Gisler E., Greuter F. and Oelhafen P., Helvetica Phys. Acta 50, 611, 1977

Gubler U., Oelhafen P., Greuter F. and Preiswerk H.P., Helvetica Physica Acta, in press

Hunderi O. and Ryberg R., J.Phys. F 4, 2084, 1974

Kliewer K.L., Phys.Rev. B15, 3759, 1977

Koyama R.Y. and Spicer W.E., Phys.Rev. B4, 4318, 1971

Norris C. and Wotherspoon J.T.M., J.Phys. F7, 1599, 1977

Norris C., Liquid Metals 1976: Inst. Phys. Conf. Ser. No 30,pp.171

Oelhafen P., Ph.D. thesis (ETH Zurich, Nr. 5767), unpublished, 1976

Read M.N. and Jennings P.J., Surface Sci. 74, 54, 1978

Sass J.K., Surface Sci. 51, 199, 1975

Schaich W.L. and Ashcroft N.W., Solid State Commun. 8, 1959, 1970

Soven P., Phys.Rev. 137A, 1706, 1965

THE MAGNETIC PROPERTIES OF ALLOYS

D.M. Edwards

Department of Mathematics, Imperial College
London SW 7 2BZ, U.K.

1. INTRODUCTION

Alloys containing transition metals exhibit a very wide range of magnetic behaviour. In these lectures it is necessary to narrow the field and I shall concentrate on ferromagnetic and nearly-ferromagnetic alloys, particularly those containing nickel and iron. Thus I shall not discuss dilute alloys with noble metal hosts, such as AuFe or CuMn, and consequently shall only mention the Anderson model in connection with quite a different system and the Kondo effect not at all. Antiferromagnetism and spin-glass behaviour will also receive scant attention. The concentration on ferromagnetic alloys in these lectures is reasonable because it is for these transition metal alloys that the observed magnetic moments have been most closely linked with theoretical work on the electronic structure at T = 0. The interpretation of the magnetic properties of pure transition metals at finite temperature is still a matter for debate[1] so little will be said about alloys at finite T. However, considerable progress has been made in the calculation of spin-wave energies and these collective elementary excitations dominate the low-temperature properties in many cases. Experimental and theoretical work on spin waves in nickel alloys will be discussed and this work throws light on the correct description of the ground state. The central theme of these lectures, from which there will be occasional diversions, is thus the interpretation in terms of electronic structure, of observed magnetic moments and spin-wave energies in ferromagnetic transition metal alloys of varying composition.

1.1. Basic ideas of itinerant electron magnetism

Magnetic moments in transition metals are due to d electrons whose spins are aligned by exchange interactions. The ferromagnetic metals Fe, Co, Ni have saturation moments of 2.2, 1.7 and 0.6 μ_B/atom respectively. The existence of these non-integral moments, together with the large observed electronic specific heat associated with partly-filled d bands, shows that the d electrons are itinerant, not localized on atoms as in the Heisenberg model of magnetic insulators. Furthermore d-like parts of the Fermi surfaces of both majority- and minority-spin electrons have been observed directly by de Haas-van Alphen measurements[2]. Clearly then, any theory of magnetism in the transition metals and their alloys must be founded in band theory. The band structure of a transition metal consists of a fairly narrow d band, with a width of several eV, and a broad overlapping s-p band. Although hybridization occurs between the d and s-p bands it is a reasonable first approximation to consider the d band alone in discussing magnetic properties. A further, more drastic, approximation which has been made in most of the existing work on magnetic alloys is to consider the d band as consisting of five independent identical sub-bands. This approximation was made in early work in non-magnetic alloys, using the tight-binding CPA discussed by Dr. Gyorffy. Some contact with reality is retained by taking the density-of-states curve for the pure metal sub-band to correspond to that calculated in a full band-structure calculation, reduced by a factor 5 of course. More recently a much more sophisticated method, the KKR-CPA method discussed by Dr. Gyorffy, has been developed and applied to non-magnetic CuNi alloys. This method will certainly be applied to magnetic alloys in the future and we shall keep this prospect in mind during these lectures. For the present, however, we shall consider the single-orbital tight-binding model and first discuss a pure metal to fix the ideas of band magnetism. In this case the Hamiltonian is that of the Hubbard model:

$$H = \sum_{ij\sigma} t_{ij} a^{\dagger}_{i\sigma} a_{j\sigma} + V \sum_{i\sigma} n_{i\sigma} + U \sum_{i} n_{i\uparrow} n_{i\downarrow} \, . \qquad (1)$$

Here $a^{\dagger}_{i\sigma}$ is the creation operator for an electron of spin σ in the orbital at site i and $n_{i\sigma} = a^{\dagger}_{i\sigma} a_{i\sigma}$ is the correspondong occupation number. Also t_{ij} is the hopping integral ($t_{ii} = 0$), V is an atomic energy level and U represents the Coulomb interaction between two electrons on the same atom. We shall treat the ground state of the model within the Hartree-Fock approximation and it is therefore necessary to regard U as an effective parameter which includes effects due not only to screening by s-p electrons but also to corelation between d electrons[3]. The single-particle part of the

Hamiltonian, the first two terms, may be written in the Bloch representation so that

$$H = \sum_{\underline{k}\sigma} (\varepsilon_{\underline{k}} + V) n_{\underline{k}\sigma} + U \sum_i n_{i\uparrow} n_{i\downarrow} \tag{2}$$

where $n_{\underline{k}\sigma}$ is the Bloch state occupation number and the band energy

$$\varepsilon_{\underline{k}} = \sum_j t_{ij} \exp [i\underline{k}.(\underline{R}_i - \underline{R}_j)] . \tag{3}$$

Here $\underline{R}_i$ is the position of the site i. The Bloch wave-functions are of the form

$$\psi_{\underline{k}}(\underline{r}) = \frac{1}{\sqrt{N}} \sum_i e^{i\underline{k}.\underline{R}_l} \phi(\underline{r} - \underline{R}_l) , \tag{4}$$

where $\phi(\underline{r})$ is an atomic orbital and N is the number of atoms. The exchange integral between Bloch states is

$$\iint \psi_{\underline{k}}^*(\underline{r}_1) \psi_{\underline{k}'}^*(\underline{r}_2) \; H_{int} \; \psi_{\underline{k}}(\underline{r}_2) \; \psi_{\underline{k}'}(\underline{r}_1) \tag{5}$$

where H_{int} corresponds to the last term of (2). This integral has the value U/N so that the intra-atomic Coulomb interaction U is often referred to as an exchange parameter. The intra-atomic exchange integral J between different orbitals on an atom may formally be included and some authors[4] make this explicit by writing U + 4J in place of the parameter U.

In the Hartree-Fock approximation the interaction term of (2) becomes

$$\begin{aligned} U \sum_i n_{i\uparrow} \langle n_{i\downarrow} \rangle &= U n_\uparrow n_\downarrow / N \\ &= \frac{1}{4} \frac{U}{N} [(n_\uparrow + n_\downarrow)^2 - (n_\uparrow - n_\downarrow)^2] \qquad (6) \\ &= - \frac{U}{4N} n^2 \zeta^2 + \text{constant} \end{aligned}$$

where $n_\uparrow$, $n_\downarrow$ are the total number of up and down spin electrons, $n = n_\uparrow + n_\downarrow$ and $\zeta = (n_\uparrow - n_\downarrow)/n$ is the relative magnetization. Thus we obtain the total energy in the form

$$E = \sum_{\underline{k}\sigma} \varepsilon_{\underline{k}} n_{\underline{k}\sigma} - \frac{U}{4N} n^2 \zeta^2 , \tag{7}$$

neglecting a constant. This is the energy expression on which the original theories of band magnetism by Slater[5] and Stoner[6] were based. The criterion for ferromagnetism at $T = 0$ is obtained by comparing the energy $E(\zeta)$, given by (7), for small ζ with $E(0)$. Thus

$$E(\zeta) = E(0) + N\int_{E_F}^{\mu_+} \varepsilon N(\varepsilon)d\varepsilon - N\int_{\mu_-}^{E_F} \varepsilon N(\varepsilon)d\varepsilon - \frac{U}{4N}n^2\zeta^2 \tag{8}$$

where $N(\varepsilon)$ is the density of states per atom of the band energies $\varepsilon_{\underline{k}}$, E_F is the paramagnetic Fermi energy and μ_+, μ_- are the up and down spin Fermi energies in the magnetized state (see fig.1(a)). Clearly

$$E(\zeta) = E(0) + \frac{n\zeta}{2}\cdot\frac{1}{2}\frac{n\zeta}{NN(E_F)} - \frac{U}{4N}n^2\zeta^2 + O(\zeta^4) \tag{9}$$

where in the second term the first factor is the number of spins turned round in going to the magnetized state and the second factor is the average energy increase for each spin reversal. Hence

$$E(\zeta) = E(0) + \frac{1}{4N}\frac{n^2\zeta^2}{N(E_F)}\left[1 - UN(E_F)\right] + O(\zeta^4) \tag{10}$$

and the energy is lowered on magnetizing if

$$UN(E_F) > 1. \tag{11}$$

This is known as the Stoner criterion for ferromagnetism. If this is satisfied the ground state corresponds either to a state of maximum spin alignment (fig. 1(b) or 1(c) depending on whether the band is greater or less than half-filled) or to a state of partial alignment with ζ corresponding to a minimum of $E(\zeta)$ (fig.1(d)). The former case is called a strong ferromagnet (e.g.1(b) occurs in Ni) and the latter case corresponds to a weak ferromagnet such as Fe. If in the ground state $\zeta \ll 1$ we have a very weak itinerant ferromagnet (e.g. $ZrZn_2$, $Ni_3A\ell$). From (6) an up spin electron sees an interaction energy $U < n_{i\downarrow} > = n_\downarrow U/N$. Thus the difference between up and down spin one-electron energies, for the same $\underline{k}$, is

$$\Delta = U(n_\uparrow - n_\downarrow)/N = Un\zeta/N. \tag{12}$$

This is the exchange splitting between the spin sub-bands, as shown in figs. 1(b), (c), (d); it is proportional to the magnetization and thus corresponds to a molecular field, of magnitude $Um/(2\mu_B^2)$ where m is the magnetic moment on an atom.

The concept of a molecular field makes the calculation of the generalized spin susceptibility $\chi(\underline{q},\omega)$ very simple in many cases. Its use corresponds to a time-dependent Hartree-Fock theory which is known to be equivalent to the random phase approximation (RPA).

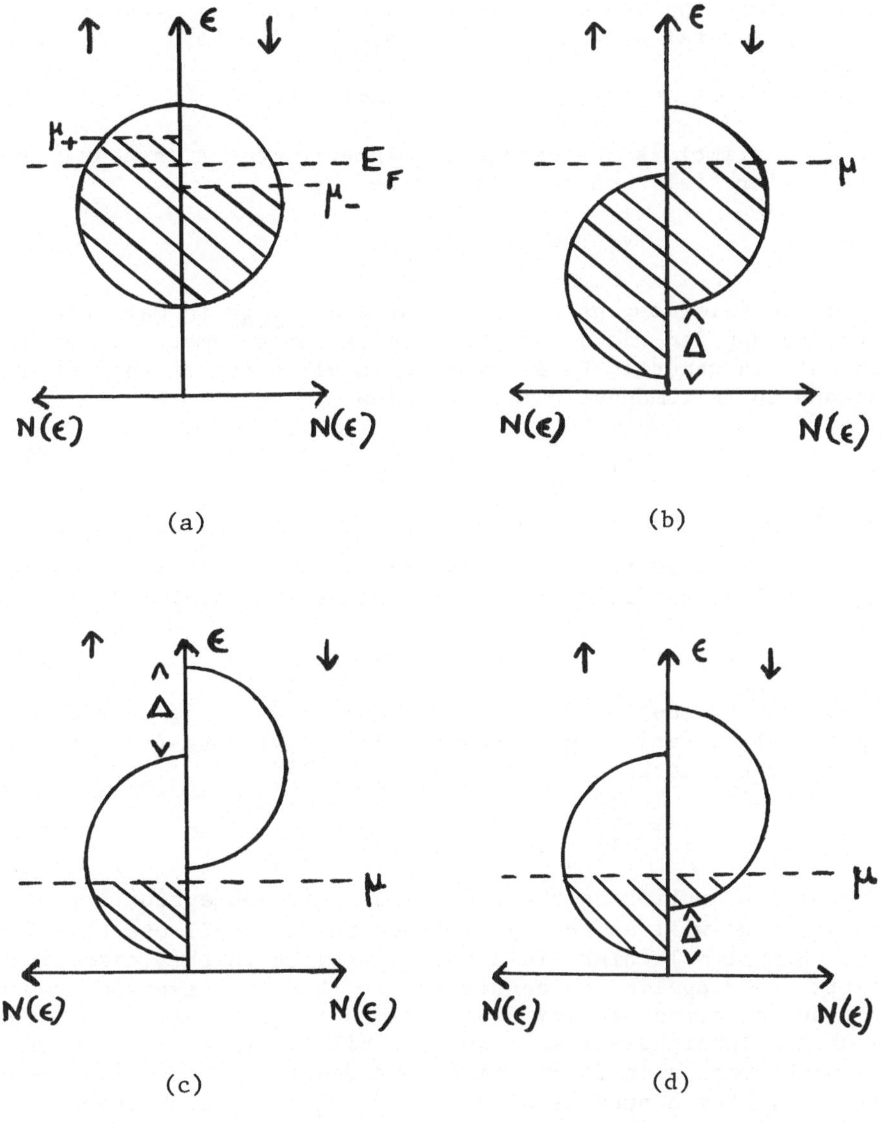

Fig. 1: Schematic densities of states in an itinerant electron ferromagnet. μ the chemical potential in the ferromagnetic state.

We shall consider in this section the case of a paramagnetic metal, where the Stoner criterion is not satisfied but where electron interaction is nevertheless important. Suppose the system is in a time-dependent applied magnetic field of magnitude $h_i \exp(i\omega t)$ at site i. The total perturbing field at the site is

$$H_i \exp(i\omega t) = (h_i + v_i \delta m_i)\exp(i\omega t) \tag{13}$$

where $\delta m_i \exp(i\omega t)$ is the change in magnetic moment on atom i and the molecular field constant is defined by

$$v_i = U_i/(2\mu_B^2). \tag{14}$$

We have specified δm_i as representing the change in magnetic moment due to the applied field and the interaction parameter U_i as site-dependent, in general, to allow later application of the following equations to ferromagnetic alloys. Now

$$\delta m_i = \sum_j \chi^0_{ij}(\omega) H_j \tag{15}$$

where $\chi^0_{ij}(\omega)$ is the Hartree-Fock, or "non-interacting", susceptibility which relates the moment induced at site i to the magnitude of the oscillating field at site j. Hence from (15) and (13),

$$\delta m_i = \sum_\ell \chi^0_{i\ell}(\omega)(h_\ell + v_\ell \delta m_\ell). \tag{16}$$

Considering the applied field to act only at site j, and defining the physical "interacting" susceptibility $\chi_{ij}(\omega)$ by (15) with H_j replaced by h_j, we have

$$\chi_{ij} = \chi^0_{ij} + \sum_\ell \chi^0_{i\ell} v_\ell \chi_{\ell j}. \tag{17}$$

This equation applies to the transverse spin susceptibility of a ferromagnet as well as to a paramagnet and we shall use this fact later. Simple molecular field theory applies in this case since a magnetic field applied perpendicular to the magnetization does not induce any electron density fluctuations to first order. In the case of the longitudinal spin susceptibility in a ferromagnet, coupling between spin fluctuations and density fluctuations must be considered. For a pure metal $v_\ell = U/(2\mu_B^2) = v$, a constant, and χ^0_{ij} and χ_{ij} depend on $\underline{R}_l - \underline{R}_j$ only. Equation (17) is then simply solved by Fourier transformation and we find

$$\chi(\underline{q},\omega) = \sum_j \chi_{ij} e^{-i\underline{q}\cdot(\underline{R}_i-\underline{R}_j)}$$

$$= \frac{\chi^0(q,\omega)}{1 - v\chi^0(\underline{q},\omega)} \tag{18}$$

where

$$\chi^0(\underline{q},\omega) = \sum_j \chi^0_{ij} e^{-i\underline{q}\cdot(\underline{R}_i-\underline{R}_j)} . \tag{19}$$

The generalized susceptibility $\chi(\underline{q},\omega)$ describes the response to an applied field $h \exp[-i(\underline{q}\cdot\underline{r} -\omega t)]$ with wave-vector $\underline{q}$ and angular frequency ω. The corresponding non-interacting susceptibility χ^0 is easily calculated by elementary perturbation theory and takes the form

$$\chi^0(\underline{q},\omega) = 2\mu_B^2 \sum_{\underline{k}} \frac{f_{\underline{k}} - f_{\underline{k}+\underline{q}}}{\varepsilon_{\underline{k}+\underline{q}} - \varepsilon_{\underline{k}} - \hbar\omega + i\eta} \tag{20}$$

where $f_{\underline{k}}$ is the ground-state occupation number of state $\underline{k}$. The term $i\eta$ ($\eta \to 0+$) arises from the usual adiabatic boundary conditions where the perturbing field is turned on by an amplitude factor $\exp(\eta t)$. If

$$v\chi^0(\underline{q},0) > 1 \tag{21}$$

then clearly, from (18), at a smaller value of v the susceptibility $\chi(\underline{q},0) \to \infty$. This implies that the paramagnetic state has become unstable against the formation of a spin density wave of wave-vector $\underline{q}$. If $\chi^0(\underline{q},0)$ has its maximum value at $q = 0$ the system has a tendency to become a normal ferromagnet and the condition (21) is just the Stoner criterion (11). This is easily seen since, introducing the Fermi function $f(\varepsilon)$ at $T = 0$ (a step function),

$$(2\mu_B^2)^{-1} \lim_{q\to 0} \chi^0(\underline{q},0) = \frac{1}{N} \lim_{q\to 0} \sum_{\underline{k}} \frac{f(\varepsilon_{\underline{k}}) - f(\varepsilon_{\underline{k}+\underline{q}})}{\varepsilon_{\underline{k}+\underline{q}} - \varepsilon_{\underline{k}}}$$

$$= \frac{1}{N} \sum_{\underline{k}} - \frac{\delta f}{\delta \varepsilon_{\underline{k}}} = \int_0^\infty \delta(\varepsilon - E_F) N(\varepsilon) d\varepsilon = N(E_F) . \tag{22}$$

If $\chi^0(\underline{q},0)$ has a maximum at the zone boundary where $\underline{q}$ is half a reciprocal lattice vector the instability is towards a simple antiferromagnet. If the maximum occurs at a general point in the zone a spin density wave, incommensurate with the lattice, will form (if (21) is satisfied) as in chromium. From (18) and (22), the uniform static susceptibility χ_s in a paramagnet is given by

$$\chi_s = \lim_{q\to 0} \chi(\underline{q},0) = \frac{2\mu_B^2 N(E_F)}{1 - UN(E_F)} . \tag{23}$$

The numerator is the Pauli susceptibility of a non-interacting system and $[1-UN(E_F)]^{-1}$ is known as the Stoner enhancement factor.

Consider now a nearly ferromagnetic metal such as palladium which has a Stoner enhancement factor of about 10. The following considerations are useful in discussing dilute PdFe and PdNi alloys. For a nearly ferromagnetic metal $\chi^o(\underline{q},0)$ has a maximum at $q = 0$ and the denominator in $\chi(\underline{q},0)$, given by (18), almost vanishes for this value of q. Thus $\chi(\underline{q},0)$ is very strongly peaked at $q = 0$ and the response to any perturbing field is dominated by long wavelength Fourier components. For small q, and a cubic crystal,

$$\chi^o(\underline{q},0) \simeq 2\mu_B^2 N(E_F)\ (1 - bq^2), \tag{24}$$

where b is a constant depending on the band structure.

Hence

$$\chi(\underline{q},0) \simeq \chi_s(1 + q^2/\kappa^2)^{-1} \tag{25}$$

where

$$\kappa^2 = \frac{1 - \bar{U}}{b\bar{U}}, \quad \bar{U} = UN(E_F). \tag{26}$$

Equation (25) is the Fourier transform of the real-space equation

$$- C\nabla^2 M(\underline{r}) + AM(\underline{r}) = H(\underline{r}), \tag{27}$$

where $M(\underline{r})$, and $H(\underline{r})$ are the spatially-varying magnetization and magnetic field and

$$A = \chi_s^{-1}, \ C = A\kappa^{-2} = bU/(2\mu_B^2) . \tag{28}$$

For a localized magnetic field $H(\underline{r}) = H\delta(\underline{r})$ the solution of (27) is easily obtained in the form

$$M(\underline{r}) = \frac{H}{(2\pi)^3}\int \frac{\exp(-i\underline{q}.\underline{r})}{A + Cq^2}\, d^3q$$
$$= \frac{H\chi_s}{(2\pi)^3}\int \frac{\exp(-i\underline{q}.\underline{r})}{1 + q^2/\kappa^2}\, d^3q$$

$$\propto H\ \chi_s r^{-1} \exp(-\kappa r)\ . \tag{29}$$

Furthermore the total induced moment is

$$\int M(\underline{r})d^3r = \chi_s\ H, \tag{30}$$

where χ_s is the susceptibility per unit volume. (Previously our susceptibilities have been defined as per atom, for convenience.) Thus the magnetic moment induced by a localized magnetic field is large, when χ_s is large, and extends to a distance of order κ^{-1}. This length is known as the correlation length and, from (26), $\kappa^{-1} \to \infty$ if the metal is on the verge of ferromagnetism ($\bar{U} \to 1$). A localized magnetic field such as that considered here arises in practice as the molecular field due to an impurity carrying a local moment, such as Fe or Co in Pd. The resultant "giant" moment is about $10\mu_B$ and is known from neutron scattering experiments[7] to extend over more than 200 Pd atoms around the impurity, corresponding to $\kappa^{-1} \sim 5$Å in Pd. The large extent of these moments results in ferromagnetic ordering extending down to very low impurity concentrations of less than 0.1 at.%. The situation in PdNi is more subtle, since isolated Ni impurities do not have a local moment, and ferromagnetism only appears for more than about 2% Ni. This will be discussed later.

Finally in this section we note that equation (27) may be derived by minimizing an energy expression

$$E = \int E(\underline{r})d^3r, \tag{31}$$

$$E(\underline{r}) = E_0 - M(\underline{r})H + \frac{1}{2}AM^2(\underline{r}) + \frac{1}{4}BM^4(\underline{r}) + \frac{1}{2}C|\nabla M(\underline{r})|^2\ ,$$

where E_0 is the energy density for $M = 0$. We have included an additional term in M^4 so that the expression may be applied to very weak ferromagnets as well as paramagnets. In the case of uniform magnetization ($\nabla M = 0$) equation (31) corresponds to equation (10), with the addition of an applied field. Clearly, from the derivation of (10) from (8), the coefficient B depends on details of the density of states $N(\varepsilon)$ near E_F. Expression (31) is familiar in Landau's theory of phase transitions, E becoming the free energy at finite T, and its applicability to itinerant electron magnets, even at T = 0, was first pointed out by Mathon[8]. On minimizing E with respect to $M(\underline{r})$ we obtain

$$-C\nabla^2 M(\underline{r}) + AM(\underline{r}) + BM^3(\underline{r}) = H(\underline{r}) \tag{32}$$

and we shall make use of this result later.

We close this section with a general remark about alloys near the critical composition for the onset of ferromagnetism. Moments will first appear on atoms with particularly favourable local environments and these will induce giant moments or "polarisation clouds" in their vicinity owing to the large correlation length in the alloy. This effect was first observed by Hicks et al[9] in CuNi alloys. Thus in general the magnetization in an alloy near the critical composition is expected to be rather inhomogeneous. NiPt alloys, which are discussed later, may be an exception.

1.2. Spin density functional theory

Recently a new approach, much more sophisticated than that involving tight-binding models and effect interaction parameters, has been made to the ground state of magnetic transition metals. This is based on the spin density functional (SDF) method introduced by Hohenberg, Kohn and Sham and developed for magnetic problems by von Barth and Hedin[10] and others. The result of this work, discussed in Dr. Gunnarsson's lectures, is that in principle the exact ground state energy and exact particle and spin densities of a many-electron system may be determined by solving a one-electron problem self-consistently. The Schrödinger equation takes the form

$$\{ -\frac{\hbar^2}{2m}\nabla^2 + v(\underline{r}) + \int \frac{e^2 n(\underline{r}')}{|r-r'|} d^3r' + v_{xc}^{(\sigma)}[n_\uparrow(\underline{r}), n_\downarrow(\underline{r})]\}\psi_{i\sigma}(\underline{r})$$

$$= \varepsilon_{i\sigma}\,\psi_{i\sigma}(\underline{r}) \tag{33}$$

with

$$n_\sigma(\underline{r}) = \sum_i |\psi_{i\sigma}(\underline{r})|^2\,\theta(\mu-\varepsilon_{i\sigma}) \tag{34}$$

and $n(\underline{r}) = n_\uparrow(\underline{r}) + n_\downarrow(\underline{r})$. Here $v_{xc}^{(\sigma)}$ is a universal functional, in general non-local (and unknown!), of the spin densities and $v(\underline{r})$ is an external potential, e.g. due to the nuclei and an applied magnetic field. The chemical potential μ is determined to give the correct number of electrons. Use of the exact functional $v_{xc}^{(\sigma)}$ would lead to the exact spin densities $n_\sigma(\underline{r})$ and the exact total energy; the latter quantity is given by the usual Hartree expression plus a functional $E_{xc}[n_\uparrow(\underline{r}), n_\downarrow(\underline{r})]$ from which $v_{xc}^{(\sigma)}$ is derived by differentiation. The one-electron energies $\varepsilon_{i\sigma}$ are Lagrangian parameters which are not guaranteed any physical significance. Thus for a pure metal, where the one-electron states are Bloch states, the energies $\varepsilon_{\underline{k}\sigma}$ are not quasi-particle energies and should not even give the exact Fermi surface.

In practice approximate forms of the exchange-correlation functional $v_{xc}^{(\sigma)}$ must be used and these are usually local functionals derived from theories of the uniform electron gas. The simplest functional is that of the Slater Xα method,

$$v_{xc}^{(\sigma)} = - \frac{3e^2}{2\pi} (3\pi^2)^{\frac{1}{3}} \alpha[n_\sigma(\underline{r})]^{\frac{1}{3}} , \tag{35}$$

and when $\alpha = 2/3$ this corresponds to the simplest potential of Kohn and Sham with correlation neglected. Callaway and Wang[11] have carried out self-consistent band calculations for pure bcc iron and fcc nickel using both the Kohn-Sham potential and a more sophisticated one, including correlation effects, due to von Barth and Hedin[10]. The only input in these calculations is the atomic number and the crystal structure. To a good approximation there is a rigid exchange splitting Δ between the up and down spin bands as in Stoner theory. However Gunnarsson[12] has pointed out that the exchange splitting is slightly larger at the top of the d band than at the bottom. Callaway and Wang find that the von Barth-Hedin potential gives a slightly better result for the magnetization than the Kohn-Sham potential, the calculated values being within 5% of those observed in Fe and Ni. For Ni the Kohn-Sham potential gives Δ = 0.88eV at the top of the d band, with a moment of $0.65\mu_B$/atom, and the von Barth-Hedin potential gives Δ = 0.63eV with a moment of $0.58\mu_B$/atom. There is quite a lot of experimental evidence, the most recent being angular-resolved photoemission measurements by Eastman, which suggests that the quasi-particle exchange splitting in Ni is little bigger than 0.3eV. It seems unlikely that using an improved functional in the SDF calculations could reduce Δ in Ni by a factor 2 and the discrepancy may be due to the necessity of including an energy-dependent self-energy, arising from dynamical processes such as virtual electron-magnon scattering, in the calculation of quasi-particle energies. This view is supported by the fact that the large SDF values of Δ seem to be essential for calculating realistic values of spin-wave energies in nickel and its alloys. This whole matter was discussed at the Toronto Conference[13] and we return to it in ¶3. Callaway and Wang calculated Fourier components of the spin density $\mu_B[n_\uparrow(\underline{r}) - n_\downarrow(\underline{r})]$, to compare with neutron magnetic form factors, and also calculated the hyperfine field at the nucleus, measured by Mössbauer and NMR measurements, from $n_\uparrow(0) - n_\downarrow(0)$. The agreement with experiment in Fe and Ni is reasonable, although the Kohn-Sham potential gives considerably better results for the hyperfine field than the von Barth-Hedin potential. Calculated Fermi surfaces are also substantially in agreement with de Haas-van Alphen measurements, although perfect agreement is not expected here even with the exact functional v_{xc}.

Calculations in an applied magnetic field, and/or with different lattice constants, make possible the theoretical evaluation of equilibrium lattice constants, bulk moduli, spin susceptibilities and the pressure derivatives of these quantities, as well as of ferromagnetic moments. Impressive general agreement with experiment is reported, for a number of magnetic and paramagnetic metals, by Andersen et al[14] and Janak and Williams[15], the latter also reporting cohesive energies. A remarkable result of this work is that calculations of lattice constant and bulk modulus for magnetic metals only agree reasonably with experiment when calculations are made in the magnetic phase. The calculated lattice constant of ferromagnetic bcc Fe is more than 2% larger than that calculated for the hypothetical paramagnetic phase; the bulk modulus ($B = -V\,dP/dV$) is 50% smaller. Much earlier Shiga and Nakamura[16] estimated the magnetovolume effect using Stoner theory and estimated an upper bound of a 3% increase in lattice constant on magnetizing. They showed that the effect was due to shifting electrons from down spin bonding states to up spin antibonding states. They also point out that the effect is much larger than that observed when Fe and Co disorder magnetically at the Curie temperature T_c. Shiga and Nakamura concluded that the magnetovolume effect may be interpreted locally as associated with magnetized atoms and that in Fe and Co disordered local moments persist above T_c, rather as in the Heisenberg model. This contrasts strongly with finite-temperature Stoner theory in which magnetic disordering occurs simply due to repopulation of one-electron states according to Fermi-Dirac statistics. In Stoner theory there are no local moments above T_c and predicted values of T_c, for Fe and Co are far too high. However, for some materials, such as antiferromagnetic Cr and very weak itinerant ferromagnets, it is believed that magnetic moments disappear above T_c and here one expects, and finds, important magnetic volume effects. Thus Cr has a negative thermal expansion coefficient at low temperatures due to the contraction associated with decreasing magnetic moments as the temperature rises. This negative magnetic contribution tends to be balanced by the normal positive thermal expansion due to the phonons and conduction electrons. Typical effects in very weak itinerant ferromagnets are a small thermal expansion coefficient below T_c, a large forced volume magnetostriction (increase of volume in applied field) a sharp decrease of the spontaneous magnetization and of T_c on applying pressure and large charges in elastic constants on going through T_c. These effects have been discussed by Wohlfarth[17] using an extension of Stoner theory and are associated with a quadratic dependence of the volume magnetostriction ($\Delta V/V$) on magnetization. They are often called invar effects since they appear in the classical fcc invar alloys FeNi (65% Fe), FePd and FePt. Here the variation of magnetic moment with composition, field, pressure and temperature may be associated with the existence of two states of fcc Fe, one with large moment and one with small or zero moment, which are close in energy. This possibility was suggested empirically by Weiss and has been substantiated by SDF calculations for pure fcc Fe and ordered Fe_3Pt. Mössbauer studies[18] indicate the co-existence of large and small hyperfine fields on Fe atoms in FeNi invar alloys. On the other hand Wohlfarth treats the invar alloys as normal very weak itinerant ferromagnets. No SDF calculations have yet been carried out for disordered alloys.

Finally we note that Shiga[19] has used alloy data to deduce the lattice parameter of hypothetical paramagnetic bcc Fe and finds it to be 3% smaller than for the ferromagnetic metal, in good agreement with theory. The magnetic contribution to the thermal expansion of the fcc invar alloy $Fe_{65}Ni_{35}$ corresponds to a lattice contraction[20] of 0.6% between $T = 0$ and $T = T_c$.

2. THE GROUND STATE OF MAGNETIC TRANSITION METAL ALLOYS

2.1 Experimental data

Some experimental measurements which have an important bearing on the electronic structure of magnetic alloys are as follows:-

1. Bulk magnetization and its dependence on applied field.

2. Magnetic (elastic diffuse) scattering of neutrons.

The scattering cross-section is proportional to

$$|F(\underline{q})|^2 \frac{1}{N} \sum_{i,j} e^{i\underline{q}.(\underline{R}_i - \underline{R}_j)} \langle (m_i - \overline{m})(m_j - \overline{m}) \rangle \qquad (36)$$

where m_i is the moment on the atom at $\underline{R}_i$, $\overline{m}$ is the mean atomic moment, $F(\underline{q})$ is the atomic form factor and $\underline{q}$ is the scattering vector. The brackets $\langle\ \rangle$ denote a configurational average. If, in an AB alloy, we suppose that each atom A has the same moment m_A and that each atom B has moment m_B (Shull-Wilkinson model) the correlation function in (36) vanishes for $i \neq j$ and (36) becomes

$$|F(\underline{q})|^2 c(1 - c)\ (m_B - m_A)^2, \qquad (37)$$

where c is the concentration of B atoms. Thus the neutron cross-section determines $|m_B - m_A|$ and hence $m_B - m_A$ if we assume $m_B > m_A$ or vice-versa. Bulk magnetization measurements give $\overline{m} = cm_B + (1-c)m_A$ so we can deduce m_A and m_B.

A more precise theory due to Marshall[21] takes account of local environment effects. For the simplest case of a dilute alloy ($c \ll 1$)

equation (37) is modified so that the cross-section is proportional to

$$|F(\underline{q})|^2 c(1-c)\{m_B - m_A + G(\underline{q})\}^2 \tag{38}$$

where

$$G(\underline{q}) = \sum_i e^{i\underline{q}.\underline{R}_i} g(\underline{R}_i) . \tag{39}$$

An equivalent result was first derived by Low and Collins[7]. Here m_A is the moment of an A atom far from a B atom, m_B is the moment of an isolated B atom and $m_A + g(\underline{r})$ is the moment of an A atom at position $\underline{r}$ relative to a B impurity. Thus $G(\underline{q})$ may be determined and hence the form of the magnetic disturbance $g(\underline{r})$ around an impurity. Clearly $\bar{m} = c[m_B + G(0)] + (1-c)m_A$ so that, since m_B and m_A are independent of c in this case,

$$\frac{d\bar{m}}{dc} = m_B - m_A + G(0). \tag{40}$$

This is the change in total moment per impurity which can be deduced from bulk magnetization measurements. It provides the $q = 0$ limit of $m_B - m_A + G(\underline{q})$ in (38) and determines its sign. $F(\underline{q})$ is almost constant over the range of q of interest and $m_B - m_A$ is deduced from the cross-section at the larger values of q where $G(\underline{q}) \rightarrow 0$. This technique has been used to investigate magnetic disturbances around impurities in Fe and Ni, giant moments around Fe, Co and Mn in Pd, and polarization clouds in a number of alloys near the critical composition.

3. Mössbauer and NMR measurements measure the hyperfine field at different nuclei. The main contribution is the Fermi contact term which is proportional to the magnetization density $\mu_B[n_\uparrow(0) - n_\downarrow(0)]$ at the nucleus.

4. Measurements of the electronic specific heat γT. $\gamma \propto N_\uparrow(E_F) + N_\downarrow(E_F)$, where $N_\sigma(E_F)$ is the alloy density of states at the Fermi level for electrons of spin σ. γ also contains an enhancement factor, due to electron-phonon interaction, which is usually unknown and may be as large as 1.5 or more.

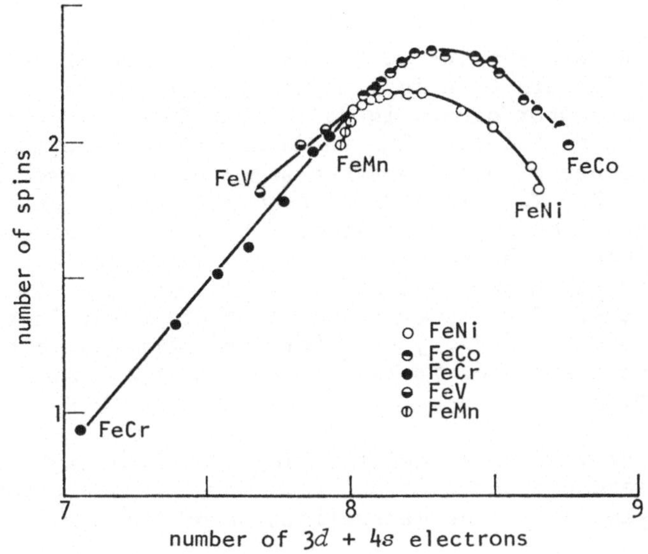

Number of uncompensated spins in b.c.c. iron alloys (Crangle and Hallam 21a).

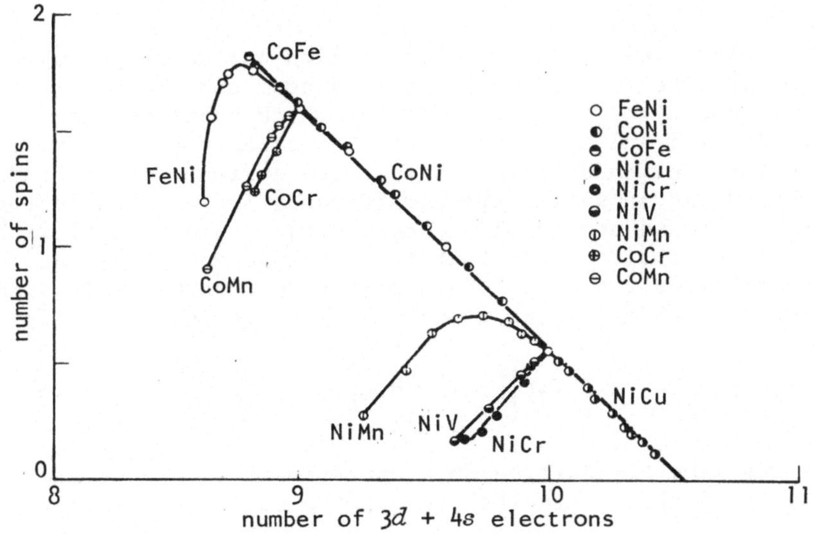

The number of magnetic electron spins per atom plotted against number of 3d + 4s electrons per atom for f.c.c. binary alloys of iron and nickel (Slater-Pauling curve; Crangle 21b).

Fig.2: Two parts of the Slater-Pauling curve.

2.2 Qualitative discussion of transition metal impurities in nickel and iron

Alloys of Ni are simpler to discuss than those of Fe owing to the simplifying feature of a full majority spin d band in the pure metal. The same factor occurs in Co so that Co alloys often resemble the corresponding Ni alloys. The variation of average moment $\bar{m}$ as a function of number of electrons per atom outside closed shells is shown in fig. 2. This is derived from bulk magnetization measurements. We note that $\underline{\text{Ni}}\text{Co}$, $\underline{\text{Ni}}\text{Fe}$ alloys lie on the straight line part of the so-called Slater-Pauling curve, whereas $\underline{\text{Ni}}\text{Cr}$ and $\underline{\text{Ni}}\text{V}$ behave quite differently and $\underline{\text{Ni}}\text{Mn}$ is somewhat anomalous. The rather sudden loss of moment in the fcc NiFe alloys at about 65% Fe, just before the martensitic transformation to the bcc structure, is associated with the invar effect mentioned in ¶1.2.

For simplicity we shall use the single-orbital tight-binding model in the present discussion but remember that the d band is in fact 5-fold degenerate. The generalization of the Hamiltonian (1) to AB alloys is

$$H = \sum_{ij\sigma} t_{ij} a^{\dagger}_{i\sigma} a_{j\sigma} + \sum_{i\sigma} V_i n_{i\sigma} + \sum_i U_i n_{i\uparrow} n_{i\downarrow} \tag{41}$$

where the atomic level V_i and the interaction parameter U_i take values V_A, U_A if site i is occupied by an A atom and V_B, U_B if by a B atom. In general t_{ij} should depend on which types of atom occupy sites i and j and this dependence is probably an essential one if A and B belong to different transition series since d band-widths in the 4d and 5d series are considerably larger than in the 3d. However since we are mainly concerned with alloys of Ni with other 3d metals we shall assume that t_{ij} is independent of the type of atom at sites i and j. The Hamiltonian then has only diagonal disorder and we may write it in the form (cf. equation (2))

$$H = \sum_{\underline{k}\sigma} \varepsilon_{\underline{k}} n_{\underline{k}\sigma} + \sum_i V_i n_{i\sigma} + \sum_i U_i n_{i\uparrow} n_{i\downarrow} \quad . \tag{42}$$

As in ¶1.1 we make the Hartree-Fock approximation and the one-electron states of spin σ are the eigenstates of the Hamiltonian

$$H_{\sigma} = \sum_{\underline{k}} \varepsilon_{\underline{k}} n_{\underline{k}\sigma} + \sum_i (V_i + U_i \langle n_{i-\sigma} \rangle) n_{i\sigma} \quad . \tag{43}$$

This may be written explicitly in one-electron form as

$$H_\sigma = \sum_{\underline{k}} \varepsilon_{\underline{k}} \, |k><\underline{k}| + \sum_i V_{i\sigma}|i><i|, \tag{44}$$

$$V_{i\sigma} = V_i + U_i < n_{i-\sigma} >. \tag{45}$$

We shall not discuss the choice of parameters V_A, V_B, U_A, U_B here since we return to this Hamiltonian in a CPA treatment for concentrated alloys in ¶2.2. For the present we discuss in a qualitative way the case of a single B impurity (B ≡ Co, Fe, Mn, Cr or V) in a pure A matrix (A ≡ Ni). We assume for the moment that $V_{i\sigma}$ is the same for all Ni atoms, being equal to $V_{A\sigma}$. We then have the familar Slater-Koster problem of an impurity with localized perturbing potentials

$$\delta_\sigma = V_{B\sigma} - V_{A\sigma} . \tag{46}$$

These potentials tend to be repulsive ($\delta_\sigma > 0$) since for alloys of Ni with elements of smaller atomic number we have $V_B > V_A$. A bound state will be split from the top of the band if

$$\delta_\sigma \, F(W) > 1 \tag{47}$$

where

$$F(\varepsilon) = \int_{-W}^{W} \frac{N(\varepsilon')d\varepsilon'}{\varepsilon - \varepsilon'} . \tag{48}$$

The band $\varepsilon_{\underline{k}}$, with density of states $N(\varepsilon)$, has been taken to lie in the energy range $(-W,W)$. For $N(\varepsilon)$ corresponding to the Ni d band (47) becomes approximately $\delta_\sigma > W/2$. Bound states of this type in transition metal alloys are called Friedel virtual bound states. The term virtual refers to the fact that they actually have a finite lifetime, with a consequent energy-width, owing to hybridization with the sp band. We list below the difference in atomic number z, and the approximate number of d electrons n_D on the atom, for a number of impurities in Ni.

Element	Ni	Co	Fe	Mn	Cr	V
z	0	-1	-2	-3	-4	-5
n_d	9.4	8.4	7.4	6.4	5.4	4.4

We distinguish two types of behaviour, depending on whether the magnetization follows or deviates from the Slater-Pauling curve (fig. 2).

Case (i) (NiCo, NiFe, NiMn):

The simple theory of the magnetization in these alloys is that the up spin d band, full in pure Ni, remains full so that the number of down spin electrons changes by z for each impurity. The moment thus increases by -z so that $d\bar{m}/dc = - z\mu_B$. Also the mean atomic number $\bar{z}$ is given by

$$\bar{z} = cz_B + (1 - c)z_A = z_A + zc \tag{49}$$

so that $d\bar{z}/dc = z$ and

$$d\bar{m}/d\bar{z} = - \mu_B \tag{50}$$

in agreement with fig.2. NiMn is anomalous at higher concentrations due to some Mn moments aligning antiparallel to the total moment. Equation (50) is sometimes thought to be a result of rigid band theory but in fact it makes no assumption about the detailed electronic structure. It assumes only that the number of sp electrons is constant and that no up spin d states split off the d band so as to appear above E_F. If we further assume local charge neutrality on each atom, the number of d holes on the impurity is 0.6 - z and on every Ni atom, even a nearest neighbour to an impurity, it is 0.6. Thus the moment on a Ni atom should be always 0.6 μ_B and those on Co, Fe, Mn impurities should be 1.6, 2.6 and 3.6 μ_B, respectively, in dilute alloys. These numbers, and the conclusion that the magnetic disturbance is confined to the impurity site, agree well with neutron-scattering data. Since screening of the change difference ze is carried out entirely by $\downarrow$ spin electrons we may determine $\delta_\downarrow$, which is equal to $\delta_0 (= V_B - V_A)$ if we make the simplifying assumption $U_B = U_A = U$, by the Friedel sum rule

$$z = \frac{5}{\pi} \tan^{-1} \left\{\frac{-\delta_0 \pi N(E_F)}{1-\delta_0 F(E_F)}\right\} . \tag{51}$$

Since the $\downarrow$ spin electrons screen the charge ze we expect that $\delta_\uparrow \sim 0$ and a solution of this type is only possible if

$$\delta_\uparrow = \delta_0 + \frac{1}{5} Uz < \delta_c, \tag{52}$$

where δ_c is the value of $\delta_\uparrow$ which just pushes an up spin bound state through E_F. In the case of Fe and Mn it is likely that δ_0 is large enough to split a bound state from the top of the down spin band. The local densities of states on the impurity, and on a nearest-neighbour (nn) Ni atom, are shown schematically in figs.3(a) and 3(b) respectively; the schematic density of states for pure Ni is that of fig.1(b). The bound state is shown as a virtual state with a finite width and is shown to extend on to nn Ni atoms. This spatial extension must occur since the unoccupied state holds 5 holes, considering the five-fold degeneracy, and for Fe only 2.6 of these can be on the Fe atom. About half the weight of the bound state must therefore be distributed on the 12 nn Ni atoms. In principle, in a cubic crystal, a bound state of only e_g or t_{2g} symmetry could exist and hold 2 or 3 holes, respectively. However, there is no evidence for this occurring. We conclude that although the magnetic disturbance is strongly localized on the impurity, due to local charge neutrality and a full up spin band, the electronic disturbance may not be so localized. The simple form of Friedel sum rule, (51), is then of doubtful validity although it appears to work quite well in practice.

Case (ii) (<u>Ni</u>Cr, <u>Ni</u>V):

For Cr, with 5.4 d electrons, the above situation with a full up spin d shell is hardly possible. A very large δ_Q would be needed to repel nearly all the down spin electrons and it would then be difficult to satisfy (52). In this case an up spin bound state passes through E_F. Friedel[22] showed that this occurrence provides a beautiful explanation of the deviation from the Slater-Pauling curve. On introducing the impurity the change in number of up spin electrons is $\delta n_\uparrow = -5$ and the change in total number of electrons is $\delta n_\uparrow + \delta n_\downarrow = z$. Hence the change of moment is

$$d\bar{m}/dc = (\delta n_\uparrow - \delta n_\downarrow)\mu_B = (-z-10)\mu_B. \tag{53}$$

The theoretical and experimental values of $d\bar{m}/dc$ for a number of impurities in Ni are given below.

Impurity	Ti	V	Cr	Mo	W
$z + 10$	4	5	6	6	6
$-\mu_B^{-1}\, d\bar{m}/dc$(expt).	4.0	5.5	4.5	5.9	6.2

If the broadened virtual bound state is not completely above E_F the observed loss of moment will be smaller than the theoretical value. Friedel suggested this is the case for Cr and pointed out that the very large residual resistance of <u>Ni</u>Cr alloys supports this hypothesis. A large increase in specific heat coefficient γ on adding Cr to Ni provides further support.

Diffuse neutron scattering measurements[7,23] show that the moments on Cr or V impurities in Ni are very small ($-0.2\mu_B$ for Cr). Most of the loss of moment occurs on surrounding Ni atoms up to a distance of 5Å. The exchange field on the impurity is thus very small and the bound state should have a small positive exchange splitting, owing to the splitting of the Ni band. Thus a bound state lies above E_F for both up and down spins. The local densities of states on the impurity, and on a nn Ni atom, are shown schematically in figs. 3(c) and 3(d) respectively. The approximate number of states in each part of the band is indicated, remembering that Cr and V atoms have about 5 d electrons. More precisely the number of unoccupied states of each spin on the impurity must be $(9.4 + z)/2$, assuming charge neutrality and zero moment. Clearly each nn Ni atom loses about 0.2 up spin electrons, compared with bulk Ni, and this deficit of electronic charge must be screened by down spin electrons. Neutron scattering shows that the magnetic disturbance extends to about two nn distances or further, and has the same spatial form, when scaled, for many impurities of this type (Cr, Nb, Mo, W, Re). Comly et al[24] concluded that the bound state extends to nn Ni atoms, as we have described, and that the resulting loss of up spin electrons cannot be screened out completely by d electrons on these atoms, leading to a disturbance of wider range. Possibly screening on n.n Ni atoms is completed by s p electrons and this results in a loss of up spin d electrons on neighbouring sites as we shall discuss for the case of non-transition metal impurities in Ni. Comparing with the perfect screening by d electrons of negatively charged Fe and Co impurities we may conclude that in a nearly full band it is easier to push d states above E_F with a repulsive potential than to pull them below with an attractive one. It is significant that the range and shape of the magnetization disturbances around non-transition metal impurities (Zn, Al, Ga, Sb, Si, Ge, Sn) are very similar to those around Cr, V, etc. We shall see later that in this case there is also a loss of up spin electrons at nn Ni sites, due to a different mechanism this time, and screening occurs as before.

Although the physical situation is clear it seems to be difficult to describe it correctly within the Hartree-Fock theory. Most authors find a large negative moment on the impurity, e.g. Demangeat and Gautier[25] find $-5\mu_B$ on Cr. This is due to a very localized up spin bound state which gives almost perfect screening of the impurity by up spin electrons. The situation may be described as an antiparallel local moment on the Cr in contrast to the true situation of no local moment. Several authors have pointed out that there are often three Hartree-Fock solutions, one with large parallel moment, one with large anti-parallel moment and one with low moment on the impurity. Jo and Miwa[26] discuss this situation for Mn in a NiCo matrix and show from energy considerations that the low moment state is unstable. This would not be surprising for Mn but, as discussed by Kanamori[27], the low moment state corresponds to the up spin bound state being just at E_F and the instability of such a solution may be general. However, as discussed later, Hennion and Hennion use such a solution in their calculations of spin wave energies in these alloys. Clearly what is required is either a very careful choice of para-

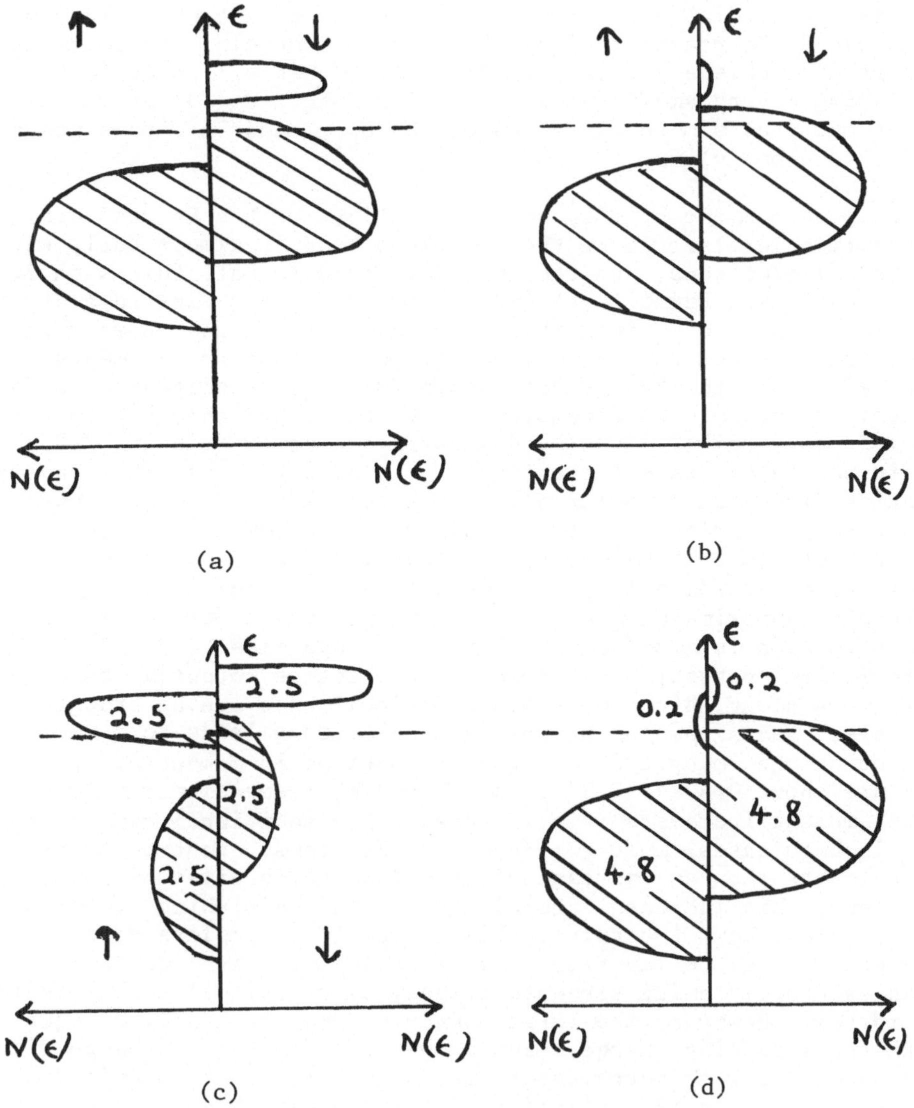

Fig. 3: Local density of states on (a) an Fe impurity in Ni. (b) A Ni atom next to the Fe impurity. (c) A Cr impurity in Ni. (d) A Ni atom next to the Cr impurity. In (c) and (d) the numbers indicate the approximate number of states in each part of the band.

meters, or an improved model, so as to obtain a correct physical solution. In principle, a SDF calculation could solve the whole problem, including the long-range screening effect which is anyway beyond the reach of simple Hartree-Fock models. Kanamori and Hasegawa[28] achieve a fairly small moment ($-0.6\ \mu_B$) on Cr in Ni only by taking a very small value for U_{Cr} such that $U_{Cr}/U_{Ni} \simeq 0.2$. They point out that a more normal value for U_{Cr} leads to a moment on Cr of about $-1.4\mu_B$.

Finally we comment briefly on dilute alloys of bcc Fe with other transition metals. In Fe the up spin d band is almost full, with about 4.7 electrons, and the down spin band is half full with 2.5 electrons. The Fermi level in the down spin band lies in a dip in the middle of the Fe density of states curve, so that down spin electrons may not screen very effectively. Thus for elements to the left of Fe in the periodic table (Mn, Cr, V) screening is done largely by up spin electrons, with states pushed above E_F in the up spin band rather like the down spin states in $\underline{Ni}$Co and $\underline{Ni}$Fe. As in the latter case screening is very effective and the magnetization disturbance is largely on the impurity. However, neutron scattering experiments show a small positive moment on Fe atoms up to a distance of 6Å or more. Clearly $d\bar{m}/dc = z\mu_B$ so that for these alloys $d\bar{m}/d\bar{z} = +\ \mu_B$ in contrast to equation (50). This corresponds approximately to the measured moments shown in fig. 2. The situation is more complicated for alloys like $\underline{Fe}$ Co and $\underline{Fe}$ Ni. Here it is impossible to screen the attractive potential by up spin electrons alone, since only 0.3 additional ones can be brought to the site, and screening by down spin electrons leads to a reduced moment on the impurity. Observed reductions in moment on the impurity are about 0.4 and 1.3 μ_B for Co and Ni respectively. Since the total moment increases (fig.2) consderable additional moment (about $2\mu_B$ in both cases) must reside on the Fe atoms. Neutron scattering experiments show an increase of moment on nn Fe sites of several per cent, this increase extending out with diminishing strength to 6Å or more. Low[7] interprets this as due to incomplete screening by d electrons on the impurity, the resulting positive charge disturbance eventually being screened largely by up spin electrons in the Fe matrix. However, the large moment of $2\mu_B$ can hardly be equated with the screening charge. Low showed earlier that a charge disturbance in a weak ferromagnet (one with holes in both spin subbands) leads to a magnetization disturbance which extends as far as the magnetic correlation length and whose magnitude is related to the high field susceptibility. (In a strong ferromagnet the correlation length is just the electric screening length). The implication seems to be that there is a large local high field susceptibility in the neighbourhood of a Ni impurity in Fe.

2.3 CPA Calculations for concentrated alloys

In the single-site CPA it is assumed that all A atoms and all B atoms are equivalent. Thus, in equations (44) and (45), $< n_{i\sigma} >$ and $V_{i\sigma}$ take just two values each, one value for A sites and one for B sites. Hence

$$V_{\lambda\sigma} = V_\lambda + U_\lambda < n_{\lambda-\sigma} >, \quad (\lambda = A, B). \tag{54}$$

The standard form of the CPA can then be applied to obtain one-electron Green functions and densities of states for each spin. These quantities are functions of $< n_{A\sigma} >$, $< n_{B\sigma} >$ and partial densities of states for A and B sites must be calculated in order to determine these occupation numbers self-consistently. This procedure was first carried out, for Ni and Fe alloys, by Hasegawa and Kanamori[28,29] who calculated the magnetization, moments on A and B atoms, and specific heat coefficient γ. Other authors have done similar calculations for paramagnetic metals in an applied field, thus obtaining susceptibilities.

The basic idea of the CPA is that the disordered electron potential is replaced by a uniform, but energy-dependent, effective potential for an "effective medium". This effective potential, in general complex, is called the coherent potential and for a ferromagnetic material there are two such potentials $\Sigma_\sigma(E)$, one for each spin. The procedure for determining the Σ_σ is to insist on a zero average t-matrix for scattering by a single atom, of A or B type, set in the effective medium. Thus the Σ_σ are determined by the equation

$$c_A \frac{V_{A\sigma} - \Sigma_\sigma}{1-(V_{A\sigma} - \Sigma_\sigma)G_\sigma} + c_B \frac{V_{B\sigma} - \Sigma_\sigma}{1-(V_{B\sigma}-\Sigma_\sigma)G_\sigma} = 0 \tag{55}$$

$$G_\sigma = \frac{1}{N} \sum_{\underline{k}} \frac{1}{E-\varepsilon_{\underline{k}}-\Sigma_\sigma(\varepsilon)} = \int_{-W}^{W} \frac{N(\varepsilon)d\varepsilon}{E-\Sigma_\sigma(E)-\varepsilon} \tag{56}$$

G_σ is the site-diagonal element of the CPA Green function. On multiplying equation (55) by G_σ we may rewrite it in the form

$$c_A G_{A\sigma} + c_B G_{B\sigma} = G_\sigma , \tag{57}$$

where $G_{\lambda\sigma}$ $(\lambda = A, B)$ is defined by

$$G_{\lambda\sigma} = \frac{G_\sigma}{1-(V_{\lambda\sigma}-\Sigma_\sigma)G_\sigma} \quad . \tag{58}$$

$G_{A\sigma}$ and $G_{B\sigma}$ are the site-diagonal elements of the Green function for an A or B atom embedded in the effective medium, and equation (57) says that the coherent potential is determined so that the average of these is equal to the site-diagonal element of the Green function for the effective medium.

The density of states per atom is given by

$$\rho_\sigma(E) = -\frac{1}{\pi}\,\mathrm{Im}\,G_\sigma(E) = c_A\rho_{A\sigma}(E) + c_B\rho_{B\sigma}(E) \tag{59}$$

where

$$\rho_{\lambda\sigma}(E) = -\frac{1}{\pi}\,\mathrm{Im}\,G_{\lambda\sigma}(E). \tag{60}$$

The quantities $\rho_{\lambda\sigma}(E)$ are the local densities of states at a λ atom, so that

$$< n_{\lambda\sigma} > = -\frac{1}{\pi}\,\mathrm{Im}\int_{-\infty}^{\mu} \frac{G_\sigma(E)\,dE}{1-[V_{\lambda\sigma}(E)-\Sigma_\sigma(E)]G_\sigma(E)} \tag{61}$$

Equations (54), (55), (56) and (61) form a set of self-consistent CPA equations for a ferromagnetic alloy in the Hartree-Fock approximation.

We refer the reader to Haswgawa and Kanamori's papers for their detailed results. However, a few points of interest are as follows:

(i) Hasegawa and Kanamori use a small value of the exchange splitting in pure Ni, corresponding to $\Delta \sim 0.35$ eV, in their calculations for Ni alloys. The best SDF calculation gives $\Delta = 0.63$ eV and Hennion and Hennion[30] use $\Delta \sim 0.8$eV in their CPA calculations for Ni alloys. On the other hand $\Delta \sim 0.35$ eV is a reasonable value for the splitting in the quasi-particle spectrum, as observed experimentally (see ¶1.2).

(ii) For fcc NiCo and NiFe the up spin band is hardly deformed since screening is done by down spin electrons as discussed in ¶2.2. Hasegawa and Kanamori find that at 50% Fe the top of the up spin band in Ni Fe crosses the Fermi level so that the alloy becomes a weak ferromagnet. This explains the deviation from the Slater-Pauling curve and a rapid increase in the specific heat coefficient with increasing Fe content. The onset of weak ferromagetism only occurs in this way when Δ in pure Ni is small. In the calculations of Hennion and Hennion the top of the up spin d band sinks further and further below E_F with increasing Fe content. The deviation from the Slater-Pauling curve must then be explained in another way, e.g. a sudden decrease of moment at Fe sites with an Fe-rich local environment; the large Fe moment is known to be unstable in fcc Fe from SDF calculations. Very inhomogeneous weak ferromagnetism would then occur.

Clearly this effect cannot be treated in single-site CPA and the invar problem is very complicated. A third possibility has been worked out by Jo[31] using a "ternary alloy" approach in which large parallel and antiparallel Fe moments occur. Jo also applies this model, which we shall discuss later, to NiMn where there is some experimental evidence for large antiparallel Mn moments. However the large magnetovolume effects in the NiFe invar alloys cannot be explained if all Fe atoms retain a large moment.

(iii) We have already pointed out (¶2.2) difficulties with NiCr. Hasegawa and Kanamori's calculation does not explain the peak in γ near the critical concentration of 12% Cr. Their up spin Cr sub-band, corresponding to the Friedel bound state in the dilute limit, lies well above E_F. This is not the case in Hennion and Hennion's calculation which (D.Moody, private communication) gives reasonable agreement with the γ data. However, as discussed in ¶2.2, their solution may be unstable and does not describe the variation of magnetization with composition very well.

(iv) The description of bcc Fe alloys is rather good. For FeCr the down spin band is hardly deformed since screening is done by up spin electrons, as discussed in ¶2.2. The differences between the behaviour of FeCo and FeNi, both as regards bulk magnetization and specific heat γ, are well accounted for. This is quite surprising when the rather subtle behaviour of the dilute alloys is considered. In fact in FeCo the CPA recovers the simple rigid band result, in which the up spin band fills with increasing Co content to give a maximum average moment of about $2.5\mu_B$ at 30% Co, while at the same time giving the correct moments on individual sites. In later work on FeCo and FeNi alloys Kanamori et al[32] used a more realistic density-of-state curve for bcc Fe and calculated the high field susceptibility dM/dH, obtaining good agreement with experiment.

We now discuss some recent developments of the CPA method for ferromagnetic alloys. Sacchetti[33] and co-workers have applied the CPA to e_g and t_{2g} sub-bands, with appropriate projected densities of states. The calculations for the two sub-bands are coupled self-consistently by Coulomb and exchange terms appearing in the Hartree-Fock atomic energy levels. Thus equation (54) is replaced by

$$V_{\lambda\sigma\mu} = V_{\lambda\mu} + U_\lambda\{ <n_{\lambda-\sigma\mu}> + \sum_{\sigma',\mu'\neq\mu} <n_{\lambda\sigma'\mu'}> \} - J_\lambda \sum_{\mu'\neq\mu} <n_{\lambda\sigma\mu'}> \tag{62}$$

where μ is a label for the five sub-bands. For a given type of atom λ, and given spin σ, $V_{\lambda\sigma\mu}$ and $<n_{\lambda\sigma\mu}>$ take two distinct values depending on whether μ is of t_{2g} (3 equivalent sub-bands) or e_g (2 equivalent sub-bands) type. There are altogether 8 coupled equations to be solved self-consistently. The main object of these calculations is to determine the variation with composition of the asphericity of the magnetic form factor, investigated experimentally by neutron Bragg scattering.

For NiMn alloys Jo, and Hennion and Hennion, have used a "ternary alloy" approach. The deviation from the Slater-Pauling curve is assumed to be due to some Mn local moments being antiparallel to the bulk magnetization. Mössbauer and lattice parameter measurements support this hypothesis. Considering the solution of the Hartree-Fock problem for a Mn atom embedded in an effective medium described by a trial coherent potential, Jo showed there are two locally stable solutions having moments parallel and antiparallel to the bulk magnetization. For the magnetic state of Ni there is only one Hartree-Fock solution. Corresponding to the two Mn solutions, which we denote by BI and BII, there are different occupation numbers $<n_{B\alpha\sigma}>$, $\alpha = I, II$, and corresponding atomic levels $V_{B\alpha\sigma}$ given by (54). The self-consistent coherent potential depends on the concentrations y and z of BI and BII atoms, where y + z = x in a $Ni_{1-x}Mn_x$ alloy. Σ_σ is determined by the equation (A ≡ Ni),

$$(1-x)\frac{V_{A\sigma}-\Sigma_\sigma}{1-(V_{A\sigma}-\Sigma_\sigma)G_\sigma} + y\frac{V_{BI\sigma}-\Sigma_\sigma}{1-(V_{BI}-\Sigma_\sigma)G_\sigma} + z\frac{V_{BII\sigma}-\Sigma_\sigma}{1-(V_{BII\sigma}-\Sigma_\sigma)G_\sigma} = 0 \tag{63}$$

where G_σ is given by (56). Jo determines the concentrations y and z of parallel and antiparallel Mn moments by equating the Hartree-Fock-CPA energies which correspond to embedding a BI or BII atom in the effective medium. Considering that $-1/\pi \ \mathrm{Im}\, d/dE \ \ln(1-V_{B\alpha\sigma}G_\sigma)$ represents the additional density of states of σ spin due to the presence of $B\alpha$ impurity atoms, the energy corresponding to $B\alpha(\alpha = I,II)$ is

$$E(B\alpha) = E_0 - \frac{1}{\pi} \text{Im} \sum_\sigma \int_{-\infty}^{\mu} (E-\mu) \frac{d}{dE} \ln (1 - V_{B\alpha\sigma} G_\sigma(E)) dE \qquad (64)$$
$$- U_B n_{B\alpha\uparrow} n_{B\alpha\downarrow} \quad .$$

The constant E_0 is independent of α .

Hennion and Hennion determine the concentration of BII atoms more simply by the empirical rule that a Mn atom is of type II if it has at least four nn Mn atoms, a random distribution being assumed. Thus $z = Px$, $y = (1-P)x$ where

$$P = \sum_{q=4}^{12} \binom{12}{q} x^q (1-x)^{12-q} \ . \qquad (65)$$

The observed maximum in the bulk magnetization occurs at $x \simeq 0.1$ (see fig.2) whereas Jo finds it at $x \simeq 0.07$ and Hennion and Hennion at $x \simeq 0.15$. Both authors find that the MnI and MnII local moments take values of $+ 4\mu_B$ and $-4\mu_B$, respectively, which are essentially independent of x. The Ni moment decreases steadily to zero at the critical concentration, $x = 0.3$, at which stage the theoretical model has the appearance of a "disordered antiferromagnet" or spin glass. Mn bands of both up and down spin, corresponding to bound states in the dilute limit, occur above E_F; these correspond to MnII and MnI atoms respectively.

Finally in this section, we mention very briefly local environment effects in ferromagnetic alloys. Methods of going beyond single-site CPA have been developed by Miwa[34] and by Brouers, Gautier and Van der Rest.[35] The idea is to consider scattering by a (z + 1)-atom cluster, consisting of a central atom and its z nearest neighbours of specified composition, embedded in an effective medium. The coherent potential of the effective medium is determined so that the site-diagonal element of the effective medium Green function is equal to the site-diagonal element (at the central site) of the cluster Green function, configurationally arranged over cluster composition. An improvement on this central site approximation is to equate the configurational average of the trace of the Green function taken on the nn shell to the site-diagonal element of the effective medium Green function (boundary site approximation). We expect the moment of a given atom to depend on its local environment when the moment is found, either experimentally or in a single-site CPA calculation, to vary considerably with composition. Thus moments on Co and Fe atoms in fcc Ni alloys are expected to be essentially independent of their environment. However, the moment of Fe in bcc FeCo alloys increases from $2.2\mu_B$ in pure Fe to about $2.9\mu_B$ in the Co rich alloys. We therefore expect that at any concentration an Fe atom in a Co rich local environment will have a larger moment than one with predominantly Fe nearest neighbours. This is borne out in calculations by Kanamori et al[3,2] using the boundary site approximation, and the configurational averages of the Fe and Co moments agree well with the single-site CPA calculations.

Local environment effects are obviously important in alloys near the critical composition. Thus, as discussed earlier, atoms with favourable magnetic environments may be the source of extensive giant polarization clouds. In this case the moment of a given atom certainly does not depend just on the nearest-neighbour composition and in the presence of such massive inhomogenity the concept of an effective medium is inappropriate. We return to this problem in ¶2.5.

2.4 Dilute alloys of non-transition metals in nickel and iron

Solute	Cu	Zn	Al	Si	Ge	Sn	Sb
Solute valence	1	2	3	4	4	4	5
$-\mu_B^{-1} d\bar{m}/dc$ (in Ni)	1.14	2.11	2.80	3.77	3.70	4.22	5.31
$-\mu_B^{-1} d\bar{m}/dc$ (in Fe)			2.27	2.28	1.36	0.97	

The above table shows the observed decrease in magnetization of Ni and Fe due to some non-transition metal impurities. It is noticeable that in Ni there is a close correlation between the loss of moment and the valence of the impurity. For Fe, however, it appears that some impurities such as Al and Si merely dilute the Fe matrix, the loss of moment being close to $2.2\mu_B$, the moment on an Fe atom. For other impurities the loss of moment is considerably less than $2.2\mu_B$ so that, assuming zero moment on the impurity, neighbouring Fe atoms must have an increased moment. This is borne out by neutron diffuse scattering measurements.[7] We shall concentrate mainly on Ni alloys since the Fe case is more complicated. Contributions to the understanding of these alloys have been made by Marshall, Mott[36] and, most recently, by Terakura and Kanamori.[37] We shall first present a simplified model, synthesizing ideas of the above authors, and then a more realistic picture due to Terakura. The fundamental justification for these models is the detailed calculation of Terakura and Kanamori.

The simplest interpretation of the Ni and Co alloy data is that the valence electrons of the impurity atoms are poured into the host d bands to fill empty down spin states. However the impurity ion must be screened by s and p electrons and, indeed, X-ray spectroscopy and Mössbauer isomer-shift measurements show that the impurity has a normal number of valence electrons. The filling of the host d band must therefore be "apparent". In the simplified model of screening due to Marshall, and Mott, the s and p valence orbitals of the impurity combine with symmetrized combinations of d orbitals of surrounding host atoms to form bonding and antibonding orbitals. Consider, for example, an impurity s orbital ϕ_s and a suitable linear combination ϕ_d of d orbitals on nearest-neighbour atoms; ϕ_d has s-like symmetry with respect to the impurity site and the situation is illustrated schematically in fig. 4(a) where only four of the twelve Ni nearest-neighbours are shown. Suppose that the bonding and antibonding states are

$$(1 + \lambda^2)^{-\frac{1}{2}} (\phi_s + \lambda\phi_d), \; (1 + \lambda^2)^{-\frac{1}{2}} (\phi_d - \lambda\phi_s) \;, \tag{66}$$

respectively, and that only the bonding state is occupied, i.e. d states on nn atoms are forced above the Fermi level due to mixing with the s orbital. Also the weight of unoccupied d states $(1 + \lambda^2)^{-1}$ equals the weight of occupied s orbital and a similar effect takes place for p orbitals. The total occupied weight of s and p orbitals must correspond to the impurity valence Z, for screening, so that a total of about Z d states must be pushed through the Fermi level on nn Ni atoms. In their much more sophisticated first-principles calculation, using a pseudo-Greenian formulation, Terekura and Kanamori indeed showedthat, when a Ni atom is replaced by an impurity such as Aℓ, the change in the number of states below E_F of s and p symmetry with respect to the impurity site is very small. Since (Z - 0.6)/2 additional states of each spin with these symmetries appear on the impurity atom, to screen it, there is a corresponding loss of occupied states on neighbouring Ni atoms, these states being mainly formed from suitable linear combinations of d functions. Terakura and Kanamori argue that this loss of electronic charge on nn Ni sites, which occurs in the one-electron theory, will be screened by down spin d electrons. The net result is that on nn Ni atoms (Z - 0.6)/2 up spin electrons are lost and replaced by down spin electrons. Together with the lost moment of $0.6\mu_B$ on the impurity site, which carries almost zero moment, the total moment lost per impurity is $Z\mu_B$. The screening of the lost charge on nn Ni sites is almost the same as occurs for impurities such as Cr, as discussed in ¶2.2. The main difference is the mechanism of the loss of up spin d states below E_F on the nn Ni atoms. In the Cr case it is due to the tail of a virtual bound state whereas in the present case it is due to the formation of antibonding states with sp electrons on the impurity. This similarity between the behaviour of impurities like Cr and impurities like Aℓ explains why the shape of the magnetization disturbance, observed with neutron diffuse scattering[7], is almost the same in both cases. It extends beyond nearest-neighbours, to a distance of 5Å. As pointed out in ¶2.2 this indicates that screening by down spin d electrons is incomplete on the nn sites. We suggest that it is completed by s and p electrons on the Ni atoms and this leads in turn to a loss of up spin d electrons on neighbouring Ni atoms, thus extending the disturbance

(a)

(b)

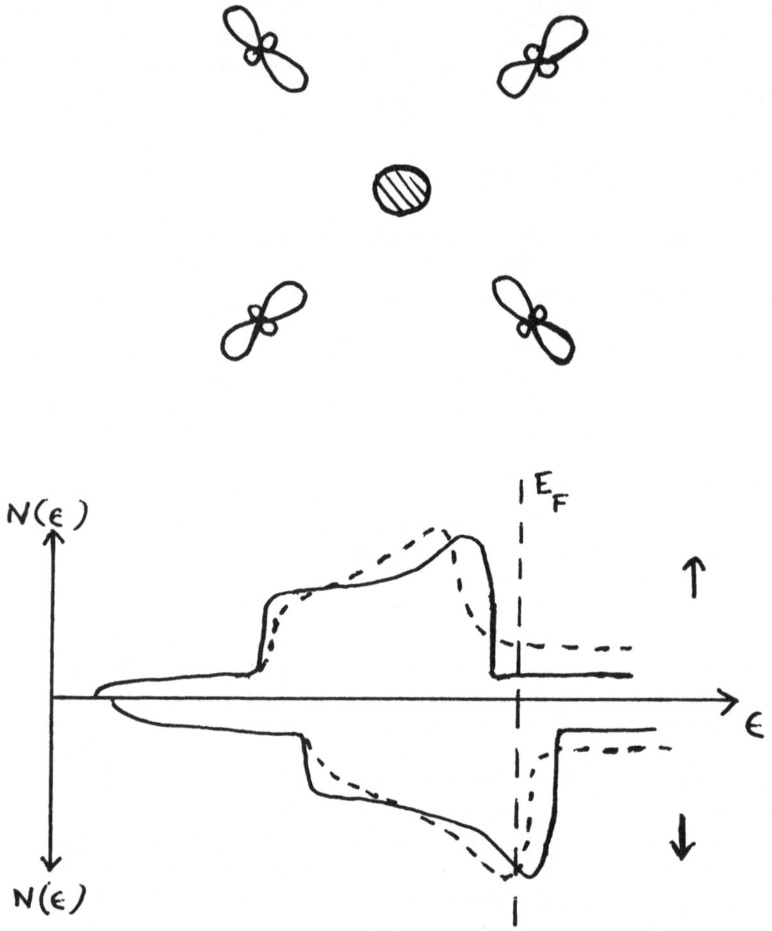

Fig.4(a) Sketch of an s-like linear combination of d orbitals on nearest-neighbour atoms to an impurity (only 4 of the 12 atoms are shown).

(b) Schematic local density of states on a nearest-neighbour Ni atom (dashed) compared with the density of states for bulk Ni (full line).

to next-nearest-neighbours to the impurity. There is one slight difference between the Cr and Aℓ case. This is due to the fact that screening on the nn Ni atoms will be carried out by down spin d electrons in states which have not been pushed upwards, i.e. linear combinations of Ni d functions having mainly d symmetry with respect to the impurity site, for the Aℓ case and linear combinations which are orthogonal to the d-like virtual bound state in the Cr case.

The simple model of Marshall and Mott suffers from the defect that the concept of impurity s and p orbitals is not well-defined, since we know that the valence electrons have a broad energy distribution in normal metals. The following improved model, in which the valence orbital is replaced by continuuum states, is due to Terakura[38]. We consider explicitly states of s symmetry about the impurity but states of p symmetry may be treated similarly.

Consider the following Anderson-like Hamiltonian:

$$H = \sum_k \varepsilon_k |k\rangle\langle k| + E_d |d\rangle\langle d| \qquad (67)$$

$$+ \Sigma_k \{ V_{dk} |k\rangle\langle d| + V^*_{dk} |d\rangle\langle k| \} .$$

Here the state $|d\rangle$ is the symmetrized orbital ϕ_d defined previously and $|k\rangle$ is a continuum state. A Hamiltonian of this type applies to electrons of each spin and for a given spin E_d is the corresponding effective atomic level on Ni sites. V_{dk} is the hybridization matrix element. If the central site is occupied by a Ni atom, instead of the impurity, the continuum states correspond to the usual nearly-free-electron-like sp band of pure Ni. The states k are then linear combinations of plane waves having s symmetry about the central site. When the central site is occupied by an impurity such as Aℓ the potential seen by the continuum states may be considered as constant apart from a spherical square well at the impurity site. The distribution of the energies ε_k is thus shifted downwards and possibly a bound state may be formed. We shall see that the downward shift pushes the hybridized d states upwards in energy. The density of states on the d orbital is given by

$$\rho_d(E) = - \pi^{-1} \, \mathrm{Im}\, G_{dd}(E) \qquad (68)$$

where

$$G_{dd}(E) = \langle d| (E^+ - H)^{-1} |d\rangle$$

$$= [E^+ - E_d - \sum_k |V_{dk}|^2 (E^+ - \varepsilon_k)^{-1}]^{-1}$$

$$= (E - E_d - \Delta + i\Gamma)^{-1} \; . \qquad (69)$$

The shift Δ and width Γ are given by

$$\Delta = P\sum_k |V_{dk}|^2 / (E-\varepsilon_k) \qquad (70)$$

and

$$\Gamma = \pi \sum_k |V_{dk}|^2 \, \delta(E-\varepsilon_k) \qquad (71)$$

where P denotes the principal part. Thus

$$\rho d(E) = \frac{1}{\pi} \frac{\Gamma}{(E-E_d-\Delta)^2+\Gamma^2} \qquad (72)$$

and the d resonance on nearest-neighbour sites is centred on $E_d+\Delta$. Suppose the central site is initially occupied by a Ni atom, the d resonance on this atom making no contribution to states of s and p symmetry, and is then replaced by an Aℓ atom. As we have noted the energy distribution of the continuum states ε_k shifts downwards, additional states appearing below E_d. This increases Δ so that the d orbital is pushed upwards. For Aℓ in Fe, where similar considerations apply, Terakura finds the increase in Δ to be about 1.5eV. The upward shift of the d level on nearest-neighbours to the impurity leads to increased hybridization above the d band, with a loss of d states below E_F. The local density of states on a Ni atom which is nearest-neighbour to the impurity is compared schematically with the density of states for bulk Ni in fig. 4(b). The net change is produced by an upward shift of d states having s and p symmetry relative to the impurity site and a smaller downward shift of all d states to produce electrical neutrality. The downward shift hardly affects the number of occupied up spin states but produces the increase in occupied down spin states necessary for screening. The hybridized d states above E_F correspond to the antibonding states of the simpler picture.

We now comment briefly on the more complicated case of Fe alloys. Terakura[39] has shown that once again, when an Fe atom is replaced by the impurity, the change in the number of states below E_F of s and p symmetry with respect to the impurity is quite small (see fig. 1 of Terakura's paper). This applies to states of both spin. In Ni, with a full up spin d band, this results in an electron deficit on nn atoms which is screened by down spin d electrons. In Fe, with both spin sub-bands incompletely filled, the loss of d states on nearest-neighbour sites is partly cancelled by other effects, a narrowing of the band which increases the number of up spin electrons and the existence of a resonance near the centre of the band which affects the number of down spin electrons in a way that is sensitive to the impurity potential. These last two effects refer to states of d symmetry with respect to the impurity and also exist in Ni alloys where they play no significant role owing to the full up spin band. Terakura has carried out detailed calculations for Fe alloys and finds good agreement with the observed dm/dc for several different impurities.

2.5 Alloys near the critical composition

Consider a binary alloy, $A_{1-x}B_x$, where B is a ferromagnetic metal, and suppose that the alloy is ferromagnetic for $x > x_o$, the critical concentration. We may distinguish two cases:

(i) No moments exist on an isolated magnetic atom; the ferromagnetic moment appears through a cooperative effect.

(ii) Isolated magnetic atoms have moments but they are aligned ferromagnetically only for $x > x_o$.

In case (i) the behaviour depends on whether $x_o \ll 1$ or is considerably larger so that the critical alloy is quite concentrated. In the former case the magnetization must be inhomogeneous, since it will be peaked up around the magnetic atoms, and we shall discuss Pd Ni with $x_o \sim 0.02$ as an example of this. In the more concentrated case the magnetization may be more homogeneous although the formation of polarization clouds in the critical region is likely. We shall discuss PtNi as an example of a homogeneous magnetic alloy.

Examples of case (ii) are CrFe and AuFe and for $x < x_o$ we have antiferromagnetic or spin glass behaviour. In these alloys $x_o \sim 0.17$ and the appearance of ferromagnetism seems to correspond to a percolation limit[40] where, with nn Fe interactions, an infinite cluster of Fe atoms first forms. Assuming predominantly nearest-neighbour hopping this corresponds to the appearance of an itinerant-electron Fe d band. Another quite different example of case (ii) is Pd Fe where x_o is less than 0.001. The difference between PdNi and PdFe is noteworthy. In discussing case (i) alloys Kato and Mathon[41] have stressed that even for x considerably less than x_o there will always be a few quite large clusters of magnetic atoms which must carry a moment. They argue that strictly no exact critical concentration x_o exists. However these isolated moments may not be aligned ferromagnetically, particularly at any practical non-zero temperature, and experimentally the critical concentration seems to be quite well-defined.

In these lectures we shall confine our attention to case (i) alloys and discuss both homogeneous and inhomogeneous alloys using a generalisation of the Landau equation (32). This phenomenological approach is particularly useful for the inhomogeneous case where single-site CPA is inappropriate.

Homogeneous magnetic alloys: In this case we neglect the $\nabla^2 M$ term in (32) and note that $A = 0$ for $x = x_0$ since the susceptibility diverges at the critical concentration. Thus we put $A = A_1(x_0-x)$ so that (32) may be written

$$M^2 = B^{-1}(H/M) + A_1 B^{-1}(x-x_0), \tag{73}$$

where B and A_1 are slowly-varying functions of x and H is a uniform applied field. Thus so-called Arrott plots of M^2 against H/M for different values of x near x_0 should be a set of nearly parallel straight lines. Deviations from linearity indicate inhomogeneity and fig.5 shows these to be small in Ni_xPt_{1-x} alloys.[42] Also from (73) we see that the magnetization M at H = 0 should vary as $(x - x_0)^{\frac{1}{2}}$ and fig. 6 shows that this holds for Ni_xPt_{1-x}. Microscopic single-site CPA calculations for these alloys have been carried out by Alben and Wohlfarth[43] and, including off-diagonal disorder to take account of differing band-widths in Ni and Pt, by Inoue and Shimizu[44]. Good agreement with experiment for M(x) is obtained.

Inhomogeneous magnetic alloys: In a system such as Ni_xPd_{1-x}, where $x_0 = 0.023$, the Arrott plots for x near x_0 curve strongly (fig.7). The first treatments of inhomogeneity using equation (32) were those of Shtrikman and Wohlfarth[45] and Yamada and Wohlfarth[46]. These general treatments are appropriate for dealing with concentration fluctuations in concentrated alloys, these fluctuations being introduced by allowing A to be a slowly-varying function of position. Thus equation (32) may be written

$$-C\nabla^2 M(\underline{r}) + A(\underline{r})M(\underline{r}) + BM^3(\underline{r}) = H. \tag{74}$$

Yamada and Wohlfarth found that the Arrott plots bend upwards at smaller values of H/M, somewhat as in fig.7, but that unlike fig.7, they very soon become parallel straight lines at higher fields. Here we describe a rather different, less macroscopic way of using equation (74) which is appropriate for alloys such as Pd Ni. The essential feature required of the alloy system is a low concentration of magnetic impurities (an isolated impurity not carrying a moment however) in a strongly paramagnetic matrix with a correlation length κ^{-1} considerably larger than the interatomic distance. Under these circumstances equation (74) may also be derived microscopically. In the matrix $A(\underline{r})$ is a positive

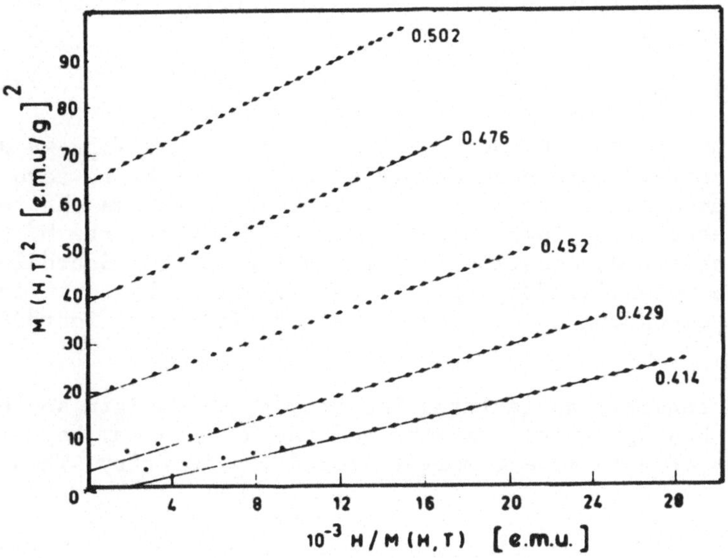

Fig.5: Arrot plots for Ni_xPt_{1-x} alloys at T = 4K and H up to 150 kOe.

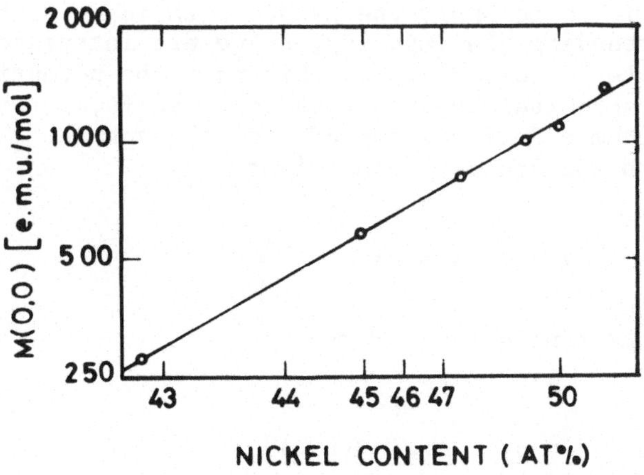

Fig.6: The spontaneous (T = O, H = O) moment of Ni_xPt_{1-x} alloys as a function of nickel content (ℓnM against ℓn $(x-x_0)$).

constant, being equal to the reciprocal of the susceptibility of the matrix, but it varies rapidly at the impurity site. We write

$$A(\underline{r}) = A_0 + \sum_i V(\underline{r} - \underline{R}_i) \tag{75}$$

where A_0 is the constant value of A in the matrix and $V(\underline{r})$ is a negative "potential" within an atomic sphere $r = a$, being zero for $r > a$. The summation is over impurity sites $\underline{R}_i$. We use the term "potential" because in a paramagnetic situation, where the M^3 term of (74) may be neglected, equation (74) is analogous to a Schrödinger equation. The potential $V(\underline{r})$ will be characterized by a single parameter corresponding to an s wave phase shift in the Schrödinger analogy.

We first consider an isolated impurity at the origin so that equation (74) has spherical symmetry and the M^3 term may be omitted in the absence of a permanent moment around the impurity. Thus putting

$$\rho = \kappa r, \quad h = H/A_0, \quad \psi = \rho M, \tag{76}$$

where $\kappa^{-1} = (C/A_0)^{\frac{1}{2}}$ is the correlation length in the matrix, we may write (74) in the form

$$-d^2\psi/d\rho^2 + \psi = h\rho \ , \quad \rho > \kappa a \ . \tag{77}$$

It is unreasonable to apply the Landau equation within the atomic sphere surrounding the impurity, so we are interested only in the exterior solution of (77). In this case the potential V(r) is completely characterised by an s wave phase shift, or equivalently, the logarithmic derivative γ of ψ on the sphere. Thus (77) must be solved with the boundary conditions

$$\left.\begin{aligned} &\frac{1}{\psi}\frac{d\psi}{d\rho} = \gamma \text{ for } \rho = \kappa a \\ &\psi \text{ is finite for } \rho \to \infty \end{aligned}\right\} \ . \tag{78}$$

Hence we find

$$M(r) = \frac{H}{A_0}\left\{1 + \frac{1-\gamma\kappa a}{1+\gamma} \ \frac{\exp[-\kappa(r-a)]}{\kappa r}\right\} \ . \tag{79}$$

The second term in brackets is the moment associated with the impurity and diverges when $1 + \gamma = 0$. Hence the condition for a local moment is $\gamma < -1$ and we suppose this is not satisfied. It would be satisfied for an Fe impurity in Pd.

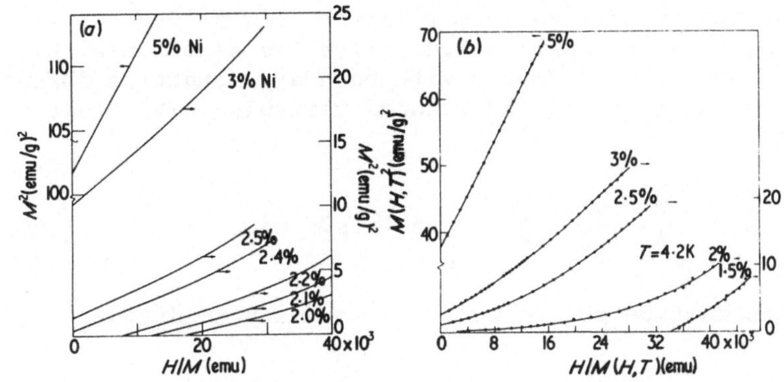

Fig. 7 : (a) Arrott plots for Pd-Ni alloys
(b) Observed Arrott plots (Beille et al[51])

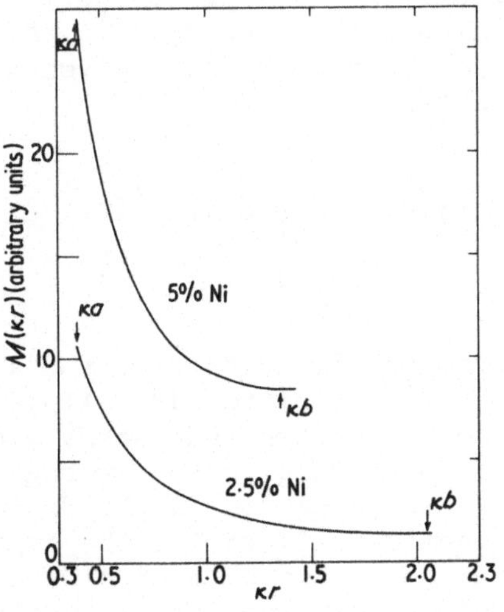

Fig. 8 : Spatial dependence M(r) for two Pd-Ni alloys.

For a finite concentration c of impurities two different procedures have been followed by Edwards et al[47,48] and Kato and Mathon[49] respectively. We first describe the approach of Edwards et al. whose philosophy is that the influence of magnetic inhomogeneity is more important than that of disorder. They therefore suppose that the impurities form a lattice and solve (74) within the Wigner-Seitz approximation i.e. solve the equation within the Wigner-Seitz sphere of radius b with boundary condition $dM/dr = 0$ for $r = b$. By introducing the reduced variables (76) equation (74) may be written

$$-\frac{d^2\psi}{d\rho^2} + \psi + \frac{B}{A_0}\frac{\psi^3}{\rho^2} = h\rho, \quad \kappa b > \rho > \kappa a. \tag{80}$$

This equation must be solved with the boundary conditions

$$\begin{aligned} \frac{1}{\psi}\frac{d\psi}{d\rho} &= \gamma \text{ for } \rho = \kappa a \\ &= \rho^{-1} \text{ for } \rho = \kappa b. \end{aligned} \tag{81}$$

This may be done numerically for a range of applied fields and impurity concentrations in both the paramagnetic and ferromagnetic regimes. The average magnetization M for a given H and c may then be calculated from the integral

$$M = \frac{3}{b^3 - a^3} \int_a^b M(r)\, r^2 dr, \tag{82}$$

the contribution to M from the impurity site $r < a$ being very small. Edwards et al [48] applied this model to PdNi alloys, taking the "impurity" to be a pair of nearest-neighbour Ni atoms and the "matrix" to correspond to pure Pd plus the average effect of single Ni impurities. A calculation based on single Ni impurities yielded insufficient magnetic inhomogeneity to fit the experimental data. Furthermore, the neutron-scattering data of Aldred et al[50] shows that the concentration of magnetic polarization clouds is much less than the concentration of single Ni atoms, although somewhat greater than that of nn Ni pairs ("doubles"). We suppose that the giant magnetization clouds surround Ni doubles. The inverse correlation length κ of the matrix (Pd plus Ni singles) deduced from the neutron data is 0.15 ± 0.03 Å^{-1} which hardly differs from the value of 0.2Å^{-1} for pure Pd, deduced from PdFe alloys. This shows that the effect of the single Ni impurities is weak. In the calculations the value $\kappa = 0.2\text{Å}^{-1}$ was taken. Also, since it is assumed that the Landau coefficient C is the same for all the alloys, the constant $A_0 = C\kappa^2$ equals the inverse susceptibility of pure Pd. The only adjustable parameters are γ and B. The concentration c of doubles was calculated using the relation

$$c = \tfrac{1}{2}x\,[\,1-(1-x)^{12}] \tag{83}$$

where x is the Ni concentration. In deriving (83) we assume a random distribution of Ni impurities and treat all larger clusters, whose concentration is very low in the range of interest, as effectively doubles. We chose $\gamma = -0.85$ to give the observed critical concentration $x = 0.0232$ for the appearance of a spontaneous magnetization. This is close to the value of -1 at which an isolated Ni double would acquire a moment. The value of B/A_0 was taken to be $5.0 \times 10^{-3}\,(\text{emu/g})^{-2}$ obtained by fitting the observed magnetization[51] for $x = 0.03$ in a field of 80 kOe. Thus only two points of the whole family of calculated curves in fig.7 are fitted. The agreement between theory and experiment is generally good. The higher observed slope of the 5% alloy compared with the others is clearly shown by these calculations. An extension of these calculations to very high fields would yield Arrott plots all of which eventually tend to be parallel with a slope given by B^{-1}, since the left hand side of (74) is ultimately dominated by the BM^3 term. The straightness of the Arrott plot of the 5% alloy can be attributed to the higher degree of homogeneity of the magnetization of this more concentrated alloy compared with those closer to the critical concentration. This is clearly shown in fig. 8.

Kato and Mathon[49] preserve the disordered nature of the "potential" in (75) in their treatment of equation (74). They linearize this equation by putting

$$\begin{aligned} M^3(\underline{r}) = [\,M_f + m(\underline{r})]^3 &= M_f^3 + 3M_f^2\, m(\underline{r}) \\ &= 3M_f^2 M(\underline{r}) - 2M_f^3 \end{aligned} \tag{84}$$

where

$$M_f = \Omega^{-1} \int M(\underline{r})\, d^3r \tag{85}$$

for a given configuration f, Ω being the volume of the crystal. Then (74) becomes

$$[\,-C\nabla^2 + A(\underline{r}) + 3BM_f^2]\;\; M(\underline{r}) = H + 2BM_f^3 \;. \tag{86}$$

The linearization procedure has been tested for special cases with the impurities arranged in a lattice. It reproduces rather accurately the form of M(r) obtained by direct integration of (74), at least when the isolated impurity has no tendency to form a local moment i.e. $\gamma + 1$ is not too small. In the disordered case, however, there will be a tendency to form local moments near clusters of Ni atoms and the linearization will be invalid for such configurations. Equations (85) and (86) should be solved self-consistently for each configuration and the configurational average $M = \langle M_f \rangle$ then calculated. Since this is not feasible the configuration dependent quantity M_f^2 is approximated by

$$M_f^2 \simeq \langle M_f^2 \rangle = M^2 + \langle \Delta M^2 \rangle \tag{87}$$

and similarly

$$M_f^3 \simeq \langle M_f^3 \rangle = M^3 + 3M \langle \Delta M^2 \rangle + \langle \Delta M^3 \rangle . \tag{88}$$

Then equation (86) becomes

$$[- C\nabla^2 + \tilde{A} + \sum_i V(\underline{r} - \underline{R}_i)]\, M(\underline{r}) = \tilde{H} \tag{89}$$

where the quantities

$$\tilde{A} = A_0 + 3BM^2 + 3B \langle \Delta M^2 \rangle \tag{90}$$

$$\tilde{H} = H + 2B\, [M^3 + 3M\langle \Delta M^2 \rangle + \langle \Delta M^3 \rangle]$$

are already independent of the configuration of impurities. Kato and Mathon make the further approximation $\langle \Delta M^3 \rangle \simeq \langle \Delta M^2 \rangle^{3/2}$. Defining the Green function $G(\underline{r}, \underline{r}';E)$ by the equation

$$[E + C\nabla^2 - \sum_i V(\underline{r}-\underline{R}_i)]\, G(\underline{r},\underline{r}';E) = \delta(\underline{r}-\underline{r}') \tag{91}$$

we have, from (89),

$$M(\underline{r}) = -\tilde{H} \int G(\underline{r},\underline{r}';-\tilde{A})\, d^3r' . \tag{92}$$

Hence

$$M = -\Omega^{-1}\, \tilde{H} \iint \langle G(\underline{r},\underline{r}')\rangle\, d^3r d^3r'$$

$$= -\tilde{H}\, G(\underline{k} = 0; -\tilde{A}). \qquad (93)$$

Also

$$G(\underline{k}, -\tilde{A}) = -\frac{1}{\tilde{A} + Ck^2 + \Sigma(\underline{k}, -\tilde{A})} \qquad (94)$$

where Σ $(\underline{k},E)$ is the self-energy arising from the disordered potential $\sum_i V(\underline{r}-\underline{R}_i)$. Combining equations (94), (93), and (90) we obtain the equation

$$[A_0 + \Sigma(0,-\tilde{A}) - 3B \langle\Delta M^2\rangle]\, M + BM^3$$
$$= H + 2B \langle\Delta M^3\rangle \qquad (95)$$

for M. The self-energy, or coherent potential, may be obtained using CPA or otherwise. Kato and Mathon use single site CPA, taking $V(\underline{r})$ to be a della function of strength $-V_0$, and fit V_0 and B to the data in the same way as Edwards et al (V_0 here replaces γ). However there is no explicit account taken of Ni^0 "doubles". It is found that near the critical concentration no self-consistent solution exists unless $\langle\Delta M^2\rangle_1 \neq 0$. In general the quantity $\langle\Delta M^2\rangle = \langle M_f^2\rangle - \langle M_f\rangle^2$ is of order Ω^{-1} and vanishes in the thermodynamic limit $\Omega \to \infty$. Kato and Mathon interpret their introduction of a finite $\langle\Delta M^2\rangle$ as rectifying faults in their linearization procedure, which fails in the presence of Ni clusters. They devise an ingenious scheme for determining $\langle\Delta M^2\rangle$ self-consistently within CPA but subsequently improve agreement with experiment by choosing $\langle\Delta M^2\rangle$ arbitrarily as a Gaussian function of x centred on x = 0.023, introducing two further parameters for the amplitude and width of the function. The calculated Arrott plots have the strong curvature of the observed ones at smaller values of H/M but all become parallel at higher fields. This indicates that the magnetization becomes rather homogeneous in moderate applied fields which does not seem to be the case experimentally. Kato and Mathon's method is attractive in that it treats disorder explicitly but is marred by the rather artificial introduction of the non-zero $\langle\Delta M^2\rangle$ term. The lattice model of Edwards et al, with its explicit treatment of strongly inhomogenous polarization clouds surrounding pairs of Ni atoms, may be closer to the true physical situation.

3. EXCITATIONS FROM THE GROUND STATE : SPIN WAVES

3.1 Introduction

Clearly from § 2 considerable progress has been made in understanding the ground state of magnetic transition metals and alloys. For pure metals quantitative first principles calculations are possible using the SDF method. On the other hand the behaviour of these materials at and above the Curie temperature T_c is still not understood very well. Do local moments disorder as in the Heisenberg model or do they melt away as in Stoner theory ? If the former is the case how do we describe the process within the band model, which one must use for the ground state ? To proceed from the known to the unknown we may start with low-lying magnetic excitations from the ground state, in particular spin waves. Long wavelength spin waves make an important, and often dominant, contribution to the temperature dependence of the magnetization M(T) at low temperatures through the well-known Bloch $T^{3/2}$ law. There is often a relation between spin wave energies and T_c. Before discussing this further it is useful to give first a macroscopic picture of a spin wave.

The macroscopic exchange energy in a ferromagnet takes the form

$$A \sum_i (\nabla n_i(\underline{r}))^2 \tag{96}$$

where n_1, n_2, n_3 are the direction cosines of the magnetization direction at position $\underline{r}$. Here A is the Bloch wall stiffness constant. A long wavelength spin wave is a macroscopic oscillation in which the magnetization precesses around the equilibrium z direction. The transverse components of magnetization vary in space and time as

$$M^+ = M_x + iM_y \propto \exp[\, i(\underline{q}.\underline{r} - \omega t)]. \tag{97}$$

The dispersion relation is

$$\hbar\omega = Dq^2 \tag{98}$$

and $D \propto A/M_s$ where M_s is the saturation moment. This macroscopic picture is valid for any ferromagnet, at zero or finite $T(< T_c)$ and with an ordered or disordered structure. Furthermore D is real, damping appearing in higher powers of the wave-vector q. In exciting a spin wave quantum (magnon), for example by neutron scattering, a single spin is flipped down. Fig. 9[52] shows the qualitative relationship between D and T_c for Fe alloys. This is confirmed by Aldreds' work[53,54] on Fe and Fe V, where it is shown that up to 400 K M(T) is determined entirely by spin wave excitations. In Ni, on the other hand[55], spin wave excitations only account for one half of the decrease in magnetization at $T_c/2$. However Fe is not purely Heisenberg-like, it resembles Ni in appearing to have unusually

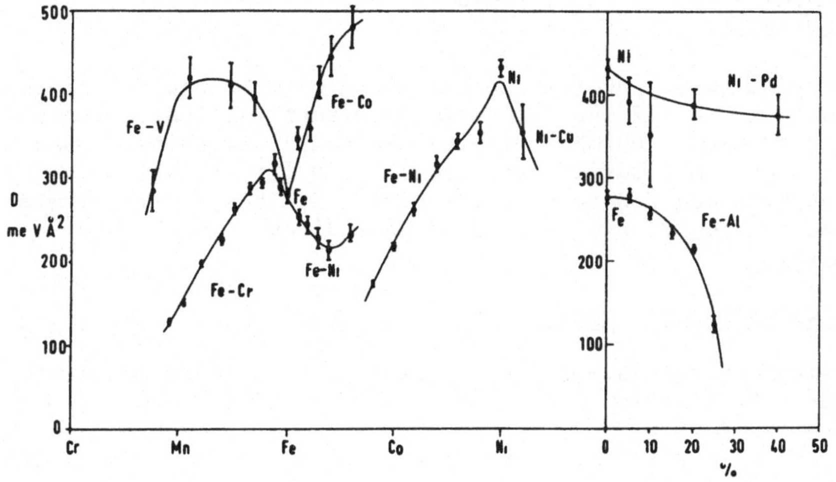

Fig 9a : Experimental values of D for ferromagnetic alloys.

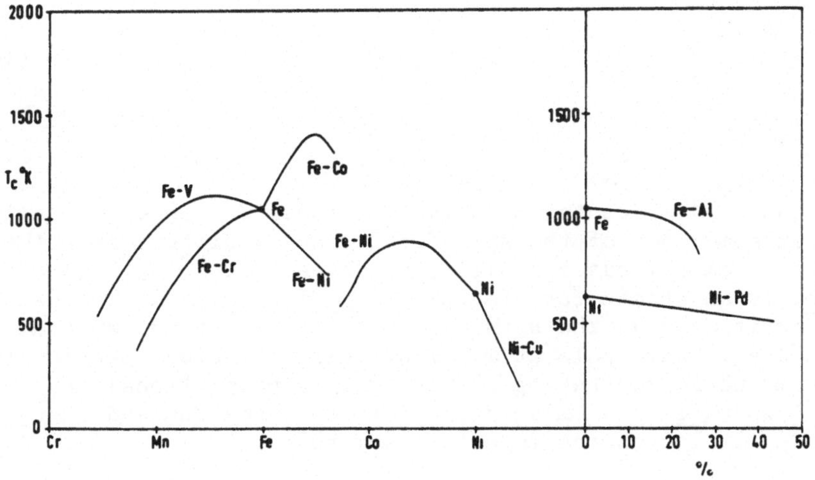

Fig. 9b : Experimental values of Curie temperature T_c.

long short-range order far above T_c.[56] Recent work by Hennion and Hennion[30] shows that NiCo alloys behave like NiFe with D varying approximately as M_s^{-1}. However in NiCr and NiV the stiffness constant D is proportional to T_c which is proportional to M_s. In the anomalous NiMn alloys D falls rapidly, faster then M_s^{-1} initially and T_c also falls.

Clearly it is of interest to calculate D in Ni and Fe alloys. In any case D is one of the following important intrinsic properties of a ferromagnet which underlie the technical properties : saturation moment M_s, Curie temperature T_c, D (or A), anisotropy coefficient K. A and K determine the width of a Bloch wall between domains. Before considering alloys we must discuss spin waves in pure metals.

3.2. Spin waves in pure metals

For simplicity consider the one-band tight-binding model with Hamiltonian (cf. equation (2))

$$H = \sum_{\underline{k}\sigma} \varepsilon_{\underline{k}}\, n_{\underline{k}\sigma} + U \sum_i n_{i\uparrow} n_{i\downarrow} . \tag{99}$$

We further consider the response function given by the Kubo formula

$$\chi_\perp(\underline{q},\omega) = \lim_{\eta\to 0+} \int_{-\infty}^{\infty} \langle\langle S^-_{\underline{q}}(t),\ S^+_{-\underline{q}} \rangle\rangle\ e^{-i(\omega-i\eta)t}\, dt \tag{100}$$

where

$$\langle\langle S^-_{\underline{q}}(t),\ S^+_{-\underline{q}} \rangle\rangle = \frac{i}{\hbar} \langle [S^-_{\underline{q}}(t),\ S^+_{-\underline{q}}] \rangle\ \theta(t) \tag{101}$$

and

$$S^-_{\underline{q}} = \sum_{\underline{k}} a^{\dagger}_{\underline{k}+\underline{q}\downarrow}\, a_{\underline{k}\uparrow} \tag{102}$$

In this chapter < > denotes a thermal average. $S^-_{\underline{q}}$ is essentially the Fourier component of the spin density and $\chi_\perp$ differs from the transverse dynamical susceptibility χ_{-+} by a factor $4\mu_B^2|F(\underline{q})|^2$, where $F(\underline{q})$ is an atomic form factor. $\chi_\perp$ is important because its singularities give the spin-flip excitations of the system, including a spin wave pole at $\omega = \omega(\underline{q})$. Also the cross-section for inelastic magnetic scattering of neutrons is proportional to $\mathrm{Im}\chi_\perp(\underline{q}, \omega)$ where $\underline{q}$ and $\hbar\omega$ correspond to the momentum and energy transfer. $\chi_\perp$ satisfies the important sum rule

$$\int_{-\infty}^{\infty} \mathrm{Im}\ \chi_\perp(\underline{q},\omega)d\omega = \frac{\pi}{\hbar}(n_\downarrow - n_\uparrow) = -\frac{2\pi}{\hbar}\langle S^z\rangle \tag{103}$$

Following Izuyama et al[57] we shall calculate $\chi_\perp$ in the random phase approximation (RPA). We require the two-particle Green function

$$\langle\langle S^{-}_{\underline{q}}(t), S^{+}_{-q}\rangle\rangle = \sum_{\underline{k}} \langle\langle \theta_{\underline{q}}(\underline{k},t), S^{+}_{-\underline{q}}\rangle\rangle \tag{104}$$

where $\theta_{\underline{q}}(\underline{k}) = a^{\dagger}_{\underline{k}+\underline{q}\downarrow}\, a_{\underline{k}\uparrow}$ and

$$\langle\langle \theta_{\underline{q}}(\underline{k},t), S^{+}_{-\underline{q}}\rangle\rangle = \frac{i}{\hbar} < [\theta_{\underline{q}}(\underline{k},t), S^{+}_{-\underline{q}}] > \theta(t) \,. \tag{105}$$

The equation of motion, using $i\hbar\, dA/dt = [A,H]$, is

$$i\hbar \frac{d}{dt} \langle\langle \theta_{\underline{q}}(\underline{k},t) , S^{+}_{-\underline{q}}\rangle\rangle$$

$$= - \delta(t) < [\theta_{\underline{q}}(\underline{k}), S^{+}_{-\underline{q}}] > + \langle\langle [\theta_{\underline{q}}(\underline{k},t), H], S^{+}_{-\underline{q}}\rangle\rangle \,. \tag{106}$$

If we write $H = H_0 + H_1$, where H_0 is the band energy and H_1 the interaction term, the commutator $[\theta_q(\underline{k}), H_0]$ is evaluated exactly. However in evaluating $[\theta_q(\underline{k}), H_1]$ we make the RPA and keep only terms of the form $n_{\underline{k}'\sigma}\, \theta_{\underline{q}}(\underline{k}'')$, replacing $n_{\underline{k}'\sigma}$ by its thermal or ground-state average $f_{\underline{k}'\sigma}$. (106) is then a closed set of linear equations and may be solved by Fourier transformation to obtain the Fourier transform of the Green function (104). Hence we find

$$\chi_{\perp}(\underline{q},\omega) = \frac{\Gamma(\underline{q},\omega)}{1 - \frac{U}{N}\Gamma(\underline{q},\omega)} \tag{107}$$

where

$$\Gamma(\underline{q},\omega) = \sum_{\underline{k}} \frac{f_{\underline{k}\uparrow} - f_{\underline{k}+\underline{q}\downarrow}}{\varepsilon_{\underline{k}+\underline{q}} - \varepsilon_{\underline{k}} + \Delta - \hbar\omega + i\eta} \tag{108}$$

and Δ is the exchange splitting in the Hartree-Fock ground state. Equation (107) for $\chi_{\perp}$ corresponds exactly to equation (18) for χ in a paramagnet and may also be derived, as was (18), by time-dependent molecular field (Hartree-Fock) theory which is equivalent to RPA. Another derivation is the following diagrammatic treatment of the two-particle Green function :

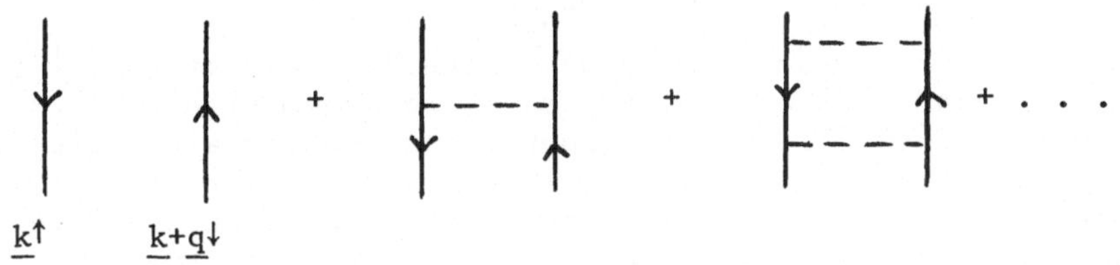

$$= \Gamma + \Gamma \frac{U}{N} \Gamma + \Gamma \frac{U}{N} \Gamma \frac{U}{N} \Gamma + \ldots \tag{109}$$

$$= \frac{\Gamma}{1 - \frac{U}{N}\Gamma} .$$

The spin-flip excitation spectrum is given by the zeros of the denominator in (107). For a given change of wave-vector $\underline{q}$ it consists of a continuum of single-particle (Stoner) excitations with excitation energies

$$\hbar\omega = \varepsilon_{\underline{k}+\underline{q}} - \varepsilon_{\underline{k}} + \Delta \tag{110}$$

and, for sufficiently small q, a state split off below the continuum. For q = 0 the split-off solution is $\omega = 0$, as is easily seen remembering that $\Delta = U(n_\uparrow - n_\downarrow)/N$. The excitation spectrum is shown schematically in fig. 10 for strong and weak ferromagnets. The gap ΔE in the single-particle spectrum for a strong ferromagnet corresponds to the gap between the top of the full up spin band and the Fermi level. The split-off states are spin waves and correspond to exciton-like excitations in which the down spin electron is bound to the up spin hole. For small q the spin wave energy $\hbar\omega$ is small and we expand the right hand side of the equation $1 = U\Gamma/N$ in powers of Δ^{-1}. We find for a cubic crystal that $\hbar\omega = Dq^2$ with

$$D = \frac{1}{3(n_\uparrow - n_\downarrow)} \sum_{\underline{k}} \{\frac{f_{\underline{k}\uparrow} + f_{\underline{k}\downarrow}}{2} \nabla^2\varepsilon_{\underline{k}} - \frac{f_{\underline{k}\uparrow} - f_{\underline{k}\downarrow}}{\Delta} |\nabla\varepsilon_{\underline{k}}|^2\}. \tag{111}$$

For multiple bands, but assuming a rigid exchange splitting Δ, we just sum over the band index n as well as $\underline{k}$. Using Green's theorem for the integrals over $\underline{k}$ space we may write (111) in the form[58]

$$D = \frac{1}{3(n_\uparrow - n_\downarrow)} \{\frac{1}{2}[M(\mu_+) + M(\mu_-)] - \frac{1}{\Delta}\int_{\mu_-}^{\mu_+} M(\varepsilon)d\varepsilon\} \tag{112}$$

where μ_+,μ_- are the up and down spin Fermi levels and $M(\varepsilon)$ is given by the surface integral

$$M(\varepsilon) = \frac{\Omega}{8\pi^3} \sum_n \int_{\varepsilon_{n\underline{k}}=\varepsilon} |\nabla\varepsilon_{n\underline{k}}| dS. \tag{113}$$

This is similar to the density of states $N(\varepsilon)$ except that $|\nabla\varepsilon_{\underline{k}}|$ appears instead of $|\nabla\varepsilon_{\underline{k}}|^{-1}$. $M(\varepsilon)$ is shown schematically in fig.11 and clearly, from (112), D is proportional to the algebraic area between the curve and the dashed line. In the limit of a very weak ferromagnet where $\Delta \to 0$ it is clear from (112) that the term in brackets tends to zero. In fact it is easy to show that $D \propto (n_\uparrow - n_\downarrow)$ i.e. proportional to the magnetization in the very

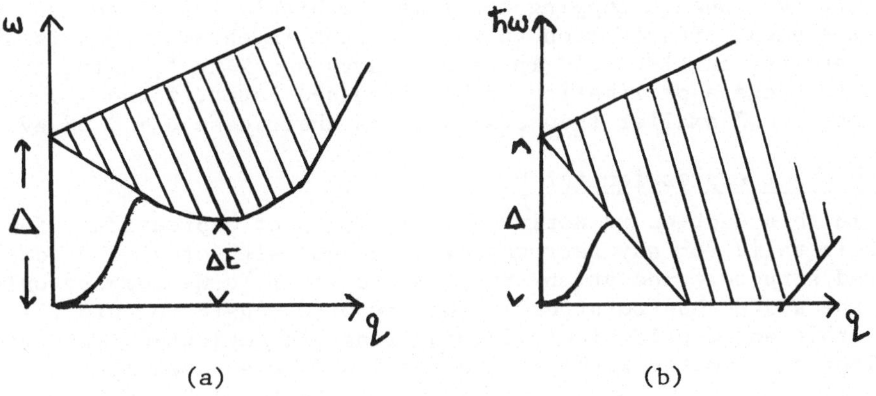

Fig.10 : Spin-flip excitation spectrum for (a) strong (b) weak ferromagnets

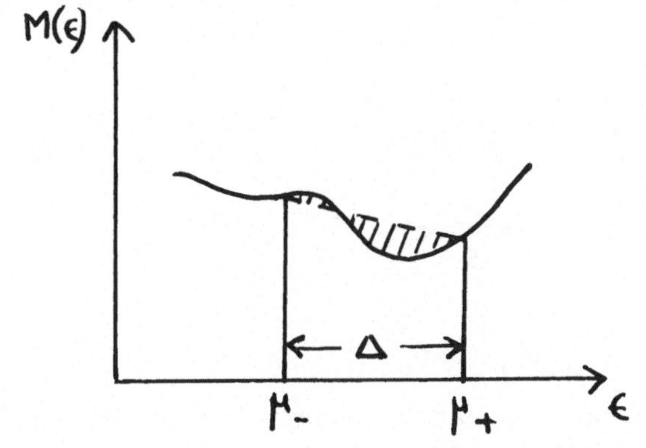

Fig.11 : Schematic M(ε) curve showing the construction for evaluating D.

weak limit. Wakoh found that equation (112) gave D much less then the observed value in Fe, and about one half the observed value in Ni, taking $\Delta = 0.47$ eV. It is necessary to take $\Delta \sim 0.8$ eV to obtain the observed value of D in Ni. Cooke et al found good agreement for Ni and Fe, with a considerably smaller value of Δ for Ni, in a more elaborate calculation assuming arbitrarily that electrons only interact when occupying the same d orbital. The consequent matrix element effects seen to be important in obtaining satisfactory results, particularly in Fe. Callaway and Wang[11], using the local exchange approximation to be discussed later, found a value of D only 10 % smaller than observed. In their case $\Delta = 0.88$ eV.

3.3 An exact expression for D

In this section we derive a formally exact expression[59] for D which is valid for any ferromagnetic system, with ordered or disordered structure and at any temperature $T(< T_c)$. We suppose only that the system has rotational symmetry with regard to spin i.e. spin-orbit and dipole-dipole interactions are neglected. Then the Hamiltonian commutes with S_0^-, the total spin step-down operator, since if ψ is any eigenstate so is $S_0^- \psi$(same total S but S^z reduced by 1) with the same energy E. Thus

$$[H,S_0^-]\psi = HS_0^-\psi - S_0^-H\psi = ES_0^-\psi - S_0^-E\psi = 0 \,. \tag{114}$$

Hence $[H,S_0^-] = 0$. We are interested in the pole $\hbar\omega = Dq^2$, for small q, of the transverse susceptibility $\chi_\perp$ defined by (100). From the equations of motion of the Green function we find

$$\chi_\perp = -\frac{2<S^z>}{\hbar\omega} + \frac{q^2}{\omega^2}\{\chi_J - \frac{1}{\hbar q}<[J_{\underline{q}}^-\,,\,S_{-\underline{q}}^+]>\} \tag{115}$$

where

$$\chi_J(\underline{q},\omega) = \int \langle\langle J_{\underline{q}}^-(t),\,J_{-\underline{q}}^+\rangle\rangle e^{-i\omega t} \tag{116}$$

with

$$\hbar q J_{\underline{q}}^- = [S_{\underline{q}}^-,H] \tag{117}$$

Thus $J_{\underline{q}}^-$ is a spin current operator, equation (117) being the Fourier transform of an equation of continuity. For q = 0 we have only the first term on the right of (115) which may be regarded as the spin wave pole for q = 0. Clearly it exhausts the sum rule (103) and, for small q, $\chi_\perp$ must be dominated by the spin wave pole that we know phenomenologically must exist. Thus for small q

$$\chi_\perp(\underline{q},\omega) = -\frac{2<S^z>(1 + O(q^2))}{\hbar\omega - Dq^2} + O(q^2)$$

$$= -\frac{2<S^z>}{\hbar\omega} - \frac{2<S^z>Dq^2}{\hbar^2\omega^2} \tag{118}$$

as $\omega \to 0$, $q/\omega \to 0$. By comparing (115) and (118) we obtain, to order q^2,

$$Dq^2 = \frac{1}{2<S^z>} \{\hbar q<[J^-_{\underline{q}},S^+_{-\underline{q}}]> - \hbar^2 q^2 \lim_{\omega\to 0} \lim_{q\to 0} \chi_J\} . \tag{119}$$

Clearly, from (115), χ_J contains the spin wave pole, but the limit in (119) is in fact finite because the residue tends to zero as $q \to 0$. Also from (115)

$$\lim_{\omega\to 0} \chi_J(\underline{q},\omega) = \frac{1}{\hbar q} <[J^-_{\underline{q}},S^+_{-q}]> \tag{120}$$

so that if the limits in (119) were reversed the terms in the curly bracket would cancel. Probably in the paramagnetic limit, where the spin wave pole disappears, the order in which the limits $\omega \to 0$, $q \to 0$ are taken is immaterial and the term in curly brackets tends to zero as ferromagnetism disappears. This may be a fundamental reason for D to tend to zero as ferromagnetism approaches paramagnetism, either in an alloy system as the concentration is varied or as $T \to T_c$. However it is not clear that in this limit D is always proportional to the magnetization as was shown to be the case in § 3.2 for a pure metal in RPA.

For a general system of electrons, with Hamiltonian

$$H = \sum_i [\frac{p_i^2}{2m} + V(\underline{r}_i)] + \sum_{i<j}\sum \frac{e^2}{r_{ij}}, \tag{121}$$

the spin density is $\frac{1}{2}\sum\delta(\underline{r} - \underline{r}_i)\sigma_i^-$ so that

$$S^-_{\underline{q}} = \frac{1}{2}\sum_i \exp(i\underline{q}.\underline{r}_i)\sigma_i^- \tag{122}$$

where σ_i^- is a Pauli operator for the ith electron. The only contribution to $[S^-_{\underline{q}},H]$ in (117) comes from the kinetic energy term in H. Then (119) gives

$$D = \frac{\hbar^2}{n_\uparrow - n_\downarrow} \{\frac{n}{2m} - \lim_{\omega\to 0} \chi_J(0,\omega)\} \tag{123}$$

where $n = n_\uparrow + n_\downarrow$ and J_0^- in χ_J is the component of $m^{-1}\ \underline{p}_i\sigma_i^-$ in the $\underline{q}$ direction.

For the one-band model with Hamiltonian (99) we take $S^-_{\underline{q}}$ as given by (102) and this commutes with the second term in the Hamiltonian. Then

$$D = \frac{1}{n_\uparrow - n_\downarrow} \{\frac{1}{2}\sum_{\underline{k}} \frac{\partial^2\varepsilon}{\partial k^2} <n_{\underline{k}\uparrow} + n_{\underline{k}\downarrow}> - \hbar^2 \lim_{\omega\to 0}\chi_J(0,\omega)\} \tag{124}$$

and

$$J_0^- = \sum_{\underline{k}} v_{\underline{k}} a^+_{\underline{k}\downarrow} a_{\underline{k}\uparrow}, \quad v_{\underline{k}} = \hbar^{-1} \frac{\partial \varepsilon}{\partial k} . \tag{125}$$

The derivatives $\partial\varepsilon/\partial k$ and $\partial^2\varepsilon/\partial k^2$ are taken in the $\underline{q}$ direction. For a pure metal we recover the RPA formula (111) when the two particle Green function in χ_J is evaluated in RPA. This means summing the ladder diagrams (109) but actually only the first diagram, corresponding to just the Hartree-Fock approximation (HFA), contributes to χ_J. The other terms cancel on carrying out the $\underline{k}$ summations arising from (125) owing to the fact that $v_{\underline{k}}$ is an odd function. This does not occur in an alloy, where $\underline{k}$ translational symmetry is lost, and then to obtain the full RPA result for D we must calculate χ_J in RPA, not just in the HFA. We see this clearly in the next section, using time-dependent Hartree-Fock theory which is equivalent to the RPA.

3.4 Spin waves in alloys : local exchange approximation

Most existing work on spin waves in transition metal alloys is based essentially on the one-band tight-binding model, taking as its starting point the Hartree-Fock-CPA theory of the ground state due to Hasegawa and Kanamori. We shall return to this model in the next section. However with the development of the more sophisticated KKR-CPA method for the ground state (see Dr. Gyorffy's lectures) it is possible to envisage calculations of D based on this. Ultimately such calculations might be combined with the SDF formalism, for self-consistent determination of the crystal potentials, thus obtaining a theory without adjustable parameters as has been done for pure Ni. As regards actual numerical calculation for alloys we are here looking somewhat into the future. Much of the work in this section and the next was done in collaboration with M.A. Rahman[60] and W.-Y.P. Fung[61].

The procedure we shall follow for calculating D in a disordered system is first to consider a particular configuration of the constituent atoms and then to take a configurational average. In fact we average $D(n_\uparrow - n_\downarrow)$ which is related to a Green function χ_J through equation (119). We then assume $\overline{D(n_\uparrow - n_\downarrow)} = \bar{D}(\overline{n_\uparrow - n_\downarrow})$, which corresponds to the macroscopic relation between D, the magnetization $M \propto n_\uparrow - n_\downarrow$ and the Bloch wall stiffness constant $A \propto DM$.

In the ground state of a particular atomic configuration, using the local exchange approximation, an electron of spin σ moves in a local potential $V_\sigma(\underline{r})$. In the KKR-CPA method for an alloy $V_\sigma(\underline{r})$ would be different within muffin-tin spheres around different types of atom. It is convenient to introduce one-electron energies $\varepsilon_{n\sigma}$ and the corresponding eigenfunctions $|n\sigma\rangle$ although they will not appear in the final result. They satisfy

$$H_\sigma|n\sigma\rangle = \varepsilon_{n\sigma}|n\sigma\rangle \tag{126}$$

where

$$H_\sigma = (\underline{p}^2/2m) + V_\sigma(\underline{r}). \tag{127}$$

In a transverse mode of wavevector $\underline{q}$ the local magnetization $M_0(\underline{r})$ precesses about the equilibrium z direction with a small cone angle $m_{\underline{q}}(\underline{r})$. Thus the deviations of the magnetization vector from equilibrium is of the form $(\delta M_x, \delta M_y, 0)$ where

$$\delta M_+ = \delta M_x + i\delta M_y = M_0(\underline{r}) m_{\underline{q}}(\underline{r}) \exp[i(\underline{q}.\underline{r} - \omega t)] \tag{128}$$

and ω is the angular frequency of the mode.
Consequently the effective one-electron Hamiltonian

$$H_0 = \underline{p}^2/2m + \frac{1}{2}[V_\uparrow(\underline{r}) + V_\downarrow(\underline{r})] + \frac{1}{2}[V_\downarrow(\underline{r}) - V_\uparrow(\underline{r})]\sigma_z \tag{129}$$

acquires a perturbing term

$$\frac{1}{4}V(\underline{r}) m_{\underline{q}}(\underline{r}) \{\exp[i(\underline{q}.\underline{r} - \omega t)]\sigma^- + HC\} \tag{130}$$

where

$$V(\underline{r}) = V_\downarrow(\underline{r}) - V_\uparrow(\underline{r}) = H_\downarrow - H_\uparrow . \tag{131}$$

The perturbed wavefunctions are calculated to first order and used to construct $\delta M_x + i\delta M_y$. On comparing with (128) we obtain the self-consistency condition in the form

$$- \mu_B \sum_{\ell n} \frac{\langle \ell\downarrow | \exp(i\underline{q}.\underline{r}) V(\underline{r}) m_q(\underline{r}) | n\uparrow \rangle}{\varepsilon_{n\uparrow} - \varepsilon_{\ell\downarrow} + \hbar\omega}$$

$$\times (f_{n\uparrow} - f_{\ell\downarrow}) \psi^{\star}_{n\uparrow}(\underline{r}) \psi_{\ell\downarrow}(\underline{r}) \exp(- i\underline{q}.\underline{r})$$

$$= m_{\underline{q}}(\underline{r}) M_0(\underline{r}) \tag{132}$$

where $f_{n\sigma}$ is the groundstate occupation number and $\psi_{n\sigma}(\underline{r})$ is the Schrödinger representation of the state-vector $|n\sigma\rangle$. Equation (132) is equivalent to an equation of the form (16), or (17), without the driving applied field. The frequencies of the modes, which include spin waves, are those values of ω for which equation (132) has a nontrivial solution $m_{\underline{q}}(\underline{r}) \neq 0$.

If we wish to regard (132) as an RPA equation, rather then time-dependent Hartree-Fock, we must construct a many-body Hamiltonian H for which H_0 is the Hartree-Fock one-electron Hamiltonian. Given $V_\uparrow(\underline{r}), V_\downarrow(\underline{r})$, and hence the electron densities $\rho_\uparrow(\underline{r}), \rho_\downarrow(\underline{r})$, this takes the form

$$H = \sum_i [\frac{p_i^2}{2m} + V_0(\underline{r}_i)] + \sum_{i \neq j}\sum \int d^3r F(\underline{r}) \delta(\underline{r} - \underline{r}_i) \delta(\underline{r} - \underline{r}_j) \tag{133}$$

where

$$V_\uparrow = V_0 + F\rho_\downarrow \ , \ V_\downarrow = V_0 + F\rho_\uparrow \tag{134}$$

or

$$V_0 = \frac{\rho_\uparrow V_\uparrow - \rho_\downarrow V_\downarrow}{\rho_\uparrow - \rho_\downarrow} \ , \ F = \frac{V_\downarrow - V_\uparrow}{\rho_\uparrow - \rho_\downarrow} \ . \tag{135}$$

An RPA treatment of the transverse dynamical susceptibility for this Hamiltonian gives equations of the type (17), and the equation corresponding to (16) is just (132).

For a long wavelength spin wave $\hbar\omega = Dq^2$ and $m_{\underline{q}}(\underline{r})$ may be expanded as

$$m_{\underline{q}}(\underline{r}) = 1 + qm_1(\underline{r}) + \ldots \tag{136}$$

since a $q = 0$ spin wave corresponds to uniform precession. (We write q as a scalar, corresponding to $|\underline{q}|$ for q in a given direction, and for a cubic or amorphous crystal D is independent of this direction, which we label x). We find a solution of (132) of this form with

$$D = \frac{\hbar^2}{n_\uparrow - n_\downarrow}\{\frac{n}{2m} - \sum_{\ell n} \langle \ell\downarrow|\frac{p_x}{m} - \frac{V(\underline{r})m_1(\underline{r})}{\hbar}|n\uparrow\rangle\langle n\uparrow|\frac{p_x}{m}|\ell\downarrow\rangle$$

$$\times \ (f_{n\uparrow} - f_{\ell\downarrow})/(\varepsilon_{\ell\downarrow} - \varepsilon_{n\uparrow})\} \tag{137}$$

where $m_1(\underline{r})$ satisfies a certain integral equation. This is of the exact form (123) and the second term in curly brackets corresponds to χ_J evaluated in RPA. Diagrammatically this means summing ladder diagrams and if, as an approximation, we neglect $m_1(\underline{r})$ this corresponds to retaining only the first diagram, without interaction lines, thus neglecting the "vertex correction ". The result is that χ_J is just evaluated in the HFA. Physically, as is clear from equation (136), neglect of $m_1(\underline{r})$ corresponds to ignoring the spatial variation of the precessional angle, or spin wave amplitude. Edwards and Rahman argued that for a pure ferromagnetic metal this neglect of the variation of the precessional angle within a unit cell was a reasonable, possibly even desirable, approximation. This corresponds to a neglect of local field effects. Callaway and Wang[61], who initiated the theory of spin waves within the local exchange approximation, used equation (137) with $m_1(\underline{r}) = 0$ to obtain their rather satisfactory value of D in Ni. Also the property $D \to 0$ as $n_\uparrow - n_\downarrow \to 0$ is preserved. However for an alloy it seems dangerous to ignore the differences between spinwave amplitudes on different atoms, thus neglecting virtual spinwave scattering processes. We shall refer to the contribution to D which arises from the term in equation (137) involving $m_1(\underline{r})$ as the 'magnon scattering" contribution. The contribution to $\overline{D}$ not arising from $m_1(\underline{r})$ corresponds to the average exchange stiffness sampled by a uniform (unscattered) spin wave and we call it the 'average exchange' contribution. For a Heisenberg model of an alloy neglect of $m_1(\underline{r})$ would correspond to the virtual crystal approximation in which the different exchange integrals are replaced by a single average exchange constant. In all existing quantitative work on alloys, from the itinerant electron standpoint, only the 'average-exchange' contribution has been considered. The successful application of the theory to nickel alloys, as discussed in § 3.6, seems to indicate that this is the dominant contribution to D, at least in some cases. In this section therefore, we shall put $m_1(\underline{r}) = 0$ but we return to a discussion of the 'magnon scattering' contribution later within the framework of the disordered Hubbard model. After making this approximation it is easy to express the χ_J term in (137) in terms of the single-particle Green function

$$G_\sigma(\underline{r},\underline{r}';E_+) = \langle\underline{r}|\frac{1}{E - H_\sigma + i\eta}|\underline{r}'\rangle = \sum_\ell \frac{\langle\underline{r}|\ell\sigma\rangle\langle\ell\sigma|\underline{r}'\rangle}{E - E_{\ell\sigma} + i\eta} . \tag{138}$$

The second term in curly brackets in (137) becomes

$$(\frac{\hbar}{m})^2 \frac{1}{\pi} \int_{-\infty}^{E_F} dE \, \mathrm{Im}\iint d^3r \, d^3r' [\frac{\partial}{\partial x} G_\downarrow(\underline{r}',\underline{r};E_+)] [\frac{\partial}{\partial x'} G_\uparrow(\underline{r},\underline{r}';E_+)] . \tag{139}$$

The final step in obtaining a formula for D is to take the configurational average of $(n_\uparrow - n_\downarrow)$ D, given by (137) and (139). In order to express D in terms of configurationally-averaged single-particle Green functions we propose the approximation

$$\overline{(\partial G_\downarrow/\partial x)(\partial G_\uparrow/\partial x')} = \overline{\partial G_\downarrow/\partial x} \; \overline{\partial G_\uparrow/\partial x'} . \tag{140}$$

This neglects the vertex correction which arises from the effective electron-electron interaction induced by averaging out the disorder The equation corresponding to (140) in the simple tight-binding CPA introduces no approximation in the formula for D since the vertex correction makes no contribution to (139)[63,64]. It is not to be expected that this result will carry over exactly in KKR-CPA (B.L. Györffy, private communication), but (140) may be a reasonable first approximation. The formula for D is

$$D = \frac{\hbar^2}{\overline{n}_\uparrow - \overline{n}_\downarrow} \{\frac{n}{2m} + (\frac{\hbar}{m})^2 \frac{1}{\pi} \int_{-\infty}^{E_F} dE \, \mathrm{Im}\iint d^3r \, d^3r' [\frac{\partial}{\partial x} \overline{G_\downarrow(\underline{r}',\underline{r};E_+)}] \times [\frac{\partial}{\partial x'} \overline{G_\uparrow(\underline{r},\underline{r}';E_+)}] \} . \tag{141}$$

Unfortunately the Green functions are only straightforward to calculate within the KKR-CPA method when $\underline{r}$ and $\underline{r}'$ lie within muffin-tin spheres (B.L. Györffy, private communication). Thus further work on the one-electron alloy problem may be required before (141) can be evaluated in practice.

3.5 Spin waves in the disordered Hubbard model

We consider the Hamiltonian (41) in the general case of both diagonal and off-diagonal disorder. This can serve as a simple model of amorphous ferromagnetic transition metals, as well as alloys. In the amorphous case the emphasis is on off-diagonal disorder where the variation of distance between neighbouring atoms is modelled by assuming a probability distribution for values of the hopping integrals t_{ij}. Clearly the RPA formula for D, in a particular configuration, is of the general form (119) with χ_J evaluated in RPA. On taking the configurational average we obtain

$$\overline{(n_\uparrow - n_\downarrow)D} = \lim_{q\to 0} q^{-1}\overline{\langle [J^-_{\underline{q}}, S^+_{-\underline{q}}]\rangle} - \overline{\langle\langle J^-_0(t), J^+_0\rangle\rangle^{HF}_{\omega=0}} + \overline{(n_\uparrow - n_\downarrow)D_{ms}} \quad (142)$$

where the first two terms are the 'average exchange' contribution, with χ_J evaluated in the HFA, and the last term stands for the 'magnon scattering' contribution which is the RPA vertex correction to χ_J^{HF}. This last term is only easy to evaluate in the case of weak scattering where we work to second order in the $\delta t_{ij}, \delta V_i$ and δU_i, these being the deviations of the corresponding parameters in (41) from their average values. In this case we find

$$\overline{(n_\uparrow - n_\downarrow)D_{ms}} = -\frac{U}{N}\sum_{\underline{q}} \frac{1}{1-\frac{U}{N}\Gamma(\underline{q},0)} \overline{|\langle\langle J^-_0(t), S^+_{-\underline{q}}\rangle\rangle^{HF}_{\omega=0}|^2} \quad (143)$$

which vanishes for a pure metal. Here U is the unperturbed interaction parameter and $\Gamma(\underline{q},\omega)$ is the 'non-interacting' susceptibility for the unperturbed system, defined by (108). The unperturbed spin wave energies $\hbar\omega^0_{\underline{q}}$ satisfy the equation $1 - U\Gamma(\underline{q},\omega^0_{\underline{q}})/N = 0$ and a straightforward expansion shows that for small $\underline{q}$

$$[1 - U\Gamma(\underline{q},0)]^{-1} \cong U(n_\uparrow - n_\downarrow)/\hbar\omega^0_{\underline{q}}\ . \quad (144)$$

Clearly then (143), which is the last term of (142), corresponds to virtual magnon scattering processes where the effective scattering matrix element is proportional to the static spin current-spin response function. We note that it is always negative. In considering[61] the difference in D between the amorphous and crystalline states of a ferromagnetic metal it is found that the 'magnon-scattering' contribution is quite important. It is also essential in the percolation limit discussed in § 3.7. However for many transition metal alloys the 'average exchange' contribution may be dominant since it reflects the way in which the electronic structure and occupation of the d band vary with composition.

For a cubic alloy with diagonal disorder only (Hamiltonian (42)), and neglecting D_{ms}, (142) becomes

$$D = \frac{1}{3(\bar{n}_\uparrow - \bar{n}_\downarrow)} \{ \frac{1}{2} \sum_{\underline{k}} \overline{< n_{\underline{k}\uparrow} + n_{\underline{k}\downarrow} >} \nabla^2 \varepsilon_{\underline{k}}$$

$$- \frac{1}{\pi} \int_{-\infty}^{E_F} dE \sum_{\underline{kk}'} \overline{G^{\uparrow}_{\underline{kk}'}(E+) G^{\downarrow}_{\underline{k}'\underline{k}}(E+)} \nabla \varepsilon_{\underline{k}} . \nabla \varepsilon_{\underline{k}'} \}, \quad (145)$$

a result first obtained by Hill and Edwards[64]. If we use the CPA we may put

$$\overline{G^{\uparrow} G^{\downarrow}} = \overline{G^{\uparrow}}\, \overline{G^{\downarrow}} , \quad (146)$$

within the $\underline{k}$-summations, as shown by Velicky[63] in investigating the conductivity. Then (145) may be written as

$$D = [6\pi(\bar{n}_\uparrow - \bar{n}_\downarrow)]^{-1} \int_{-\infty}^{E_F} dE \int d\varepsilon \, M(\varepsilon) \mathrm{Im}[G^{\uparrow}_{\varepsilon}(E+) - G^{\downarrow}_{\varepsilon}(E+)]^2 \quad (147)$$

where

$$G^{\sigma}_{\varepsilon_{\underline{k}}}(E) = [E - \varepsilon_{\underline{k}} - \Sigma_{\sigma}(E)]^{-1} \quad (148)$$

is the one-particle CPA Green function and $M(\varepsilon)$ is given by (113) without the summation over band index. This result was obtained by Fukuyama[65], using a diagrammatic method, and by Hill and Edwards. Riedinger and Nauciel-Bloch[4] considered a multi-orbital model and showed that, if one makes an equivalent orbital approximation, equation (147) is valid when the summation over band index is included in (113). In the work of Fukuyama, and Riedinger and Nauciel-Bloch, no 'magnon-scattering' contribution D_{ms} ever appears. This is because they introduce the CPA at an early stage so that the spin wave is excited in a uniform medium. More specifically Riedinger and Nauciel-Bloch use RPA equations of the form (17) for the transverse dynamical susceptibility and carry out suitable Fourier transformations. They then take the configurational average and factorize terms involving $\chi^0(\underline{q})\chi(\underline{q})$ as $\overline{\chi^0(\underline{q})}\ \overline{\chi(\underline{q})}$. Thus non-interacting susceptibilities are averaged at the outset. In our method the RPA formula for $(n_\uparrow - n_\downarrow)D$ is obtained for a given configuration, and is then averaged to give (142). Finally we mention some model calculations[66] of D in partially ordered alloys. It is found that D is quite sensitive to the degree of long-range order, as observed in Pd_3Fe and Ni_3Fe.

3.6 Numerical calculations of D in nickel alloys

The first numerical calculations of D in alloys to go beyond rigid band theory were those of Hill and Edwards[64], using the CPA-RPA equation (147). However they did not use realistic $M(\varepsilon)$ curves so that their conclusions were only qualitative. They found that in alloys with weak scattering, such as NiCo, the variation of D with composition was a smoothed version of the curve calculated using rigid band theory. Subsequent evaluations of (147), using $M(\varepsilon)$ calculated for a realistic five-orbital Ni d band, have been made by Riedinger and Nauciel-Bloch[4] and Hennion and Hennion[30]. The latter authors also report measurements of D in alloys of Co,Fe, Mn, Cr, V, Mo, Ru with Ni. We refer the reader to their papers for detailed results and merely make a few comments here :

(i) For the strong ferromagnets NiCo, NiFe, NiMn the factor $(n_\uparrow - n_\downarrow)^{-1}$ seems to play a dominant role in the variation of D with the concentration c of impurity. In fact for NiFe it is found experimentally that $D \propto M^{-1}$ quite accurately for $c \leqslant 0.5$. For NiCo the drop of D with increasing x is rather less rapid than M^{-1} and for NiMn much more rapid, even assuming that NiMn remains on the Slater-Pauling curve which it does not. If $D \propto M^{-1}$ it means that the Bloch wall stiffness constant $A \propto DM$ remains essentially constant and that the inertia of the system is merely increased by adding local moments on impurity atoms. To account for M(c) in NiMn alloys Hennion and Hennion use the 'ternary alloy' CPA approach and the change in D(c) from that calculated assuming a strong ferromagnet seems to be largely due to the M^{-1} factor. The calculated D(c) curves for NiCo and NiMn agree rather well with experiment. For NiFe the agreement is comparatively poor. This may be because in the calculations NiFe is a strong ferromagnet for all c, no mechanism being included for the deviation from the Slater-Pauling curve as the invar region is approached for $x > 0.5$. Experimentally in this region D(c) drops sharply in a parabolic manner; this appears to support Wohlfarth's picture of invar as a fairly homogenous weak itinerant ferromagnet with $D \propto M \propto (c_0 - c)^{1/2}$.

(ii) In the alloys NiCr, NiV, NiMo it is found experimentally that $D \propto M$ for all c. In the theoretical calculations too $D \propto M$ although the variation of both D and M is not in very good agreement with experiment. The critical concentrations c_0 are in reasonable agreement with experiment and the main discrepancy is the $(c_0 - c)^{1/2}$ behaviour of D and M in the theory. This is characteristic of a homogeneous CPA effective medium and the different observed behaviour is presumably associated with giant polarization clouds which are known to exist for c near c_0. As mentioned in § 2.2 there is some doubt about the stability of Hennion and Hennion's Hartree-Fock-CPA ground state in these alloys.

(iii) In all the calculations rather large values of the interaction parameters U were used ; thus for pure Ni the exchange splitting was taken as 0.8 eV. SDF calculations[11] give $\Delta = 0.63$ eV and experimentally the quasi-particle splitting seems to be about 0.3 eV. Use of smaller Δ values leads to values of D which are much too small and, for example in the case of Cr, a wrong qualitative behaviour of D(c). The difficulty may be partly due to inadequacy of the model, which gives a rigid exchange splitting in pure Ni. However Callaway and Wang[11], in a much more sophisticated but difficult calculation, find D slightly less then observed with $\Delta = 0.88$ eV. It is difficult to believe that good values of D can be obtained in Ni and Ni alloys with Δ as small as 0.3 eV. The explanation may be as follows[13].

Consider the simplest case of a single Hubbard band (99) with all spins up in the ground state (corresponding to n holes/atom in the Ni d band). Suppose the interaction parameter U is taken so that the Hartree-Fock exchange splitting $\Delta = Un$ corresponds to that calculated in a SDF calculation. The quasi-particle exchange splitting Δ_{eff} is found to be less then Δ, as required, owing to electron-magnon effects in the self-energy $\sum_{\downarrow}(\underline{k},\omega)$. We want to understand why we should use Δ rather than Δ_{eff} in calculating D. The reason is that in a more exact theory than RPA D is given by

$$D = \frac{1}{3n}\left(\frac{1}{2}\sum_{\underline{k}} \nabla^2 \varepsilon_{\underline{k}} - \sum_{\underline{k}} \frac{|\nabla \varepsilon_{\underline{k}}|^2}{\sum_{\downarrow}(\underline{k},\omega = \varepsilon_{\underline{k}})}\right) \tag{149}$$

where the summations are over occupied up spin states. This corresponds to the exact formula (124) with a vertex correction neglected in χ_J. In the RPA formula for D, used in the numerical calculations, $\sum_{\downarrow}$ takes its Hartree-Fock value Δ. When electron-magnon interaction effects are included $\sum_{\downarrow}(\underline{k},\omega = \varepsilon_{\underline{k}})$ takes values somewhat smaller than Δ but larger than the quasi-particle exchange splitting Δ_{eff}. This is because, for a given $\underline{k}$, Δ_{eff} satisfies the Dyson equation

$$\Delta_{eff} = \sum_{\downarrow}(\underline{k},\omega = \varepsilon_{\underline{k}} + \Delta_{eff}) \tag{150}$$

and owing to the ω dependence of $\sum_{\downarrow}$ this is smaller than $\sum_{\downarrow}(\underline{k},\omega = \varepsilon_{\underline{k}})$ which appears in the spin wave formula. (Fig.2 of reference 67 makes this clear). The physical reason for using different exchange splittings in different contexts is that D, related to the Bloch wall stiffness constant, is really a static (ground-state) quantity whereas the quasi-particle spectrum observed in photoemission and other experiments depends on dynamics not contained in the SDF formalism. If the effective U is really as large as SDF calculations predict then the top of the up spin band in SDF calculations for NiFe alloys would sink further and further below the Fermi level as the Fe concentration increases. Thus the deviation from the Slater-Pauling curve as the invar region is approached cannot be due to the continuous onset of weak ferromagnetism. Rather the ground state must change discontinuously, possibly with Fe atoms in an Fe-rich environment suddenly losing or greatly reducing their moments. However the top of quasi-particle band would presumably always remain near E_F owing to the electron-magnon effect as before.

3.7 The percolation limit

It was mentioned in § 2.5 that in alloys like $Cr_{1-x}Fe_x$ and $Au_{1-x}Fe_x$ the onset of ferromagnetism at the critical concentration $x_0 \sim 0.17$ seems to be associated with percolation between Fe atoms. The Fe d band can be modelled by a dilute Hubbard model with hopping only occurring between a fraction x of the atoms. Clearly D should tend to zero as $x \to x_0$ from above since there can be no long wavelength exchange stiffness in the absence of an infinite magnetized cluster. Is this result given correctly by our formulae for D ? To answer this we transform the exact expression (119) for D. We consider the case of a disordered Hubbard model and in fact the same general result also holds for the Heisenberg model. Consider the disordered crystal as lying between two crystal planes $x = \pm L$ and apply periodic boundary conditions. (Finally we may let $L \to \infty$). Then it is possible to derive the following formula for D (Edwards, unpublished) :

$$D = -\frac{\hbar^2}{2\langle S^z\rangle_{ic}} \lim_{\omega\to 0} \mathrm{Re} \int_{-\infty}^{\infty} e^{-i(\omega-i\eta)t} \langle\langle K_{X'}^-(t), K_X^+\rangle\rangle \, dt, \quad (X' \neq X) \qquad (151)$$

Here $\langle S^z\rangle_{ic}$ is the total spin in the infinite cluster, being proportional to $xP(x)$ where $P(x)$ is the fraction of Fe atoms in the infinite cluster. Also K_X^- is a local spin current operator such that

$$J_0^- = -\frac{1}{L} \sum_X K_X^- . \qquad (152)$$

The sum is over planes of atoms labelled by X and for simplicity we consider a simple cubic lattice with unit lattice constant. It may be shown that the expression (151) is independent of X and X' so long as $X \neq X'$; this corresponds to a continuity condition. For the Hubbard model

$$K_X^- = \frac{iL}{\hbar} \sum_{Y_i Z_i} t_{XY_iZ_i, X+1Y_iZ_i} \left(a^\dagger_{XY_iZ_i\downarrow} a_{X+1,Y_iZ_i\uparrow} - a^\dagger_{X+1,Y_iZ_i\downarrow} a_{XY_iZ_i\uparrow} \right) \qquad (153)$$

where the sum over Y_i, Z_i is over all atoms in a given plane X and the hopping integral $t_{XY_iZ_i, X+1Y_iZ_i}$ takes the value t or 0 depending on whether or not both sites XY_iZ_i and $X+1,Y_iZ_i$ are occupied by magnetic atoms. For the dilute Heisenberg model the Hamiltonian is

$$H = -\sum_{i<j}\sum J_{ij} \, \underline{S}_i . \underline{S}_j \qquad (154)$$

where J_{ij} takes the value J or 0 depending on whether or not the nearest neighbour atoms i and j are both magnetic. In this case

$$K_x^- = \frac{i}{\hbar} 2LS \sum_{Y_i Z_i} J_{XY_iZ_i, X+1Y_iZ_i} (S^-_{XY_iZ_i} - S^-_{X+1,Y_iZ_i}). \qquad (155)$$

From (151) it is straightforward to derive the well-known result[68] (for the dilute Heisenberg model)

$$D(x) = \frac{2S}{LxP(x)} \sigma(x). \qquad (156)$$

Here $\sigma(x)$ is the conductance of a classical resistor network in which exchange bonds J_{ij} are replaced by conductances with the same values i.e. J or 0. $\sigma(x)$ is calculated for a cube of side L.

Clearly from equation (151), $D\langle S^z\rangle_{ic} = 0$ if the concentration x is less than the percolation limit since there will be no connection between arbitrary planes X,X' i.e. the exchange stiffness constant A is zero. It may be shown that if the usual formula (124) for D in the Hubbard model is evaluated in RPA, then it is equal to (151) with the Green function there evaluated in RPA. This is only true if the vertex correction in χ_J is retained i.e. the magnon-scattering contribution to D is treated correctly. As before $D\langle S_z\rangle_{ic}$ will vanish so that in principle the RPA can deal correctly with the percolation limit; but one must do better than using CPA.

REFERENCES

1. Transition Metals 1977 : Institute of Physics Conference Series No. 39 pp 623-629.

2. A.V. Gold, J. Low Temp. Physics 16, 3 (1974).

3. D.M. Edwards, Physica 91B, 3 (1977).

4. e.g. R. Riedinger and M. Nauciel-Bloch, J. Phys.F. 5, 732 (1975).

5. J.C. Slater, Phys. Rev. 49, 537, 931 (1936).

6. E.C. Stoner, Proc. Roy. Soc. A165, 372 (1938).

7. G.G. Low, Advances in Physics 18, 371 (1969).

8. J. Mathon, Proc. Roy. Soc. A306, 355 (1968).

9. T.J. Hicks, B.D. Rainford, J.S. Kouvel, G.G. Low and J.B. Comly, Phys. Rev. Lett. 22, 531 (1969).

10. U. von Barth and L. Hedin, J. Phys. C. 5, 1629 (1972).

11. J. Callaway and C.S. Wang, Physica 91B, 338 (1977).

12. O. Gunnarsson, J. Phys. F.6, 587 (1976).

13. D.M. Edwards, Transition Metals 1977 : Institute of Physics Conference Series No. 39 pp 279-281.

14. O.K. Andersen, J. Madsen, U.K. Poulsen, O. Jepsen and J. Kollăr, Physica 86-88B, 249 (1977).

15. J.F. Janak and A.R. Williams, Phys. Rev. 14B, 4199 (1976).

16. M. Shiga and Y. Nakamura, J. Phys. Soc. Japan 26, 24 (1969).

17. E.P. Wohlfarth, J. Phys. C 2, 68 (1969), Phys. Lett. 28A, 569 (1969).

18. Y. Nakamura, M. Shiga and N. Shikazono, J. Phys. Soc. Japan 19, 1177 (1964).

19. M. Shiga, AIP Conf. Proc. No. 18, 463 (1974).

20. M. Hayase, M. Shiga and Y. Nakamura, J. Phys. Soc. Japan 30, 729 (1971).

21. W. Marshall, J. Phys. C. 1, 88 (1968).

21a J. Crangle and G.C. Hallam, Proc. Roy. Soc. A272, 119 (1963).

21b J. Crangle, article in "Electronic Structure and Alloy Chemistry of the Transition Metals" ed. P.A. Beck (Interscience).

22. J. Friedel, Nuovo Cim. Suppt. 8, 287 (1958).

23. J.W. Cable and R.A. Medina, Phys. Rev. B13, 4868 (1976).

24. J.B. Comly, T.M. Holden and G.G. Low, J. Phys. C 1, 458 (1968).

25. C. Demangeat and F. Gautier, J. Phys.C. Met. Phys. Suppl. No.3S 291 (1970).

26. T. Jo and H. Miwa, J. Phys. Soc. Japan 40, 706 (1976).

27. J. Kanamori, J. de Physique 35, C4-131 (1974).

28. H. Hasegawa and J. Kanamori, J. Phys. Soc. Japan 33, 1599 (1972).

29. H. Hasegawa and J. Kanamori, J. Phys. Soc. Japan 33, 1607 (1972).

30. M. Hennion and B. Hennion, J. Phys. F 8, 287 (1978) and to be published. The same value of Δ was used in reference 4. Also B. Hennion, M. Hennion and F. Kajzar, Proc. Symp. IAEA, Vienna 1977.

31. T. Jo, J. Phys. Soc. Japan 40, 715 (1976).

32. J. Kanamori, H. Akai, N. Hamada and H. Miwa, Physica 91B, 153 (1977).

33. G. Frollani, F. Menzinger and F. Sacchetti, Phys. Rev. B11, 2030 (1975).

34. H. Miwa, Progr. Theor. Phys. 52, 1 (1974).

35. F. Brouers, F. Gautier and J. van der Rest, J. Phys. F. 5, 975 (1975).

36. N.F. Mott, Advances in Physics 13, 325 (1964).

37. K. Terakura and J. Kanamori, Progr. Theor. Phys. 46, 1007 (1971).

38. K. Terakura, J. Phys. F. 7, 1773 (1977).

39. K. Terakura, J. Phys. F. 6, 1385 (1976).

40. B.R. Coles, Physica 91B, 167 (1977).

41. T. Kato and J. Mathon, J. Phys. F. 6, 221 (1976).

42. J. Beille, D. Bloch and M.J. Besnus, J. Phys. F. 4, 1275 (1974).

43. R. Alben and E.P. Wohlfarth, Phys. Lett. 49A, 271 (1974).

44. J. Inoue and M. Shimizu, J. Phys. Soc. Japan 42, 1547 (1977).

45. S. Shtrikman and E.P. Wohlfarth, Physica 60, 427 (1972).

46. H. Yamada and E.P. Wohlfarth, Phys. Stat. Sol.(b) 58, K151 (1973).

47. D.M. Edwards, J. Mathon and E.P. Wohlfarth, J. Phys. F. 3, 161 (1973).

48. D.M. Edwards, J. Mathon and E.P. Wohlfarth, J. Phys; F. 5, 1619 (1975).

49. T. Kato and J. Mathon, J. Phys. F. 6, 1341 (1976).

50. A.T. Aldred, B.D. Rainford and M.W. Stringfellow, Phys. Rev. Lett. 24, 897 (1970).

51. J. Beille, D. Bloch and M. Besnus, J. Phys. F. 4, 1275 (1974).

52. E.P. Wohlfarth, article in "Quantum Theory of Atoms, Molecules, Solid State" (Academic Press 1966) p. 485.

53. A.T. Aldred and P.H. Froehle, Intern. J. Magnetism 2, 195 (1972).

54. A.T. Aldred, Intern. J. Magnetism 2, 223 (1972).

55. A.T. Aldred, Phys. Rev. B11, 2597 (1975).

56. J.W. Lynn, Phys. Rev. B11, 2624 (1975).

57. T. Izuyama, D.J. Kim and R. Kubo, J. Phys. Soc. Japan 18, 1025 (1963).

58. S. Wakoh, D.M. Edwards and E.P. Wohlfarth, J. de Physique 32, C1-1073 (1971).

59. D.M. Edwards and B. Fisher, J. de Physique 32, C1-697 (1971).

60. D.M. Edwards and M.A. Rahman, J. Phys. F. 8, 1501 (1978).

61. D.M. Edwards and W.-Y.P. Fung, J. Phys.F., in the press.

62. J. Callaway and C.S. Wang, J. Phys. F. 5, 2119 (1975).

63. B. Velicky, Phys. Rev. 184, 614 (1969).

64. D.J. Hill and D.M. Edwards, J. Phys. F. 3, L162 (1973).

65. H. Fukuyama, AIP Conf. Proc. No. 10, 1127 (1973).

66. I. Takahashi and D.M. Edwards, J. Phys. F., to be published.

67. D.M. Edwards and J.A. Hertz, J. Phys. F. 3, 2191 (1973).

68. e.g. A. Brooks Harris and S. Kirkpatrick, Phys. Rev. B16, 542 (1977).

THE ELECTRONIC AND COHESIVE PROPERTIES OF DISORDERED SIMPLE METALS

R. Evans

H.H. Wills Physics Laboratory, University of Bristol

Bristol BS8 1TL, U.K.

1. INTRODUCTION

Simple metals are those without d or f electrons in the conduction band and whose core levels lie at energies well below the conduction band. For these metals it is reasonable to assume that the core electrons are rigidly fixed to the nuclei so that an elemental metal can be well-represented by a two component system of N ions and NZ 'conduction' electrons where Z is the usual chemical valence. The alkalis, Be, Mg, Zn, Cd, Hg, Al, Ga, In, Sn, Tl and Pb are usually described as simple metals. The outer electronic structure of these metals is characterized by extended s and p like states and, to some extent, can be said to be nearly-free-electron like. The alkaline earths Ca, Sr and Ba have unfilled d bands and their hybridisation with the nearly-free s-p bands has significant effects on the electronic structure in the neighbourhood of the Fermi energy. Consequently, the alkaline earths can be considered intermediate between simple and transition metals. The latter, of course, have unfilled d states at the Fermi energy.

For simple metals it is often possible to model the interaction between a conduction electron and an ion by a weak pseudopotential. This simplification implies that the electronic structure, various electronic and electrical properties and the cohesive properties of elemental, perfect crystalline metals can be usefully calculated using low-order perturbation theory with the pseudopotential as an expansion parameter. The interaction between the conduction electrons and screening of the ions can be straightforwardly incorporated into such calculations. While the pseudopotential model proved extremely fruitful for understanding the band structure and the measured Fermi surfaces of simple metals (e.g. Harrison 1966,

Cohen and Heine 1970) its usefulness as an ab-initio scheme for calculating the electronic structure of perfect crystals has been largely superseded by what are essentially selfconsistent field theories which treat the core and conduction electrons on the same footing and solve the appropriate one-electron Schrödinger equation using non-perturbative methods (e.g. Slater 1974). These theories do not involve pseudopotentials and can be applied to transition and noble metals as well as simple metals. They have also been used to calculate certain cohesive and electronic properties for perfect crystalline metallic elements and ordered intermetallic compounds.

Unlike the pseudopotential approach the self-consistent field theories are not readily amenable to the study of disordered metals. It is only the imposition of periodic (Bloch) boundary conditions which permits the solution of the appropriate one-electron equation and thereby allows one to make progress with these theories. For a defected crystal, a crystal with a surface, a crystal undergoing lattice vibrations, a disordered alloy or a liquid metal, Bloch's theorem does not apply and it is usually difficult to produce even a sensible approximate solution of the self-consistent field equations. On the other hand, we shall show in these lectures that provided we work to low (usually second) order in the pseudopotential, the problems introduced by disorder are quite tractable and it is possible to examine the physical properties of a wide variety of disordered systems using the pseudopotential approach.

Since we cannot do justice to all aspects of this far-ranging subject in a few lectures we have deliberately chosen to describe only a few applications of the pseudopotential model. We will be primarily concerned with the cohesive, structural and thermodynamic properties of simple metals. The notes are arranged as follows : in § 2 we introduce the concept of a weak pseudopotential and in § 3 we describe the calculation of the electronic structure of perfect crystals, disordered crystalline alloys and liquid metals using screened pseudopotentials and low order perturbation theory. In § 4 we formulate the pseudopotential theory of energetics and outline the application of this to perfect crystals at temperature T = OK, the thermodynamics of crystals at finite temperature, liquid metals and static crystals with point defects. The limitations of this (linear response) treatment for point defects and surfaces are described. The theory is re-formulated in a real-space representation in § 5 and the effective pairwise inter-ionic potentials which appear in this formulation are discussed in some detail. Recent applications of such pairwise potentials to the calculation of the thermodynamics and structure of liquid metals are described. A discussion of defect energetics in terms of pairwise potentials is also included in this section. In § 6 we extend the theory to binary alloys and consider order-disorder problems for both static and vibrating crystals. Work on the structural and thermodynamic properties of liquid alloys is also described. The question of the validity of low order perturbation theory for alloys is considered. Finally in §7 we mention some recent work on the energetics of point defects and some new ideas concerning the theoretical determination of crystal structures.

We stress that these are lecture notes. It is not our intention to present a comprehensive review of any particular topic. Where it is convenient we refer the reader to relevant review articles rather than original papers.

2. PSEUDOPOTENTIAL TREATMENT OF THE CONDUCTION ELECTRON-ION INTERACTION

The standard one-electron theory of pure metals assumes that the total potential experienced by a single electron can be written as a sum of contributions from each ion :

$$V(\underline{r}) = \sum_{\ell=1}^{N} v(\underline{r} - \underline{R}_\ell) \tag{2.1}$$

where N is the total number of ions and $\underline{R}_\ell$ denotes the position of the ℓth ion. The electronic structure of the metal is then determined by solving the Schrödinger equation :

$$H|\psi_n\rangle = E_n|\psi_n\rangle \tag{2.2}$$

where

$$H = -\nabla^2 + V(\underline{r}) \tag{2.3}$$

and the eigenfunctions ψ_n and eigenvalues E_n refer to a particular static configuration of the ions. For a perfect crystal the $\underline{R}_\ell$ refer to perfect lattice sites and $E_n \equiv E(\underline{k})$, $\psi_n \equiv \psi_{\underline{k}}(\underline{r})$ where $\underline{k}$ is the Bloch wavevector. The solution of (2.2) then – constitutes the 'band structure problem'. For a disordered system we are required to solve (2.2) for various configurations of the ions for which $\underline{k}$ is not a good quantum number.

In all problems we need to define exactly what is meant by the one-electron potential associated with each ion and this is discussed at more length in Appendix 1. For the present we can think of V as a Hartree-like selfconsistent potential. In general $V(\underline{r})$ will exhibit screened Coulomb behaviour outside the ion cores due to screening by other conduction electrons. Within each ion core V will be strongly attractive.

The valence (conduction) electron wavefunctions oscillate rapidly in the ion cores in order to orthogonalize themselves to the atomic-like core eigenfunctions. The valence eigenvalue spectrum, however, does not depend on the details of these oscillations ; it is determined by matching logarithmic derivatives of the wave function at some suitable core radius. Thus if one is only concerned with the valence eigenvalues the true potential V can be replaced by a pseudopotential V^{ps}_{sc} (the subscript sc implies suitable screening) which is defined to be equal to V outside the ion cores in the interstitial region, but which is much weaker than V inside the cores. V^{ps}_{sc} must have the same logarithmic derivatives at the core radius and hence the same valence eigenvalues as V, but no eigenstates of lower energy. The corresponding pseudowavefunctions will then be smooth functions with no nodes in the cores. If the resultant pseudopotential is everywhere 'small' then it should be possible to treat it as a weak perturbation acting on an otherwise free-electron gas. This obviously leads to considerable simplifications.

There are many different techniques for constructing pseudopotentials, see e.g. Harrison (1966), Cohen and Heine (1970) and we will not describe these in any detail here. Almost all prescriptions involve fitting to experimental data of one kind or another. (An exception is the powerful and elegant prescription of Rasolt and Taylor (1975)). The screened pseudopotential associated with each ion v^{ps}_{sc} is not necessarily a local function of $\underline{r}$. All that is required of V^{ps}_{sc} is that its eigenvalues are the same as the valence eigenvalues of the true potential. Consequently V^{ps}_{sc} is an operator which may depend on angular momentum or energy or may exhibit other non-local behaviour. Indeed any operator of the form

$$W^{ps} = V + \sum_c |\psi_c\rangle\langle F_c| \qquad (2.4)$$

is a valid pseudopotential (Austin, Heine and Sham 1962). Here V is the true potential and the summation runs over all core eigenfunctions ψ_c. The F_c are completely arbitrary functions. It is straightforward to show that

$$(-\nabla^2 + W^{ps})|\psi_n^{ps}\rangle = E_n|\psi_n^{ps}\rangle \qquad (2.5)$$

where the eigenvalues are identical to the valence eigenvalues of the true potential (see (2.2)) provided the true and pseudowavefunctions are related by

$$|\psi_n\rangle = |\psi_n^{ps}\rangle - \sum_c |\psi_c\rangle\langle\psi_c|\psi_n^{ps}\rangle \qquad (2.6)$$

W^{ps} is clearly non-local. An alternative prescription is to construct a 'bare' model pseudopotential with a simple functional form inside the ion core and a Coulomb tail outside. The parameters of the model are adjusted so that the model wavefunction matches smoothly to the true wavefunction at the core radius. This 'bare' pseudopotential is then screened in a linear approximation (Appendix 1). It is often necessary to have different model potentials for different angular momenta and energies (Abarenkov and Heine 1965, Shaw 1968). Clearly the more complicated the pseudopotential the less useful it becomes.

For most of the simple metals it is possible to construct weak pseudopotentials characterized by a small number of parameters, and often v^{ps}_{sc} can be treated as a local function of $\underline{r}$. For noble and transition metals the situation is much more complicated. Any pseudopotential for these metals must be extremely energy dependent and 'strong' so that there is no great advantage in working with pseudowavefunctions etc. This is most easily expressed in the language of scattering theory. If we approximate the true potential by a muffin-tin model i.e. assume V is spherically symmetrical inside a given radius which should be greater than that of the ion core, and average the potential outside to some constant value, E_0, we can describe the scattering properties of each ion (muffin-tin) in terms of phase shifts. In most of the simple metals the phase shifts are small modulo π for all angular momenta. This means that the core of the ion constitutes only a small scattering cross-section for the valence electrons throughout the relevant range of energies. Such behaviour is easily modelled by a weak pseudopotential. In the noble metals the d phase shift exhibits resonant behaviour i.e. it increases rapidly from ~ 0 to $\sim \pi$ at an energy of a few eV above the energy zero E_0. In transition metals similar resonant behaviour occurs at higher energies. The scattering cross-sections are very large at resonant energies and cannot be evaluated using perturbation theory. Any attempt to model such resonances in terms of a simple pseudopotential rapidly runs into difficulties and it is usually better to work directly with the true (strong) potential or equivalently the logarithmic derivatives or phase shifts of the latter (e.g. Ziman 1970). We should note that the distinction between simple and non-simple metals is not precise. The alkaline earths Ca, Sr and Ba have excited atomic d states which are close to the s^2 ground state. In the metallic state these elements are characterized by fairly large d phase shifts at energies close to the Fermi energy (Ratti and Evans 1973) and a pseudopotential treatment of these is not straightforward. Under high pressure Cs becomes 'a transition-metal' i.e. the phase shifts for energies below the Fermi energy are similar to those obtained for early transition metals (Stocks et. al 1972). Several of the four and five valent metals have fairly large and energy dependent s and p phase shifts and it is difficult to construct reliable, weak pseudopotentials for these elements.

For simplicity we will ignore complications of non-locality etc. and assume that the screened pseudopotential of a single ion is a local function of $\underline{r}$ so that the total pseudopotential can be written as (see Appendix 1) :

$$V^{ps}_{sc}(\underline{r}) = \sum_{\ell=1}^{N} v^{ps}_{sc}(\underline{r} - \underline{R}_\ell) \tag{2.7}$$

We will further assume that $v^{ps}_{sc} \equiv v^{ps}_{sc}(r)$ is spherically symmetrical.

3. ELECTRONIC STRUCTURE IN TERMS OF PSEUDOPOTENTIALS

3.1 Perfect Crystals

Provided V_{sc}^{ps} is everywhere weak it is reasonable to expand the pseudo Bloch function in plane waves :

$$\psi_{\underline{k}}^{ps}(\underline{r}) = \sum_{\underline{g}} \alpha_{\underline{k}-\underline{g}} \exp(i(\underline{k}-\underline{g}).\underline{r}) \tag{3.1}$$

and presume that only a few coefficients $\alpha_{\underline{k}-\underline{g}}$, where $\underline{g}$ is a reciprocal lattice vector, will make a significant contribution. The band structure is then given by solving

$$\det|((\underline{k}-\underline{g})^2 - E(\underline{k}))\delta_{\underline{g},\underline{g}'} + \Omega^{-1}V_{sc}^{ps}(\underline{g}' - \underline{g})| = 0 \tag{3.2}$$

where

$$\begin{aligned} V_{sc}^{ps}(\underline{g}' - \underline{g}) &= \int d\underline{r} \exp(i(\underline{g}'-\underline{g}).\underline{r})\, V_{sc}^{ps}(\underline{r}) \\ &= v_{sc}^{ps}(|\underline{g}'-\underline{g}|)S(\underline{g}'-\underline{g}) \end{aligned} \tag{3.3}$$

and Ω is the total volume. $v_{sc}^{ps}(q)$ is the screened form factor of a single ion defined by

$$v_{sc}^{ps}(q) = \int d\underline{r} \exp(i\underline{q}.\underline{r})v_{sc}^{ps}(r) \tag{3.4}$$

and $S(\underline{q})$ is the structure factor :

$$S(\underline{q}) = \sum_{\ell=1}^{N} \exp(i\underline{q}.\underline{R}_{\ell}) \tag{3.5}$$

For a perfect lattice this vanishes unless $\underline{q}$ is a reciprocal lattice vector i.e. $S(\underline{q}) = N\delta_{\underline{q},\underline{g}}$. Thus the band structure is determined by the magnitude of the form factor at the first few reciprocal lattice vectors. For many simple metals this description of the band structure appears to be quite accurate - at least for energies near the Fermi energy. Indeed the form factors are often fitted to measured Fermi surface data or electrical transport coefficients (the latter depend on integrals of the form factor over the Fermi surface) ; see Cohen and Heine (1970) for details. For strong - scattering metals where pseudopotential theory is inappropriate one must solve the Schrödinger equation (2.2) involving the true potential. Various well-known techniques are available e.g. the APW and KKR methods. All methods treat the scattering problem at a single ion (muffin-tin) exactly and then solve the appropriate multiple scattering problem. The first step entails the calculation of the phase shifts (or the single-site transition matrix) corresponding to the potential v while the second is possible because of the imposition of periodic boundary conditions.

3.2 Disordered Metals

Since we cannot define an $E(\underline{k})$ relation for a disordered metal it is better to focus attention directly on the configuration averaged density of states $n(E)$ rather than on wavefunctions and eigenvalues. (Throughout this section we have in mind disordered crystalline alloys and liquid metals).

We define the Green operator, corresponding to the Schrödinger equation (2.2), for arbitrary complex E as

$$G(E) = (E - H)^{-1} = \sum_n \frac{|\psi_n\rangle\langle\psi_n|}{E - E_n} \tag{3.6}$$

The spectral operator is then defined by

$$\rho(E) = -\frac{1}{2\pi i} \lim_{\eta\to 0} (G(E + i\eta) - G(E - i\eta)) = \sum_n |\psi_n\rangle\langle\psi_n| \, \delta(E - E_n) \tag{3.7}$$

where E is now real. It is convenient to use the momentum representation for these operators since after averaging they will either have the full symmetry of the lattice (crystalline alloys) or be translationally invariant (liquid metal). We denote by $|\underline{k}\rangle$ a suitable configuration independent set of states and introduce the spectral function

$$\rho(\underline{k},E) = \langle\underline{k}|\rho(E)|\underline{k}\rangle = -\frac{1}{\pi} \text{Im } G(\underline{k},E^+) = \sum_n |\langle\underline{k}|\psi_n\rangle|^2 \, \delta(E - E_n) \tag{3.8}$$

where $G(\underline{k},E) = \langle\underline{k}|G(E)|\underline{k}\rangle$ and $E^+ = E + i0$. For a perfect crystal ψ_n is a Bloch function and $\rho(\underline{k},E)$ is an array of delta functions in reciprocal space separated by reciprocal lattice vectors. The relationship between E and the positions of these delta functions defines the band structure $E(\underline{k})$. In a disordered system $\rho(\underline{k},E)$ is generally a continuous function of $\underline{k}$ for a given E and no precise $E(\underline{k})$ can be defined.

For each $\underline{k}$ the spectral function satisfies the sum rule

$$\int_{-\infty}^{\infty} dE\rho(\underline{k},E) = \sum_n |\langle\underline{k}|\psi_n\rangle|^2 = 1 \tag{3.9}$$

since the $|\psi_n\rangle$ form a complete set.

The configuration averaged density of states per ion is

$$n(E) = N^{-1} \langle\sum_n \delta(E - E_n)\rangle \tag{3.10}$$

which, from (3.8) is

$$n(E) = N^{-1} \sum_{\underline{k}} \langle\rho(\underline{k},E)\rangle \tag{3.11}$$

We see that any calculation of the density of states requires some prescription for $\mathcal{G}(E) = \langle G(E)\rangle$.

The standard approach uses perturbation theory i.e. we divide the Hamiltonian into a reference part H_0 plus a perturbation V_1 :

$$H = H_0 + V_1 \tag{3.12}$$

H_0 is taken to be configuration independent but the choice depends on the particular problem. For example in a crystalline alloy with a small concentration of impurities H_0 could correspond to the perfect pure crystal while for a concentrated alloy H_0 might refer to a perfect crystal in which some suitable average of the different ionic potentials is located at every lattice site. For a liquid metal, however, it is difficult to think of any choice other than $H_0 = -\nabla^2$, the kinetic energy operator.

In general we have

$$\begin{aligned} \mathcal{G}(E) &= \langle(E - H_0 - V_1)^{-1}\rangle \\ &= G_0 + \langle G_0V_1G_0\rangle + \langle G_0V_1G_0V_1G_0\rangle + \ldots \end{aligned} \tag{3.13}$$

where $G_0 = (E - H_0)^{-1}$ is configuration independent. It is useful to introduce an invariant function, the irreducible self-energy operator $\sum(E)$ defined by

$$\mathcal{G}(E) = (G_0(E)^{-1} - \textstyle\sum(E))^{-1} \tag{3.14a}$$

Since $\mathcal{G}, G_0$ and $\sum$ are diagonal in momentum representation this is equivalent to ★

$$\mathcal{G}(\underline{K},E) = (G_0(\underline{k},E)^{-1} - \textstyle\sum(\underline{k},E))^{-1} \tag{3.14b}$$

$\sum(E)$ is a complex quantity which measures the energy shift arising from disorder and describes the damping of the electronic states. If $\sum(E)$ is calculated in some approximate fashion by summing

★ This is valid for a single band situation.

a certain set of diagrams in a perturbation expansion the $G(E)$ obtained from (3.14) automatically contains the infinite sum of 'reducible' terms associated with the particular 'irreducible' diagrams contained in $\Sigma(E)$. The leading terms in the perturbation expansion for $\Sigma(E)$ are given by

$$\begin{aligned}\Sigma(E) &= G_0(E)^{-1} - G(E)^{-1} \\ &= E - H_0 - [G_0(1 + \langle V_1\rangle G_0 + \langle V_1 G_0 V_1\rangle G_0 + \ldots)]^{-1} \\ &= \langle V_1\rangle + (\langle V_1 G_0 V_1\rangle - \langle V_1\rangle G_0 \langle V_1\rangle) + \mathcal{O}(V_1^3) \qquad (3.15)\end{aligned}$$

To lowest order the electronic structure of the disordered system is to be calculated using the configuration independent Green operator at a shifted energy $E - \langle V_1\rangle$. This approximation does not allow for any broadening of the electronic states due to disorder i.e. the averaged spectral function

$$\langle\rho(\underline{k},E)\rangle = -\frac{1}{\pi}\,\mathrm{Im}\, G(\underline{k},E^+) \qquad (3.16)$$

still exhibits delta function behaviour. This implies that the decay time of an electron in state $|\underline{k}\rangle$, which is inversely proportional to the width of the spectral function (e.g. Ehrenreich and Schwartz 1976), is infinite in this order of perturbation theory. The incorporation of second order terms gives a finite width to the spectral function and ensures that the lifetime of the states is finite.

The configuration averaging which is involved in the second order term can usually be performed. In the case of a pure liquid metal we take $H_0 = -\nabla^2$, so that G_0 is the free particle propagator and $V_1 = V$, and find that $\langle V_1 G_0 V_1\rangle$ involves the usual pairwise distribution function for the ions in the liquid. This function can be extracted from diffraction experiments or is readily modelled in various ways (see § 5.3). Higher order terms in the expansion of $\Sigma(E)$ require the higher order distribution functions of the liquid. As these are not available from experiment and cannot be accurately modelled in any straightforward manner, a theory of this kind rapidly becomes intractable if the series is not rapidly convergent. Indeed if V is the true (strong) electron - ion potential then (3.13) or (3.15) will not converge and the series must be formally re-summed to infinite order using scattering theory approaches (e.g. Ballentine 1975). If, on the other hand, we replace V by V^{ps}_{sc} and the latter quantity is, in some sense, weak it is reasonable to concentrate on the leading terms in the perturbation series. We then find (Ballentine 1975)

$$\Sigma(\underline{k},E) = N v^{ps}_{sc}(q=0) + \frac{N}{(2\pi)^3}\int d\underline{q}\,\frac{|v^{ps}_{sc}(q)|^2 a(q)}{E - (\underline{k}+\underline{q})^2} \qquad (3.17)$$

where a(q) is the liquid structure factor which is given as a configuration average :

$$a(q) = N^{-1} < \sum_{\ell,\ell'} \exp(i\underline{q}.(\underline{R}_\ell - \underline{R}_{\ell'}))> - N\delta_{q,0} \tag{3.18}$$

The first term in (3.17) is simply the average pseudopotential which can be eliminated by changing the energy scale. It is clear that n(E) departs from free electron behaviour by terms depending on $|v^{ps}_{sc}|^2$ a(q). The form factor is largest for q = 0 (Appendix 1) but here a(q) is very small. Conversely, in many simple metals, $|v^{ps}_{sc}(q)|^2$ is small for q near the principal maximum of the structure factor. Thus the product tends to be small for all values of q and the second order perturbation theory predicts very small ($\sim$ few percent) deviations from free electron behaviour. This is confirmed by the results of detailed calculations based on a more sophisticated approximation for $\sum(E)$ which replaces the denominator in (3.17) by $E - (\underline{k} + \underline{q})^2 - \sum(\underline{k} + \underline{q}, E)$ (Ballentine 1975). The approximation corresponds to replacing the free particle propagator G_0 by the 'full' propagator G and can be viewed as a properly self-consistent second order perturbation theory.

Recent photoemission spectroscopy experiments on liquid In, Al, Sn and Hg (Norris 1977) cannot easily be reconciled with an almost free-electron density of states and in liquid Ga, Tl, Pb and Bi there is evidence for pronounced structure in n(E) at energies below the Fermi energy (Wotherspoon 1977). The interpretation of photoemission data is, of course, a difficult and delicate business (n(E) is not measured directly) and it would be premature to abandon the pseudopotential approach on this evidence alone - especially when we consider how fruitful this approach has proved for understanding the electrical transport properties of simple liquid metals (e.g. Faber 1972). We might speculate that in these polyvalent metals states at energies below the Fermi energy are not well-described by simple pseudopotential theory, while those at, or close to the Fermi energy, are. This would be the case if the s and p phase shifts of the true potential showed strong variation with energy at low energies but remained fairly constant at higher energies. As mentioned in § 2 there is some evidence for such behaviour in the polyvalent metals.

For strongly scattering ions (especially the noble and transition metals) it is necessary, as in the case of the perfect crystal, to work with the transition operation t rather than the ionic potential v. The self-energy $\sum(E)$ can be expanded in powers of t (Ballentine 1975) and one proceeds, in principle, as above by seeking to truncate this multiple scattering expansion at low order using self-consistency arguments. Various theories exist (see e.g. the review by Watabe 1977) but there are few calculations based on realistic models of the liquid metal so it is difficult to ascertain what progress has been achieved. It is still a major challenge for theorists to produce a reliable and tractable approximation scheme for n(E) in a strong-scattering, topologically disordered metal.

As we mentioned earlier, in disordered crystalline alloys where we have an underlying periodicity, there are more options available for the choice of a reference Hamiltonian. We will not discuss this matter in any detail since the subject will be covered in Dr. Györffy's lectures. It suffices to say that for random, substitutional alloys the coherent potential approximation (CPA) has proved highly successful for the calculation of the electronic structure. The CPA is applicable to both strong and weak scattering situations since it is based on a t matrix approach. It is, however, strictly a single-site approximation which ignores all correlations between sites and cannot easily be extended to discuss ordering effects. Here we concentrate on binary alloys of simple metals and develop the pseudopotential model for these systems.

We assume a static lattice with an alloy Hamiltonian of the form

$$H = H_0 + V_1$$

where

$$V_1(\underline{r}) = \sum_A v_{sc}^A(|\underline{r} - \underline{R}_A|) + \sum_B v_{sc}^B(|\underline{r} - \underline{R}_B|) \tag{3.19}$$

where $H_0 = -\nabla^2$, v_{sc}^A is the screened pseudopotential of a single A ion and v_{sc}^B the corresponding quantity for a B ion. Provided both v_{sc}^A and v_{sc}^B are 'weak' then the total pseudopotential V_1 will be 'small' and we might usefully calculate the self-energy by computing the leading terms of (3.15). The first order contribution is

$$\langle V_1\rangle_c = \sum_{\ell=1}^{N} (c\, v_{sc}^A + (1-c)v_{sc}^B) = \sum_{\ell=1}^{N} \bar{v}_{sc} \tag{3.20}$$

where $\langle\ \rangle_c$ denotes a configuration average, c is the atomic concentration of species A and the summation runs over all lattice sites. $\bar{v}_{sc}$ is referred to as the virtual crystal potential. The second order contribution can also be simplified and we find in momentum representation :

$$\Sigma(\underline{k},E) = N\bar{v}_{sc}(q=0) + \frac{1}{\Omega}\sum_{\underline{q}\neq 0} \frac{\langle|V_1(\underline{q})|^2\rangle_c}{E - |\underline{k}+\underline{q}|^2} \tag{3.21}$$

where

$$\begin{aligned} V_1(\underline{q}) &= \int d\underline{r}\, \exp(i\underline{q}.\underline{r})V_1(\underline{r}) \\ &= v_{sc}^A(q)S_A(\underline{q}) + v_{sc}^B(q)S_B(\underline{q}) \end{aligned} \tag{3.22}$$

and we have introduced a structure factor for each species :

$$S_A(\underline{q}) = \sum_A \exp(i\underline{q}.\underline{R}_A) \qquad S_B(\underline{q}) = \sum_B \exp(i\underline{q}.\underline{R}_B)$$

$V_1(\underline{q})$ is more conveniently expressed in terms of the virtual crystal potential and the total structure factor $S(\underline{q})$:

$$V_1(\underline{q}) = \bar{v}_{sc}(q)S(\underline{q}) + (v^A_{sc}(q) - v^B_{sc}(q))((1-c)S_A(\underline{q}) - cS_B(\underline{q})) \quad (3.23)$$

For a Bravais lattice it follows that if $\underline{q} = \underline{g}$, a reciprocal lattice vector, then $\exp(i\underline{q}.\underline{R}_\ell) = 1$ for all lattice vectors so that $S(\underline{q}) = N$, $S_A(\underline{q}) = cN$ and $S_B(\underline{q}) = (1-c)N$. $V_1(\underline{q})$ then reduces to $N\bar{v}_{sc}(g)$. If $\underline{q}$ is not a reciprocal lattice vector then $S(\underline{q}) = 0$ but $S_A(\underline{q})$ and $S_B(\underline{q})$ are non-zero. Consequently (3.21) can be written as

$$\sum(\underline{k},E) = N\bar{v}_{sc}(q=0) + \frac{N^2}{2\Omega} \sum_{\underline{g}\neq 0} \frac{|\bar{v}_{sc}(g)|^2}{E - (\underline{k}+\underline{g})^2}$$

$$+ \frac{1}{(2\pi)^3} \int d\underline{q} \frac{|v^A_{sc}(q) - v^B_{sc}(q)|^2}{E - (\underline{k}+\underline{q})^2} \langle |(1-c)S_A(\underline{q}) - cS_B(\underline{q})|^2 \rangle_c \quad (3.24)$$

The first term merely shifts the energy scale while the second term is the second order perturbation theory contribution from the perfect lattice in which the virtual crystal potential is located at each site. The third term contains all information regarding the distribution of A and B ions in the alloy. This only depends on the difference between the ionic pseudopotentials i.e. $|v^A_{sc} - v^B_{sc}|$.

The configuration average is easily performed for a completely random alloy (Harrison 1966, Inglesfield 1969a) and reduces to $Nc(1-c)$. Ordering effects can, however, be straightforwardly incorporated into the theory by introducing suitable order parameters (e.g. the Bragg-Williams long-range order parameter) - see § 6.2.

One obvious improvement can be made to the above theory and that is to incorporate the virtual crystal potential into the reference Hamiltonian i.e. define

$$H_0 = -\nabla^2 + \sum_{\ell=1}^{N} \bar{v}_{sc} \quad (3.25)$$

so that the corresponding perturbation potential is now

$$V_1 = (1-c)\sum_A (v^A_{sc} - v^B_{sc}) - c\sum_B (v^A_{sc} - v^B_{sc}) \quad (3.26)$$

The Green operator G_0 can be calculated exactly, without recourse to perturbation theory, by computing the band structure of the virtual crystal. The self-energy then takes the form of an expansion in $v^A_{sc} - v^B_{sc}$. The linear term vanishes since $\langle V_1 \rangle_c = 0$ while the second order contribution depends on $|v^A_{sc} - v^B_{sc}|^2$ and involves a configuration average of the same kind as in (3.24) (Stern 1966). A perturbation theory of this kind will be valid provided $|v^A_{sc} - v^B_{sc}| \ll$ band width of the virtual crystal.

These examples should illustrate how pseudopotential theory allows several problems, which would otherwise appear intractable, to be tackled in a realistic fashion. In particular it allows one to perform the configuration averages which play an integral role in disordered systems.

Our discussion has been based on one-electron theory. We have assumed the existence of effective one-electron potentials and screened pseudopotentials. Such potentials will depend on the distribution of conduction electrons in the metal ; the potential due to a single Na ion in liquid Na will be different from that in crystalline Na and the potential of a Cu ion in CuZn alloys will, in general, depend on the concentration and distribution of the ions. Any proper theory of metals must contain a prescription for the conduction electron screening. Consequently the problems of the electron-electron interaction, self-consistency etc. must be examined in more detail. These topics are discussed in Appendix 1. In the next section we develop the theory of the cohesive properties of simple metals within the pseudopotential framework. This also makes use of some of the results of Appendix 1.

4. THEORY OF THE COHESIVE PROPERTIES OF SIMPLE METALS

For simplicity we consider a pure metal at a temperature T. We will generalize to a binary alloy in § 6.

4.1 From a two-component to a quasi one-component system

We suppose that the metal has a volume Ω and contains N ions of valence Z so that the total number of conduction electrons is NZ. The basic Hamiltonian for the system of (pseudo) ions and conduction electrons is

$$\mathcal{H} = \underbrace{\sum_i \frac{p_i^2}{2m}}_{H_e} + \underbrace{\frac{e^2}{2}\sum_{i\neq j}\frac{1}{|\underline{r}_i-\underline{r}_j|}}_{H_{ee}} + \underbrace{\sum_{i,\ell} v^{ps}(|\underline{r}_i-\underline{R}_\ell|)}_{H_{ei}} + \underbrace{\sum_\ell \frac{P_\ell^2}{2M}}_{H_k} + \underbrace{\frac{1}{2}\sum_{\ell\neq\ell'} W(|\underline{R}_\ell-\underline{R}_{\ell'}|)}_{H_{ii}} \qquad (4.1)$$

where $\{\underline{r}_i\}$ and $\{\underline{R}_\ell\}$ denote electronic and ionic coordinates, $\{p_i\}$ and $\{P_\ell\}$ the corresponding momenta and m and M the masses. v^{ps} is the bare pseudopotential of a single ion. For simplicity it is assumed to be a local function of r. Inside the core it is assumed small and outside it will behave like $-Ze^2/r$. The parameters of v^{ps} should be chosen so that it has the same valence eigenvalues as the true potential of the free ion. (The relationship between v^{ps} and v^{ps}_{sc} is discussed in Appendix 1). W is the bare ion-ion interaction which is taken to be pairwise. For large separations W will be Coulombic but at short range it should include contributions from van der Waals attraction and Born-Mayer repulsion between the ion cores. Provided the ion cores are small and the ions are not highly polarizable both of these contributions will be insignificant for the separations of interest in normal metals so we simply set $W(R) = Z^2e^2/R$.

The Helmholtz free energy F corresponding to the Hamiltonian $\mathcal{H}$ is

$$F = -\beta^{-1}\ln Q_N \tag{4.2}$$

where $\beta = 1/K_BT$ and Q_N is the canonical partition function :

$$Q_N = \mathrm{Tr}\,\exp(-\beta\mathcal{H}) \tag{4.3}$$

The trace runs over both electronic and ionic states of the system but if we make the adiabatic approximation these can be separated :

$$Q_N = \mathrm{Tr}_{ion}\,\mathrm{Tr}_e\,\exp(-\beta\mathcal{H}) \tag{4.4}$$

where Tr_e refers to a complete set of electronic states corresponding to a particular ionic configuration. Equation (4.4) can be rewritten as

$$Q_N = \mathrm{Tr}_{ion}\,\exp(-\beta(H_K+H_{ii}))\{\mathrm{Tr}_e\,\exp(-\beta(H_e + H_{ee} + H_{ei}))\} \tag{4.5}$$

where we have separated terms which are independent of the electronic coordinates. The term in brackets is just $\exp[-\beta F'(\underline{R}_1,\ldots,\underline{R}_N)]$ where $F'(\underline{R}_1,\ldots,\underline{R}_N)$ is the free energy of an electron system interacting in the presence of an external potential described by H_{ei}. Provided we can evaluate F' (within some approximation scheme) the electronic degrees of freedom no longer explicitly appear and the problem is reduced to that of a quasi one-component system of pseudo-ions with an effective Hamiltonian given by

$$H_{ion} = H_K + H_{ii} + F' \tag{4.6}$$

In order to perform the trace over ionic coordinates i.e. evaluate thermodynamic properties, F' must be a relatively simple function of these coordinates. We shall see that this is indeed the case if F' is calculated to second order in v^{ps}.

4.2 The free energy of the electron system

We begin by considering a homogeneous interacting electron gas of density $n_0 = NZ/\Omega$, in the presence of a uniform compensating charge background. The potential of the latter is

$$V_+(\underline{r}) = -e^2 \int \frac{d\underline{r}' \, n_0}{|\underline{r} - \underline{r}'|} \tag{4.7}$$

Let the free energy of this uniform system be F_{eg} (this energy includes the contribution from the electrostatic self-energy of the positive background). We now perturb this system by replacing the potential V_+ by the total bare electron-ion pseudopotential $V^{ps}(\underline{r}) = \sum_\ell v^{ps}(|\underline{r} - \underline{R}_\ell|)$. The change in potential is

$$\delta V(\underline{r}) = V^{ps}(\underline{r}) - V_+(\underline{r}) \tag{4.8}$$

The free energy F' of the final system of electrons and pseudo-ions can be obtained using the Hellman-Feynman theorem :

$$F' = F_{eg} - \frac{e^2}{2} \iint \frac{d\underline{r} d\underline{r}' n_0^2}{|\underline{r} - \underline{r}'|} + \int_0^1 d\lambda \int d\underline{r} \delta V(\underline{r}) n(\underline{r};\lambda) \tag{4.9}$$

where $n(\underline{r};\lambda)$ is the electron density in a system in which the external potential is $V_+(\underline{r}) + \lambda \delta V(\underline{r})$. Until now we have not required V^{ps} to be weak so the theory could equally well apply in strong-scattering situations. In order to proceed, however, we need to evaluate $n(\underline{r};\lambda)$. If $\delta V(\underline{r})$ is small we can use linear response theory to calculate this quantity :

$$n(\underline{r};\lambda) = n_0 + \int d\underline{r}' \chi(|\underline{r} - \underline{r}'|) \lambda \delta V(\underline{r}') \tag{4.10}$$

where χ is the interacting density response function of the homogeneous electron gas - for density n_0 (e.g. Pines and Nozières 1966). ($n(\underline{r};\lambda)$ is strictly the pseudo-charge density if δV refers to the pseudopotential as in (4.8)). Fourier transforming we have :

$$n(\underline{q};\lambda) = \begin{cases} \lambda \chi(q) \delta V(\underline{q}) & \underline{q} \neq 0 \\ NZ & q = 0 \end{cases}$$

and substituting into (4.9) we obtain

$$F' = F_{eg} - \frac{e^2}{2} \iint \frac{d\underline{r} d\underline{r}' n_0^2}{|\underline{r} - \underline{r}'|} + \frac{1}{\Omega} \int_0^1 d\lambda \left(NZ \delta V(q=0) + \sum_{\underline{q} \neq 0} \lambda \chi(q) |\delta V(\underline{q})|^2 \right)$$

$$= F_{eg} - \frac{e^2}{2} \iint \frac{d\underline{r} d\underline{r}' n_0^2}{|\underline{r} - \underline{r}'|} + n_0 \delta V(q=0) + \frac{1}{2\Omega} \sum_{\underline{q} \neq 0} \chi(q) |\delta V(\underline{q})|^2 \tag{4.11}$$

$\delta V(\underline{q})$ can then be expressed in terms of the pseudopotential and the structure factor $S(\underline{q})$:

$$\delta V(\underline{q}) = v^{ps}(q)S(\underline{q}) = V^{ps}(\underline{q}) \qquad \underline{q} \neq 0$$
$$Nw_c(q = 0) \qquad q = 0 \qquad (4.12)$$

where

$$w_c(q) \equiv v^{ps}(q) + \frac{4\pi}{q^2} Ze^2 \qquad (4.13)$$

is the non-Coulombic part of the bare pseudopotential. The free energy can then be written as

$$F' = F_{eg} - \frac{e^2}{2} \iint \frac{d\underline{r}d\underline{r}' n_0^2}{|\underline{r} - \underline{r}'|} + U_1 + U_2 \qquad (4.14)$$

where $U_1 = Nn_0 w_c(q = 0)$ and

$$U_2 = \frac{1}{2\Omega} \sum_{\underline{q} \neq 0} \chi(q) |v^{ps}(q)|^2 |S(\underline{q})|^2 \qquad (4.15)$$

The second term in (4.14) is the negative of the self-energy of the uniform positive background. This is precisely cancelled by the uniform q = 0 contribution to the bare ion-ion interaction H_{ii} leaving an ion-ion contribution :

$$U_M = \frac{1}{2\Omega} \sum_{\underline{q} \neq 0} \frac{4\pi Z^2 e^2}{q^2} \left(|S(\underline{q})|^2 - N \right) \qquad (4.16)$$

Thus, in this linear response approximation, the effective Hamiltonian for the pseudo-ions is

$$H_{ion} = H_K + U_M + F_{eg} + U_1 + U_2 \qquad (4.17)$$

This result is valid to second order in δV and constitutes the starting point for almost all studies of the structure and thermodynamic properties of simple metals. Since the terms depending on the position coordinates of the ions U_M and U_2 involve only the quantity $|S(\underline{q})|^2$ it is usually straightforward to perform the ensemble average over the ions ; only pairwise correlations enter the theory. U_M is usually called the 'Madelung term' and U_2 the 'band structure term' (Harrison 1966).

At temperatures appropriate to metals (including simple liquid metals near their melting points) the free energy of the uniform electron gas can be approximated by the ground state energy. The correction terms are $\sim - 5\pi^2/12\ (K_B T/E_F)^2$. Thus we can write

$$F_{eg} \approx NZu_{eg}(n_0) \qquad (4.18)$$

where $u_{eg}(n_0)$ is the ground state energy per electron. u_{eg} contains exchange and correlation contributions in addition to the kinetic energy (see Appendix 1). The density response function $\chi(q)$ can also be approximated by its zero temperature limit and this function is discussed in Appendix 1. The detailed behaviour of $\chi(q)$ depends on the treatment of exchange and correlation effects but realistic approximation schemes are now available.

Thus, given some prescription for the pseudopotential v^{ps}, it is a relatively simple task to evaluate H_{ion}. Before describing specific applications of this model it is useful to investigate the consequences of going beyond second order in δV in the expansion of F'. For finite q the electron density can be expanded as

$$n(\underline{q};\lambda) = \lambda\chi(q)\delta V(\underline{q}) + \lambda^2\sum_{\underline{k}}\chi^{(2)}(\underline{q},\underline{k})\delta V(\underline{k})\delta V(\underline{q}-\underline{k}) + \ldots \tag{4.19}$$

where $\chi^{(2)}$ is the second-order response function. It is clear that the term in λ^2 introduces three-body correlations when substituted into (4.9). The higher-order terms involve higher-order correlations. As the second and higher order response functions are not well understood and it is difficult to calculate the corresponding correlation functions, the usefulness of this approach rapidly diminishes if linear response proves inadequate.

4.3 Applications of second-order pseudopotential theory

For a perfect crystal at T = OK, $S(\underline{q}) = N\delta_{\underline{q},\underline{g}}$ and (4.17) gives for the total energy per ion

$$\frac{U}{N} = Zu_{eg}(n_0) + n_0 w_c(q=0) + \frac{U_M}{N} + \frac{U_2}{N} \tag{4.20}$$

where

$$\frac{U_M}{N} = \frac{1}{2\Omega}\sum_{\underline{q}\neq 0}\frac{4\pi}{q^2}Z^2e^2(N\delta_{\underline{q},\underline{g}} - 1)$$

and

$$\frac{U_2}{N} = \frac{N}{2\Omega}\sum_{\underline{g}\neq 0}\chi(g)|v^{ps}(g)|^2$$

We have ignored any contribution from the zero-point motion of the ions. The Madelung contribution can be easily evaluated using an Ewald procedure, (Harrison 1966) and we find

$$\frac{U_M}{N} = \frac{\alpha e^2 Z^{5/3}}{2r_s(n_0)} \tag{4.21}$$

where r_s, the electron radius, is defined by

$$\frac{4\pi}{3}r_s^3 = n_0^{-1} \tag{4.22}$$

and α is the Ewald constant for the particular lattice under consideration. (The values of α are well-known for the simple lattices). The 'band structure' contribution U_2/N can be computed directly but the summation over reciprocal lattice vectors is not rapidly convergent since $\chi(g) \sim g^{-2}$ and $v^{ps}(g) \sim g^{-2}$ for large g. (We will return to this point in § 7). In order to obtain accurate energies one needs to include more than 100 reciprocal lattice vectors. By calculating U/N for various crystal structures one can attempt to understand why a particular metal adopts a particular crystal structure (e.g. Heine and Weaire 1970). Energy differences between structures are, however, rather small, typically 10^{-4} or 10^{-5} Ryd per ion. Consequently the calculations are rather sensitive to the choice of pseudopotential and exchange and correlation contribution to $\chi(q)$. Since the energy differences are so small it is not too surprising to find that including 3rd order terms in v^{ps} can change the conclusions of a purely 2nd order calculation (McLaren and Sholl 1974). The 3rd order term does not make a substantial contribution to the binding energy of the metal. The latter is given by

$$E_{cohesive} = -\left(\frac{U}{N} + \frac{\text{sum of ionziation energies}}{Z}\right) \tag{4.23}$$

and calculations of this quantity are discussed by Heine and Weaire (1970). The pressure p and bulk modulus B_T are easily evaluated in the 2nd order theory :

$$\Omega_0 p = \Omega_0 p_0 + n_0 w_c(q=0) + \frac{U_M}{3} + \frac{1}{2\Omega_0}\sum_{\underline{g}\neq 0}\left(F(g) + \frac{k_F}{3}\frac{\partial F(g)}{\partial k_F} + \frac{g\partial}{3\partial g}F(g)\right)$$

with

$$p_0 = n_0^2\,\frac{du_{eg}(n_0)}{dn_0} \tag{4.24}$$

where $\Omega_0 = \Omega/N$ is the atomic volume and $F(q) \equiv \chi(q)|v^{ps}(q)|^2$. The band structure term in (4.24) involves the derivative of $F(g)$ w.r.t. the average electron density n_0 since on changing the volume of the system we must change the electron density in order to maintain change neutrality. The Fermi wavevector is defined as usual by $k_F^3 = 3\pi^2 n_0$. The bulk modulus is given by $B_T = -\Omega(\partial p/\partial\Omega)$. If the zeroth Fourier coefficient of the pseudopotential $w_c(q = 0)$ is adjusted so that (4.24) yields a pressure ~ 0 the calculated bulk moduli are in rather good agreement with experiment for many simple metals (Ashcroft and Langreth, 1967).

For a crystal at finite temperature we must perform the appropriate ensemble average over the ionic coordinates. From equations (4.2 - 4.6) we have

$$F = -\beta^{-1}\mathrm{Tr}_{ion}\exp(-\beta H_{ion})$$

or

$$F = U - TS_{ion}$$

$$= \langle H_{ion} \rangle - TS_{ion} \tag{4.25}$$

where U is the internal energy and S_{ion} is the ionic contribution to the entropy. The brackets < > denote an ensemble average over the ions. Using (4.17) and (4.18) the free energy can be written as

$$F = NZu_{eg}(n_0) + Nn_0 w_c(q=0) + \langle U_M \rangle + \langle U_2 \rangle + \langle H_k \rangle - TS_{ion} \tag{4.26}$$

If the excursion of the ions from their perfect crystalline sites can be described by a linear combination of phonon coordinates the last two terms in (4.26), the kinetic energy and entropy of the ions, can be readily expressed in terms of Brillouin zone integrals of phonon frequencies $\omega_\lambda(\underline{q})$ where λ denotes polarization. The average of the structure factor $\langle|S(\underline{q})|^2\rangle$ which enters $\langle U_M\rangle$ and $\langle U_2\rangle$ can also be evaluated in terms of $\omega_\lambda(\underline{q})$ (if the one-phonon approximation is employed). Thus, provided one has determined the phonon frequencies the calculation of the free energy and its derivatives is straightforward but somewhat tedious.

The calculation of phonon frequencies within the 2nd order pseudopotential model has been described by many authors (e.g. Harrison 1966, Wallace 1972). The force constants depend on $F(q)$ and the direct ion-ion interaction W(q). Relatively few calculations of the free energy and its derivatives have been carried out (see Wallace 1972 for a review of some of these) and almost all authors have made the additional approximation of replacing $\langle|S(\underline{q})|^2\rangle$ by its static, equilibrium value. Animalu (1967) has discussed the temperature induced fcc-bcc phase transition in Ca and Sr within this framework.

A computationally simpler and physically more appealing scheme for calculating free energies was introduced by Stroud and Ashcroft (1972). This makes use of the so-called Gibbs-Bogoliubov inequality. This relates the free energy F of the 'true' system with Hamiltonian $H_k + \Phi$, where Φ is the potential energy, to that of a 'reference' system at the same density and temperature but with Hamiltonian $H_k + \Phi_0$. The inequality states

$$F \leqslant F_0 + \langle \Phi - \Phi_0 \rangle_0 \tag{4.27}$$

where F_0 is the free energy of the reference system and $\langle\ \rangle_0$ indicates an ensemble average over the reference system. The aim of this approach is to find a reference system which forms a realistic approximation to the true system and for which F_0 and the ensemble average can be easily calculated. By varying the parameter(s) of such a reference system the r.h.s. of (4.27) can be minimized. The minimum value is then used as an estimate of the true free energy F, i.e.

$$F \approx \text{Min}\,(\langle H_k\rangle_0 - TS_0 + \langle\Phi\rangle_0)$$

where S_0 is the entropy of the reference system. This means the free energy has exactly the same form as (4.26) but the averages now refer to the reference system and the entropy S_{ion} is replaced by S_0. Two obvious choices of reference system present themselves ; the Debye and Einstein models. In each case $\langle H_k\rangle_0$ and S_0 are simple functions of the characteristic temperature Θ and $\langle|S(\underline{q})|^2\rangle_0$ has a simple form. The minimization procedure leads to physically acceptable values of Θ provided the temperature is such that the mean displacement of the ions from their equilibrium sites is a small fraction of the interatomic distance. Θ shows only a weak dependence on density and a smooth, almost constant, behaviour with temperature. For details see Stroud and Ashcroft (1972) who used the Debye model to calculate the free energy of Na and Jones (1973) who used the Einstein model to calculate several thermodynamic properties of Li, Na, K and Al. This work was primarily aimed at understanding melting in simple metals and therefore required a theory for the free energy of the liquid phase (see below). We should note that these approximation schemes can be viewed as crude versions of the usual self-consistent phonon approximation (e.g. Wallace 1972).

The pseudopotential model is readily applicable to liquid metals. The ensemble average over the ionic coordinates can be performed classically at liquid metal temperatures so that

$$\text{Tr}_{ion}\exp(-\beta H_{ion}) \rightarrow \frac{1}{N!h^{3N}}\int d\underline{R}_1 \ldots d\underline{R}_N \int d\underline{P}_1 \ldots d\underline{P}_N \exp(-\beta H_{ion})$$

$$= \left(\frac{2\pi M}{h^2\beta}\right)^{3N/2} \frac{1}{N!}\int d\underline{R}_1 \ldots d\underline{R}_N \exp[-\beta\Phi(\underline{R}_1 \ldots \underline{R}_N)]$$

where Φ is the potential energy given by the sum of the last four terms in (4.17). The internal energy of the liquid metal is given by

$$U = \frac{3}{2}Nk_BT + F_{eg} + Nn_0w_c(q=0) + \langle U_M\rangle + \langle U_2\rangle \tag{4.28}$$

where the first term is the classical kinetic energy and the brackets now refer to a configuration average defined by

$$\langle f\rangle \equiv \frac{\int d\underline{R}_1 \ldots d\underline{R}_N\, f(\underline{R}_1 \ldots \underline{R}_N)\exp(-\beta\Phi)}{\int d\underline{R}_1 \ldots d\underline{R}_N \exp(-\beta\Phi)} \tag{4.29}$$

$\langle U_M \rangle$ and $\langle U_2 \rangle$ involve the liquid structure factor a(q) which is defined in (3.18) so that (4.28) can be written as

$$\frac{U}{N} = \frac{3}{2}k_B T + Zu_{eg}(n_0) + n_0 w_c(q=0) + \frac{1}{4\pi^2}\int_0^\infty dq 4\pi Z^2 e^2 (a(q) - 1)$$

$$+ \frac{1}{4\pi^2}\int_0^\infty dq q^2 F(q) a(q) \tag{4.30}$$

where we have converted summations over $\underline{q}$ to integrals. The 'band structure' term in (4.30) includes contributions from all values of q - unlike the corresponding term for a perfect crystal. This means the pseudopotential is sampled at small q, where it is large, as well as reciprocal lattice vectors, where it is small. However, a(q) is small where q is small so this region does not make a dominant contribution to the total energy. This is analagous to the density of states problem (see (3.17)). The pressure and bulk modulus can also be evaluated by differentiating the partition function w.r.t. volume (e.g. Hasegawa and Watabe 1972). Given some means of evaluating the liquid structure factor and its density dependence it is straightforward to calculate U,p and B_T. Some of the first calculations used experimental data (from X-ray or neutron diffraction) for a(q). This was unsatisfactory since a 'proper' theory should contain a prescription for this quantity in order to maintain self-consistency. It is well-known that such considerations are very important in the thermodynamics of liquids. In the next section we shall see that a(q), or more accurately its Fourier transform the radial distribution function g(R), can be calculated directly from pseudopotential theory using a molecular dynamics or a Monte Carlo simulation. An elegant approximation scheme for the thermodynamic properties was introduced by Stroud and Ashcroft (1972) and Jones (1973) who again made use of the Gibbs-Bogoliubov inequality (4.27).

The natural reference system for a liquid metal, or indeed any dense fluid is hard-spheres. The structure factors of most liquid metals can be crudely modelled by a hard-sphere system at the same density with a suitably chosen diameter d (Ashcroft and Lekner 1966). In the Percus-Yevick approximation both F_{hs} and $a_{hs}(q)$ are available in convenient forms so that it is computationally straightforward to minimize the r.h.s. of (4.27) w.r.t. the hard-sphere diameter and thereby obtain an estimate of the free-energy. We will return to this topic in the next section since it is rather more convenient to work in a real-space representation.

As a final example of the application of second-order pseudopotential theory we briefly consider defect problems. For simplicity we restrict discussion to a static lattice i.e. T = OK. In this case the total energy takes the form

$$U = ZNu_{eg}(n_0) + Nn_0 w_c(q{=}0) + \frac{1}{2\Omega}\sum_{\underline{q}\neq 0}\frac{4\pi}{q^2}Z^2e^2(|S(\underline{q})|^2 - N)$$

$$+ \frac{1}{2\Omega}\sum_{\underline{q}\neq 0} F(q)|S(\underline{q})|^2 \qquad (4.31)$$

This result is valid for any arrangement of the ions provided the volume Ω is kept constant. Consequently defect energetics are more easily formulated under constant volume conditions. In all cases we are required to evaluate the structure factor $S(\underline{q})$ appropriate to the defect under consideration. The defects might be extended e.g. dislocations and stacking faults or point defects e.g. vacancies and interstitials. Here we focus attention on the simplest point defect, a monovacancy, and calculate the vacancy formation energy E_v.

We consider a single vacancy located at the origin. Ignoring any relaxation of the ions around the vacancy, the structure factor of the defect lattice is

$$S_v(\underline{q}) = S_p(\underline{q}) - 1 \qquad (4.32)$$

where

$$S_p(\underline{q}) = \sum_{\ell=1}^{N+1} \exp(i\underline{q}.\underline{R}'_\ell)$$

is the structure factor of a perfect lattice of $N + 1$ sites in which the lattice spacing is reduced by a factor μ^{-1} (where $\mu^3 = 1 + N^{-1}$) over that in the original perfect lattice of N sites. Harrison (1966) showed that the Madelung energy associated with this defected lattice differs from that of the original lattice by $-2U_M/3N$ where U_M/N is the Madelung energy per ion of the perfect crystal and is given by (4.21). The change in 'band structure' energy is also easily calculated. Since $S_p(\underline{q})$ is only non-zero at reciprocal lattice vectors $\underline{g}'$ we find

$$\frac{1}{2\Omega}\sum_{\underline{q}\neq 0} F(q)|S_v(\underline{q})|^2 = \frac{N^2}{2\Omega}\sum_{\underline{g}'\neq 0} F(g') + \frac{1}{4\pi^2}\int_0^\infty dq\, q^2 F(q) \qquad (4.33)$$

with $g' \equiv |\underline{g}'| = \mu|\underline{g}|$ where $\underline{g}$ is a reciprocal lattice vector of the original lattice. Now

$$F(g') \approx F(g) + (\mu - 1)g\,\frac{\partial F(g)}{\partial g}$$
$$= F(g) + \frac{g}{3N}\frac{\partial F(g)}{\partial g}$$

so the constant volume formation energy is

$$E_v = -\frac{2}{3}\frac{U_M}{N} + \frac{1}{4\pi^2}\int_0^\infty dq\, q^2 F(q) + \frac{1}{2\Omega_0}\sum_{\underline{g}\neq 0}\frac{g}{3}\frac{\partial F(g)}{\partial g} \qquad (4.34)$$

where terms $O(1/N)$ have been neglected. The effects of an expansion in volume associated with vacancy formation can be included in a rather straightforward fashion in this model (e.g. Evans 1977). The results of calculations based on (4.34) are in reasonable agreement with experiment for the alkali metals, but for polyvalent metals (4.34) often predicts negative formation energies (A critical review of the theory and calculations is given by Evans (1977)). We should note that relaxation effects can be systematically taken into account in this model and it seems to be well accepted that relaxation reduces the formation energy by <0.1eV for the simple metals. The measured formation energies lie in the range 0.3 eV (Rb and Cs) to 0.8 eV (Mg). Since E_v in (4.34) depends on the cancellation of terms of differing sign it is not surprising that the results are sensitive to the choice of pseudopotential and exchange and correlation contribution to $\chi(q)$. The sensitivity is especially pronounced in the polyvalent metals where each term is large in magnitude.

This treatment of the vacancy formation energy is based on linear response theory. One might imagine that the presence of a vacancy would severely distort the conduction electron density in its neighbourhood so that the linear theory becomes invalid. Replacing a uniform positive background by a lattice of pseudo-ions with a vacancy introduces a perturbation potential $\delta V(\underline{r})$ which is relatively large near the vacancy. On Fourier transforming and using (4.12) and (4.32) we have

$$\delta V(\underline{q}) = \begin{cases} N w_c(q{=}0) & q = 0 \\ N v^{ps}(g) & \underline{q} = \underline{g} \neq 0 \\ -\, v^{ps}(q) & \underline{q} \neq \underline{g} \neq 0 \end{cases} \tag{4.35}$$

i.e. the perturbation involves the low q Fourier components of the pseudopotential. The latter make a substantial contribution to the second term in (4.34). This argument suggests that linear screening (second order perturbation theory) may be inadequate and one should seek an alternative, non-perturbative technique. Some progress in this direction has recently been made (see Evans 1977) using a procedure similar to that developed by Lang and Kohn (1970) for the surface energy. We return to this in § 7.

The problem of calculating the surface energy of a crystalline metal or, indeed, the surface tension of liquid metal is of course closely related to the defect problem. It is possible to calculate the surface energy of a metal at T = OK using second-order pseudopotential theory; equation (4.31) is also valid for a crystal with a surface. Such calculations (Finnis 1975) yield surface energies which are roughly 30 % of the experimental estimate for Na and < 10% of the experimental estimate for Al. The surface energy and surface tension of liquid metals can also be formulated within the linear screening theory (Evans and Kumaravadivel 1976) and the results calculated by these authors are again much smaller than those obtained from experiment. Linear screening is completely inadequate for the surface problem for the reasons given above in connection with the point defect problem i.e. δV will have large (small q) Fourier components. Considerable effort has gone into improving the theory of the surface energy (for a review see Lang 1973 and Gunnarsson 1978) and we mention some more recent work in § 7.

5. REAL SPACE REPRESENTATION AND EFFECTIVE PAIRWISE POTENTIALS

The theory described in § 4 was formulated in reciprocal space since this is the natural representation for a linear response approach. It is often instructive to transform to a real space picture and think in terms of effective pairwise potentials. This is particularly useful for liquid state and defect problems because, in these fields, powerful theoretical and computational techniques have been developed which use empirical pairwise potentials as input data.

5.1 Transformation of the effective Hamiltonian

Equation (4.17) can easily be transformed to a real space representation. The structure dependent terms can be written as :

$$U_M + U_2 = \frac{1}{2\Omega} \sum_{\underline{q}} \phi(q;n_0) \sum_{\ell \neq \ell'} \exp(i\underline{q}.(\underline{R}_\ell - \underline{R}_{\ell'})) + \frac{N}{2\Omega} \sum_{\underline{q}\neq 0} F(q) - \frac{N(N-1)}{2\Omega} \lim_{q\to 0} \phi(q;n_0) \tag{5.1}$$

where

$$\phi(q;n_0) = \frac{4\pi Z^2 e^2}{q^2} + F(q) = \frac{4\pi Z^2 e^2}{q^2} + \chi(q)\,|v^{ps}(q)|^2 \tag{5.2}$$

The n_0 in $\phi(q;n_0)$ indicates that $\chi(q)$ depends on the average electron density n_0. The limit in the last term of (5.1) can be evaluated :

$$\phi(q = 0;n_0) = 2Zw_c(q=0) - Z^2(\chi_0(q=0)^{-1} - U_{xc}(q=0)) \tag{5.3}$$

where $\chi_0(q)$ is the non-interacting Lindhard function and $U_{xc}(q)$ is the exchange and correlation contribution to the effective electron-electron interaction (see Appendix 1). The term in brackets is proportional to the bulk modulus of the uniform electron gas

$B_e(n_0)$ via the compressibility sum rule (see (A.22)) so that

$$\phi(q = 0;n_0) = 2Zw_c(q=0) + \Omega_0^2 B_e(n_0) \tag{5.4}$$

The effective Hamiltonian for the ions then takes the form

$$H_{ion} = H_k + Nu(n_0) + \frac{1}{2} \sum_{\ell \neq \ell'} \phi(|\underline{R}_\ell - \underline{R}_{\ell'}|;n_0) \tag{5.5}$$

with

$$\begin{aligned} u(n_0) &= Zu_{eg}(n_0) + n_0 w_c(q=0) - \phi(q=0;n_0)/2\Omega_0 + \frac{1}{2\Omega} \sum_{q \neq 0} F(q) \\ &= Zu_{eg}(n_0) - \frac{\Omega_0}{2} B_e(n_0) + \frac{1}{4\pi^2} \int_0^\infty dq q^2 F(q) \end{aligned} \tag{5.6}$$

and

$$\phi(R;n_0) = \frac{1}{(2\pi)^3} \int d\underline{q} \exp(- i\underline{q}.\underline{R}) \phi(q;n_0) \tag{5.7}$$

We have assumed that F_{eg} and $\chi(q)$ take their zero temperature values. $u(n_0)$ is structure independent and is sometimes referred to as the self-energy of a pseudo-ion. The effective pairwise potential $\phi(R;n_0)$ combines the direct Coulomb repulsion between the ions and the effects of polarization of the ion by the conduction electron screening cloud.

If the bare pseudopotential is non-local the physical content of ϕ remains the same but $F(q)$ must be replaced by a more complicated energy-wave number characteristic (e.g. Kumaravadivel and Evans 1976). Taking the q = 0 limit then becomes difficult and, in general, there is no simple, explicit expression for $u(n_0)$.

If we go beyond linear screening i.e. we retain the term in λ^2 in (4.19), the real space representation of H_{ion} becomes considerably more complicated :

$$\begin{aligned} H_{ion} = H_k &+ Nu'(n_0) + \frac{1}{2} \sum_{\ell \neq \ell'} \phi'(|\underline{R}_\ell - \underline{R}_{\ell'}|;n_0) \\ &+ \frac{1}{3} \sum_{\ell \neq \ell' \neq \ell''} \phi^{(3)}(\underline{R}_\ell;\underline{R}_{\ell'};\underline{R}_{\ell''};n_0) \end{aligned} \tag{5.8}$$

Both the self-energy and the pair potential are modified and an explicit three-body potential $\phi^{(3)}$ appears. For a discussion of these modifications see Hasegawa (1976).

5.2 The form of the calculated pairwise potentials

$\phi(R;n_0)$ has been calculated by many authors for many simple metals. It is well-known that the pair potentials, especially in the region of the first minimum, are quite sensitive to the choice of pseudopotential and to the form of the exchange and correlation correction U_{xc} in the response function. Some consensus of what are 'good' pseudopotentials and what constitutes a reasonable treatment of exchange and correlation has emerged (e.g. Kumaravadivel and Evans 1976) but in the polyvalent metals in particular, ϕ is very sensitive to U_{xc}. This sensitivity results from the strong cancellation between the direct Coulomb repulsion and the polarization term. In Al for $R \sim 5$ a.u, the Coulomb part is ~ 3 Ryd but the resultant $\phi(R;n_0)$ is only $\sim 10^{-3}$ Ryd. We will comment further on this in § 7.

The depth of the first minimum in ϕ does not appear to vary in any systematic way from element to element. It is usually in the range $1 - 4 \times 10^{-3}$ Ryd but varies somewhat with density (Kumaravadivel and Evans 1976). It does not correlate with the observed cohesive energies. The latter are typically an order of magnitude larger than the depths of the potential minima. This implies that most of the binding energy of a metal resides in the structure independent self-energy $u(n_0)$ and this is confirmed by detailed calculations which show that for both crystals and liquids the pairwise contribution to the internal energy is only a small contribution to the internal energy (e.g. Heine and Weaire 1970, Price 1971).

The short-range repulsive force calculated from ϕ is relatively insensitive to U_{xc}, pseudopotential and density and exhibits a systematic variation from element to element (e.g. Kumaravadivel and Evans 1976). In the alkalis the first minimum in ϕ shifts to progressively larger separations as the atomic number increases (see Figure 1). This reflects the increasing 'size' of the metallic atoms. We should note that the first zero of the pair potential is much larger than the diameter of the ion core in metals. For example, in Na the diameter of the core is < 4 a.u. while the zero of the potential occurs at ~ 6 a.u. The repulsive force in the metal arises from the incomplete screening, at small R, of the direct Coulomb interaction between the ions. This can be contrasted with the rare gases where the short range forces are due to repulsion between closed electron shells. The repulsive part of ϕ 'softens' on going from Li to Cs i.e. as the average electron density is reduced.

The pair potentials for polyvalent metals are considerably 'harder' than those for the alkalis but are still 'softer' than a typical Lennard-Jones 6 - 12 potential for a rare-gas solid or liquid (see Figure 2).

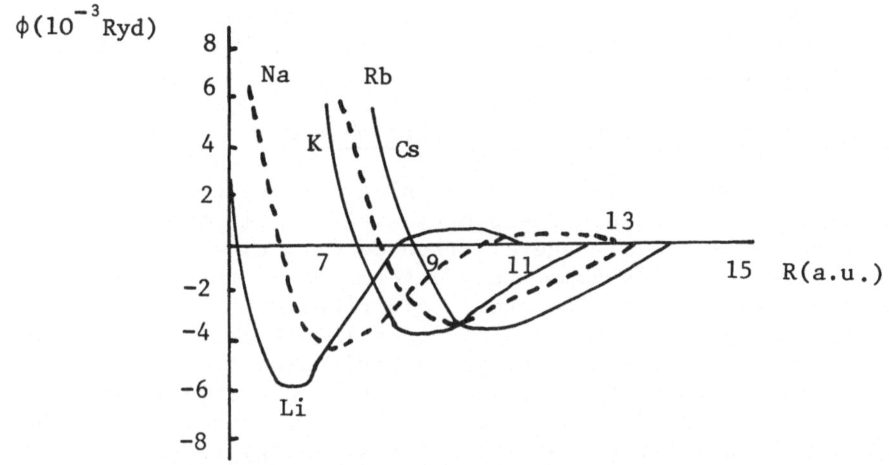

Figure 1 : Pair potentials calculated for the liquid alkali metals (from Kumaravadivel and Evans 1976).

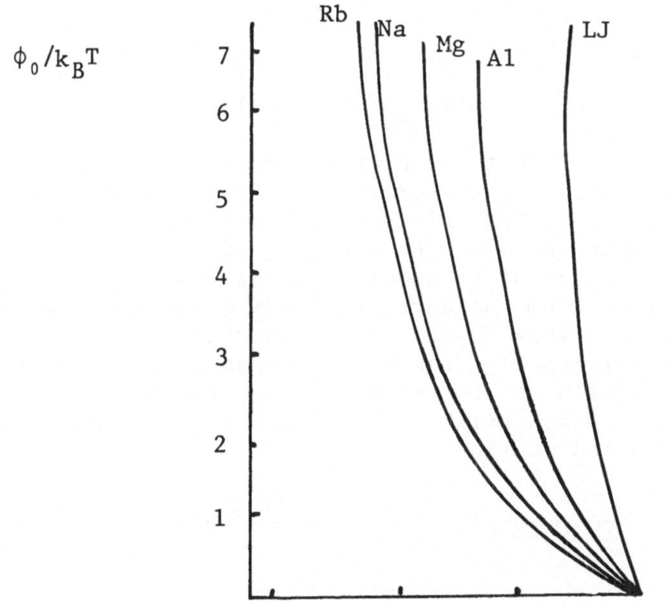

Figure 2 : The repulsive force part ϕ_0 of the pairwise potentials calculated from pseudopotential theory (from Kumaravadivel and Evans 1976). R_{min} denotes the position of the first minimum in ϕ and $\phi_0(R) = \phi(R) - \phi(R_{min})$. T is the melting temperature. Curve LJ is calculated for a 'typical' Lennard-Jones 6-12 potential.

The systematic variation of the size and softness of metallic atoms has important consequences for the entropy and the structure factor in liquid metals.

5.3 Thermodynamics and structure of liquid metals

The Hamiltonian of (5.5) is a convenient starting point for the calculation of liquid state properties. If we fix the ion density ρ (and hence the average electron density $n_0 = \rho Z$) the structure of the liquid is completely determined by the effective pairwise potential $\phi(R;n_0)$. The atomic distribution functions can be calculated using the standard techniques (Monte-Carlo, molecular dynamics, diagrammatic expansions, perturbation theories) which were originally developed for simple insulating (rare-gas) liquids (e.g. Hansen and McDonald 1976). The thermodynamic functions can be derived after the usual fashion. The internal energy is

$$\frac{U}{N} = \frac{3}{2} k_B T + u(n_0) + \frac{\rho}{2} \int d\underline{R} g(R) \phi(R;n_0) \tag{5.9}$$

where g(R) is the radial distribution function evaluated for the particular density under consideration. g(R) and the liquid structure factor a(q) are related via

$$a(q) - 1 = \rho \int d\underline{R} \exp(i\underline{q}.\underline{R})(g(R) - 1) \tag{5.10}$$

Equation (5.9) differs from that for simple insulating fluids due to the presence of the self-energy term $u(n_0)$. As mentioned above the dominant contribution to U/N arises from the latter. Other constant volume quantities such as the entropy and the specific heat will have the same form as in simple liquids - apart from very small contributions from the entropy of the uniform electron gas, which can safely be ignored. Quantities which depend on volume derivatives differ significantly from the corresponding quantities obtained for the simple fluids. For example, the pressure can be calculated using the standard trick of scaling the variables in the partition function and we find :

$$p = \rho k_B T + \rho n_0 \frac{du(n_0)}{dn_0} - \frac{\rho^2}{2} \int d\underline{R} g(R) \left(\frac{R}{3}\frac{\partial}{\partial R} - n_0 \frac{\partial}{\partial n_0}\right) \phi(R;n_0) \tag{5.11}$$

This result differs from the familiar virial equation of state because of the appearance of the two density derivatives. A liquid metal or indeed, any metallic system, is not in equilibrium under the influence of pairwise forces alone ; the additional terms reflect the inherent two component nature of the metal.

Over the last few years there has been a lot of work on the pseudopotential theory of the thermodynamics of liquid metals. This subject has recently been reviewed by Ashcroft and Stroud (1978) and Evans (1978) so here we will only describe one scheme which has been widely used in calculations of the free energy and its derivatives. This is the scheme based on the Gibbs-Bogoliubov inequality which we referred to earlier.

We write the effective Hamiltonian of (5.5) as $H_{ion} = H_k + \Phi$ and consider a reference system at the same density and temperature consisting of hard spheres of diameter d. The corresponding potential function is

$$\Phi_0 = \frac{1}{2} \sum_{\ell \neq \ell'} \phi_{hs}(|\underline{R}_\ell - \underline{R}_{\ell'}|) \tag{5.12}$$

where

$$\phi_{hs}(R) = \infty \qquad R \leqslant d$$
$$\qquad\qquad 0 \qquad R > d$$

It then follows from (4.27) that the free energy satisfies the inequality

$$F \leqslant F_{hs} + Nu(n_0) + \frac{N\rho}{2} \int d\underline{R} g_{hs}(R)\phi(R;n_0) \tag{5.13}$$

since for the hard sphere fluid $g_{hs} = 0$ for $R \leqslant d$. F_{hs} is the free energy of a hard sphere fluid of density ρ and temperature T. The diameter d is then used as a variational parameter to minimize the r.h.s. of (5.13) and the minimum is taken as an approximation for the true free energy F. It is computationally more efficient to carry out this procedure in reciprocal space since (a) $\phi(R;n_0)$ must be calculated by a Fourier transform i.e. (5.7) and (b) $a_{hs}(q)$ is available in analytical form in the Percus-Yevick approximation. Thus, in practice, one minimizes

$$F_{hs} + \frac{N}{4\pi^2} \int_0^\infty dq 4\pi Z^2 e^2 (a_{hs}(q) - 1) + \frac{N}{4\pi^2} \int_0^\infty dq q^2 F(q) a_{hs}(q)$$

which are the only structure dependent contributions to the r.h.s. of (5.13). The resultant estimate of the free energy satisfies

$$\left(\frac{\partial F}{\partial d}\right)_{\Omega,T} = 0$$

or equivalently

$$\left(\frac{\partial F}{\partial \eta}\right)_{\Omega,T} = 0 \tag{5.14}$$

where $\eta = \pi d^3\rho/6$ is the packing fraction of the hard spheres. The entropy is given by

$$\begin{aligned} S_{ion} &= -\left(\frac{\partial F}{\partial T}\right)_\Omega \\ &= -\left(\frac{\partial F}{\partial T}\right)_{\Omega,\eta} - \left(\frac{\partial F}{\partial \eta}\right)_{\Omega,T} \left(\frac{\partial \eta}{\partial T}\right)_\Omega \end{aligned} \tag{5.15}$$

but the second term vanishes by virtue of (5.14). In the present approximation the only explicit temperature dependence in F is in F_{hs} so the entropy then takes on a particularly simple form

$$S_{ion} = -\left(\frac{\partial F_{hs}}{\partial T}\right)_{\Omega,\eta} \tag{5.16}$$

This is just the entropy of the hard sphere fluid of density ρ in which the diameter is fixed via the minimization procedure. It then follows that the internal energy and the pressure (Watabe and Young 1974) have simple forms in this scheme :

$$\frac{U}{N} = \frac{3}{2} k_B T + u(n_0) + \frac{\rho}{2} \int d\underline{R} g_{hs}(R)\phi(R;n_0) \tag{5.17}$$

$$p = \rho k_B T + \rho n_0 \frac{du(n_0)}{dn_0} - \frac{\rho^2}{2} \int d\underline{R} g_{hs}(R)\left(\frac{R}{3}\frac{\partial}{\partial R} - n_0 \frac{\partial}{\partial n_0}\right)\phi(R;n_0) \tag{5.18}$$

Thus the variational scheme is equivalent to replacing the true radial distribution function in the formally exact equations (5.9) and (5.11) by its hard sphere equivalent - with the diameter determined by the minimization procedure.

The excess entropies, defined as $S_{ion} - S_{ideal}$, where S_{ideal} is the entropy of the ideal gas, as calculated from (5.16) for many simple metals (Kumaravadivel and Evans 1976) are in good qualitative agreement with experiment. One finds

$$S_{ion} - S_{ideal} = - Nk_B \mathscr{s}(\eta) \tag{5.19}$$

where $\mathscr{s}$ is solely a function of the packing fraction. The minimization procedure yields packing fractions which are close to 0.45 for most metals at temperatures near their melting points. This implies that the excess entropy of all liquid metals should be roughly constant and equal to $- k_B \mathscr{s}(0.45) \sim - 4k_B$ per ion. The experimental values lie roughly in the range − 3.3 to − 4.6 k_B per ion.

It is well-known that the hard sphere structure factors evaluated with $\eta \approx 0.45$ are in fairly good agreement with experiment, at least in the neighbourhood of the first peak, for many simple metals. Consequently, it is encouraging that the pseudopotential theory actually predicts packing fractions of this magnitude. This implies that the repulsive part of the calculated pairwise potentials must be realistic. If the repulsive potentials were grossly wrong they would lead to values of η very different from 0.45 and produce non-physical structure factors.

More sophisticated theories which take into account the effect of 'softness' in the repulsive potential have been applied to the calculation of liquid metal structure factors (see Kumaravadivel and Evans 1976 and Evans 1978). The softer the repulsive potential the more damped are the oscillations in a(q) at large q. Thus in the light of § 5.2 we might expect the alkali metals to show very damped structure factors while the polyvalent metals should exhibit more pronounced large q oscillations. This trend is found in the measured structure factors. The small q behaviour of a(q) is also of interest in liquid metals since it should reflect the influence of the long-range oscillations in $\phi(R;n_0)$. A discussion of this problem has recently appeared (Evans and Schirmacher 1978).

5.4 Defect energetics in the real-space representation

As in § 4.3 we consider a static, defected lattice at T = OK. The total energy of (4.31) then transforms to

$$U = Nu(n_0) + \frac{1}{2} \sum_{\ell \neq \ell'} \phi(|\underline{R}_\ell - \underline{R}_{\ell'}|;n_0) \tag{5.20}$$

which is, as stressed earlier, valid for any arrangement of ions. This real space representation is appealing because it casts defect problems into a form which is similar to that which has been studied at length by those workers who model metals in terms of empirical pairwise potentials (see e.g. Johnston 1973 and Heald 1977). Such studies are primarily concerned with relaxation effects and interactions between defects rather then the calculation of point defect formation energies. Furthermore, they are frequently directed at transition and noble metals for which the pseudopotential model is inappropriate. Nevertheless it is clear that the computational techniques which have been developed for treating these problems can be readily applied to the model defined by (5.20) - provided one works at constant volume and is careful about the interpretation of the results.

The constant volume vacancy formation energy is easily obtained in the real space representation :

$$E_v = -\frac{1}{2} \sum_{\ell \neq 0} \left[\phi(R;n_0) + \frac{R}{3} \frac{\partial \phi(R;n_0)}{\partial R} \right]_{R=|\underline{R}_\ell|} \tag{5.21}$$

where the summation is over all ions. The first term is due to the breaking and re-making of pairwise bonds in the formation process while the second takes into account the shortening of the interatomic distances which occurs under constant volume conditions. If we ignore relaxation of the ions around the defect so that $\underline{R}_\ell$ refers to perfect lattice sites, the first term is equal to the pairwise contribution to the cohesive energy (see (4.23)) while the second is proportional to the virial term in the pressure since

$$\Omega_0 p = n_0 \frac{du(n_0)}{dn_0} - \frac{1}{2} \sum_{\ell \neq 0} \left(\frac{R}{3} \frac{\partial}{\partial R} - n_0 \frac{\partial}{\partial n_0} \right) \phi(R;n_0) \Big|_{R=|\underline{R}_\ell|} \tag{5.22}$$

The virial term is in general non-zero and can be larger than the first contribution to E_v (Minchin et al 1974). This again reflects the fact that the metal is not in equilibrium under the influence of pairwise forces alone. If we were to assume that the total energy could be represented as

$$U = \frac{1}{2} \sum_{\ell \neq \ell'} \phi_{em}(|\underline{R}_\ell - \underline{R}_{\ell'}|) \tag{5.23}$$

where $\phi_{em}(R)$ is some empirical density independent pairwise potential then the virial of this potential would vanish and the corresponding unrelaxed constant volume formation energy would be identical to the cohesive energy. Experimentally the vacancy formation energies of metals are typically about one third of the cohesive energies so relaxation would have to bring about enormous reductions of energy if this empirical model were to be appropriate for metals. For rare-gas solids such a model may be much more realistic since the vacancy formation energy is roughly 70 or 80% of the cohesive energy. The formation volume associated with a vacancy depends on the volume derivative of E_v and therefore involves derivatives of $\phi(R;n_0)$ w.r.t. n_0 as well as w.r.t. R. It is not permissible to use the harmonic lattice statics formalism, as is conventionally applied with empirical potentials, for calculating formation volumes (see Evans 1977 and references therein).

The real space representation has one important disadvantage when used in connection with defect problems : it tends to make these problems appear easier than they really are. Although the electronic aspects of the problem are in a sense properly treated (i.e. within the linear response approximation) they are hidden away and therefore easily ignored or forgotten. If linear screening is inadequate for a particular defect situation, as suggested in §4.3, then the present description in terms of pairwise potentials becomes invalid.

6. COHESIVE PROPERTIES OF BINARY ALLOYS

Understanding the structure, stability and thermodynamic properties of alloys is a major goal of the electron theory of metals. Starting with the pioneering work of Mott and Jones in the 1930's on the Hume-Rothery rule for the α-β phase boundary in brass-type alloys, physicists have attempted to formulate a theory for the energetics which will give some microscopic insight into phase diagrams, ordering phenomena, etc. This is certainly a rich field but one in which relatively little progress has been achieved. (For an instructive review of work before 1970 see Heine and Weaire 1970).

Here we will generalize the pseudopotential model to binary alloys and discuss some applications.

6.1 The effective Hamiltonian for the ions

We consider a binary alloy A_cB_{1-c} consisting of ions of valence Z_A and Z_B. The basic Hamiltonian for the system of (pseudo) ions and conduction electrons is

$$H = \sum_i \frac{p_i^2}{2m} + \frac{e^2}{2} \sum_{i \neq j} \frac{1}{|\underline{r}_i - \underline{r}_j|} + \sum_{i,\ell} v_\ell(\underline{r}_i - \underline{R}_\ell) + \sum_\ell \frac{P_\ell^2}{M_\ell} + \frac{e^2}{2} \sum_{\ell \neq \ell'} \frac{Z_\ell Z_{\ell'}}{|\underline{R}_\ell - \underline{R}_{\ell'}|} \tag{6.1}$$

where the symbols have the same meaning as in (4.1) except that v_ℓ refers to the bare pseudopotential at position $\underline{R}_\ell$; P_ℓ and M_ℓ refer to the ℓth ion and Z_ℓ is the valence of the ion at $\underline{R}_\ell$. (We have assumed the bare ion-ion interaction H_{ii} is purely Coulombic). Following the arguments of § 4.1 we obtain an effective Hamiltonian for the ions of the form given in (4.6) but we must now determine the free energy F' of the electron system interacting in the presence of an external potential

$$V(\underline{r}) = \sum_A v^A(|\underline{r} - \underline{R}_A|) + \sum_B v^B(|\underline{r} - \underline{R}_B|) \tag{6.2}$$

where v^A and v^B are the bare pseudopotentials of the A and B ions respectively. The calculation of F' follows as in § 4.2 except at concentration c the uniform compensating positive background is taken to have a density

$$n_0 = \frac{N\bar{Z}}{\Omega} = \frac{N}{\Omega}(cZ_A + (1 - c)Z_B) \tag{6.3}$$

where N is the total number of ions and Ω is the volume of the alloy. The perturbation potential is now

$$\delta V(\underline{q}) = V(\underline{q}) - V_+(\underline{q}) = v^A(q)S_A(\underline{q}) + v^B(q)S_B(\underline{q}) - V_+(\underline{q})$$

which reduces to

$$\delta V(\underline{q}) = \begin{cases} v^A(q)S_A(\underline{q}) + v^B(q)S_B(\underline{q}) & \underline{q} \neq 0 \\ N(cw^A(q{=}0) + (1 - c)w^B(q{=}0)), & q = 0 \end{cases} \tag{6.4}$$

where w^A and w^B are the non-Coulombic parts of the bare pseudopotential (see (4.13)). F' can then be written as in (4.14) but now

$$U_1 = Nn_0(cw^A(q{=}0) + (1 - c)w^B(q{=}0))$$

and

$$U_2 = \frac{1}{2\Omega} \sum_{\underline{q} \neq 0} \chi(q)\,|v^A(q)S_A(\underline{q}) + v^B(q)S_B(\underline{q})|^2 \tag{6.5}$$

and F_{eg} and $\chi(q)$ refer to the uniform electron gas with density n_0 given by (6.3). Thus the effective Hamiltonian has the form

$$H_{ion} = H_k + U_M + F_{eg} + U_1 + U_2$$

with a Madelung term

$$U_M = \frac{1}{2\Omega} \sum_{\underline{q}\neq 0} \frac{4\pi e^2}{q^2} \left(|Z_A S_A(\underline{q}) + Z_B S_B(\underline{q})|^2 - N(cZ_A^2 + (1-c)Z_B^2) \right) \quad (6.6)$$

We have converted the original three component (electrons and A and B bare ions) problem to an effective two-component problem. Taking the trace over ionic coordinates now involves averaging w.r.t. both configurational and thermal disorder. We concentrate first on the configurational aspects of the problem and consider a static substitutionally disordered crystal. Later we consider the effects of lattice vibrations.

6.2 Crystalline alloys

If we neglect the vibrations of the ions the Helmholtz free energy of the alloy is

$$F = F_{eg} + U_1 + \langle U_M \rangle_c + \langle U_2 \rangle_c - TS_c \quad (6.7)$$

where $\langle\ \rangle_c$ denotes a configuration average only and S_c is the configurational entropy of the alloy. The averages involved in the Madelung and 'band structure' terms have been discussed in § 3.2 and we find

$$\langle U_2 \rangle_c = \frac{N^2}{2\Omega} \sum_{g\neq 0} \chi(g) |\bar{v}(g)|^2$$

$$+ \frac{1}{2(2\pi)^3} \int d\underline{q}\, \chi(q) |v^A(q) - v^B(q)|^2 \langle |(1-c)S_A(\underline{q}) - cS_B(\underline{q})|^2 \rangle_c \quad (6.8a)$$

where $\bar{v}(q) = cv^A(q) + (1-c)v^B(q)$ and $\underline{g}$ is a reciprocal lattice vector. Similarly we have

$$\langle U_M \rangle_c = \frac{N^2}{2\Omega} \sum_{g\neq 0} \frac{4\pi \bar{Z}^2 e^2}{g^2}$$

$$+ \frac{1}{2(2\pi)^3} \int d\underline{q} \frac{4\pi e^2}{q^2} \left[(Z_A - Z_B)^2 \langle |(1-c)S_A(\underline{q}) - cS_B(\underline{q})|^2 \rangle_c - \right.$$

$$\left. - N(cZ_A^2 + (1-c)Z_B^2) \right] \quad (6.8b)$$

For a completely random distribution of ions these results reduce to

$$\frac{\langle U_2\rangle_r}{N} = \frac{N}{2\Omega}\sum_{\underline{g}\neq 0}\chi(g)|\bar{v}(g)|^2 + \frac{c(1-c)}{4\pi^2}\int_0^\infty dq q^2\chi(q)|v^A(q) - v^B(q)|^2 \quad (6.9a)$$

and

$$\frac{\langle U_M\rangle_r}{N} = \frac{1}{2\Omega}\sum_{\underline{q}\neq 0}\frac{4\pi\bar{Z}^2e^2}{q^2}(N\delta_{\underline{q},\underline{g}} - 1) \quad (6.9b)$$

Note that while there is no term in $|Z_A - Z_B|^2$ for the random case there will be contributions of this type for less disordered situations.

In order to specify the degree of order it is often useful to introduce a long range order parameter of the kind used in the Bragg-Williams theory of order-disorder transitions. At a given stoichiometric composition we suppose there is an ordered state in which all the A ions occupy the α sub-lattice sites and all the B ions occupy the β sub-lattice sites. A given configuration of the alloy system may be specified by the fraction δ_α of the α sites occupied by A atoms, or equivalently by the order parameter

$$\eta = \frac{\delta_\alpha - c}{1 - c} \quad (6.10)$$

For perfect order $\eta = 1$ and for the random state $\eta = 0$. It is, of course, possible to introduce further order parameters but this further complicates the configuration averaging. Leung (1978) has calculated $\langle U_2\rangle_c$ and $\langle U_M\rangle_c$ for the one order parameter case and finds a free energy of the form

$$F = F_{eg} + U_1 + U_M^{(0)} + U_2^{(0)} + \eta^2(U_M^{(1)} + U_2^{(1)}) - TS_c(\eta) \quad (6.11)$$

where $U_M^{(1)}$ depends on $(Z_A - Z_B)^2$ and $U^{(1)}$ depends on $|v^A - v^B|^2$ and on the crystal structure. This is of the same form as the free energy obtained in the usual Bragg-Williams theory (e.g. Guggenheim 1952) and therefore may be analysed in the standard fashion. The order-disorder transition temperature T_c is given by

$$k_BT_c = -\gamma(U_M^{(1)} + U_2^{(1)}) \quad (6.12)$$

where γ is a numerical factor depending on the particular stoichiometric concentration of the alloy. Leung (1978) applied this analysis to the order-disorder transitions in Mg_3Cd and $MgCd_3$. Since these alloys are homovalent, ordering is solely determined by the difference in pseudopotentials $|v^{Mg} - v^{Cd}|^2$. Using local pseudopotentials adjusted to fit the non-local ones of Animalu and Heine (1965), Leung calculates transition temperatures of 421 K for Mg_3Cd and 438 K for $MgCd_3$. These should be compared with the experimental values of 424 K and 357 K respectively.

Inglesfield (1969a,b,c) had earlier presented a rather successful account of the phase diagram, crystal structure and order-disorder behaviour of the CdMg, CdHg, MgHg alloy systems by considering only energy differences between the completely ordered and completely disordered alloys. Inglesfield's papers afford excellent examples of the richness of this subject and show how far one can progress with this rather simple theory (see also Hafner 1976,1977a). The alloys considered by Inglesfield are very special, however. Not only are they homovalent, their constituent elements have very similar atomic volumes and come from the same column in the periodic table. These features make the alloys prime candidates for study in the pseudopotential scheme. Whereas volume differences can be accounted for and only make the calculations more complicated, large differences in electronegativity, as (roughly) measured by $|v^A - v^B|$, might lead to a breakdown of the perturbation theory.

To understand this we examine (6.8a) or (6.9a). The term involving the virtual crystal contribution $\bar{v}$ only requires this quantity at non-zero reciprocal lattice vectors $\underline{g}$. We expect, in general, both $v^A(g)$ and $v^B(g)$ to be small so that $\overline{v(g)}$ is also small and the appropriate contribution to the 'band structure' energy is similar to that in a perfect crystal. The disorder introduces an additional contribution which depends on an integral of $|v^A - v^B|^2$ over <u>all</u> wave vectors q. However, if $|v^A - v^B|$ is large for small values of q, as will be the case if the constituents have different valences or come from different parts of the periodic table, it may be inappropriate or inaccurate to calculate the relevant contribution in the linear response approximation. (The situation is analogous to that discussed in § 4.3 for the case of a monovacancy). A similar argument suggests that calculating the alloy self-energy $\sum(\underline{k},E)$ using second-order perturbation theory (see (3.24)) may be inaccurate for these alloys where v^A and v^B are very different.

The effects of lattice vibrations have been considered by Leung et al. (1976) and Leung (1978). These authors have generalized the Gibbs-Bogliubov procedure, which we described in § 4.3, to a binary alloy. As a reference system they chose independent Einstein oscillators i.e. an A or B type ion is assumed to oscillate independently about its lattice point in a harmonic potential well with a frequency corresponding to Einstein temperatures Θ_A or Θ_B. The thermal averages over such a reference system are then straightforward. The configuration averaging is performed <u>after</u> the thermal averaging and is equivalent to that described above. The entropy of the reference system is the sum of thermal and configurational contributions. Leung et al. used the theory to investigate the excess entropy and heat of alloying in the Mg-Cd system while Leung (1978) considered the effect of lattice vibrations on the order-disorder transition temperature T_c. He found that lattice vibrations lowered T_c to 380 K and 363 K for Mg_3Cd and $MgCd_3$ respectively. The more pronounced lowering of T_c in $MgCd_3$ is related to the fact that the calculated Einstein temperatures are smaller in $MgCd_3$ than in Mg_3Cd.

6.3 Real space representation for alloys

Just as in the case of pure metals it is often useful to transform the effective Hamiltonian for the ions into a real space representation. Clearly H_{ion} must be of the form

$$H_{ion} = H_k + U_0 + \frac{1}{2} \sum_{\ell \neq \ell'} \phi_{\ell\ell'}(|\underline{R}_\ell - \underline{R}_{\ell'}|; n_0) \tag{6.13}$$

where U_0 is structure independent and

$$\phi_{\alpha\alpha'}(q; n_0) = \frac{4\pi}{q^2} Z_\alpha Z_{\alpha'} e^2 + \chi(q) v^\alpha(q) v^{\alpha'}(-q) \tag{6.14}$$

with $\alpha, \alpha' =$ A or B. The pairwise potentials $\phi_{AA}, \phi_{AB}, \phi_{BB}$ vary with concentration due to their dependence on the average electron density n_0. (The bare pseudopotentials are not usually allowed to depend on concentration but see Ashcroft and Stroud 1978) .

In order to treat order-disorder problems in crystalline alloys it is convenient to re-write (6.13) in terms of mean and difference pairwise potentials, i.e.

$$H_{ion} = H_k + U_0 + \frac{1}{2} \sum_{\ell \neq \ell'} \phi_m(|\underline{R}_\ell - \underline{R}_{\ell'}|) + \frac{(1-c)^2}{2} \sum_{\ell \neq \ell'}^{(AA)} \phi_d(|\underline{R}_\ell - \underline{R}_{\ell'}|)$$

$$+ \frac{c^2}{2} \sum_{\ell \neq \ell'}^{(BB)} \phi_d(|\underline{R}_\ell - \underline{R}_{\ell'}|) - c(1-c) \sum_{\ell \neq \ell'}^{(AB)} \phi_d(|\underline{R}_\ell - \underline{R}_{\ell'}|) \tag{6.15}$$

where the mean potential is

$$\phi_m(q) = \frac{4\pi}{q^2} \bar{Z}^2 e^2 + \chi(q) |\bar{v}(q)|^2$$

and the difference potential is

$$\phi_d(q) = \frac{4\pi}{q^2} (Z_A - Z_B)^2 e^2 + \chi(q) |v^A(q) - v^B(q)|^2$$

Order-disorder transitions for Hamiltonians with the same structure as (6.15) are well-studied (e.g. Guggenheim 1952). As we have seen in § 6.2 all the ordering aspects of the alloy problem are governed by $\phi_d(R)$ or its Fourier transform. These functions are discussed in some detail by Inglesfield (1969a,b,c) and Heine and Weaire (1970).

6.4 Liquid alloys

The Hamiltonian of (6.13) has recently been employed in many calculations of the thermodynamic properties of liquid alloys (for reviews see Ashcroft and Stroud 1978 and Young 1977). Almost all work is based on the generalization of the variational method, described in § 5.3, to binary alloys. The reference system is chosen to be a mixture of hard-spheres of diameters d_1 and d_2. For such a system the partial structure factors and free energy are available in closed form in the Percus-Yevick approximation. More accurate approximations for the free energy are also available. Most theories assume the collision diameter for spheres of different types is $1/2(d_1 + d_2)$ - the so-called 'additivity requirement'. This choice of reference system would appear to restrict application of the theory to those alloys in which the repulsive part of ϕ_{AB} lies intermediate to ϕ_{AA} and ϕ_{BB}. There is good evidence (Ashcroft and Stroud 1978 and Young 1977) that the calculated pairwise potentials for alloys of simple metals satisfy this requirement. For those alloys in which ϕ_{AB} is not intermediate to ϕ_{AA} and ϕ_{BB} it will be necessary to find another more suitable reference system.

The Helmholtz free energy satisfies the inequality

$$F \leqslant F_{hs}^{alloy} + U_0 + \left\langle \frac{1}{2} \sum_{\ell \neq \ell'} \phi_{\ell\ell'}(|\underline{R}_\ell - \underline{R}_{\ell'}|; n_0) \right\rangle_{hs} \tag{6.16}$$

F_{hs}^{alloy} is the free energy of the hard-sphere reference system at the same density and temperature as the true system ; it includes the classical entropy of mixing. The r.h.s. of (6.16) is minimized w.r.t. the diameters d_1 and d_2. This is again most easily carried out in reciprocal space (see § 5.3) so that one minimizes, at each concentration

$$F_{hs}^{alloy} + \frac{N}{4\pi^2}\int_0^\infty dq 4\pi e^2 \sum_{\alpha,\alpha'} (c_\alpha c_{\alpha'})^{1/2} Z_\alpha Z_{\alpha'} (a_{\alpha\alpha'}^{hs}(q) - 1)$$

$$+ \frac{N}{4\pi^2}\int_0^\infty dq q^2 \chi(q) \sum_{\alpha,\alpha'} (c_\alpha c_{\alpha'})^{1/2} v^\alpha(q) v^{\alpha'}(-q) a_{\alpha\alpha'}^{hs}(q)$$

where $a_{\alpha\alpha'}^{hs}(q)$ is a hard-sphere partial structure factor and c_α is the concentration of α. The structure factors are defined in general by

$$a_{\alpha\alpha'}(q) = (N_\alpha N_{\alpha'})^{-1/2} \left\langle \sum_{\ell,\ell'} \exp(i\underline{q}.(\underline{R}_\ell^{(\alpha)} - \underline{R}_{\ell'}^{(\alpha')})) \right\rangle - (N_\alpha N_{\alpha'})^{1/2} \delta_{\underline{q},0} \tag{6.17}$$

where N_α and $N_{\alpha'}$ are the total numbers of α and α' ions respectively.

The minimum value then serves as an estimate of F.

The procedure described above has been used to calculate the atomic volume, enthalpy and entropy of formation for a few alloy systems (Hafner 1977b). Since such calculations necessarily involve the computation of very small energy differences it is not surprising that the results are sensitive to small changes in the input information e.g. pseudopotentials and densities. Nevertheless, the results are encouraging. Young and co-workers (see Young 1977) have studied the entropies of mixing of many alloy systems using simplified versions of this model. The important but difficult problem of phase separation in alloys has also been treated within the general framework of this model (see Ashcroft and Stroud 1978). Considering the complexity of these problems (neither phase separation nor the thermodynamical properties of mixtures of simple insulating liquids are particularly well-understood) the progress in understanding liquid alloys has been remarkable.

We should note that the alloy Hamiltonian (6.13) can be used directly in computer simulations of liquid alloys. Given the pairwise potentials ϕ_{AA}, ϕ_{AB} and ϕ_{BB} the partial radial distribution functions $g_{\alpha\alpha'}(R)$ and hence $a_{\alpha\alpha'}(q)$ can be computed as functions of concentration. The thermodynamic properties can also be obtained. The partial distribution functions obviously contain a great deal of information concerning the interionic forces and their concentration dependence. Changes in electronic structure which occur on alloying will influence these forces and be reflected in the behaviour of the partial structure factors. Since the latter can be extracted from neutron diffraction experiments (in those cases where suitable isotopes exist) it should be possible to test theories of interionic potentials rather directly. Recently Jacucci et al. (1978) have calculated the partial structure factors of equimolar NaK using pairwise potentials generated in a pseudopotential scheme. The results are qualitatively similar to those obtained in a hard-sphere model of the same system and there are interesting asymmetries associated with the principal peaks. Unfortunately, there do not appear to be any experimental results for this alloy.

7. POSSIBLE IMPROVEMENTS ON THE THEORY AND SOME RECENT DEVELOPMENTS

In spite of its many successes the pseudopotential theory of cohesion and structure has several important limitations - leaving aside its restriction to simple metals. We have discussed two examples, point defects and heterovalent alloys, where the simple second-order perturbation theory fails or becomes suspect. In both cases the difficulties are associated with the occurrence of large Fourier components of the pseudopotential at small wavevectors.

These arise because the difference δV between the total pseudopotential and the potential due to the initial uniform positive background is 'large' so that the linear screening approximation for the charge density, i.e. (4.10) is invalid. Under such circumstances we must solve for $n(\underline{r};\lambda)$ in a proper non-linear, self-consistent scheme. The Kohn-Sham theory (see Appendix 1) is the obvious choice but it is difficult - even within the simplified framework of the pseudopotential model - to solve the relevant self-consistent field equations for an arbitrary infinite array of ions. We recall that the effective one-body potential of the Kohn-Sham theory is given by

$$V_{eff}[n;\underline{r}] = V_H(\underline{r}) + \frac{\delta E_{xc}[n]}{\delta n(\underline{r})} \tag{7.1}$$

where $E_{xc}[n]$ is the exchange-correlation functional and $V_H(\underline{r})$ is the total electrostatic potential. In the pseudopotential model

$$V_H(\underline{r}) = e^2\int\frac{d\underline{r}'n(\underline{r}')}{|\underline{r}-\underline{r}'|} + \sum_{\ell} v^{ps}(|\underline{r}-\underline{R}_{\ell}|) \tag{7.2}$$

For certain problems it is convenient to add and subtract the potential due to a selected neutralizing positive background $n_+(\underline{r})$ and write

$$V_H(\underline{r}) = V_0(\underline{r}) + \delta V'(\underline{r}) \tag{7.3}$$

where

$$V_0(\underline{r}) = e^2\int d\underline{r}'\,\frac{(n(\underline{r}') - n_+(\underline{r}'))}{|\underline{r}-\underline{r}'|}$$

and

$$\delta V'(\underline{r}) = \sum_{\ell} v^{ps}(|\underline{r}-\underline{R}_{\ell}|) + e^2\int d\underline{r}'\,\frac{n_+(\underline{r}')}{|\underline{r}-\underline{r}'|}$$

If $n_+(\underline{r})$ is sufficiently simple it is possible to solve the Kohn-Sham equations, using the local density approximation for $E_{xc}[n]$, with the potential $V_0(\underline{r})$. If, in addition, $\delta V'$ is 'weak' then it might be possible to treat this additional potential, which contains all the ionic contributions, in perturbation theory. This procedure (Gunnarsson 1978) was employed by Lang and Kohn (1970) in their calculations of the surface energy ; $n_+(\underline{r})$ was then the positive background of the jellium surface. Manninen et al. (1975) used the same scheme to calculate monovacancy formation energies. In this case $n_+(\underline{r})$ corresponds to a spherical hole, whose radius is equal to the Wigner Seitz radius, in an otherwise uniform positive background of density n_0. Working to first order in $\delta V'$ these authors calculated surface energies and vacancy formation energies which were in qualitative agreement with experiment for most of the metals which they considered. Later Finnis (1975) pointed out that $\delta V'$ is still substantial for the surface of polyvalent metals and showed that the inclusion of terms $\mathcal{O}(\delta V')^2$ in the surface energy could destroy the good agreement between theory and experiment. Evans and Finnis (1976) made the same points for the vacancy problem. The second-order contributions are, however, difficult to calculate since they require the density response function of the non-uniform electron gas. Recently Monnier and Perdew (1978) have re-examined the surface energy problem. They argue that if $\delta V'$ is large then it is better to include some suitable average of this potential in the Kohn-Sham procedure and treat the remainder by perturbation theory. The average must have the same symmetry as $V_0(\underline{r})$. In practice Monnier and Perdew add a parametrized potential V_v to V_0, solve the Kohn-Sham equations in this potential, treat $\delta V' - V_v$ in first order perturbation theory and minimize the total energy w.r.t. the parameters of V_v. This variational procedure appears to give realistic surface energies for almost all the simple metals and does not seem to have the deficiencies of the earlier work. Manninen and Nieminen (1978) have adapted the variational procedure to the vacancy formation problem and they calculate rather satisfactory formation energies. (These authors also refer to other recent work on the energetics of vacancies). For an instructive review of recent work on the energetics of light interstitials e.g. H and He, in metals see Stott 1978. The theoretical techniques which are being applied to these problems are very similar to those outlined above i.e. the screening of the impurity is treated non-linearly but the lattice is handled in perturbation theory.

For concentrated disordered alloys the need to perform a configuration average seriously impedes progress. Although the CPA will provide a good approximation to the averaged density of states of the alloy when the difference between the ionic potentials $v^A - v^B$ is large, it does not contain a prescription for these potentials. Self-consistency i.e. calculating potentials from charge densities and vice versa, has not yet been properly incorporated into the theory of strong-scattering alloys.

One key requirement for a theory of the energetics of metals is that it can predict correct crystal structures. As we mentioned in § 4.3 the present pseudopotential approach has only been partially successful in this respect. Since the energy differences between structures are so small uncertainties or inaccuracies in any ingredient of the theory can lead to substantial errors in the total energy difference. Thus it is sensible to attempt to isolate the contribution to the total energy which is most 'structure dependent'. With this aim in mind we re-formulate the theory in terms of screened pseudopotentials rather than bare pseudopotentials and re-group the various contributions to the total energy.

The theory of § 4 involves the response of the electron gas to the perturbation associated with the bare pseudo-ions. Consequently the total energy contains a bare ion-ion Madelung term U_M and a term U_2 which involves the re-distribution of the interacting electrons. The effects of electron-electron interaction, screening, as well as band structure, are lumped together in this term. It is instructive to decompose U_2 into an effective one-electron contribution plus a correction term. From (4.11, 4.12 and 4.15) we have

$$U_2 = \frac{1}{2\Omega} \sum_{\underline{q}\neq 0} \chi(q)\,|v^{ps}(\underline{q})|^2$$
$$= \frac{1}{2\Omega} \sum_{\underline{q}\neq 0} n_1(\underline{q})\,v^{ps}(-\underline{q}) \tag{7.4}$$

where $n_1(\underline{q})$ is the electron density in the linear screening approximation. We can, however, introduce the total screened pseudopotential V^{ps}_{sc} which gives a prescription for the electron density in terms of the non-interacting density response (see Appendix 1 and (A.15) in particular) :

$$n_1(\underline{q}) = \chi_0(q)\,V^{ps}_{sc}(\underline{q}) \tag{7.5}$$

The screened pseudopotential is an effective one-body potential and is related to the bare pseuopotential by the analogue of (7.1) :

$$V^{ps}_{sc}(\underline{q}) = V^{ps}(\underline{q}) + U_{ee}(q)\,n_1(\underline{q}) \tag{7.6}$$

where $U_{ee}(q) = 4\pi e^2/q^2 + U_{xc}(q)$. Substituting (7.5) and (7.6) into (7.4) we find

$$U_2 = \frac{1}{2\Omega} \sum_{q\neq 0} [\chi_0(q)\,|V^{ps}_{sc}(\underline{q})|^2 - U_{ee}(q)\,|n_1(\underline{q})|^2] \tag{7.7}$$

The first term in (7.7) is the effective one-electron contribution to the total energy and is therefore associated with purely 'band-structure' effects. The second term corrects for the double counting of the electron-electron interaction which occurs in the first term. (Incidentally the formulation of U_2 as given by (7.7) is the same as that originally presented by Harrison 1966).

We now add the Madelung term and re-group. It is clear that $U_2 + U_M$ can be written as in (5.1) with the pairwise potential decomposed as

$$\phi(q;n_0) = \phi_{no}(q;n_0) + \phi_{1e}(q;n_0) \tag{7.8}$$

where

$$\phi_{no}(q;n_0) = \frac{4\pi}{q^2} Z^2 e^2 - U_{ee}(q)\,|\chi_0(q)\,v^{ps}_{sc}(q)|^2 \tag{7.9}$$

represents a residual pairwise potential between 'neutral pseudo-atoms' and

$$\phi_{1e}(q;n_0) = \chi_0(q)|v_{sc}^{ps}(q)|^2 \tag{7.10}$$

is the pairwise potential arising from the effective one-electron 'band structure' of these pseudo-atoms. It is easy to check that the q = 0 limits of both ϕ_{no} and ϕ_{1e} are finite. $\phi_{no}(q;n_0)$ is positive and $\phi_{1e}(q;n_0)$ is negative.

The 'neutral object' pairwise potential requires further comment. In the absence of exchange and correlation ($U_{xc} = 0$) the sum of $\phi_{no}(|\underline{R}_\ell - \underline{R}_{\ell'}|;n_0)$ over all pairs of ions is just the difference in electrostatic energies between a lattice of positive point ions of charge Z and a lattice in which the positive charge has the same density distribution as the electrons. It is unlikely that such an electrostatic energy difference would be very crystal structure dependent. Indeed the Madelung energy, which is difference in electrostatic energy between the lattice of point ions and that of a uniform positive charge distribution, is extremely insensitive to structure for perfect crystals (e.g. Harrison 1966). Including exchange and correlation should not drastically alter these conclusions. Consequently we might expect the effective one-electron contribution to the total energy to be the most important in determining which crystal structure a particular metal will adopt for a given atomic volume. We will discuss this further below but first we examine the behaviour of $\phi_{no}(R;n_0)$ and $\phi_{1e}(R;n_0)$. This is illustrated for Al in Figure 3. ϕ_{no} is strongly repulsive at short distances and has weak oscillations for large R. ϕ_{1e} is strongly attractive for small R and exhibits pronounced Friedel oscillations at large R. The sum of these gives the familiar pairwise potential $\phi(R;n_0)$. We note that while there is strong cancellation between ϕ_{no} and ϕ_{1e} at small R these individual contributions are only one order of magnitude greater than the resultant potential ϕ. This should be contrasted with the standard division of the pairwise potential into the direct Coulomb and 'band structure' contributions where, as we saw in § 5.2, cancellations of three orders of magnitude are involved.

Preliminary calculations for several simple metals confirm that the one-electron contribution is much more structure sensitive than the neutral object contribution. This means that it is worthwhile searching for a more convenient representation of the one-electron contribution to the total energy. For a perfect crystal the latter can be written as

$$\frac{U_{1e}}{N} = \frac{N}{2\Omega}\sum_{\underline{g}\neq 0} \chi_0(g)|v_{sc}^{ps}(g)|^2 \tag{7.11}$$

which converges rather slowly (see § 4.3) so that it is difficult

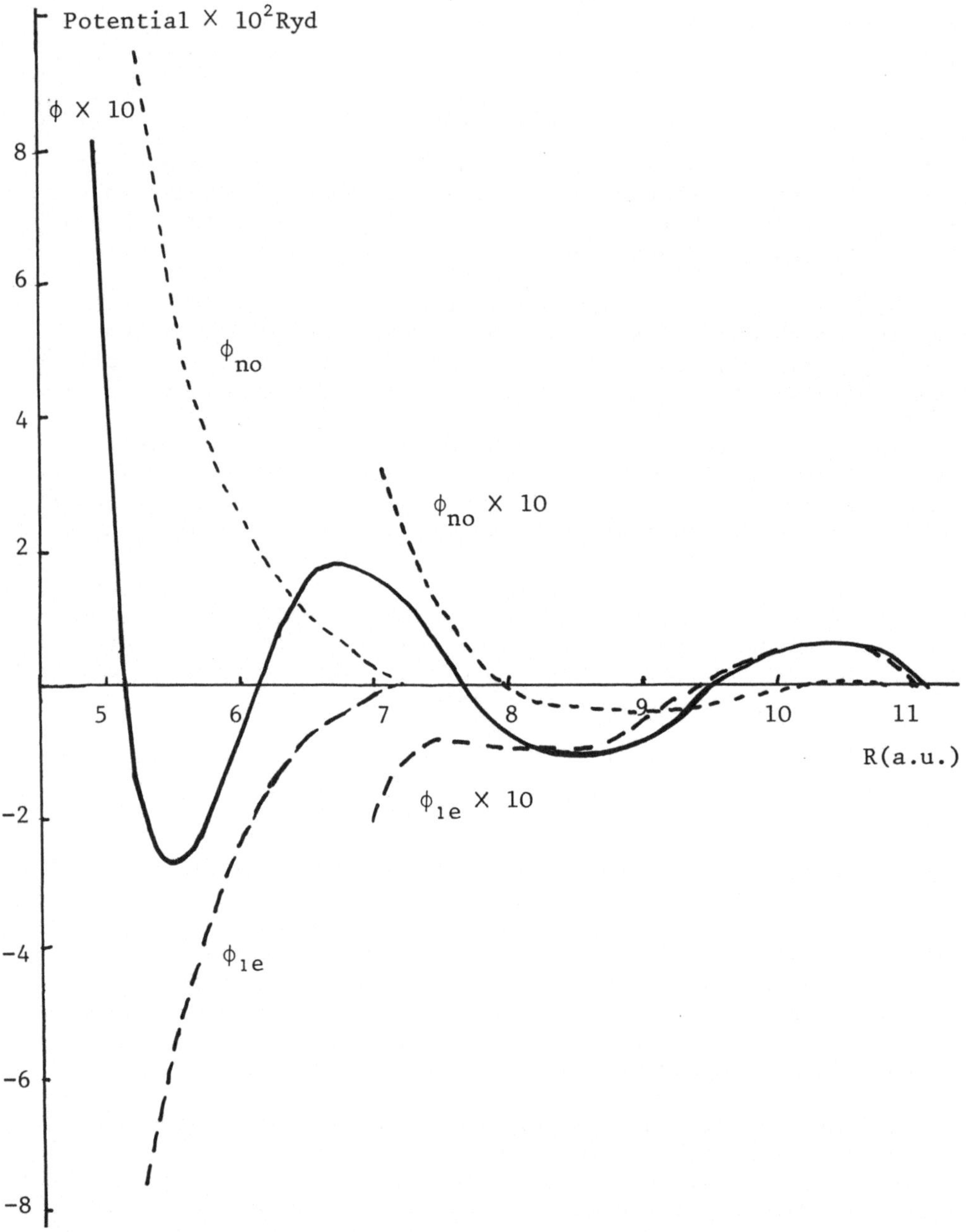

Figure 3 : The pair potential for Al. ϕ_{no} is the 'neutral object' contribution, ϕ_{1e} is the one-electron contribution and the solid curve is $\phi = \phi_{no} + \phi_{1e}$. The calculations were carried out using an Ashcroft empty core pseudopotential with core radius = 1.2 a.u. and the Vashishta-Singwi form of U_{xc}. The atomic volume is 126.9 a.u.

to ascertain exactly what it is that makes a particular crystal structure energetically favourable. Using an analytical representation of the Lindhard function χ_0, we have been able to re-express U_{1e} in a physically more transparent fashion which emphasizes the importance of the first few reciprocal lattice vectors and the diameter of the free-electron Fermi sphere $2k_F$ in determining crystal structures. This work will be reported elsewhere but the approach appears to be very useful for understanding the phase stability of certain binary alloys. In particular it lends support to the Stroud and Ashcroft (1971) explanation of the Hume-Rothery rule for the α-β phase boundary in CuZn and related alloys.

We should stress that if the effective one-electron contribution is the most important structure sensitive term in this second-order perturbation theory it is likely that the corresponding term in a fully non-linear theory will also be structure sensitive. In fact we can re-group terms in the Kohn-Sham theory in a similar fashion to that described above. The total ground state energy (T = OK) of a system of electrons and ions can be written as (see Appendix 1)

$$U = U_{es} + E_{xc}[n] + \sum_i \varepsilon_i - \int d\underline{r}\, n(\underline{r}) V_{eff}[n;\underline{r}] \tag{7.12}$$

where U_{es} is the total electrostatic energy, $E_{xc}[n]$ is the exchange correlation functional, $V_{eff}[n;\underline{r}]$ is the effective one-body potential given by (7.1) and (7.2) and ε_i are the eigenvalues of the Schrödinger equation with V_{eff} as potential (see A.2). The electrostatic energy for a pure metal of valence Z is

$$U_{es} = \frac{e^2}{2} \iint d\underline{r}\, d\underline{r}' \frac{n(\underline{r})n(\underline{r}')}{|\underline{r} - \underline{r}'|} + \frac{e^2}{2} \sum_{\ell \neq \ell'} \frac{Z^2}{|\underline{R}_\ell - \underline{R}_{\ell'}|} + \int d\underline{r}\, n(\underline{r}) V^{ps}(\underline{r}) \tag{7.13}$$

thus we can re-write (7.12) as

$$U = \sum_i \varepsilon_i + E_{xc}[n] - \int d\underline{r}\, n(\underline{r}) \frac{\delta E_{xc}[n]}{\delta n(\underline{r})} + \frac{e^2}{2} \sum_{\ell \neq \ell'} \frac{Z^2}{|\underline{R}_\ell - \underline{R}_{\ell'}|} - \frac{e^2}{2} \iint \frac{d\underline{r}\, d\underline{r}'\, n(\underline{r})n(\underline{r}')}{|\underline{r} - \underline{r}'|} \tag{7.14}$$

This can be further simplified if we make the local density approximation for $E_{xc}[n]$ (see Appendix 1) :

$$U = \sum_i \varepsilon_i + \frac{e^2}{2} \sum_{\ell \neq \ell'} \frac{Z^2}{|\underline{R}_\ell - \underline{R}_{\ell'}|} -$$

$$- \frac{1}{2} \iint d\underline{r} d\underline{r}' n(\underline{r}) n(\underline{r}') [\frac{e^2}{|\underline{r} - \underline{r}'|} + 2\delta(\underline{r} - \underline{r}') \frac{d\varepsilon_{xc}}{dn} |_{n(\underline{r})}] \quad (7.15)$$

where $\varepsilon_{xc}(n)$ is the exchange and correlation energy per electron of a uniform electron gas of density n. The first term is just the sum of the one-electron eigenvalues while the sum of the remaining terms has precisely the same physical content as the 'neutral object' contribution which we discussed above. The difference lies in the fact that the electron density $n(\underline{r})$ is to be evaluated non-linearly so that we can no longer associate a spherical charge distribution with each ion, as was the case previously. Consequently the 'neutral object' contribution cannot now be written as a sum of pairwise contributions. Nevertheless we might expect this contribution to be relatively insensitive to the crystal structure - we now have in mind closely packed perfect crystals. Thus by concentrating on the one-electron contribution, now calculated in the fully non-linear scheme, we might still be able to expose those physical features which determine the crystal structure.

The above argument is not restricted to simple metals. We could equally well consider noble and transition elements by taking Z to refer to all the electrons outside the core i.e. including the outer d shell. It seems reasonable to attempt to explain the crystal structures of say the transition metals by comparing the quantity $\int^{E_F} dE\, E\, n(E)$ for different crystal structures at the same atomic volume. Such a procedure (e.g. Pettifor 1970) is, in fact, known to give a rather good account of the observed crystal structure across the transition metal series. The Cambridge group (e.g. Pettifor 1977, Heine 1978) has other evidence to support this viewpoint.

It would also make sense to study the crystal structure of disordered transition metal alloys in this approximation since all that is required is the configuration averaged density of states n(E). As we mentioned earlier the CPA is probably accurate enough for such purposes. A more ambitious programme which uses the above division of the total energy into effective one-electron and 'neutral object' contributions has been developed to investigate lattice dynamics in transition metals (Pickett and Györffy 1976).

APPENDIX 1

(a) The Kohn-Sham Prescription for the One-Electron Potential

The effective one-electron potential V in (2.3) has to be specified. For this potential we have in mind the effective potential which enters the Kohn-Sham (1965) theory of the inhomogeneous electron gas. This theory is described in other lectures at this school (Gunnarsson 1978) so here we simply state the results which we require.

Hohenberg and Kohn (1964) proved that the total ground state energy U (at temperature T = OK) of any system of electrons and compensating positive charge could be written as

$$U = U_{es} + G[n] \tag{A.1}$$

where U_{es} is the total electrostatic energy and G[n] is an unknown but unique functional of the electron density $n(\underline{r})$. Kohn and Sham used this result to transform the many-electron problem into an effective one-electron problem. In their theory the exact electron density is given by

$$n(\underline{r}) = \sum_{i=1}^{N_e} |\psi_i(\underline{r})|^2 \tag{A.2a}$$

where the ψ_i are the eigenfunctions of an effective one-electron Schrödinger equation :

$$(-\nabla^2 + V_{eff}[n;\underline{r}])\psi_i(\underline{r}) = \varepsilon_i\psi_i(\underline{r}) \tag{A.2b}$$

The summation in (A.2a) extends over the N_e lowest lying orthonormal solutions of (A.2b). N_e is the total number of electrons in the system. The effective potential has the form

$$V_{eff}[n;\underline{r}] = V_H(\underline{r}) + \frac{\delta E_{xc}[n]}{\delta n(\underline{r})} \tag{A.3}$$

where $V_H(\underline{r})$ is the total electrostatic (Hartree) potential and the functional $E_{xc}[n]$ is defined by

$$E_{xc}[n] \equiv G[n] - T[n] \tag{A.4}$$

The functional T[n] is just the kinetic energy of the non-interacting system defined by equations (A.2) and it is straightforward to show

$$T[n] = \sum_i \varepsilon_i - \int d\underline{r}\, n(\underline{r}) V_{eff}[n;\underline{r}] \tag{A.5}$$

where the second term corrects for the double counting involved in the eigenvalue summation. $E_{xc}[n]$ is the exchange and correlation energy which is another unique functional of $n(\underline{r})$. All the difficulties of the many-body problem are concealed in this functional.

The simplest and most widely employed approximation for $E_{xc}[n]$ is the local density approximation. This assumes

$$E_{xc}[n] = \int d\underline{r}\, n(\underline{r})\, \varepsilon_{xc}(n(\underline{r})) \qquad (A.6)$$

where $\varepsilon_{xc}(n)$ is the exchange and correlation energy per electron of a uniform electron gas of density n. In this approximation

$$V_{eff}[n;\underline{r}] = V_H(\underline{r}) + \mu_{xc}(n(\underline{r})) \qquad (A.7)$$

where $\mu_{xc}(n) = d/dn(n\varepsilon_{xc}(n))$, and V_{eff} is clearly a local function of $\underline{r}$. Equations (A.2) then constitute a Hartree-like self-consistent field theory, the solution of which yields an approximate ground state density and total energy. Although the eigenvalues ε_i of this formalism are strictly only parameters in a variational procedure for the total energy of the system, it is common practice in band structure calculations to identify the ε_i with the band structure (see Callaway 1977 for an instructive discussion of this procedure). In other words it is conventional to identify V in (2.3) with V_{eff} of (A.7) so that for a metal consisting of N nuclei of atomic number Z_N

$$V(\underline{r}) = -e^2 \sum_{\ell=1}^{N} \frac{Z_N}{|\underline{r} - \underline{R}_\ell|} + e^2 \int \frac{d\underline{r}'\, n(\underline{r}')}{|\underline{r} - \underline{r}'|} + \mu_{xc}(n(\underline{r})) \qquad (A.8)$$

should constitute a realistic one-electron potential.

(b) Screening Pseudopotentials and the Effective Electron-Electron Interaction

Although one could attempt to construct screened pseudopotentials which have the same valence eigenvalues as those given by $V(\underline{r})$ in (A.8) a more standard and useful procedure is to introduce a 'bare' pseudopotential and then screen this in a linear approximation i.e. we write

$$V_{sc}^{ps}(\underline{r}) = \sum_{\ell=1}^{N} v^{ps}(|\underline{r} - \underline{R}_\ell|) + \int d\underline{r}'\, U_{ee}(|\underline{r} - \underline{r}'|)\, n_1(\underline{r}') \qquad (A.9)$$

where v^{ps} is the bare pseudopotential associated with each ion and n_1 is the (pseudo) electron density evaluated to first order in v^{ps}. U_{ee} is an effective electron-electron interaction which has an exchange and correlation contribution U_{xc} as well as the direct Coulomb contribution. Equation (A.9) is analogous to (A.3). $\sum_\ell v^{ps}(|\underline{r} - \underline{R}_\ell|)$ acts as an 'external' potential for the conduction electrons while the second term represents the screening of this potential. Fourier transforming (A.9) we obtain

$$V^{ps}_{sc}(\underline{q}) = V^{ps}(\underline{q}) + U_{ee}(q)n_1(\underline{q}) \tag{A.10}$$

where $V^{ps}(\underline{q}) = v^{ps}(q)S(\underline{q})$ and n_1 is to be calculated using linear response theory i.e.

$$n_1(\underline{q}) = \chi(q)V^{ps}(\underline{q}) \tag{A.11}$$

$\chi(q)$ is the interacting density response function (see § 4.2) of the uniform electron gas whose density is equal to the mean density of conduction electrons. Clearly this will be a realistic description of the screening provided V^{ps} is in some sense 'weak'. From (A.10) and (A.11) it then follows that

$$V^{ps}_{sc}(\underline{q}) = \frac{v^{ps}(q)}{\varepsilon_{el}(q)} S(\underline{q}) \tag{A.12}$$

where

$$\frac{1}{\varepsilon_{el}(q)} = 1 + U_{ee}(q)\chi(q)$$

Thus we can associate a screened pseudopotential v^{ps}_{sc} with each ion where $v^{ps}_{sc}(q) = v^{ps}(q)/\varepsilon_{el}(q)$. $\varepsilon_{el}(q)$ is usually referred to as the <u>electron</u> dielectric function. Since $\chi(q)$ depends on the mean electron density, the screened pseudopotential varies with atomic volume. In an alloy the screened pseudopotential of each species will vary with concentration. Some typical bare and screened pseudopotentials are sketched in Figure 4. ε_{el} is not to be confused with the true dielectric function $\varepsilon(q)$. The latter is defined by

$$\delta V_H(\underline{q}) \equiv \delta V_{ext}(\underline{q})/\varepsilon(q) \tag{A.13}$$

i.e. it relates the infinitesimal change in the total electrostatic potential δV_H to the change in whatever external potential V_{ext} is perturbing the electron system. Since

$$V_H(\underline{r}) = V_{ext}(\underline{r}) + e^2\int\frac{d\underline{r}'n(\underline{r}')}{|\underline{r} - \underline{r}'|}$$

it follows that $\delta V_H(\underline{q}) = \delta V_{ext}(\underline{q}) + 4\pi e^2/q^2\delta n(\underline{q})$ where $\delta n(\underline{q}) = \chi(q)\delta V_{ext}(\underline{q})$ and therefore

$$\frac{1}{\varepsilon(q)} = 1 + \frac{4\pi e^2}{q^2}\chi(q) \tag{A.14}$$

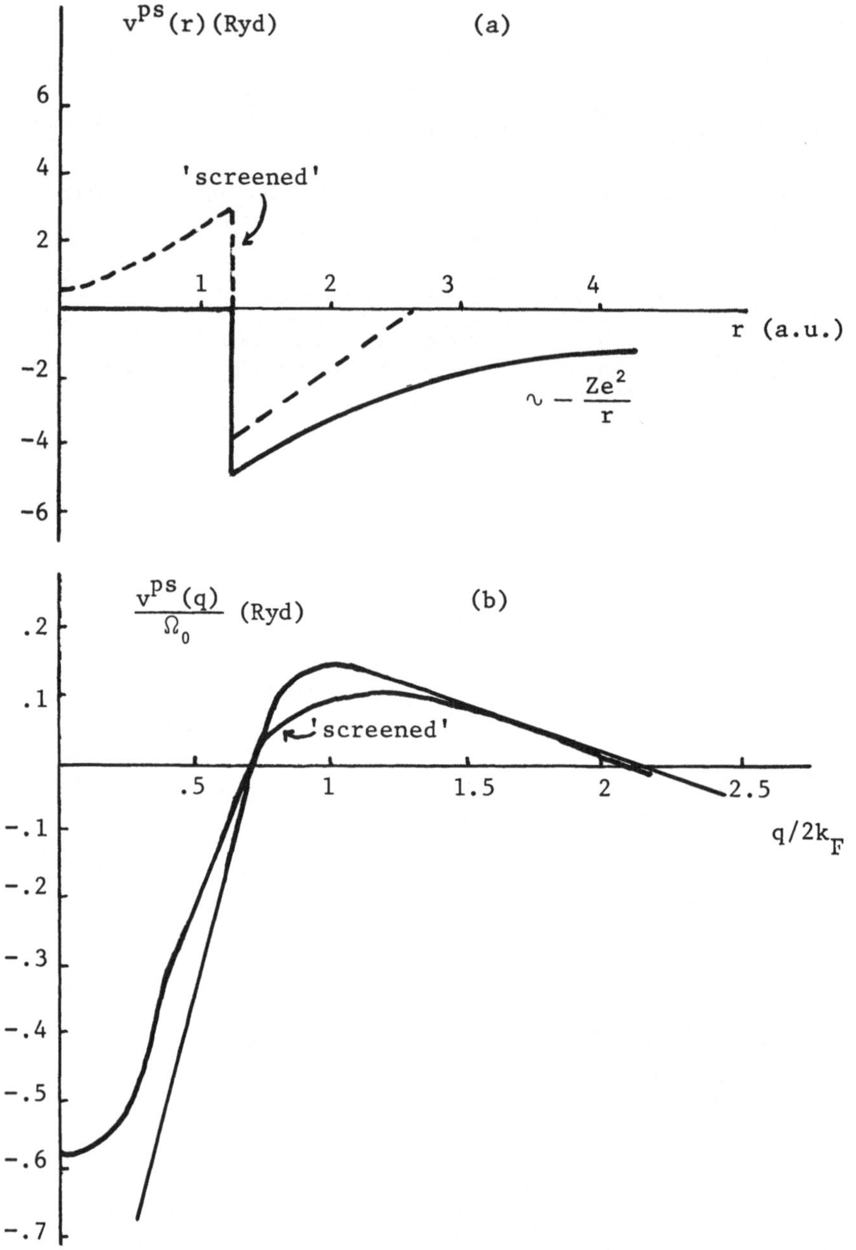

Figure 4 : Bare and screened pseudopotentials for Al. In (a) we have drawn the Ashcroft empty core potential in real space and indicated how this might appear after linearly screening. In (b) we show the corresponding results in reciprocal space. Note that $v^{ps}_{sc}(q=0)/\Omega_0 = -2/3\ k_F^2$. The first two reciprocal lattice vectors have $g/2k_F \doteq 0.78$ and 0.88 respectively.

If we write $U_{ee}(q) = 4\pi e^2/q^2 + U_{xc}(q)$ it is clear that the electron dielectric function contains an additional exchange and correlation contribution. Only in the random phase approximation (Hartree theory) where $U_{xc} = 0$, are the two dielectric functions the same.

We still have to specify U_{xc}. If V^{ps}_{sc} in (A.9) is to be an effective one-body potential then we require that

$$\chi_0(q)V^{ps}_{sc}(\underline{q}) = n_1(\underline{q}) = \chi(q)V^{ps}(\underline{q}) \tag{A.15}$$

where $\chi_0(q)$ is the <u>non-interacting</u> density response function of a uniform electron gas. V^{ps}_{sc} should induce the same density in a non-interacting linear response approach as the bare pseudopotential V^{ps} does in the fully interacting scheme. Combining (A.10), (A.11) and (A.15) we find the well-known relationship between χ, χ_0 and U_{ee} i.e.

$$\chi(q) = \chi_0(q)(1 - \chi_0(q)U_{ee}(q))^{-1} \tag{A.16}$$

At zero temperature $\chi_0(q)$ is simply the Lindhard function

$$\chi_0(q) = \frac{-k_F}{4\pi^2}\left(1 + \frac{(4k_F^2 - q^2)}{4k_F q} \ln\left|\frac{q + 2k_F}{q - 2k_F}\right|\right) \tag{A.17}$$

where k_F is the Fermi wavevector and we have used Rydberg units.

Thus we see that U_{ee} as introduced in (A.9) is the same effective electron-electron interaction which enters the conventional theory of the uniform electron gas (e.g. Pines and Nozières 1966). Consequently estimates of U_{xc} can be obtained from the various approximate theories which are currently available. The Hubbard (1958) approximation only includes exchange

$$U_{xc}(q) = \frac{-2\pi e^2}{q^2 + k_F^2} \tag{A.18}$$

More recent theories also include correlation effects. A rather simple functional form has been suggested by Singwi and co-workers :

$$U_{xc}(q) = -\frac{4\pi e^2}{q^2} A[1 - \exp(-B(q/k_F)^2)] \tag{A.19}$$

where A and B are parameters which have a rather weak dependence on the electron density. The Vashishta and Singwi (1972) version of the parameters appears to yield a $\chi(q)$ which is realistic for the metallic range of densities. Another realistic form of U_{xc} is that due to Geldart and Taylor (1970).

We should also note that U_{xc} is related to the functional $E_{xc}[n]$ of the Kohn-Sham theory. Comparing (A.3) for the effective one-body potential with (A.9) it is easy to show

$$U_{xc}(|\underline{r} - \underline{r}'|) = \frac{\delta^2 E_{xc}[n]}{\delta n(\underline{r})\delta n(\underline{r}')}\Big|_{n_0} \tag{A.20}$$

where n_0 is the density of the uniform gas. If we make a local density approximation to $E_{xc}[n]$, as in (A.6), it follows from (A.20) that $U_{xc}(r)$ is a delta function i.e.

$$U_{xc}(q) = \frac{d^2}{dn^2}(n\varepsilon_{xc}(n))\Big|_{n_0} \tag{A.21}$$

is a constant independent of q. This approximation will be inaccurate at finite wave vectors since the local density approximation cannot describe short-wavelength density fluctuations. At q = 0, however, (A.21) will be exact and is essentially a statement of the electron gas compressibility sum rule (Pines and Nozières 1966)*. This relation is useful for testing prospective theories of U_{xc} and ε_{xc}. The full statement of the sum rule (including the kinetic energy) is

$$\chi_T^{-1} = B_e(n_0) = - n_0^2(\chi_0(q{=}0)^{-1} - U_{xc}(q{=}0)) = n_0^2 \frac{d^2}{dn^2}(nu_{eg}(n))\Big|_{n_0} \tag{A.22}$$

where $u_{eg}(n) = 3k_F^2/5 + \varepsilon_{xc}(n)$ is the total energy per electron of the uniform gas and χ_T is the isothermal compressibility. Equation (A.22) is a consequence of the general result (Hohenberg and Kohn 1964) :

$$-\left(\frac{1}{\chi(q)} + \frac{4\pi e^2}{q^2}\right) = K(q) \tag{A.23}$$

with

$$K(|\underline{r} - \underline{r}'|) = \frac{\delta^2 G[n]}{\delta n(\underline{r})\delta n(\underline{r}')}\Big|_{n_0}$$

ACKNOWLEDGEMENTS

The author is grateful to Drs. P. Lloyd and B.L. Györffy for several illuminating discussions on the subject matter of these notes.

---

* Recently (R. Taylor, J.Phys.F. : Metal Phys., 8, 1699 (1978)) it has been argued that the 'local density approximation' to $U_{xc}(q)$ should be accurate throughout the important region $q \leqslant 2K_F$.

REFERENCES

Abarenkov, I., and Heine, V., 1965, Phil. Mag., 12, 529.

Animalu, A.O.E., 1967, Phys. Rev., 161, 445.

Animalu, A.O.E., and Heine V., 1965, 12, 1249.

Ashcroft, N.W., and Langreth, D.C., 1967, Phys. Rev., 155, 682.

Ashcroft, N.W., and Lekner, J., 1966, Phys. Rev., 145, 83.

Ashcroft, N.W., and Stroud, D., 1978, in Sol. St. Phys. (to appear).

Austin, B.J., Heine, V., and Sham, L.J., 1962, Phys. Rev., 127, 276.

Ballentine, L.E., 1975, Adv. Chem. Phys., 31, 263.

Callaway, J., 1977, in 'Electrons in Finite and Infinite Structures' ed. P. Phariseau and L. Scheire (NATO ASI Series, B24, Plenum) p. 321.

Cohen, M.L., and Heine, V., 1970, Sol. St. Phys., 24, 38.

Ehrenreich, H., and Schwartz, L.M., 1976, Sol. St. Phys., 31, 149.

Evans, R., 1977, in 'Vacancies '76', ed. R.E. Smallman and J.E. Harris, (London : Metals Soc.)p. 30.

Evans, R., 1978, in 'Microscopic Structure and Dynamics of Liquids' ed. A.J. Dianaux and J. Dupuy (NATO ASI Series, B 33, Plenum) p. 153.

Evans, R., and Finnis, M.W., 1976, J. Phys.F., 6, 483.

Evans, R., and Kumaravadivel, R., 1976, J.Phys.C., 9, 1891.

Evans, R., and Schirmacher, W., 1978, J. Phys.C., 11, 2437.

Faber, T.E., 1972, 'Introduction to the Theory of Liquid Metals' (London : Cambridge University Press).

Finnis, M.W., 1975, J.Phys.F., 5, 2227.

Geldart, D.J.W., and Taylor, R., 1970, Can. J. Phys., 48, 167.

Guggenheim, E.A., 1952, 'Mixtures' (Oxford : Oxford University Press).

Gunnarsson, O., 1978, this volume.

Hafner, J. 1976, J. Phys. F., 6, 1243.

Hafner, J., 1977a, Phys. Rev. B., 15, 617.

Hafner, J., 1977b, in 'Liquid Metals 1976' ed. R. Evans and D.A. Greenwood (Inst. of Phys. Conf. Series No. 30) p. 102.

Hansen, J.-P., and McDonald, I.R., 1976, 'Theory of Simple Liquids' (New York,: Academic Press).

Harrison, W.A., 1966, 'Pseudopotentials in the Theory of Metals' (New York : Benjamin).

Hasegawa, M., 1976, J. Phys. F., 6, 649.

Hasegawa, M., and Watabe, M., 1972, J. Phys. Soc. Japan, 32, 14.

Heald, P.T., 1977, in 'Vacancies '76', ed. R.E. Smallmann and J.E. Harris (London : Metals Soc.), p. 11.

Heine, V., 1978, in 'Transition Metals 1977' ed. M.J.G. Lee, J.P. Perz and E. Fawcett (Inst. of Phys. Conf. Series No. 39) p. 722.

Heine, V., and Weaire, D., 1970, Sol. St. Phys., 24, 250.

Hohenberg, P.C., and Kohn, W., 1964, Phys. Rev., 136, B 864.

Hubbard, J., 1958, Proc. Roy. Soc., A243, 336.

Inglesfield, J.E., 1969a, J. Phys. C., 2, 1285.

Inglesfield, J.E., 1969b, J. Phys. C., 2, 1293.

Inglesfield, J.E., 1969c, Acta Met., 17, 1395.

Jacucci, G., McDonald, I.R., and Taylor R., 1978, J. Phys. F., 8, L121.

Johnston, R.A., 1973, J. Phys.F., 3, 295.

Jones, H.D., 1973, Phys. Rev. A, 8, 3215.

Kohn, W., and Sham, L.J., 1965, Phys. Rev., 140, A 1133.

Kumaravadivel, R., and Evans, R., 1976, J. Phys. C., 9, 3877.

Lang, N.D., 1973, Sol. St. Phys., 28, 224.

Lang, N.D., and Kohn, W., 1970, Phys. Rev. B., 1, 4555.

Leung, C.H., 1978, to be published.

Leung, C.H., Stott, M.J., and Young, W.H., 1976, J. Phys.F., 6, 1039.

Manninen, M., and Nieminen, R., 1978, to be published.

Manninen, M., Nieminen, R., Hautojärvi, P., and Arponen, J., 1975, Phys. Rev. B, 12, 4012.

McLaren, R.E., and Sholl, C.A., 1974, J. Phys. F., 4, 2172.

Minchin, P., Meyer, A., and Young, W.H., 1974, J. Phys. F., 4, 2117.

Monnier, R., and Perdew, J.P., 1978, Phys. Rev. B., 17, 2595.

Norris, C., 1977, in 'Liquid Metals 1976' ed. R. Evans and D.A. Greenwood (Inst. of Phys. Conf. Series No. 30) p. 171.

Pettifor, D.G., 1970, J. Phys. C., 3, 367.

Pettifor, D.G., 1977, Calphad, 1, 305.

Pickett, W.E., and Györffy, B.L., 1976, in 'Superconductivity in d and f Band Metals' ed. D.H. Douglass (New York : Plenum) p. 251.

Pines, D., and Nozières, P., 1966, 'Quantum Liquids' (New York : Benjamin).

Price, D.L., 1971, Phys. Rev. A, 4, 358.

Rasolt, M., and Taylor, R., 1975, Phys. Rev. B., 11, 2717.

Ratti, V.K., and Evans, R., 1973, J. Phys. F., 3, L 238.

Shaw, R.W., 1968, Phys. Rev., 174, 769.

Slater, J.C., 1974, 'Quantum Theory of Molecules and Solids' Vol.IV. (New York : McGraw-Hill).

Stern, E.A., 1966, Phys. Rev., 144, 545.

Stocks, G.M., Gaspari, G.D., and Györffy, B.L., 1972, J. Phys. F., 2, L 123.

Stott, M.J., 1978, in 'Properties of Atomic Defects in Metals' ed. N.L. Peterson and R.W. Siegel (Published in J. Nuc. Mat. 69 and 70) p. 157.

Stroud, D., and Ashcroft, N.W., 1971, J. Phys. F., 1, 113.

Stroud, D., and Ashcroft, N.W., 1972, Phys. Rev. B, 5, 371.

Vashishta, P., and Singwi, K.S., 1972, Phys. Rev. B, 6, 875.

Wallace, D.C., 1972, 'Thermodynamics of Crystals' (New York : Wiley).

Watabe, M., 1977, in 'Liquid Metals 1976' ed. R. Evans and D.A. Greenwood (Inst. of Phys. Conf. Series No. 30) p. 288.

Watabe, M., and Young, W.H., 1974, J. Phys. F., 4, L 29.

Wotherspoon, J., 1977, private communication.

Young, W.H., 1977, in 'Liquid Metals 1976' ed. R. Evans and D.A. Greenwood (Inst. of Phys. Conf. Series No. 30) p.1.

Ziman, J.M., 1970, Sol. St. Phys., 26, 1.

A PSEUDOPOTENTIAL APPROACH TO SOME PROBLEMS IN BINARY ALLOYS OF SIMPLE METALS

R. Taylor

Division of Physics, National Research Council of Canada

Ottawa, Ontario K1A OR6 , Canada

ABSTRACT

Some recent calculations of binary alloy properties are described. The importance of choosing a good pseudopotential is stressed and it is pointed out that a wide variety of physical properties can be calculated successfully without any experimental input. Specializing to the alloy problem, homovalent simple metal alloys appear to be well described by pseudopotential techniques but finite concentrations of non-homovalent impurities appear to present some difficulties beyond the scope of standard pseudopotential approaches.

1. INTRODUCTION

The concept of the pseudopotential has been with us for about twenty years. When first introduced into the literature pseudopotentials gave tremendous insight into the reasons for the nearly-free-electron behaviour of metals but from a computational point of view there was very little saving in effort. However the conception of the parametrized model potential pioneered by Heine and Abarenkov (1964) changed all this. In the past fifteen years vast numbers of pseudopotentials have been spawned, with parameters adjusted to fit virtually any experimental property that one can think of. This effort has certainly helped in our understanding of many physical properties and has often provided useful interpolation formulae (e.g. Fermi surface fits). But, when confronted with the problem of calculating a specific physical property, the worker all too often finds himself confronted with a choice of a number of reasonable-looking pseudopotentials all giving different answers. The reason for this lies in the fact that the pseudopotential is only useful if it can be treated as a weak potential that can be used in a perturbation scheme. Hence it is certainly possible to choose a number of pseudopotentials all of which would give the same answer when summed to convergence but when only used to first or second order perturbation theory would give quite different results. Convergence then is of critical importance. In this connection the work of Shaw (1968) is of great significance. He attempted to optimize the Heine-Abarenkov model potential by matching the square well depth to the Coulombic tail at the square well radius for each angular momentum value. The Shaw optimized model potential is considered by many to be the most reliable choice but it is still not clear whether or not, even in this case, the terms beyond second order perturbation theory, (when calculating energies) are significant.

Another problem that plagues the would-be user of a pseudopotential is the confusion of effects that might or might not be included. There are those who claim that a simple local model potential is completely adequate for virtually any calculation, but always lurking in the back-ground are such questions as the importance of non-locality, energy-dependence, depletion holes and effective masses. All of these effects complicate the problem, give rise to the possibility of more adjustable parameters and by their very presence defeat the concept of simplicity that makes the pseudopotential so attractive. Hence it would be desirable to know just how important these effects are in various situations in addition to being able to place confidence in the results generated by the pseudopotential of one's choice.

In this lecture I will outline one possible scheme for determining pseudopotential parameters. This is an entirely first principles scheme involving no experimental information other than the density and crystal structure. The resulting pseudopotentials are non-local and energy-independent with no accompanying problems of depletion holes and effective masses. Hence they are relatively easy to use and yet are capable of giving very accurate results for a wide variety of physical properties when properly screened. The latter part of the paper will be devoted to some calculations of properties of simple metal alloys that I and my colleagues have investigated.

2. CHOICE OF PSEUDOPOTENTIAL

To investigate the convergence of a given pseudopotential within perturbation theory there are two rather obvious approaches. The first is simply to proceed to the next order and calculate the contribution of this term directly. This has been done by a number of authors (e.g. Brovman, Kagan and Kholas 1972, Bertoni et al 1974 and Hasegawa 1976) with great difficulty. Not too surprisingly, they found that third order contributions to phonons and pair potentials could not be ignored, thereby raising serious questions about the usefulness of attempting quantitative calculations via pseudopotential theory. To proceed further appears to be a virtually intractable problem with diminishing returns.

The second approach was suggested by Rasolt and Taylor (1975) and Dagens, Rasolt and Taylor (1975), collectively referred to as DRT. These authors suggested that one should select a problem for which the answer can be calculated exactly by other means and compare the answer with that generated by pseudopotential theory. The problem selected by DRT was that of the charge density induced by an isolated ion embedded in an infinite electron gas. This can be solved to any desired accuracy using a Hartree-Fock (HF) approach with, say, Gaspar-Kohn-Sham exchange. DRT found that in the cases studied (Li, Na, K, Mg, Ca and Al) the first order perturbation theory result using a carefully constructed pseudopotential gave good agreement with the HF calculation. But it was not good enough. When considering the phonon problem, higher order terms turned out to be quite important as one would expect from the work of the earlier mentioned authors. Hence DRT proposed a novel way of determining the pseudopotential parameters. They suggested that the pseudopotentials be adjusted in order to force the first order perturbation theory result to agree with the full HF result outside the ionic core region. By this technique all multiple scattering events at a single ion site are effectively summed whilst conserving the simplicity of pseudopotential theory. In terms of interionic potentials this means that all contributions to the two-body interaction are summed when the pseudopotential is used to second order in perturbation theory whilst N-body ($N \geqslant 3$) forces are ignored. DRT showed that the phonons generated by these adjusted pseudopotentials when screened with the Geldart and Taylor (1970) dielectric function, agreed very well with experiment. Hence one can place a great deal of confidence in the pair potentials generated by them. Subsequently Cohen and Klein (1975) and Cohen et al (1976) have investigated thermodynamic properties of K and Na by computer simulation with great success. Calculated liquid structure factors of the alkali metals (Jaccuci, Klein and Taylor 1977 and Jaccuci, McDonald and Taylor 1978) and aluminium (Michler, Hahn and Schofield 1976) yielded good agreement with experiments. Defect properties such as point defect energies (Jaccuci and Taylor 1978) and dynamics of self-diffusion (Da Fano and Jacucci 1977) in the alkalis also appear to correlate very well with experiment. Finally, the calculated transport properties, electrical resistivity (Shukla and Taylor 1976 and Taylor, Leavens and Shukla 1976), thermal resistivity (Leavens 1977) and thermoelectric power (Leavens and Taylor 1978) of Na and K all agree remarkably well with experiment. Hence pseudopotentials generated by the DRT procedure are capable of giving accurate results for a wide variety of properties without any adjustable parameters. This of course is exactly what one feels a pseudopotential should be able to do and clearly the procedure of fitting to a single experimental property should no longer be considered a useful end in itself.

3. SOME APPLICATIONS TO BINARY ALLOY PROBLEMS

Before turning our attention to the alloy problem we should at least address ourselves to the obvious suggestion that the fitting of pseudopotentials to phonon dispersion should be adequate to determine pair potentials. The answer to this is that this procedure only determines the first and second derivatives of a pair potential at a few selected points. For any defect problem one needs to know both the magnitude and derivatives of a pair potential at all relative separations and hence such a procedure is just not adequate, particularly when considering a problem like diffusion in a dilute alloy.

To describe lattice dynamical properties of a binary alloy, with constituents A and B, one requires three pair potentials, $v_{AA}(r)$, $v_{BB}(r)$ and $v_{AB}(r)$. Those can all be determined by straightforward application of second order perturbation theory and the relevant formulae are given by Rasolt and Taylor (1975) for the case of a non-local, energy-independent pseudopotential. Having constructed these potentials it is then necessary to use some form of computer simulation to calculate the desired physical property. In the following paragraphs I would like to discuss briefly some applications of pseudopotential pair potentials to binary alloys.

(i) Dilute NaAu and LiAg alloys

In each of these alloys the host lattice is the alkali metal and the noble metal can be considered to be an isolated impurity. The unusual feature of these two systems is that Au diffuses anomalously fast in Na whereas Ag in Li diffuses only slowly. In view of the apparent similarity of the two alloys it is clearly of interest to see whether a pseudopotential-based approach to pair potentials can account for the difference. One might very well ask if it is meaningful to describe a noble metal ion by a pseudopotential. The answer to this is that, in the dilute limit, d-band effects are of no significance for a noble metal in an alkali metal host. Hence it can be regarded as another simple metal and be treated accordingly. This was the approach of Schober et al (1975) who calculated NaAu and LiAg pair potentials in addition to the host metal pair potentials to which they are compared in Figure 1. Particularly striking is the very deep minimum in the NaAu potential, well inside the bcc nearest neighbour distance, strongly suggesting that the Au ion would like to sit interstitially in the Na lattice. This was confirmed by direct calculation. The reason for the deep potential well is that a Au^+ ion has a greater affinity for electrons and can screen itself at a shorter distance than can a Na^+ ion as evidenced by the fact that the electron density parameter r_s

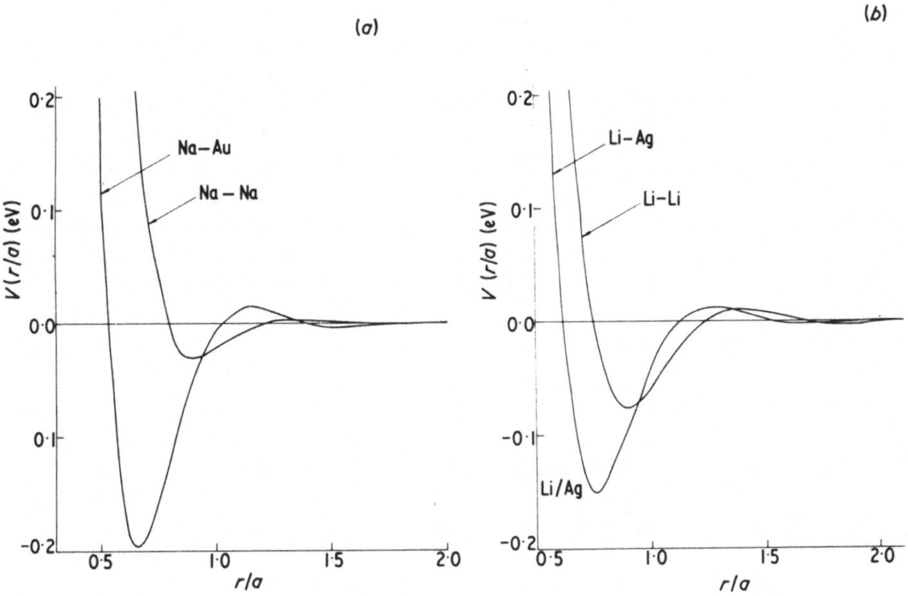

Figure 1 : Interionic potentials in electron-volts as functions of (r/a), a being the b.c.c. lattice parameter. (a) V(r) for Na-Au and Na-Na. (b) V(r) for Li-Ag and Li-Li.

is considerably larger for Na metal than for Au metal. Using the Harwell DEVIL computer simulation program and considering a number of possible diffusion mechanisms, Schober et al calculated an activation energy, for diffusion of Au in Na, of 0.12 eV as compared with the observed value of 0.10 eV (Barr et al 1971).

Looking at Figure 1, the LiAg case is not nearly so clearcut reflecting the fact that Li has a much stronger electron affinity than Na and is therefore not so distinctly different from the impurity. Schober et al found that in this case the Ag would enter the Li lattice substitutionally and the resulting calculated diffusion energy turned out to be 0.53 eV as compared to the experimental value of 0.56 eV (Mundy and McFall 1973). In view of the fact that these results were generated via a static calculation, the agreement with experiment is probably better than the authors had a right to expect. But they certainly illustrate the fact that a good understanding of physical processes can be obtained from the pseudopotential approach.

(ii) Phonons in Rb K

Recently Kamitakahara and Copley (1978) have used inelastic scattering of neutrons to study [ξξ0]L and [ξ00]T phonons of $Rb_{1-c}K_c$ at three different concentrations (c = 0.06, 0.18 and 0.29). A particularly interesting feature of these measurements is the presence of a local mode associated with both phonon branches at high frequencies. This mode grows in intensity with increasing c and appears to arise from the motion of the K ions. With an eye to providing a good test of theory Jacucci, Klein and Taylor (1978) constructed pair potentials for the Rb K system and embarked on a Molecular Dynamics simulation of the $Rb_{71}K_{29}$ alloy. A typical example of these results is illustrated in figure 2 where $S(\underset{\sim}{Q},\omega)$ is plotted for $\underset{\sim}{Q} = (2,2,1)\ (2\pi/a)$. The two peaks are clearly evident and reference to Kamitakahara and Copley (1978) shows that they are positioned correctly although the relative amplitudes are not quite correct. In figure 3 the phonon dispersion curves are illustrated where it can be seen that the computed points agree very well with experiment.

(iii) The Li Mg alloy system

I now want to turn my attention to a non-homovalent alloy to examine the validity of the pseudopotential pair potential description of such a system. Using the DRT pseudopotentials Beauchamp, Taylor and Vitek (1975) constructed pair potentials for a wide range of concentrations of Mg in Li. The potentials corresponding to 25 at % Mg are plotted in figure 4. The computed elastic constants showed good agreement with experiment and the heat of mixing showed the correct concentration dependence. However the calculated short range order parameters uncovered a serious flaw in the pair potentials. Experimentally the Li Mg alloy is completely disordered but the theory predicts a strong degree of ordering. Upon examination of the potentials in figure 4 one can see that the calculated Li-Li interaction is very similar to the Li-Mg interaction whereas the Mg-Mg interaction is much more repulsive in the near-neighbour region. Hence, although a Li ion shows no preference for either a Mg or another Li ion, a Mg ion clearly shows a preference for a Li ion and a strong degree of ordering would take place if these were the actual forces. It is most unlikely that the pseudopotentials themselves are at fault because of the method of construction. The problem almost certainly lies in the fact that with both Li^+ and Mg^{++} ions present the electron density is far from uniform. Quite conceivably, in the immediate neighbourhood of a Mg ion, the electron charge is sufficiently depleted that a Li ion can no longer effectively screen itself giving rise to a stronger Li Mg repulsion than was calculated. It was suggested, therefore, by the authors that the screening of a given ion could be significantly influenced by the type of ion in the near neighbour region. This of course gives rise to a three-body force which is beyond the scope of a pair potential second order perturbation theory description. One concludes that non-homovalent alloys must be treated much more carefully and results based on a pseudopotential description for finite concentration should be regarded with suspicion.

The foregoing remarks apply only to finite concentrations. If one considers a pure simple metal, the electron concentration

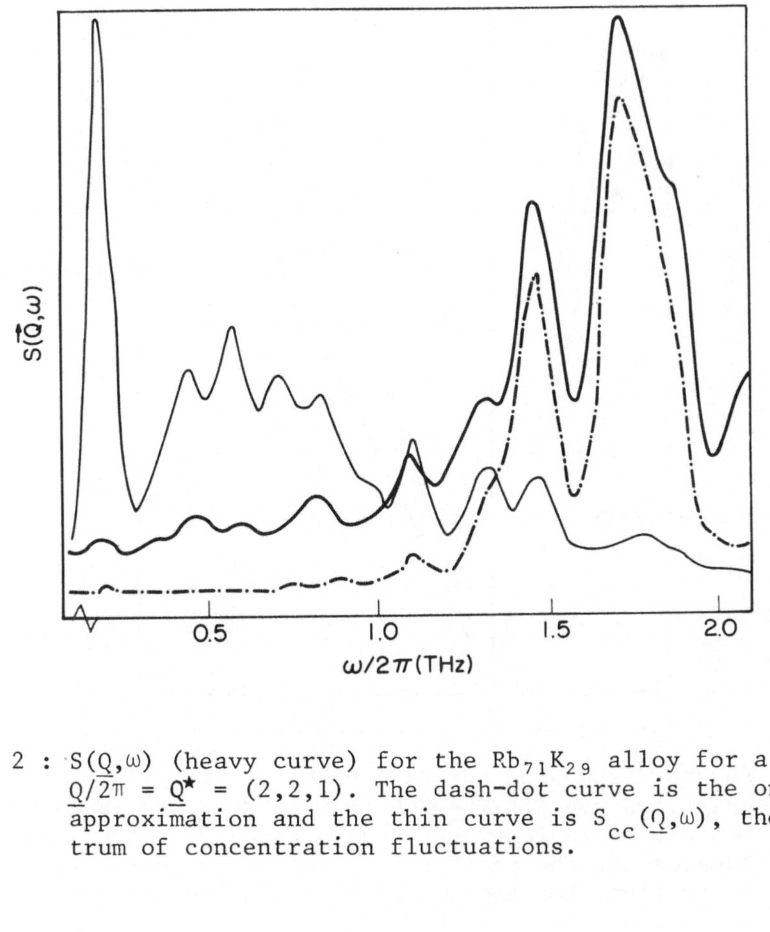

Figure 2 : $S(\underline{Q},\omega)$ (heavy curve) for the $Rb_{71}K_{29}$ alloy for a $\underline{Q}/2\pi = \underline{Q}^{\star} = (2,2,1)$. The dash-dot curve is the one-phonon approximation and the thin curve is $S_{cc}(\underline{Q},\omega)$, the spectrum of concentration fluctuations.

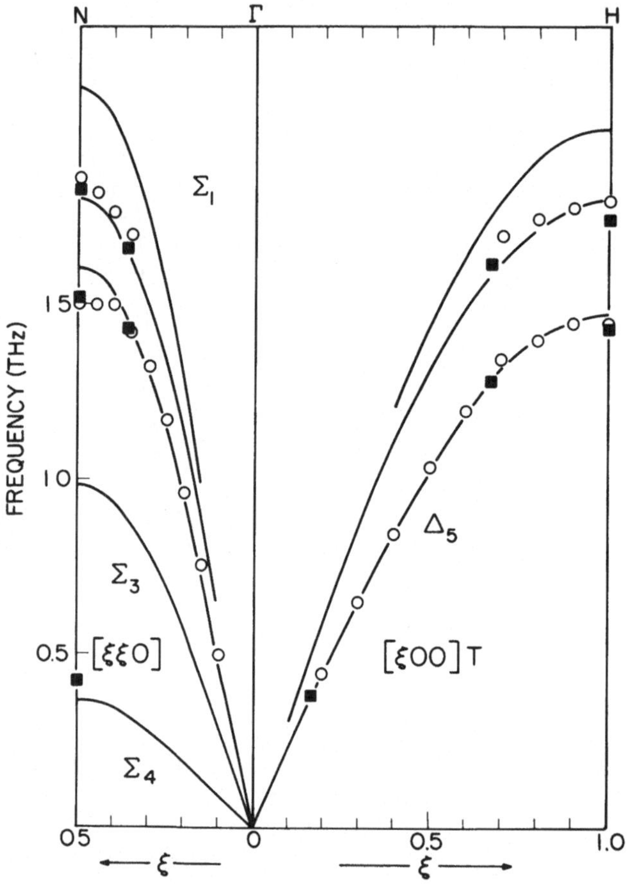

Figure 3 : Phonon dispersion curves for the $Rb_{71}K_{29}$alloy. The calculations are for T = 80K (a = 5.506Å) and the experimental points (open circles) are for T = 10 K. The solid squares are the results of a Molecular Dynamics computer simulation. For the Σ_3 and Σ_4 modes the solid lines are a quasi-harmonic calculation for pure Rb at the alloy lattice parameter. For the Σ_1 and Δ_5 modes the solid lines are the quasi-harmonic calculation for. in order of increasing magnitude, pure Rb, pure K and k in Rb all at the alloy lattice parameter.

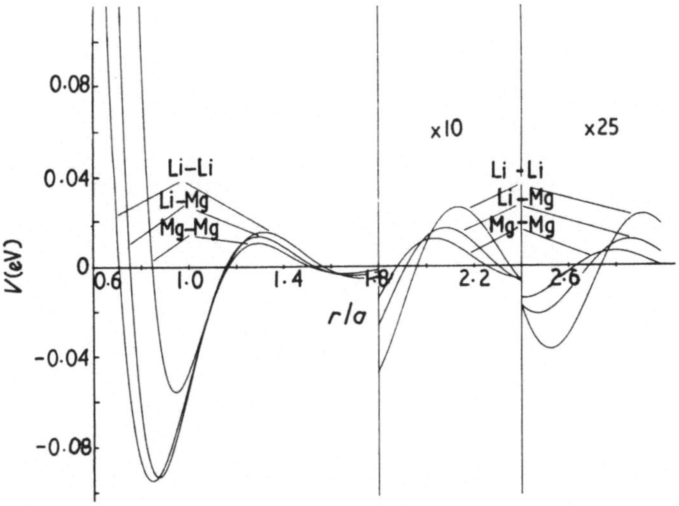

Figure 4 : Interionic potentials for Li-Li, Li-Mg and Mg-Mg corresponding to 25 at % Mg.

therein is roughly uniform. Hence, if one places an impurity in this environment, one can expect linear screening to give a good answer. Only, when the concentration becomes sufficiently high that impurity-impurity interactions are significant, will the problems that beset Beauchamp et al manifest themselves.

(iv) Dilute Be alloys

Very recently Kaufmann et al (1977) have reported the results of implanting a large number of different metallic ions in a Be host. They find that the implanted ions tend to occupy substitutional sites or either tetrahedral or octahedral interstitial sites. They can correlate their date by a Miedema (1973) plot and from there make predictions concerning the site preferences of unmeasured ions. One particular prediction of this plot is that Li will occupy a tetrahedral interstitial site but in fact it turns out to be substitutional (Kaufmann 1978). This is just the sort of problem that might well be tackled profitably from the point of view of pair potentials. In this lecture I shall report some preliminary calculations carried out by M.S. Duesbery at NRC and which we hope to publish at some later time.

In figure 5 are plotted Be-Be, Be-Al and Be-Li potentials. These are of course not homovalent impurities but since we are in the dilute limit we feel that there should be no problem using them. The similarity between the Be-Be and Be-Al potentials is so striking that it is clear that we could predict that Al would go substitutionally into Be. This is in fact observed and also is predicted by the Miedema plot. Also by inspection, and remembering the Na-Au

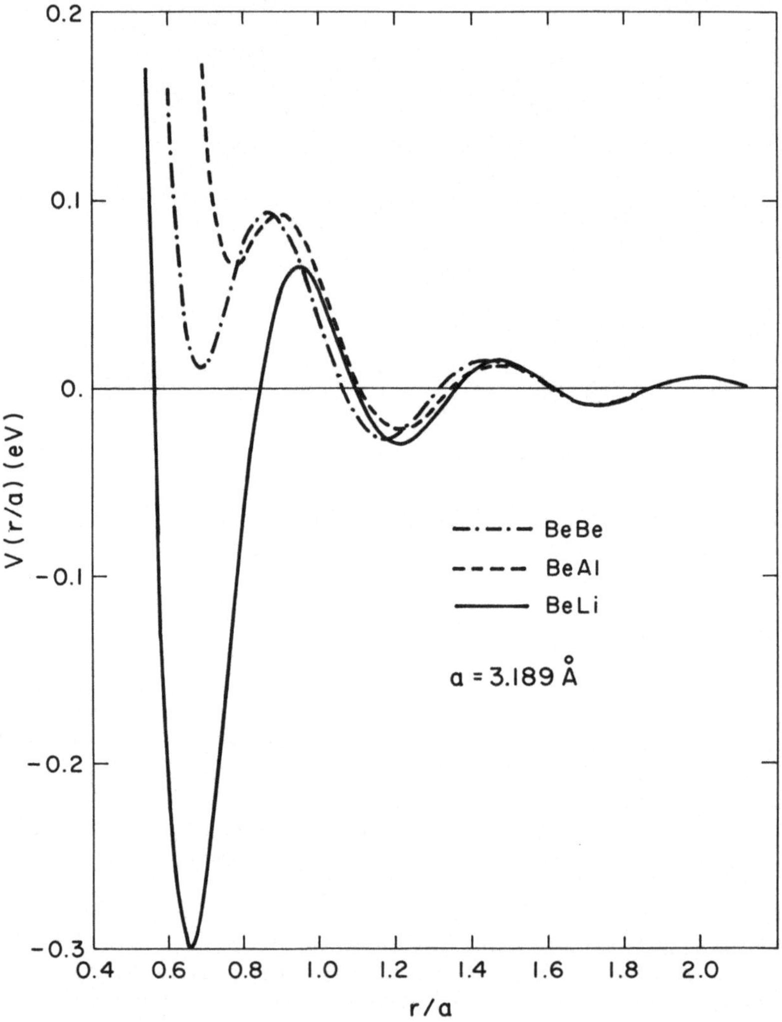

Figure 5 : Interionic potentials for Be-Be, Be-Al and Be-Li in the limit of zero concentration of impurities in Be.

potential, it seems equally clear that we would predict an interstitial site for Li, as does the Miedema plot. This is indeed the case although we favour the octahedral site rather than the tetrahedral site predicted by the Miedema plot. How can this be reconciled with the experimental observation ?

As Kaufmann (1978) has pointed out to us we must keep in mind the energetics of the particle scattering. With a heavy ion impinging on the Be lattice large amounts of energy cannot be transferred in any single scattering event. Hence a heavy ion (i.e. anything heavier than B) will travel a long distance from the region of radiation damage before thermalizing and choosing a site. But Li can loose some substantial fractions of its energy in single collisions which raises the possibility of it trapping on a vacancy site created during the damage process. Our calculations indicate that just as for the Na-Au case Li diffuses rapidly in Be but there is also an attractive interaction with a vacancy site. Then, once in the vacancy site, far and away the most stable configuration is for the Li ion to occupy the host lattice site. Thus we find that our potentials are entirely consistent with experiment. In future we hope to examine other dilute Be alloys.

4. SUMMARY

In this brief review of some of our alloy calculations I have first tried to make the point that one should, if at all possible, test the convergence of a pseudopotential before using it blindly to calculate some property. Once having obtained satisfactory convergence by a procedure analogous to that of DRT one finds that a large number of calculated properties can be found to agree well with experiment. Thus one can confidently tackle still further problems with the hopes of getting useful results. In particular for alloy problems we have enjoyed good success in describing lattice dynamic properties, diffusion processes and site occupancies. However serious problems arise when dealing with finite concentrations of non-homovalent impurities. The validity of pseudopotential techniques in these particular cases appears to be questionable.

Quite possibly in some of the examples considered we could have obtained similar results with other pseudopotentials but by choosing them, where possible, in the way that we have, we are able to place a great deal more confidence in the meaning of our results.

ACKNOWLEDGEMENTS

I would like to thank Drs. E.N. Kaufmann and M.S. Duesbery for some very useful discussions concerning the Be alloys.

REFERENCES

Barr, L.W., Le Claire, A.D. and Smith, F.A., 1971 Atomic Transport in Solids, ed. A. Lodding, and T. Lagerwall (Verlag) p. 336.

Beauchamp, P., Taylor R. and Vitek, V., 1975, J.Phys.F. : Metal Phys., 5, 2017.

Bertoni, C.M., Bortolani, V., Calendra, C., and Nizzoli, F., 1974, J. Phys. F. : Metal Phys., 4, 19.

Brovman, E.G., and Kagan, Y., and Kholas, A., 1972, Sov. Phys. - JETP, 34, 394.

Cohen, S.S. and Klein, M.L., 1975, Phys. Rev. B12, 2984.

Cohen, S.S., Klein, M.L., Duesbery, M.S., and Taylor, R., 1976, J. Phys. F. : Metal Phys., 6, 337; ibid, 6, L271.

Da Fano, A., and Jacucci, G., 1977, Phys. Rev. Letters, 39, 950.

Dagens, L., Rasolt, M., and Taylor, R., 1975, Phys. Rev. B11, 2726.

Geldart, D.J.W., and Taylor, R., 1970, Can. J. Phys., 48, 167.

Hasegawa, M., 1976, J. Phys. F. : Metal Phys., 6, 649.

Heine, V., and Abarenkov, I., 1964, Phil.Mag. 9, 451.

Jacucci, G., Klein, M.L., and Taylor, R., 1977, Solid State Commun., 24, 685.

Jacucci, G., Klein, M.L., and Taylor, R., 1978, Phys. Rev. B to be published.

Jacucci, G., McDonald, I.R., and Taylor, R., 1978, J. Phys. F. : Metal Phys., 8, L121.

Jacucci, G., and Taylor, R., 1978, J. Phys.F. : Metal Phys. submitted.

Kamitakahara, W.A., and Copley, J.R.D., 1978, Phys. Rev. B., to be published.

Kaufmann, E.N., 1978 private communcation.

Kaufmann, E.N., Vianden, R., Chelikowsky, J.R., and Phillips, J.C., 1977, Phys. Rev. Letters.

Leavens, C.R., 1977, J. Phys.F. : Metal Phys., 7, 163.

Leavens, C.R., and Taylor, R., 1978, Thermoelectricity in Metallic Conductors, ed. F.J. Blatt, and P.A. Schroeder, (Plenum) p. 131; J. Phys. F. : Metal Phys., 8, 1969.

Michler, E., Hahn, H., and Schofield, P., 1976, J. Phys. F. : Metal Phys., 6, L319.

Miedema, A.R., 1973, J. Less. Com. Met., 32, 117.

Mundy, J.N. and McFall, W.D., 1973, Phys. Rev. B7, 4363.

Rasolt, M., and Taylor, R., 1975, Phys. Rev. B11, 2717.

Schober, H., Taylor, R., Norgett, M.J., and Stoneham, A.M., 1975, J. Phys. F. : Metal Phys., 5, 637.

Shaw, R.W., 1968, Phys. Rev., 174, 769.

Shukla, R.C., and Taylor, R., 1976, J. Phys. F. : Metal Phys., 6, 531.

Taylor, R., Leavens, C.R., and Shukla, R.C., 1976, Solid State Commun., 19, 809.

DETERMINATION OF THE DISTORTION FIELD IN BINARY ALLOYS

K. Werner

Institut für Festkörperforschung der Kernforschungsanlage Jülich,
D-5170 Jülich, F.R.G.

ABSTRACT

The distortion field induced by a substitutional defect in a simple metal has been calculated from lattice statics, using Shaw's optimized model potential in combination with a screening function suggested by Vashishta and Singwi. Using this procedure, interactions of arbitrary range can be taken into account. In order to test the validity of the theory, the diffuse elastic cross-section, which is directly related to the Fourier transform of the distortion field, was measured experimentally with the help of a neutron scattering technique. The system under investigation was Al containing a few at% Mg. A comparison of the experimental data with the predictions of the pseudopotential concept gave quantitative agreement for a limited, but nevertheless continuous, range of momentum transfer vector. Furthermore, qualitative agreement was obtained throughout the entire experimental range. Supplementary virtual force model calculations gave, even by simulating forces upto the fourth nearest neighbour shell, no satisfactory description of the experimental cross-section.

INTRODUCTION

If a defect is substitutionally inserted into a pure host lattice a perturbation is produced : firstly because the ionic radii may be different and secondly because the charge configuration is changed in the neighbourhood of the defect. The lattice responds to the defect induced forces by establishing new equilibrium positions for the atoms surrounding the defect.

In Fig. 1 the situation is illustrated. The atom ℓ (whose posi-

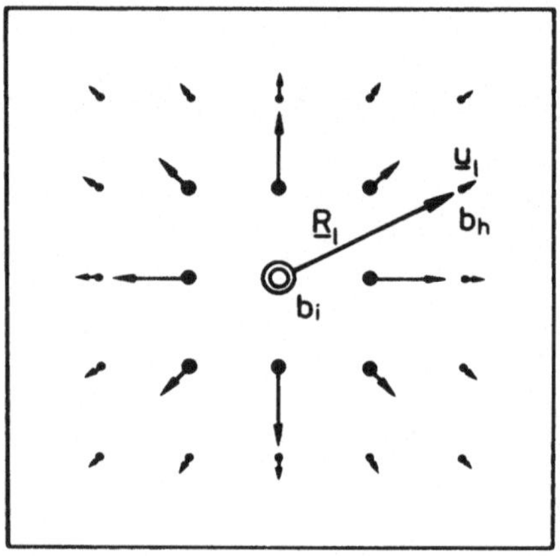

Fig. 1 : Effect of a substitutional impurity on the host lattice. The impurity (scattering length b_i) displaces the atom ℓ (equilibrium position R_ℓ, scattering length b_h) by a vector $\underline{u}_\ell$.

tion in the unperturbed matrix is $\underline{R}_\ell$, with a scattering length b_h) is displaced by a vector $\underline{u}_\ell$ owing to the presence of the impurity atom (positioned at the origin, scattering length b_i).

The aim of this lecture is to show how one can determine experimentally and theoretically the defect induced distortion of the host matrix in the immediate vicinity of the impurity. The experimental technique used is diffuse neutron scattering and to describe the measured cross section theoretically the pseudopotential concept is applied.

In order that a system is experimentally accessible and can be treated theoretically, certain restrictions have to be applied. The experimental requirements are : randomly distributed solute atoms, a large difference $|b_h - b_i|$ in the neutron scattering lengths between host and impurity, measurable lattice distortions, available elastic constants for the matrix, the incoherent cross-section is known and is small, small absorption and that single crystals can be obtained. To apply the pseudopotential concept, one needs simple (s,p - bond) metals having no covalent bonding and in which the lattice distortions are small. A number of systems fulfil these requirements, e.g. Al Mg, Al Ga, Al Li, Al Ge or Pb Sn. I shall discuss the Al Mg system on which measurements were performed[1] using the diffuse elastic neutron scattering spectrometer (DENS) at the KFA Jülich . The calculations were made by Alexis Baratoff (IBM Zürich).

DIFFUSE-ELASTIC NEUTRON SCATTERING

The basic arrangement for a neutron scattering experiment is shown in Fig. 2. The neutrons, coming from a neutron source (e.g. a reactor), pass through a collimation system and are then scattered by a specimen ; the primary beam being absorbed by a beam stop.

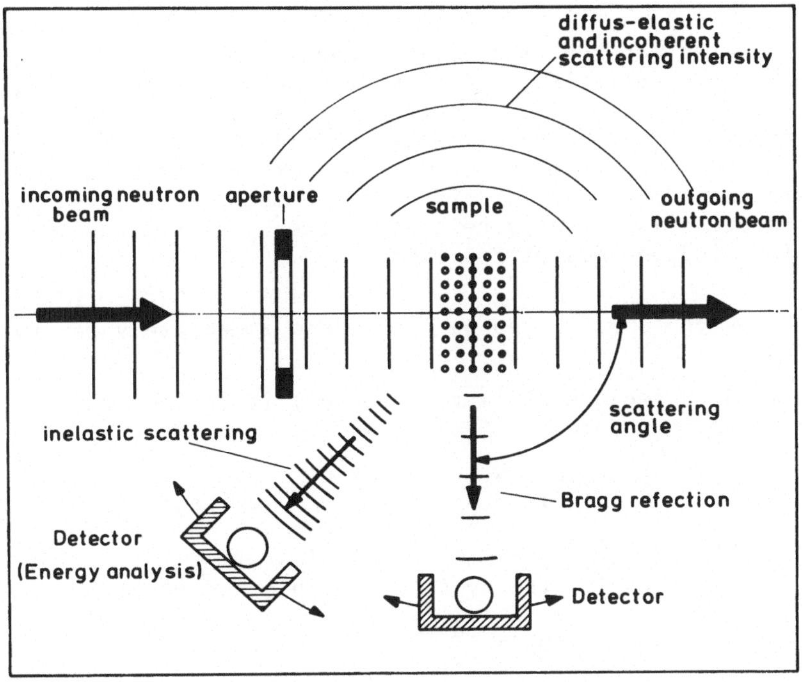

Fig. 2 : Basic arrangement for a neutron scattering experiment.

The crucial scattering processes are shown schematically in the figure, i.e.

1. Bragg reflection, which gives information about the atomic configuration of the whole crystal ;
2. inelastic scattering, produced by atomic vibrations ;
3. incoherent scattering, owing to spin and isotopic incoherence ;
4. coherent diffuse-elastic scattering, arising from the defect induced distortion field.

The first contribution was eliminated by using single crystals and measuring the scattered intensity far away from the well defined Bragg peaks. Time-of-flight analysis was used to discriminate against inelastic scattering ; the difference between the elastic scattering from an alloy sample and a similar one of pure aluminium was taken in order to eliminate multiphonon scattering and the incoherent scattering of Al.

The cross-section for the remaining coherent diffuse-elastic neutron scattering is given by[2]

$$\left(\frac{d\sigma}{d\Omega}\right)_{coh,el,dis} = c_i(1-c_i)\,|b_i e^{-W_i} - b_h e^{-W_h} - \bar{b}e^{-\bar{W}}\, i\underline{Q}\tilde{\underline{u}}(\underline{Q})|^2 := \left(\frac{d\sigma}{d\Omega}\right)_D \tag{1}$$

b_i and b_h are the coherent scattering lengths, $\bar{b}$ their average, c_i is the defect concentration, exp (— W) the Debye-Waller factor, $\underline{Q}$ the momentum transfer of the neutron to the target and $\tilde{\underline{u}}(\underline{Q})$ is the Fourier transform of the distortion field. The expression is valid for a low concentration ($c_i \ll 1$) of weakly distorting randomly distributed defects. In this case the measured diffuse-elastic cross-section is directly related to the Fourier transform of the defect induced distortion field.

The value of $(d\sigma/d\Omega)_D$ can be determined immediately for two values of $\underline{Q}$:

1. If $\underline{Q} = 1/2\underline{G}$ ($\underline{G}$ reciprocal lattice vector) it follows that $\tilde{\underline{u}}(\underline{Q}) = 0$ and therefore

$$\left(\frac{d\sigma}{d\Omega}\right)_D^{Q=1/2\underline{G}} = c_i(1-c_i)\,|b_i e^{-W_i} - b_h e^{-W_h}|^2 = \left(\frac{d\sigma}{d\Omega}\right)_{Laue} \tag{2}$$

i.e. the pure Laue monotonic scattering.

2. When $\underline{Q} \to 0$, elasticity theory is applicable and it follows that

$$i\underline{Q}\tilde{\underline{u}}(\underline{Q}) = \frac{\Delta a}{ac_i}\, f(c_{ij}) \tag{3}$$

where $\Delta a/uc_i$ is the measured lattice parameter change and $f(c_{ij})$ are linear combinations of the elastic constants of the host lattice.

In the case of Al Mg one has

$$\left(\frac{d\sigma}{d\Omega}\right)^{Al\ Mg}_{Laue} = 35 \text{ mb/sterad per defect atom}$$

$$\left(\frac{d\sigma}{d\Omega}\right)^{Al\ Mg}_{Q\to 0\ ,Q||[110]} = 12.8 \text{ mb/sterad per defect atom}$$

EXPERIMENTAL RESULTS

In Fig. 3 the coherent diffuse-elastic cross-section is plotted as a function of the momentum transfer $\underline{Q}||[110]$. Since this is the close-packed direction in the fcc-lattice, one would expect the structure of $\tilde{\underline{u}}(\underline{Q})$ to be most pronounced in this direction. The hatched areas represent the values given above for $\underline{Q} \to 0$ and $\underline{Q} = 1/2\underline{G}_{220}$.

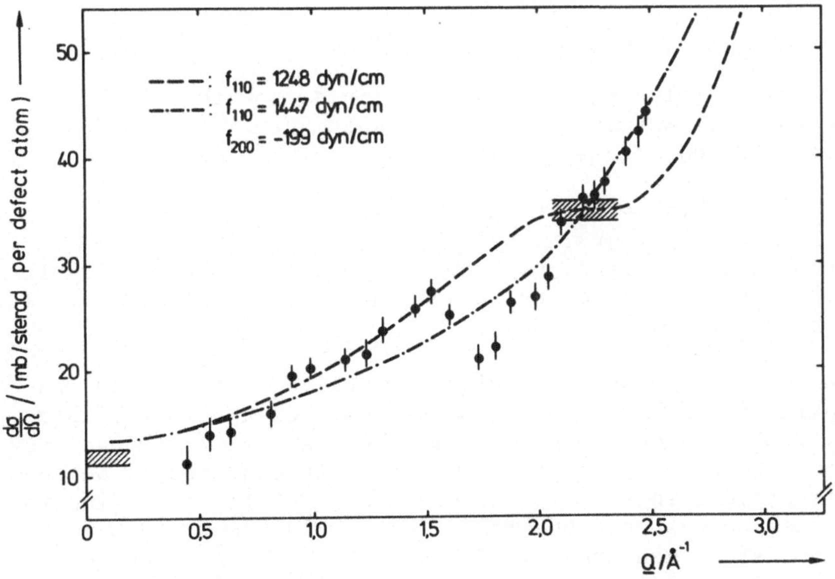

Fig. 3 : Coherent diffuse-elastic cross-section in the [110]-direction.

⧫ experimental results

– – Nearest neighbour (NN) Kanzaki model

–·– NN and next NN Kanzaki model.

The data points were obtained from two independent measurements made at a temperature of 725 K, using Al Mg single crystals containing 2.1 and 3.6 at% Mg. To obtain the cross-section in absolute units a vanadium callibration was used[3].

One notices that the experimental data agree with both the long wavelength value and the value of the Laue scattering at one half of the reciprocal lattice vector. Since the spectrometer has a high resolution ($\leqslant$ 1 mb) the maximum at $\underline{Q} \approx 1.5$ Å^{-1}, the secondary minimum at $\underline{Q} \approx 1.7$ Å^{-1} and the finite slope at $\underline{Q} = 1/2G_{220}$ are statistically significant.

THEORETICAL DESCRIPTION

a) Kanzaki-Force Model

Until now the phenomenological Kanzaki-force model[4] has been widely used to describe the defect induced distortion field[1,3]. In this case the actual effect of an impurity atom in the real lattice is simulated by virtual forces acting on the nearest neighbours of a single atom in the pure host lattice.

One starts from the basic equation of lattice statics

$$\tilde{\underline{u}}(\underline{q}) = \tilde{\underline{\underline{\phi}}}^{-1}(\underline{q})\ \tilde{\underline{K}}(\underline{q}) \tag{4}$$

The dynamical matrix $\tilde{\phi}(\underline{q})$ can be determined from the Al phonon spectrum (Born v. Karman fit)[5], $\tilde{\underline{K}}(\underline{q})$ is the Fourier transform of the defect induced forces :

$$\tilde{\underline{K}}(\underline{q}) = \sum_{\ell=1}^{\infty} \underline{K}_{\ell}(\underline{R}_{\ell})\ \exp(i\underline{q}\underline{R}_{\ell}) \tag{5}$$

This can be obtained by simulating the forces only up to the first (or second, third ...) nearest neighbours and fitting the resulting parameters to the measured data.

The results are shown in fig. 3 where the dashed line was computed taking forces only up to the nearest neighbourshell into account. The one free parameter which occurs was determined from experimental change in lattice parameter. It can clearly be seen that this curve does not fit the measured profile. Including forces up to second nearest neighbours, and adjusting their magnitude so that the slope measured at the point $1/2\underline{G}_{220}$ is produced, yields the structureless dash-dotted curve. This result is even worse.

In Fig. 4 the resulting curves for forces up to the 3rd and 4th neighbour shells are shown. Even in the last case, where three independent parameters are fitted to the data, the measured shape cannot be successfully reproduced.

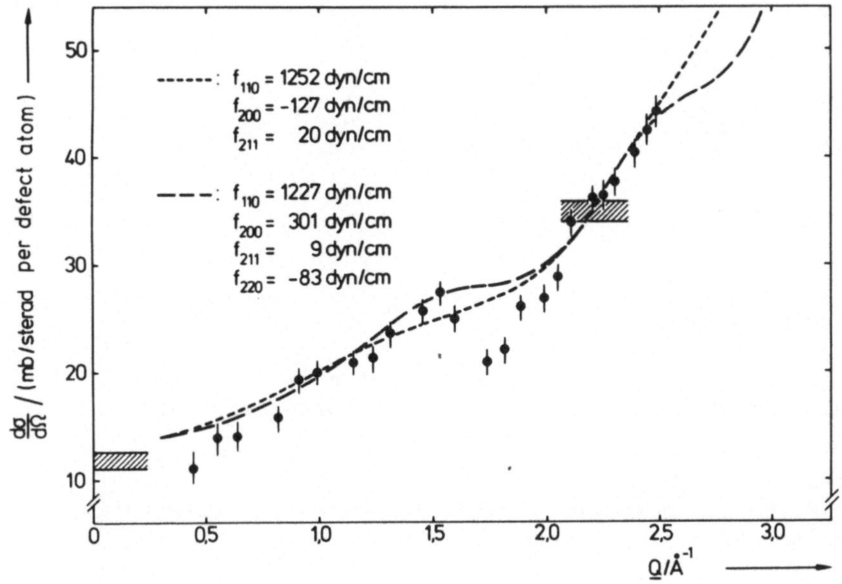

Fig. 4 : Coherent diffuse-elastic cross-section in the [110]-direction
experimental data
----- Kanzaki model including forces up to the 3rd neighbour shell
– – – Kanzaki model including forces up to the 4th neighbour shell

b) Pseudopotential Concept

Equation (4) is the basic eqaution, which for convenience is written down again :

$$\underline{\tilde{u}}(\underline{q}) = \underline{\underline{\tilde{\phi}}}^{-1}(\underline{q})\ \underline{\tilde{K}}(\underline{q}) \qquad (4)$$

Rather then trying to find phenomenological expressions for $\underline{\tilde{K}}(\underline{q})$ one can begin from another point of view : If one succeeds in developing an effective interatomic potential $V(\underline{q})$, $\underline{K}(\underline{q})$ and $\underline{\underline{\phi}}^{-1}(\underline{q})$ can be found and consequently the distortion field calculated a priori from (4)[6] :

$$\underline{\tilde{K}}(\underline{q}) = \frac{i}{\Omega} \sum_{\underline{G}} (\underline{q} + \underline{G})\ \Delta\tilde{V}(\underline{q})(\underline{q} + \underline{G}) \qquad (6)$$

with

$$\Delta\tilde{V}(\underline{q}) = \tilde{V}_{ih}(\underline{q}) - \tilde{V}_{hh}(\underline{q}) \tag{7}$$

and

$$\phi^{\ell}_{\alpha\alpha'} = \{\frac{V'_{hh}(R_\ell)}{R_\ell} - V''_{hh}(R_\ell)\}\hat{R}_{\ell\alpha}\hat{R}_{\ell\alpha'} - \frac{V'_{hh}(R_\ell)}{R_\ell}\delta_{\alpha\alpha'} =$$

$$= (K_T - K_R)\hat{R}_{\ell\alpha}\hat{R}_{\ell\alpha'} - K_T\delta_{\alpha\alpha'} \tag{8}$$

Here Ω is the atomic volume, K_T and K_R the tangential and radial force constants respectively.

In order to find the interatomic potential $\tilde{V}(\underline{q})$ the pseudopotential concept is used[7]. All together three steps have to be made :

First step :
Define a "free ion potential" which describes the interaction between an ion and a single conduction electron.

Second step :
Screen the free ion potential, incorporate exchange and correlation effects and create a "screened ion potential" which describes the interaction between an ion and the whole electron gas.

Third step :
Develop the interatomic potential as a sum of the Coulombic repulsion between two ions and the bandstructure contribution ; the latter being calculated to second (or third ...) order in the screened ion potentials using perturbation theory.

Shaw's nonlocal optimized model potential (NLOP)[8] was chosen for the free ion potential, i.e. $W_{NLOP} = - A_\ell = Z/R_\ell$ if $r < R_\ell$ for states with non-zero projection on the core states, and $W_{NLOP} = - Z/r$ otherwise. In Fig. 5 the NLOP is shown.

The bare ion NLOP's were screened linearly using the mean electron density of Al. To take into consideration the correlation and exchange effects, the formula suggested by Vashishta and Singwi (VS) was used[9]. The combination of the NLOP together with the VS screening function yields a good fit to the phonon spectrum of Al[10].

The interatomic potential $\tilde{V}(\underline{q})$ was calculated to second order in $W_{NLOP,VS} := W_{ij}$:

$$\tilde{V}(\underline{q}) = \frac{4\pi Z^\star_i Z^\star_h}{q^2}\,[1 - F_{Nij}(q)], \tag{9}$$

where

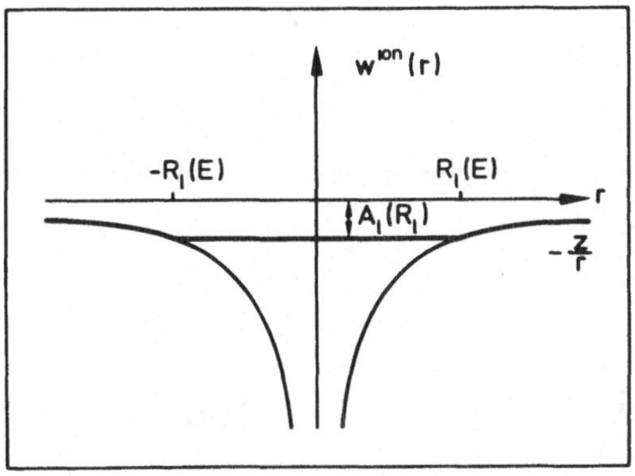

Fig. 5 : Nonlocal optimized model potential in real space.

$$F_{Nij}(q) = -\frac{1}{2N} \sum_{K}^{K_F} \sum_{q \neq 0} \frac{|\langle \underline{K} + \underline{q}|W_i|\underline{K}\rangle\langle \underline{K}|W_j|\underline{K} + \underline{q}\rangle|}{1/2K^2 - 1/2(\underline{K} + \underline{q})^2} \tag{10}$$

is the so called "energy wavenumber characteristic", with i and j representing Al and Mg respectively. In Fig. 6 the energy wavenumber characteristics are shown together with the screened formfactors as a function of q/K_F.

The resulting forces (equation(6)) and the dynamical matrix (equation (8)) are shown in Fig. 7. In the upper part of the figure, a curve calculated from the Kanzaki force model (nearest neighbours) is shown for comparison with the force calculated using the pseudopotential approach. The calculated dispersion curve for $\underline{Q} \| [110]$ is shown with the measured experimental data[11] in the lower part of the figure.

In Fig. 8 the final result, the diffuse elastic cross- section, is plotted with the experimental data shown previously in figures 3 and 4. At small $\underline{Q}$ values there is a considerable discrepancy which is presumably due to the inadequency of second-order perturbation theory : the form factors shown in Fig. 6 are not small compared with the Fermi energy in this region. For $Q \geqslant 1.5$ Å^{-1}

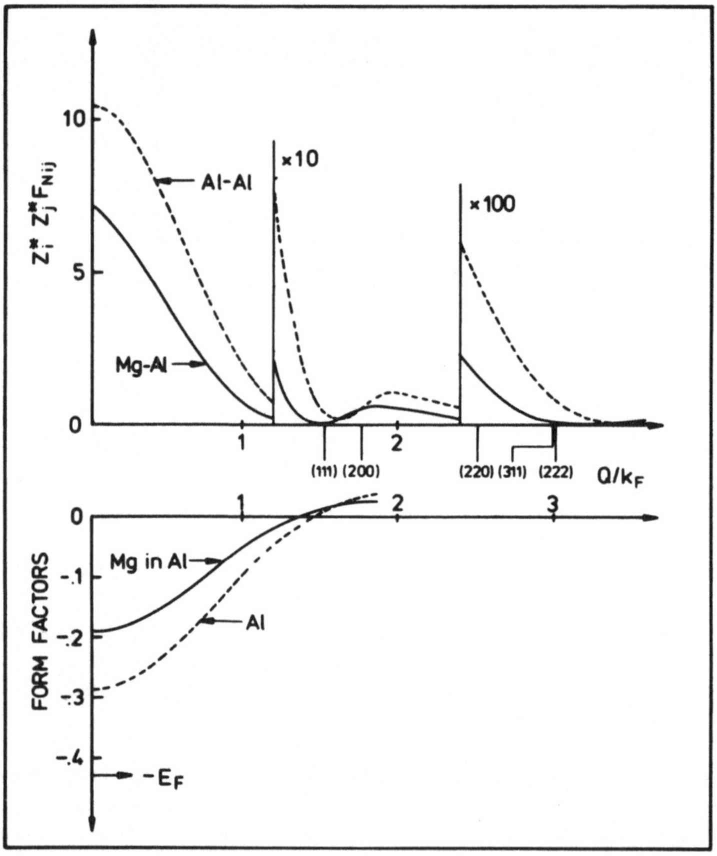

Fig. 6 : Screened model potential form factors and energy wave-number characteristics for Al-Al and Al-Mg.

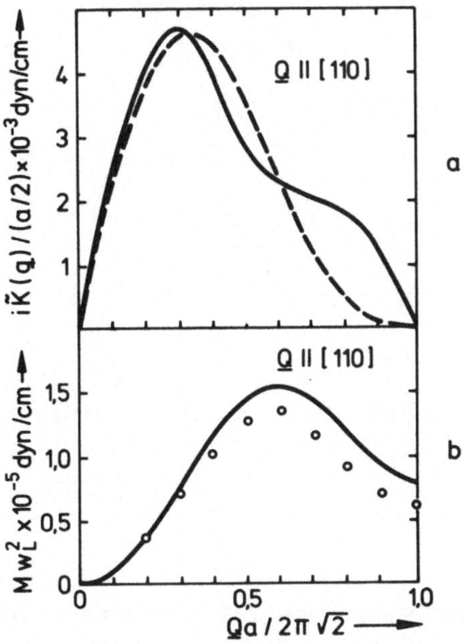

Fig. 7 : a) Defect induced forces $\tilde{\underline{K}}(\underline{q})$ for AlMg.
Solid line : result using the pseudopotential concept.
Dashed line : NN Kanzaki model.
b) Experimental data[11] and dispersion curve calculated from pseudopotential theory.

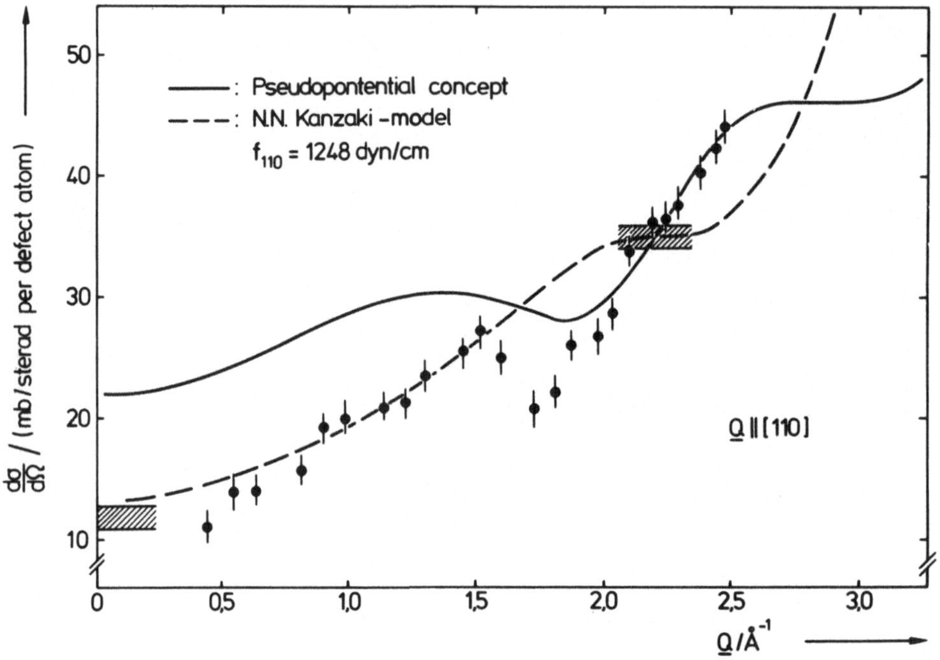

Fig. 8 : A comparison of the experimental diffuse-elastic cross-section with the theoretical cross-section calculated using a NN Kanzaki model and the pseudopotential concept.

good agreement with the measurement is obtained. In particular, the secondary minimum and the slope at $1/2G_{220}$ are well reproduced. This is in agreement with Fig. 6, where one can see that for $\underline{Q} \geqslant 1.5$ Å^{-1} the electronic perturbation is weak.

This work clearly demonstrates the potential of diffuse elastic neutron scattering for investigation long-range defect-host interactions. The most direct interpretation is in Q-space, a description which is also convenient for theories of alloys based on pairwise interactions. Previous pseudopotential calculation[12-14] of defect properties dealt with quantities which can be expressed in terms of integrals over $\underline{Q}$ and for $\underline{q} \rightarrow 0$ contributions. By contrast our calculation provides an encouragung test over a continous, albeit limited range of $\underline{Q}$, and may thus help one decide which defect or alloy properties might be well described by model potential perturbation theory.

REFERENCES

1. K. Werner, Diss. Uni. Bochum, Jül.-1519, (1978).

2. W. Schmatz, in "Treatise on Materials Science and Technology", ed. H. Hermans, Academic Press New York, 2 (1973) 105.

3. G.S. Bauer, E. Seitz and W. Just, J. Appl. Cryst. 8 (1975) 162.

4. H. Kanzaki, Phys. Chem. Sol. 29 (1957) 24.

5. G. Gilat and R.M. Nicklow, Phys. Rev. 143 (1966) 487.

6. R. Benedek, Thesis, Cornell University (1972).

7. V. Heine, Sol. State Phys. 24 (1970) 1.

8. R.W. Shaw, Jr., Phys. Rev. 174 (1968) 769.

9. P. Vashishta and K.S. Singwi, Phys.Rev. B6 (1972) 875.

10. P.V.S. Rao, J. Phys. Chem. Solids 35 (1974) 669.

11. R. Stedman and G. Nilsson, Ark. Rys. 30 (1965) 564.

12. Z.D. Popovic, J.P. Carbotte and G.R. Piercy, J. Phys. F4 (1974), 351.

13. R. Evans and M.W. Finnis, J. Phys. F6 (1976) 483 ; M.D. Whitmore ibid 1259.

14. R. Benedek and A. Baratoff, Sol. Stat. Commun. 13 (1973) 385.

METALLIC GLASSES: BULK AND SURFACE PROPERTIES

H.-J. Güntherodt, H. Beck, P. Oelhafen, K.P. Ackermann, M. Liard, M. Müller, H.U. Künzi, H. Rudin and K. Agyeman

Institut für Physik, Universität Basel

CH-4056 Basel, Switzerland

1. INTRODUCTION

Metallic glasses open a new, exciting, and rapidly expanding field of considerable scientific as well as technological interest. The unique combination of metallic and glassy properties leads to new materials with outstanding mechanical, magnetic and chemical properties. When these desirable features are considered alongside their manufacturing and economic advantages, many possible applications of metallic glasses can be envisaged. For more information the reader is referred to other review papers and conference reports. Pertinent references up to 1977 are listed in reference 1. More recent reviews are covered by the reference 2.

This review deals mainly with the physical properties of metallic glasses with emphasis on some new developments which have not yet appeared in the literature. We will draw heavily on our own unpublished results. These new developments include recent experiments on the static ionic structure by the EDXD (Energy Dispersive X-ray Diffraction) and EXAFS (Extended X-ray Absorption Fine Structure) methods, new information on the ionic dynamics, electron spectroscopy such as XPS, UPS and AES, optical reflectivity spectra, field electron emission experiments and ion scattering from the surface. One thing we will not attempt - and we admit this with apologies - is a summary of all the interesting work which has been carried out with an application point of view. There the interest has been focused mainly on high yield strength, high saturation magnetization, zero magnetostriction and so on.

Amorphous metals can be prepared by a variety of methods: 1. Evaporation, 2. Sputtering, 3. Chemical deposition, 4. Electrodeposition and finally 5. Rapid quenching from the liquid state. In this paper we are concerned only with amorphous metals prepared by the last method, the so-called metallic glasses or glassy metals. For a long time the field of metallic glasses was restricted to few canonical materials such as Pd-Si or Fe-B glasses, or similar alloys containing several components. Today, the following families of metallic glasses are known: T-N (e.g. $Fe_{80}B_{20}$, $Pd_{80}Si_{20}$), N-N (e.g. $Mg_{70}Zn_{30}$, $Ca_{70}Mg_{30}$), T_L-T_E (e.g. $Ni_{60}Nb_{40}$, $Cu_{60}Zr_{40}$), RE-N (e.g. $La_{70}A\ell_{30}$, $Ce_{70}A\ell_{30}$), RE-T_L (e.g. $Gd_{70}Co_{30}$, $Gd_{70}Ni_{30}$), U-T (e.g. $U_{70}Fe_{30}$, $U_{70}Cr_{30}$) and R-T_L (e.g. $Ta_{55}Ir_{45}$, $Ta_{55}Rh_{45}$), where T: transition metal, N: polyvalent metal, T_L: late transition metal, T_E: early transition metal, RE: rare earth , U: Uranium and R: refractory metal.

There are two main techniques to achieve rapid quenching from the melt: 1. the piston and anvil method, in which a liquid droplet is squeezed between a rapidly moving piston and a fixed anvil, 2. the melt spinning method, in which a molten alloy is extruded through an orifice by means of gas pressure to form a molten jet. The jet impinges on a chilled rotating wheel and solidifies to form a continuous glassy ribbon. Two new preparation techniques have been described recently. These are electric field atomisation from the melt (3) and arc furnace quenching (4). In the first of these methods, a high velocity ionized vapour is produced by electric field evaporation from melts in vacuum. The stream consists of atoms, clusters and small droplets. It is highly directional and can be controlled electrically. The stream has a velocity up to 20 km/s , which leads to very high cooling rates. This method seems to have great potential in surface studies. The second technique takes advantage of the high temperatures obtained in an arc furnace and is suitable for preparation of refractory metal glasses.

2. IONIC STRUCTURE

2a Static Structure

The most direct information on atomic arrangements in the amorphous state has been obtained from conventional X-ray and neutron scattering experiments in terms of the structure factor or the interference function. The structure of the alloys is described in terms of partial structure factors. The Fourier transform of the structure factor is related to the radial distribution function (RDF).

Models based on the dense random packing of hard spheres (DRPHS) have been quite successful in describing the arrangement of atomic species in metallic glasses. Recently, the structure of a binary metallic glass, $Mg_{70}Zn_{30}$, has been described by a relaxed model structure of 800 atoms (5). $Mg_{70}Zn_{30}$ seems to be an example of a "model metallic glass", containing only simple metals which can be treated in a pseudopotential approach. In contrast to earlier work, von Heimendahl (5) used realistic pair potentials calculated from general nonlocal pseudopotential theory and periodic boundary conditions. The calculated pair distribution functions are in good agreement with recent X-ray and neutron scattering results on $Mg_{70}Zn_{30}$ (6).

New information on the structure can be extracted from EDXD and EXAFS data. We will first describe the EDXD method (7). In the conventional angular-scan diffraction experiment, the X-ray source is monochromatic, so that the diffraction vector, k , is proportional to sin Θ , where Θ is the diffraction angle. In the EDXD technique, white X-ray radiation is used as a source, while Θ is fixed during the measurement. In this case, k is proportional to E, the energy of an incident photon. The diffracted photons are directed to an energy sensitive solid state detector. One of the major advantages of the EDXD method is the high statistical accuracy of the photon count, since the total intensity of white X-rays is higher than characteristic X-rays. Furthermore, all energy channels are counted in parallel thus nearly eliminating the effect of source fluctuations. Also the use of high energy X-rays, often up to 40-45 keV , makes it possible to determine the structure factor at high k values up to 25 $Å^{-1}$. The absence of a mechanical scanning system makes the entire system simpler, therefore less costly, and ensures excellent reproducibility. The main application of this new technique lies in real time study of the kinetics of structural relaxation and in the study of compositional short range order of metallic glasses.

Next, we will describe the EXAFS technique (8,9). Conventional X-ray and neutron diffraction provide a statistically averaged description of the arrangements of the atoms. EXAFS has the unique capability of probing the individual atomic environment of a particular constituent in the presence of many others. In an EXAFS experiment, the observation of interest is the modulation of the absorption coefficient on the high energy side of the X-ray absorption edge of a constituent atom. For such measurements the ratio of incident to transmitted intensity is recorded as a function of photon energy above the absorption edge. Using procedures described in detail in the literature (see ref.8), the EXAFS as a function of finalstate electron momentum can be extracted from the raw data and

Fourier transformed to real space. From this it is possible to obtain local structural information, particularly concerning the near neighbour environment in multicomponent systems.

2b. Dynamics of the Ions

Since many of the known metallic glasses are ferromagnetic, up to now only information on the spin wave spectra (10) has been published. Recently, some attention has been directed towards studies of phonon spectra of metallic glasses.

Dynamical structure factors S(k,ω) for longitudinal and transverse waves for a series of k vectors have recently been calculated (5). The same model structure mentioned in section 2a has been used to calculate this inelastic neutron scattering cross section by the equation of motion technique. The results are shown in Fig.1 for longitudinal and transverse excitations respectively. The most interesting feature of the spectra is clearly the k-dependence of the peak positions. A representation of this is shown in Fig.2 . This "dispersion relation" shows a linear behavior at small k-values,

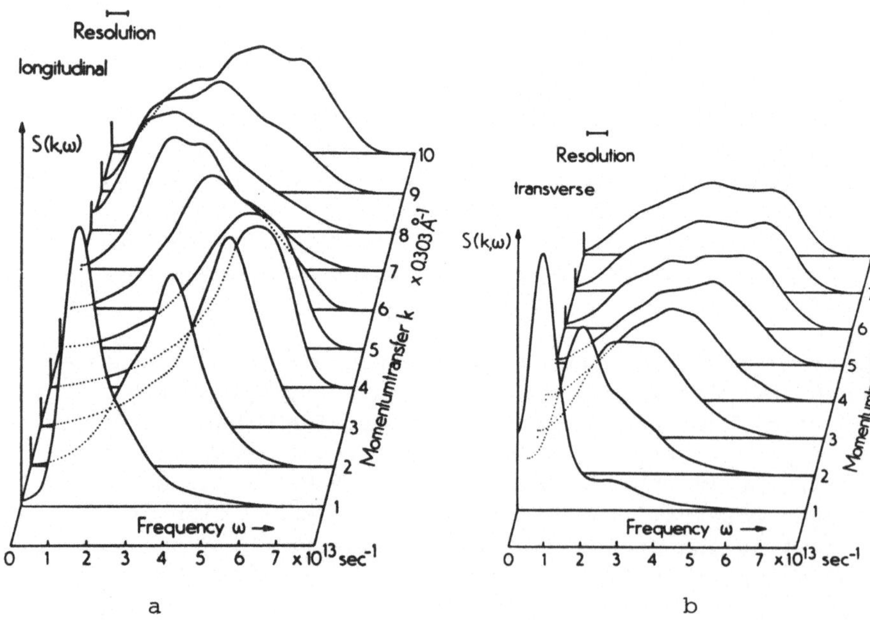

Fig.1 : S(k,ω) for longitudinal (a) and transverse (b) excitations.

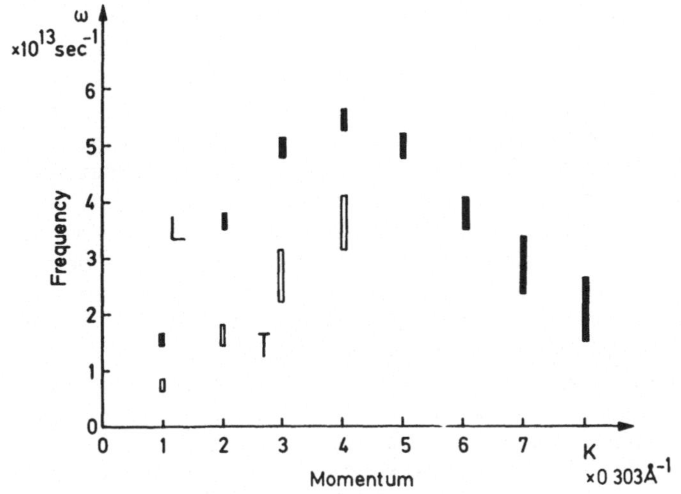

Fig.2 : Dispersion relation.

followed by a maximum. The identification of a minimum following the maximum is not really possible, because the S(k,ω) spectra become too broad. It turns out that the "dispersion relation" is quite similar to that for rotons in liquid He and for magnons in ferromagnetic metallic glasses. Inelastic neutron scattering experiments are under way in order to study S(k,ω) in the wave vector range on the right hand side of the maximum (11). The linear ω(k) relation can be tested by Brillouin scattering from metallic glasses (12) and by sound velocity data. The sound velocity can be measured or calculated by using known bulk elastic constants (Young's modulus and Poisson's ratio). Brillouin scattering experiments of glassy $Mg_{70}Zn_{30}$ have shown a linear ω(k) relation for small k-values of surface phonons (13). Since they scale to the bulk phonons, a similar linear ω(k) relation for small wave vectors is expected. The sound velocities extracted from light scattering experiments and elastic constants are in reasonable agreement (14).

3. GLASS FORMING ABILITY

The question of glass forming ability has been of great interest. Every experimental group working in the field of metallic glass preparation has its own recipe on how to find alloy concentrations forming stable metallic glasses. Basically one consults the phase diagrams of the alloys. Nowadays, there are so many new metallic glasses known, that one can start to ask, whether there is any underlying general principle. Recently Nagel and Tauc (15) have attempted to interpret the stability of metallic glasses in terms of the properties of the electron gas, in a manner reminiscent of Hume - Rothery's classical work on the stability of crystalline electron phases. It seems very difficult to find experimental evidence for the minimum of the density of states at the Fermi energy E_F which they proposed. There is even some evidence that the density of states argument is based on a misleading interpretation (see section 7). However, they postulated that in stable metallic glasses the diameter $2k_F$ of the Fermisphere is about equal to the position k_p of the first peak in the static structure factor S(k). This seems to be fulfilled in most of the known families of metallic glasses. The condition $2k_F \sim k_p$ is also very important for the explanation of negative temperature coefficients (NTC) of the electrical resistivity of noncrystalline metals at room temperature and at higher temperatures in terms of structural disorder scattering. The different concentration ranges for the known families of metallic glasses where glass forming is observed are in good agreement with the concentration ranges where NTC occur. A recent paper (16) has elucidated the condition $2k_F \sim k_p$ for the glass forming ability from a different approach leading to consequences on the phonon dispersion curves and, possibly, on the phase diagrams. An appealing aspect of this criterion $2k_F \sim k_p$ is its independence of details: it merely links the "basic inverse lengths" of electronic and ionic structure. As mentioned in section 2, mechanical stability and ionic dynamics can best be discussed in terms of pairwise ionic interactions. The stability problem is investigated by analyzing the quenching process in three steps:

(i) We start from a liquid alloy A_xB_{1-x} characterized by a set of partial pair correlation functions (PCF) $g_{\alpha\beta}(R)$. Except for details these are determined by the short range (hard sphere-like) part of the ionic interactions.

(ii) The main effect of rapid quenching (in a strongly simplified view) consists in "freezing" the liquid configuration at a given instant. Relaxation of energetically unfavorable neighbour positions will change the details but not the essential behavior of $g_{\alpha\beta}$. The mean static energy of such a system is

$$\langle E \rangle = 1/2 \sum_{\alpha\beta} x_\alpha x_\beta \int d^3R \, g_{\alpha\beta}(R) \, \phi_{\alpha\beta}(R) \; . \qquad (1)$$

(iii) The stability is checked by imposing small displacements on each particle, e.g. in the form of a plane wave:

$$\vec{R}_n \longrightarrow \vec{R}'_n = \vec{R}_n + \vec{A}(q) \cdot e^{i\vec{q}\cdot\vec{R}_n} \; . \qquad (2)$$

The system will be stable if the average variation of the energy

$$< \delta E > \equiv 1/2 \int d^3r \sum_{\alpha\beta} x_\alpha x_\beta \, \delta g_{\alpha\beta}(r) \phi_{\alpha\beta}(R) > 0 \qquad (3)$$

is positive (contributions from kinetic energy and entropy are neglected here).

We discuss the implications of (3) in terms of g, an average PCF, and a corresponding averaged pair potential ϕ . Fig. 3a shows these quantities. They were obtained for Au_xSi_{1-x} , but the following features are general.

(i) The distances between successive peaks of g (i.e. between consecutive neighbours of a given ion) are about the same and are given by

$$\lambda_p \approx 2\pi/k_p \ . \qquad (4)$$

(ii) The distances between consecutive minima of ϕ are well represented by

$$\lambda_F \approx 2\pi/2k_F \ . \qquad (5)$$

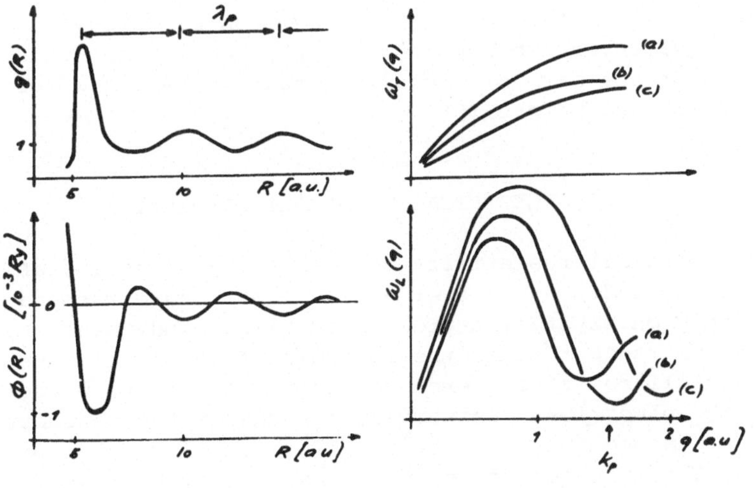

a b

Fig. 3 : a) Pair correlation function g(r) and pair potential $\phi(r)$ of metallic glasses where $2k_F \sim k_p$
b) ω_T and ω_L for Au_xSi_{1-x}: The three curves are for x = 1 (a) , x = 0.8 (b) and x = 0.6 (c) .

The position of the first neighbour is mainly determined by the "hard core" of the ions, which was not taken into account in Fig. 3a. However, the stability requirement (3) is best fulfilled if the structure of the quenched liquid (characterized by g(R)) places most of the successive neighbours of a given ion in the consecutive minima of ϕ ; i.e. when $\lambda_F \approx \lambda_p$, or $2k_F \approx k_p$. If this is not the case the ions will rearrange in order to reach an energetically more favorable state: the metallic glass is unstable.

For more quantitative statements the stability is checked by calculating the second energy derivatives

$$\omega_i^2(q) \equiv < \frac{d^2E}{\delta A_i(q)\,\delta A_i(-q)} > \tag{6}$$

with respect to longitudinal (i=L) and transverse (i=T) displacement amplitudes.In a crystal $\omega_i(q)$ would be the harmonic phonon frequencies. In the disordered system it is the fourth frequency moment of the dynamical structure factor. ω_T and ω_L are shown in Fig. 3b for Au concentrations $x = 1$, 0.8 , 0.6 . ω_L has a low lying part for $q_o \approx 2k_F$. In the light of our stability arguments this is interpreted as follows: at $x = 0.8$, where $q_o \approx k_p$, the wave length of these displacements coincides with the most probable distance between various neighbours. They do not change the overall structure. If, on the other hand, $q_o \neq k_p$ these displacements of "low energy" rearrange particles and can drive the system away from its original structure. Again, the qualitative behavior of $\omega_{T,L}$ is consistent with computer simulations of the harmonic dynamic structure factor of glassy $Mg_{70}Zn_{30}$ discussed before (5).

4. ELECTRICAL TRANSPORT PROPERTIES

The electrical resistivity of metallic glasses is, at a first glance, almost independent over a wide temperature range and then drops sharply on crystallization. A detailed study over a larger concentration range reveals the following facts (17): The observation of a small positive, zero or negative temperature coefficient (NTC) of the electrical resistivity depends on the concentrations and can be changed continuously on alloying. Of great importance are the NTC's. The main success of the Ziman theory has been the explanation of the NTC's of liquid metals by structural disorder scattering. The magnitude of the electrical resistivity as well as its temperature and concentration dependence in the glassy and liquid states are comparable. This suggests that the theory developed for liquid metals is also applicable to metallic glasses in the high temperature range.

In terms of such a theory the electrical resistivity of metallic glasses can be explained by the following model. Metallic glasses consist of randomly distributed ions and conduction electrons. The electrical resistivity can be calculated in a single-site approximation by evaluating the scattering properties of the conduction electrons from non-overlapping muffin-tin potentials. The electrical resistivity of a binary metallic glass A_xB_{1-x} is given by

$$\rho = c(k_F) \int_0^{2k_F} |U(K)|^2 K^3 dK \tag{7}$$

where

$$|U(K)|^2 = c_A |t_A|^2 \left(1-c_A + c_A S_{AA}\right) + c_B |t_B|^2 \left(1-c_B + c_B S_{BB}\right) + c_A c_B \left(t_A^* t_B + t_A t_B^*\right)\left(S_{AB} - 1\right) \tag{8}$$

where c_A, c_B are the concentrations, and S_{AA}, S_{BB} and S_{AB} are the partial structure factors of the alloy. For transition and rare earth ions t_A and t_B represent atomic t-matrices, whereas for simple metals these quantities can be replaced by pseudopotentials. Eq.(7) is valid for temperatures higher than the Debye temperature, whereas for low T the static structure factors have to be replaced by integrals over the dynamical structure factors, see Ref. 18.

There are two quantities determining the electrical resistivity. The first contribution arises from the structure factors and the second contribution from the scattering properties of the conduction electrons. In this concept NTC's arise when $2k_F \sim k_p$, corresponding to about 1.5 to 2 conduction electrons per ion. It is the variation of S(k) with temperature which determines the temperature dependence of the electrical resistivity. As the temperature is increased the height of the first peak of S(k) is reduced and therefore the electrical resistivity decreases. The crucial problem in these calculations, for glasses containing RE and T is the determination of the number n_c of conduction electrons. For T_L-N and RE-N systems one can make use of n_c-values determined by experience and by simple theoretical estimates. T_L-T_E, RE-T_L and R-T_L are more difficult to interpret.

A careful analysis shows that the T_E in the metallic glasses $Ni_{60}Nb_{40}$ and $Cu_{60}Zr_{40}$ provide a larger number of conduction electrons than the T_L. Problems arise when we come to the RE-T metallic glasses. Fig.4 shows the electrical resistivity of glassy and liquid $Gd_{67}Co_{33}$ alloys indicating NTC's. Since pure Co and Gd have $2k_F$

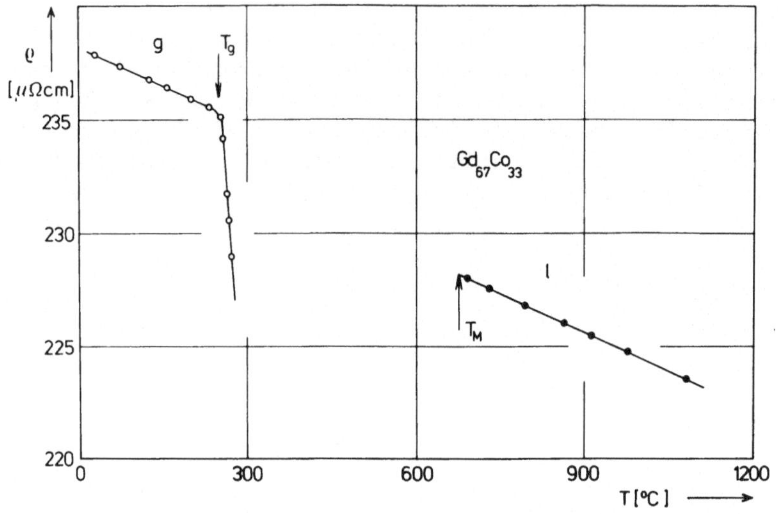

Fig.4 : Electrical resistivity of glassy and liquid $Gd_{67}Co_{33}$.

less than k_p it is unexpected that in terms of a simple alloying behavior $2k_F$ becomes equal to k_p in Gd-Co alloys. The strong evidence that NTC in the liquid state can indeed be explained by the Ziman theory in terms of $2k_F \sim k_p$, suggests, that "charge transfer" in both, glassy and liquid, Gd-Co alloys is responsible for the validity of this condition. This is in good agreement with a recent value of the normal Hall coefficient extracted from a Hall coefficient versus magnetic susceptibility plot (19).

A linear relation (20) between the crystallization temperature and the temperature coefficient of the electrical resistivity has been found, an indication of the importance of the $2k_F \sim k_p$ criterion for both the glass forming ability and the temperature coefficient of the electrical resistivity. Fig.5a shows the electrical resistivity and the temperature coefficient of $(Fe_{0.5}Ni_{0.5})_{100-y}B_y$ alloys. The correlation between the crystallization temperature and the temperature coefficient is shown in Fig.5b . The crystallization temperature increases with decreasing temperature coefficient. These results provide strong evidence that an electronic contribution to the glass forming of metallic glasses exists.

The electrical resistivity of metallic glasses at low temperature has been treated by various authors. In the framework of Baym's (18) expression (see above) simple models for the dynamical structure

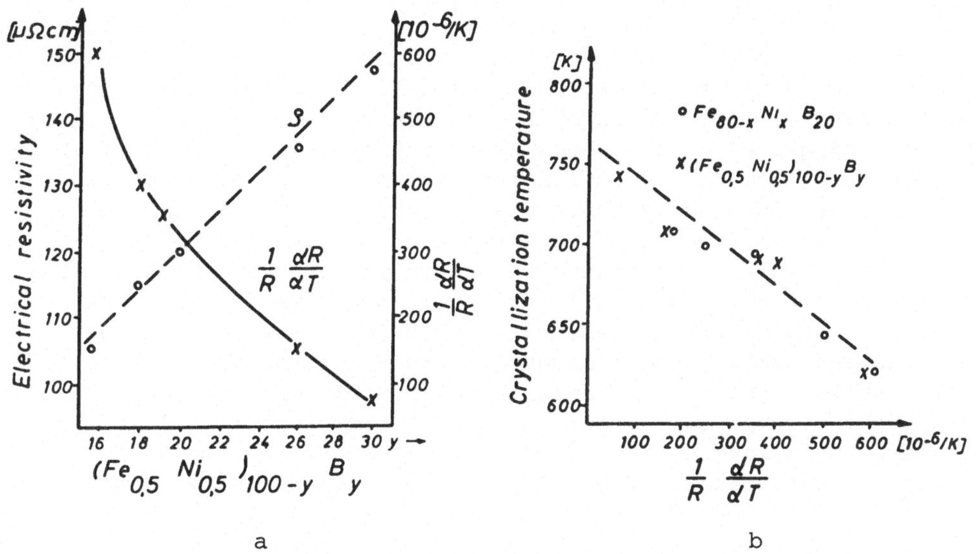

Fig.5 : a) Electrical resistivity and temperature coefficient of $(Fe_{0.5}Ni_{0.5})_{100-y}\ B_y$.
b) Correlation between the crystallization temperature and the temperature coefficient of ternary Fe-Ni-B alloys.

factor have been used (21,22). Different approaches have been developed for explaining resistivity minima and NTC's at low temperatures (23,24). It seems, however, to be difficult to obtain concrete numerical values by these methos.

5. MAGNETIC PROPERTIES

This section is presented only for the sake of completeness, since this field has been reviewed in more detail in other papers (25,26). The most valuable property of metallic glasses is ferromagnetism. A great deal of experimental and theoretical work has been done on ferromagnetic metallic glasses. Apart from ferromagnetism, metallic glasses can show para- and diamagnetism.

Magnetic properties of Pd-Si and Gd-Co glasses are discussed in some detail in this section. Pd-Si alloys show diamagnetism in the glassy state and extrapolate continuously into the liquid state where paramagnetism is observed. In the liquid state we can investigate alloys over the entire concentration range. From the behavior of the magnetic susceptibility, χ , as a function of concentration and temperature, information on the magnetic properties of the glassy state can be gained. Fig.6 shows the temperature dependence of χ for Si in the high temperature solid and liquid states (27). For crystalline Si, χ is diamagnetic, only weakly temperature dependent and changes at the melting point to paramagnetic values in the liquid state. This result demonstrates in a nice way, a semiconductor-metal transition upon melting. The liquid state data are in excellent agreement with the calculated χ values assuming four conduction electrons and taking into account electron-electron interaction. These results represent the first experimental confirmation of the widely used assumption of four conduction electrons for Si in liquid and amorphous alloys. Fig.7 shows χ of liquid Pd-Si alloys as a function of concentration. We observe a drastic decrease of the large paramagnetic χ of pure liquid Pd by adding liquid Si. For the plotted

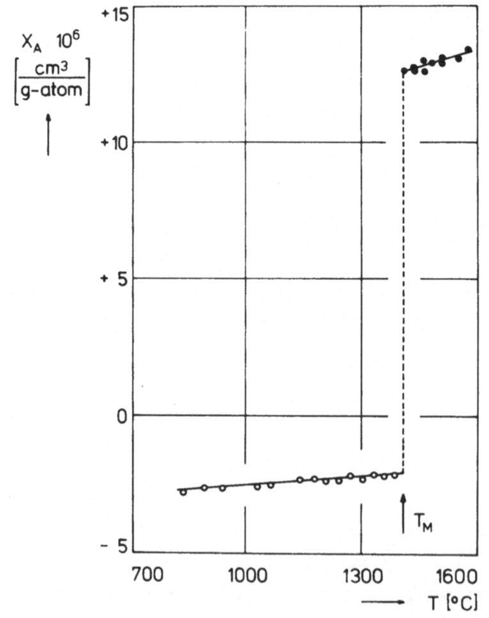

Fig.6 : Magnetic susceptibility of solid and liquid Si at high temperatures.

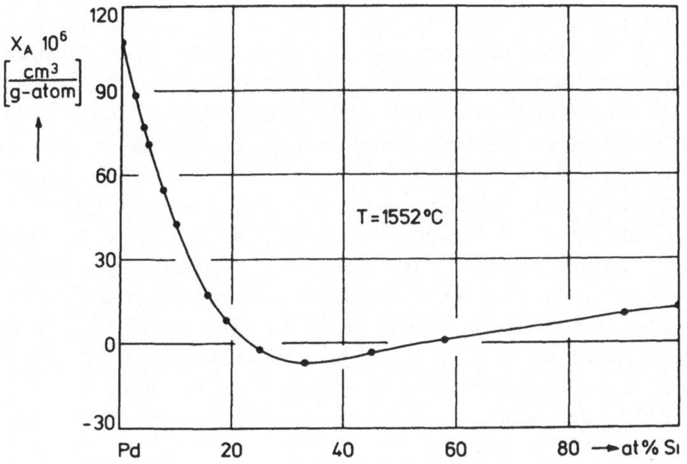

Fig.7 : Magnetic susceptibility of liquid Pd-Si alloys.

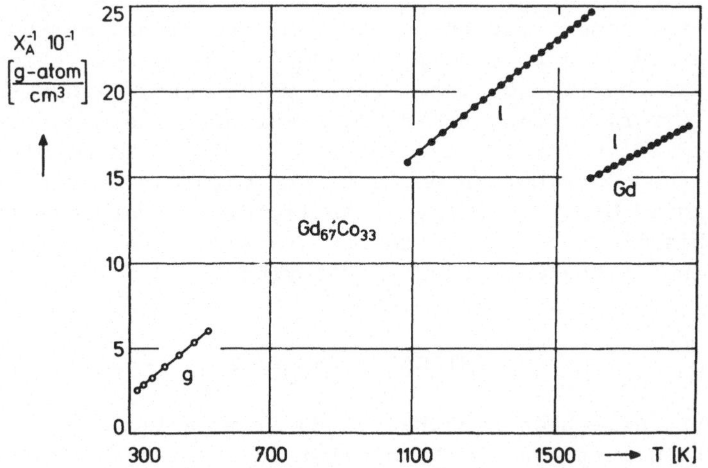

Fig.8 : Inverse magnetic susceptibility of glassy and liquid $Gd_{67}Co_{33}$ and liquid Gd.

isotherm at 1552°C, χ becomes diamagnetic in a concentration range from 25 to 55 at.% Si. The temperature coefficient of χ^{-1} changes sign from positive to negative at a Si concentration of 10 at.% . The magnetic properties of Pd-Si alloys can be understood qualitatively by Stoner theory taking into account the true density of states and the effective exchange of the Pd d electrons in the alloy. These quantities are not yet known for the liquid state. For the following discussion we use photoemission measurements on Pd-Si metallic glasses (see section 7). The following results are obtained: For glassy $Pd_{81}Si_{19}$, the observed intensity, which is related to the density of states, is lower at the Fermi energy, E_F , than the intensity for pure Pd at E_F . However, the intensity for the alloy is still a factor of 3 larger than the expected intensities for s and p states for monovalent noble metals. The resulting shoulder on the high energy side of the d band in the alloys still has d character. Thus, the drastic decrease of the large paramagnetic χ of pure liquid Pd on alloying with Si can be associated with the shift of E_F towards regions of lower density of states.

The inverse magnetic susceptibility of glassy and liquid $Gd_{67}Co_{33}$ (28) is shown in Fig.8 . The χ^{-1} values for the glassy (g) state show a continuous extrapolation into the liquid (ℓ) state. The inverse magnetic susceptibility of pure liquid Gd is shown for comparison. From the slope of the Curie-Weiss plot we have calculated the magnetic moments for the alloy. We obtain $\mu_{eff} = 6.70\,\mu_B$ in the glassy state and $\mu_{eff} = 6.86\,\mu_B$ in the liquid state. By taking into account the concentrations we are able to determine the magnetic moment per Gd atom to be $\mu_{eff} = 8.2\,\mu_B$ in the glassy state and $\mu_{eff} = 8.4\,\mu_B$ in the liquid state. These values are, within the experimental error, in reasonable agreement with the magnetic moment of pure liquid Gd, which is given by $\mu_{eff} = 8.6\,\mu_B$. In conclusion it follows that the contribution of Co to the magnetic moment of the glassy and liquid $Gd_{67}Co_{33}$ alloy is negligible or very small. The reason for this might be due to charge transfer resulting in a partial filling of the Co d-band on alloying.

6. OPTICAL PROPERTIES

We report reflection spectra for Pd-Si metallic glasses (29) and compare these with data obtained for Au-Si alloys prepared by getter sputtering in argon (30). The spectral reflectivity of an unpolished splat of $Pd_{81}Si_{19}$ in the photon energy range from 0.03 to 12 eV is shown in Fig.9 . For comparison the spectrum of pure crystalline Pd (31) is also presented. In the glassy state of $Pd_{81}Si_{19}$ the reflectivity decreases with increasing photon energy

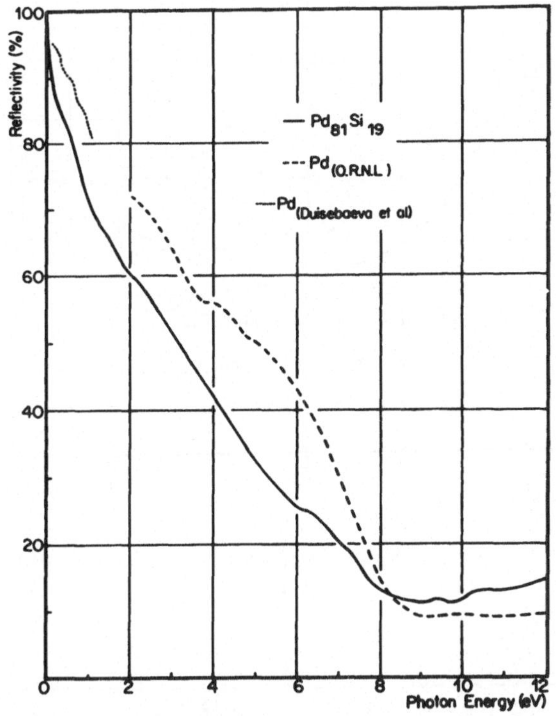

Fig.9 : Optical reflectivity of glassy $Pd_{81}Si_{19}$ and crystalline Pd.

and reaches a minimum around 9 eV. Structure appears in the low energy range at 1.5 and 2.3 eV. More pronounced structures are observed at 6.4, 7.3, 9.4 and at 10.3 eV. The reflectivity of the glassy $Pd_{81}Si_{19}$ alloy is lower than that of pure crystalline Pd up to 8 eV. The experimental data of Pd have been discussed in the literature (e.g. Ref.32) in terms of Drude theory in the range up to 0.16 eV and in terms of interband transitions at higher energies. The reflectivity of glassy $Pd_{81}Si_{19}$ is similar to the data published for amorphous $Au_{81}Si_{19}$ (30). Of particular interest is the observation that the measured reflectivity spectrum is virtually unchanged in going from the glassy to the crystalline state of $Pd_{81}Si_{19}$.

Fig. 10 shows the computed real and imaginary part of the dielectric function $\varepsilon = \varepsilon_1 + i\,\varepsilon_2$. The inset shows the data between 0.3 and 2.4 eV and the other curves are drawn in the energy range 2 to 12 eV. At first glance it seems very tempting to interprete the curves over the entire, or a large, energy range by Drude-like behavior, but a careful analysis shows that this is not possible.

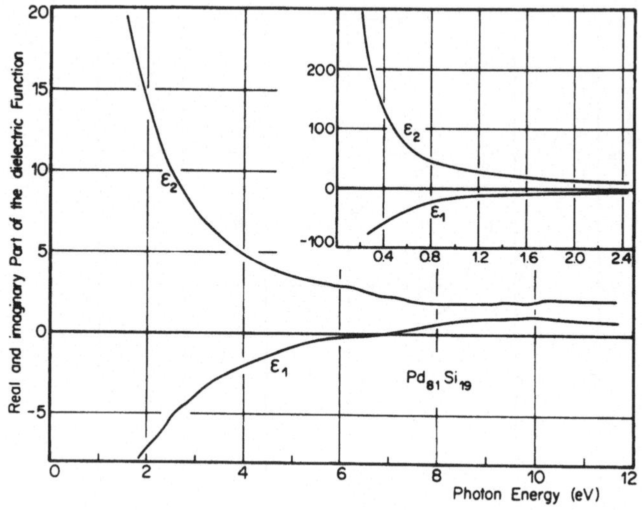

Fig.10 : Real and imaginary part of the dielectric function of glassy $Pd_{81}Si_{19}$.

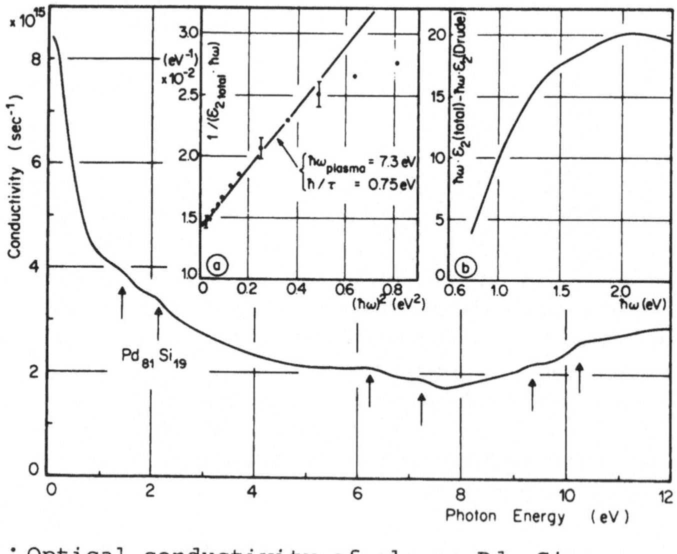

Fig.11 : Optical conductivity of glassy $Pd_{81}Si_{19}$

Inset a: $1/\hbar\omega\ \varepsilon_{2\,tot}$ versus $(\hbar\omega)^2$ of glassy $Pd_{81}Si_{19}$

Inset b: Interband absorption in the low-energy range.

The spectral reflectivity of glassy $Pd_{81}Si_{19}$ shows, apart from the lower values and the precise position in energy of the structures, a rough overall resemblance to the one for pure crystalline Pd with its contributions of intra- and interband transitions. Hence, it seems reasonable to explain the optical data of the glassy alloy $Pd_{81}Si_{19}$ based on the analysis of crystalline Pd but taking into account the alloying with Si . An interpretation very similar to the one for $Au_{81}Si_{19}$ (30) seems inadequate. We try to analyse the experimental reflectivity data of $Pd_{81}Si_{19}$ in the following way:

a) Intraband (Drude) range: Below the onset of interband transitions and within the Drude approximation, the imaginary part of the complex dielectric constants is given by

$$\varepsilon_2 = \frac{\omega_p^2}{\omega\tau(\omega^2+1/\tau^2)}$$

where ω_p is the free-electron plasma frequency, τ is the electronic relaxation time and $\gamma = \hbar/\tau$ is the damping. According to the inset a of Fig. 11, the quantity $1/\hbar\omega\varepsilon_2$ indeed varies linearly with $(\hbar\omega)^2$ up to 0.65 eV for $Pd_{81}Si_{19}$ and one can determine the Drude parameters ω_p and τ : $\hbar\omega_p = 7.3$ eV and $\gamma = 0.75$ eV . The comparable data for pure crystalline Pd (32) in the energy range up to 0.16 eV are $\hbar\omega_p = 6.5$ eV and $\gamma = 0.16$ eV . From the plasma frequency we have calculated the effective number of conduction electrons per formula unit $N^* = N/m^*/m = 0.6$. This value is in reasonable agreement with the number 0.95 deduced from Hall coefficient measurements (33), considering the uncertainty in m* and in the carrier concentration deduced from a free-electron Hall coefficient formula. Fig. 11 shows the calculated optical conductivity based on ε_2 as a function of energy. The optical conductivity extrapolated towards zero frequency yields $\sigma_o = 8.5 \cdot 10^{15} sec^{-1}$, corresponding to an electrical resistivity of 106 μΩcm in reasonable agreement with the DC resistivity of 80 μΩcm (33). The optical resistivity is somewhat larger than the DC resistivity as usually observed.

b) Interband transitions: Fig.11 shows clearly that the Drude range up to 0.65 eV is followed by the onset of interband transitions. The different structures are indicated by arrows. A precise assignment of all interband transitions of the alloy is not yet possible. This will require detailed band structure calculations for the metallic glass Pd-Si. Due to the similarity of the reflectivity spectra of the glassy alloy and crystalline Pd one could base an interpretation on the electronic structure of crystalline Pd. However, one must take into account the possibility that in an alloy, and especially in the amorphous state, the interband transitions can be weighted differently, are subject to broadening effects and can even occur at different energies due to a relaxation of the k-selection rule. According to theoretical calculations for crystalline Pd, the Fermi energy E_F lies about 0.35 eV below the top of the d band. Additional information for such an explanation can be gained by experimental photoemission studies comparing crystalline Pd and the glassy Pd-Si alloy. The inset b of Fig.11 shows the interband absorption given by the difference of the total $\varepsilon_2 \hbar\omega$ and the Drude contribution $\varepsilon_{2_D} \hbar\omega$. We see an absorption treshold at 0.65 eV, followed by a shoulder at 1.3 eV and a maximum at 2.1 eV. By restricting our discussion to the low-energy range, we find structures at 0.65, 1.3, 2.1 and 6.2 eV. The respective values in pure Pd are observed at 0.16, 0.35, 1.0 and 4.8 eV. In what follows, we discuss the observed differences in the positions of the corresponding structures in the glassy alloy and pure Pd. The onset of interband transitions at 0.16 eV in crystalline Pd is given by transitions from the d band to p states at E_F (L point in Λ direction). In the glassy alloy, the onset occurs at 0.65 eV. This higher value is due to a larger difference between the d states, from which transitions occur, and E_F. Such a larger difference can be attributed to a rise of E_F and simultaneous lowering of these d states on alloying with Si. A similar shift of 0.65 eV is observed in the UPS spectra (see Fig.16) by comparing crystalline Pd and glassy $Pd_{81}Si_{19}$. The shoulder at 1.3 eV and the maximum at 2.1 eV in the interband absorption of glassy $Pd_{81}Si_{19}$ are shifted by 1 eV compared to crystalline Pd. A shift of the same amount is observed for the intensity maximum in the UPS spectra of $Pd_{81}Si_{19}$ and crystalline Pd (see Fig.17). The broad maximum of the interband absorption of crystalline Pd at about 4.8 eV corresponds to transitions between the bottom of the d band and E_F, probably in the Δ and Σ directions. The corresponding maximum at 6.2 eV for glassy $Pd_{81}Si_{19}$ alloys can be explained by taking into account the distance in energy between the bottom of the d band and E_F in the glassy alloy. According to the UPS measurements this distance is approximately 6.2 eV.

Although the experimental reflectivity data of $Pd_{81}Si_{19}$ and $Au_{81}Si_{19}$ are quite similar, a detailed analysis leads to different interpretations. The Au-Si alloys are characterized by extremely short electronic relaxation times. This very high damping of 4 eV leads to a strong decrease of the typical optical reflectivity of pure Au on alloying with Si in the range up to 2.4 eV. The Drude behavior in this range is consistent with high values of γ, N_{eff} and ω_p. This drastic change is in agreement with the large increase of the electrical resistivity of Au on alloying with Si. For energies higher than 2.4 eV, interband transitions are observed. By contrast, data on glassy $Pd_{81}Si_{19}$ alloy can be interpreted in a similar way as pure Pd. The Drude behavior holds only in a limited energy range up to 0.65 eV. The results are consistent with lower values of γ , N_{eff} and ω_p . For energies above 0.65 eV interband transitions occur.

The optical reflection spectra for the metallic glasses $Mg_{70}Zn_{30}$, $Pd_{84}Si_{16}$, $Pd_{80}Si_{20}$ and $Gd_{67}Co_{33}$ are shown in a limited energy range from 1.5 to 6 eV in Fig.12 . In the infrared range the spectra obtained for $Mg_{70}Zn_{30}$ are in good agreement with the Drude behavior based on a free-electron model. The deviations at higher energies can be explained in terms of corrections to the Drude formula proposed by Helman and Baltensperger (34) for liquid simple metals. The reflectivity of glassy $Gd_{67}Co_{33}$ is dominated by interband transitions in a similar manner as we have discussed for the case of glassy Pd-Si alloys.

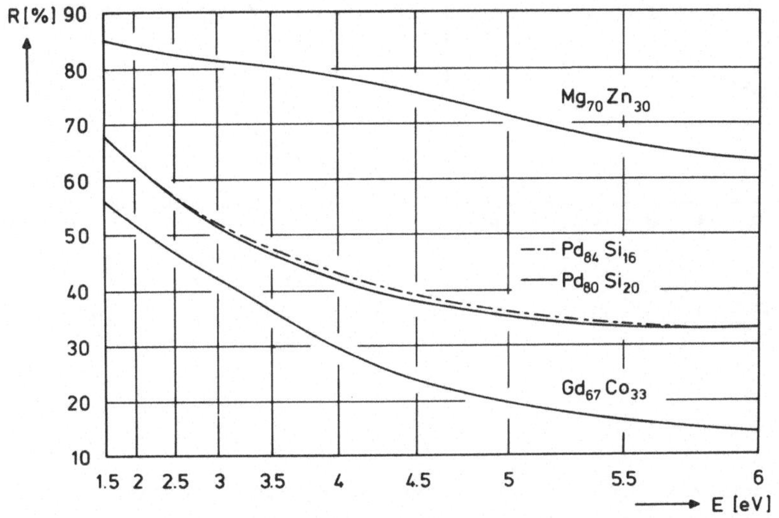

Fig.12 : Optical reflectivity of metallic glasses in a limited energy range.

7. ELECTRON SPECTROSCOPY

In this section we restrict ourselves to the presentation of new experimental results obtained by XPS (35) and UPS (36) measurements. The detailed discussion will be presented elsewhere. Fig.13 shows the valence band of METGLAS 2826 A ($Fe_{32}Ni_{36}Cr_{14}P_{16}B_6$) and 2826 ($Fe_{40}Ni_{40}P_{14}B_6$) from high resolution X-ray photoemission spectroscopy. These data are quite similar to those for the binary glassy $Fe_{80}B_{20}$ alloy. Of particular interest is the observation

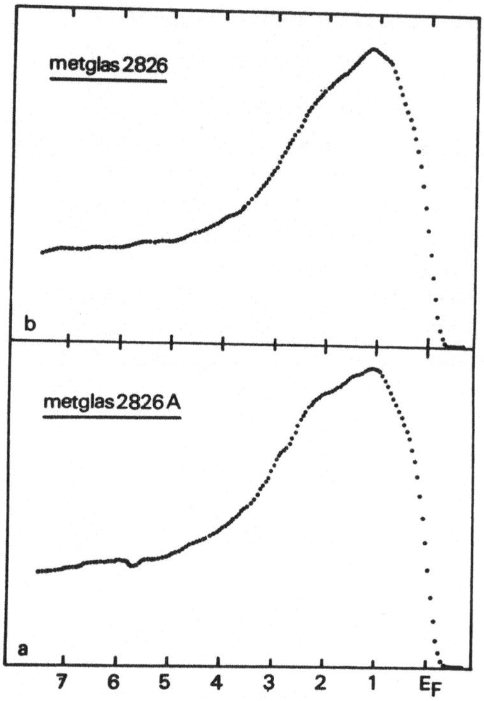

Fig.13:
Valence bands of METGLAS 2826 A (a) and 2826 (b) .

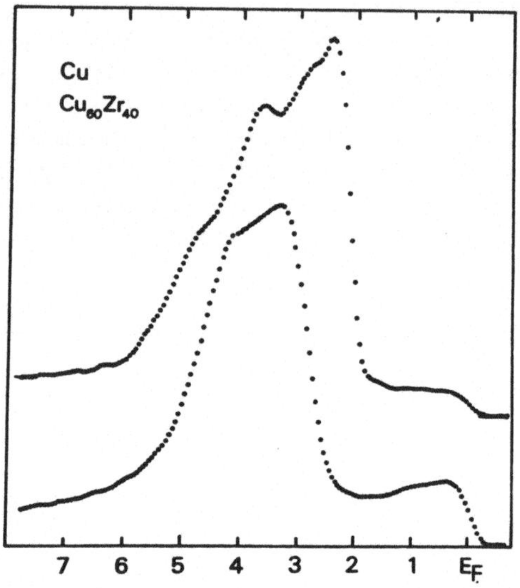

Fig.14:
Valence bands of glassy $Cu_{60}Zr_{40}$ and Cu.

that the measured density of states in all three cases is virtually unchanged in going from the glassy to the crystalline state. Data for glassy $Cu_{60}Zr_{40}$ are shown in Fig.14 , in which the density of states of pure Cu is shown for comparison. The density of states of glassy $Cu_{60}Zr_{40}$ can, to a very good approximation, be made up by a superposition of the Cu and Zr data. Hence, the density of states at E_F increases on alloying with Zr. In high resolution XPS measurements on glassy $Pd_{80}Si_{20}$, there is an indication of a small shoulder at E_F . This shoulder becomes more pronounced in UPS measurements. Fig.15 shows the two types of measurement for glassy $Pd_{81}Si_{19}$. The UPS data were obtained using a He II (40.8 eV) source. The photon energy for the XPS measurements was 1253,6 eV . UPS measurements using a He I (21.2 eV) source are shown for several Si concentrations in Fig.16 . This figure shows clearly, that the distance in energy between E_F and the onset of the high density of states d-band increases with increasing Si content. For glassy $Pd_{82}Si_{18}$ this distance

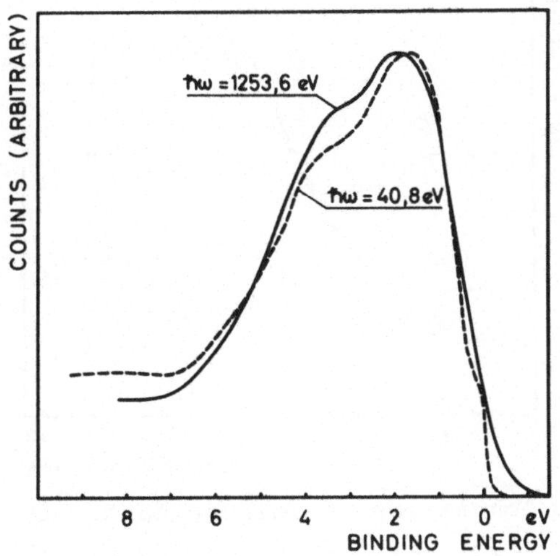

Fig.15 : Valence bands of glassy $Pd_{81}Si_{19}$ measured by XPS and UPS.

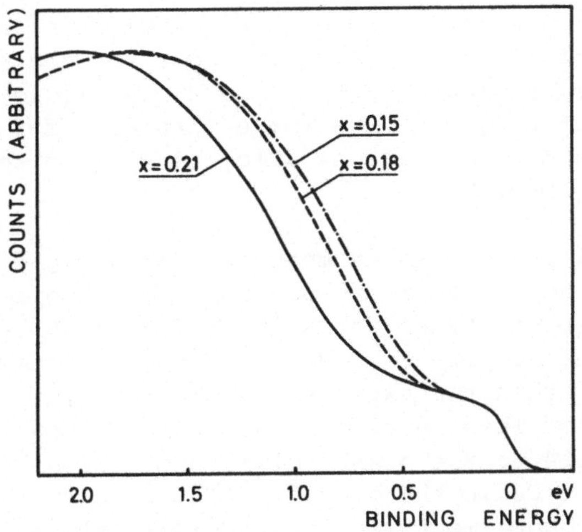

Fig.16 : UPS spectra of glassy $Pd_{1-x}Si_x$ obtained with He I source.

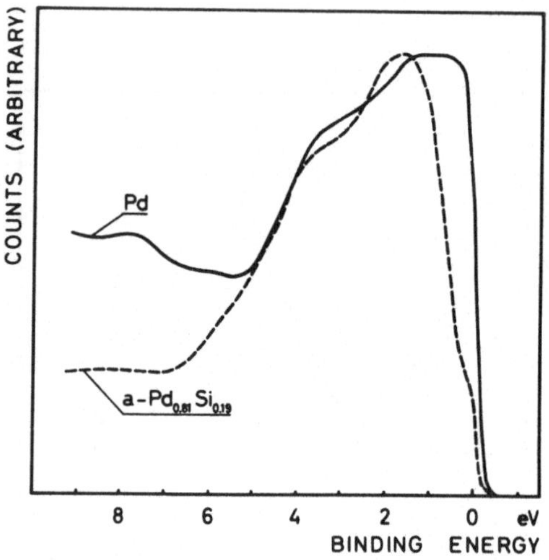

Fig.17 : Comparison of valence bands of crystalline Pd and glassy $Pd_{81}Si_{19}$.

is roughly 0.65 eV , which coincides with the onset of interband absorption in the optical reflectivity spectra. A comparison between the valence bands of pure Pd and glassy $Pd_{81}Si_{19}$ is shown in Fig.17 . The valence band of glassy $Pd_{81}Si_{19}$ is characterized by the following features: A shoulder occurs at E_F , the d band is lowered and in the energy range of 2 to 5 eV the curve fits that for pure Pd . The bottom of the d-band of the glassy alloy is shifted to higher energy values compared to pure Pd .

These careful photoemission measurements using different wavelength from He I , He II and X-ray sources yield the following results: The density of states at E_F in the glassy $Pd_{81}Si_{19}$ alloys is reduced compared to pure Pd . However, these values are still a factor of 3 larger than the expected values for s,p states in monovalent noble metals. This is also clear from a comparison of the Cu spectra in Fig.14 to the glassy alloy spectra in Fig.15. The resulting shoulder at E_F in the alloys still has d character. This finding is in good agreement with theoretical calculations of crystalline Pd (37), where E_F lies 0.35 eV below the upper d band edge and a shoulder of empty d states is indicated.

8. OTHER PROPERTIES

We would like to mention several other interesting properties. For more detailed information, the reader is referred to the original literature.

Non-crystalline superconductors prepared by quench condensation were first studied over twenty years ago by Buckel and Hilsch (38). More recently melt-spinning and splat-cooling techniques have been employed. A breakthrough has been reported recently. For a long time the highest reported superconducting transition temperature of metallic glasses was about 6 K. Recently Johnson (39) has been able to find transition temperatures up to 9 K for $Mo_{80}P_{10}B_{10}$. These values come very close to the expected maximum from Collver and Hammond's work (40).

It has been known for a long time, that electrons emitted by field emission from a crystalline tungsten tip show characteristic pattern which can be interpreted in terms of crystallinity. Thus it is expected that field emission pattern from glassy metals cannot show structures because of the absence of planes and that the work function should be uniform over the surface. Such experiments have been reported in references 40 and 41.

Of similar interest as electron spectroscopy is the Ion Scattering Spectroscopy (ISS). This method reveals information of the first monolayer of the surface. Fig.18 shows the ISS spectrum of a glassy $Pd_{81}Si_{19}$ alloy. More details will be published elsewhere.

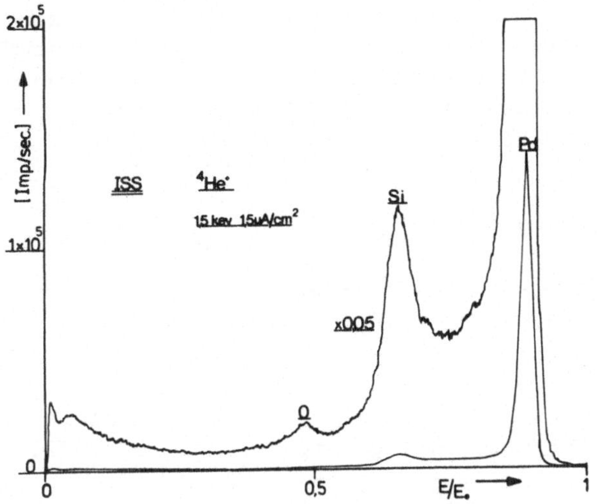

Fig.18 : ISS spectrum of glassy $Pd_{81}Si_{19}$.

9. CONCLUSION

In this review, we have tried to indicate the types of experiments and theories on the physical properties of metallic glasses presently known. Space precluded a detailed discussion of all the properties, so emphasis was placed on the ionic structure, glass forming ability, electrical transport, optical properties and electron spectroscopy. It is now becoming apparent that some of the traditional experimental techniques that have been so useful in work on crystalline materials can also be helpful in understanding the physics of metallic glasses. Though theoretical explanations for some of the data on metallic glasses can, at times, be quite difficult, progress made in the past few years does give rise to some optimism.

ACKNOWLEDGEMENTS

The unpublished results of optical reflectivity and XPS measurements were obtained as part of a stimulating collaboration with P. Wachter, Y. Baer, A. Schlegel and E. Cartier, ETH Zurich. Other XPS and the UPS measurements were performed in collaboration with H.D. Polaschegg and K. Berresheim, Leybold-Heraeus Köln. Further information on the static structure and the dynamics of ions came as a result of a very productive collaboration with J. Suck, ILL Grenoble and P. Fischer, A. Furrer, EIR Würenlingen. We are grateful to Dr. L. v. Heimendahl, Dr. H. Hillmann, Dr. H.R. Hilzinger, Prof. T. Egami, Dr. J. Wong, Dr. H. Heinrich and Prof. W.L. Johnson for sending their papers prior to publication. The authors are very grateful for stimulating discussions with many colleagues in the field of glassy and liquid metals. Financial support of the Swiss National Science Foundation, the "Kommission zur Förderung der wissenschaftlichen Forschung", the "Eidgenössische Stiftung zur Förderung Schweizerischer Volkswirtschaft durch wissenschaftliche Forschung" and the Research Center of Alusuisse is gratefully acknowledged.

REFERENCES

(1) H.-J. Güntherodt in "Festkörperprobleme" (Advances in Solid State Physics), Vol. XVII, ed. by J. Treusch (Vieweg, Braunschweig), p.25, 1977.

(2)

a. J.J. Gilman in "Crystal growth and materials", ed. by E. Waldis and H.J. Scheel (North-Holland Publishing Company), p.728, 1977.

b. "Metallic Glasses" ed. by J.J. Gilman and H.J. Leamy (American Society for Metals, Metals Park, Ohio), 1978.

c. P.Chaudhari and D. Turnbull, Science, 1978.

d. R. Cahn, "The Metallurgist", p.309, June 1978.

e. Proc. of the Symposium on "Structure and Properties of Amorphous Metals", Jizaka, Fukushima, Japan (Suppl. to the Science Reports of the Research Institutes, Tohoku University, Series A), 1978.

f. Proc. 3rd Int. Conf. on Rapidly Quenched Metals, Brighton, 1978.

g. H. Warlimont, Z. Metallkunde 69, 212, 1978.

(3) R. Clampitt, M.G. Scott, K.L. Aithen and L. Gowland in ref. 2f.

(4) S. Davis, M. Fischer, B.C. Giessen and D. Polk in ref. 2f.

(5) L. von Heimendahl, to be published.

(6) to be published.

(7) T. Egami, R.S. Williams and Y. Waseda in ref. 2f.

(8) T.M. Hayes, J.W. Allen, J. Tauc, B.C. Giessen and J.J. Hauser, Phys.Rev. Letters 40, 1282, 1978.

(9) J. Wong, F.W. Lytle, R.B. Gregor, H.H. Liebermann, J.L. Walter and F.E. Luborsky in ref. 2f.

(10) J.A. Tarvin, G. Shirane, R.J. Birgeneau and H.S. Chen, in "Transition Metals", ed. by M.J.G. Lee, J.M. Perz and E. Fawcett (Conf.Ser. No 39, The Institute of Physics, Bristol and London, 1978), p.514 .

(11) to be published.

(12) P.H. Chang, A.P. Malozemoff, M. Grimsditch, W. Senn and G. Winterling, to be published.

(13) M. Grimsditch and G. Güntherodt, to be published.

(14) to be published.

(15) S.R. Nagel and J. Tauc, Phys. Rev. Letters 35, 380, 1975.

(16) H. Beck and R. Oberle in ref. 2f.

(17) H.-J. Güntherodt and H.U. Künzi in ref. 2b.

(18) G. Baym, Phys.Rev. 135, A1691, 1964.

(19) to be published.

(20) H. Hillmann and H.R. Hilzinger in ref. 2f.

(21) P.J. Cote and L.V. Meisel in ref. 2f.

(22) K. Froböse and J. Jäckle, J. Phys. F, 7, 2331, 1977.

(23) R.W. Cochrane, R. Harris, J.O. Ström-Olsen and M.J. Zuckermann, Phys.Rev. Letters 35, 676, 1975.

(24) C.C. Tsuei, Sol. State Com. 27, 691, 1978.

(25) F.E. Luborsky, in: Ferromagnetic Materials, E.P. Wohlfarth, ed. (North-Holland Publ. Co., Amsterdam) ch.XX, 1978 and J. of Magnetism and Magnetic Materials 7, 143, 1978.

(26) M. Müller and H.-J. Güntherodt, to be published.

(27) M. Müller, H. Beck and H.-J. Güntherodt, to be published.

(28) M. Liard, M. Müller, K.P. Ackermann, B. Delley, K. Agyeman, H.U. Künzi, H. Rudin and H.-J. Güntherodt in ref. 2f.

(29) A. Schlegel, P. Wachter, K.P. Ackermann, M. Liard and H.-J. Güntherodt, to be published.

(30) E. Hauser, R.J. Zirke, J. Tauc, J.J. Hauser and S.R. Nagel, Phys.Rev. Letters 40, 1733, 1978.

(31) Z.H. Duisebaeva, M.I. Korsunskii and G.P. Motulevich, Opt. Spectroscopy 34, 307, 1973; 37, 82, 1974.
R.C. Vehse, E.T. Arakawa and M.W. Williams, Phys.Rev. B1, 517, 1970.

(32) J. Lafait in "Transition Metals", ed. by M.J.G. Lee, J.M. Perz and E. Fawcett (Conf. Ser. No 39, The Institute of Physics, Bristol and London, 1978), p.130 .

(33) H.-J. Güntherodt, H.U. Künzi, M. Liard, M. Müller, R. Müller and C.C. Tsuei in "Amorphous Magnetism II" (Proc. of the 2nd International Symposium on Amorphous Magnetism, RPI, Troy, New York), edited by R.A. Levy and R. Hasegawa, (Plenum Press, New York, 1977), p.257.

(34) J.S. Helman and W. Baltensperger, Phys. Cond. Matter 5, 60, 1966; 15, 346, 1973.

(35) E. Cartier, Y. Baer, M. Liard and H.-J. Güntherodt, to be published.

(36) P. Oelhafen et al., to be published.

(37) F.M. Mueller, A.J. Freeman, J.O. Dimmock and A.M. Furdyna, Phys.Rev. B1, 4617, 1970.
O.K. Andersen, Phys.Rev. B2, 883, 1970.
N.E. Christensen, Phys.Rev. B14, 3446, 1976.

(38) W. Buckel and R. Hilsch, Z. Phys. 138, 109, 1954; 146, 27, 1956.

(39) W.L. Johnson in ref. 2f.

(40) M.M. Collver and R.H. Hammond, Phys.Rev. Letters 30, 92, 1973.

(41) R.M.J. Cotterill and F. Kragh, Sc. and J. of Metallurgy 6, 191, 1977.

(42) H. Heinrich, T. Haag and J. Geiger, to be published.

ON THE THEORY OF DISORDERED SYSTEMS: CPA CALCULATION OF $(SN)_x$ WITH HYDROGEN IMPURITIES AND HARTREE-FOCK THEORY OF SURFACE STATES OF THREE-DIMENSIONAL CRYSTALS

J. Ladik and M. Seel

Lehrstuhl für Theoretische Chemie
Friedrich-Alexander-Universität Erlangen-Nürnberg
852 Erlangen, F.R.G.

ABSTRACT

A procedure to solve the coherent potential approximation (CPA) equation for energy and k-dependent self-energies is described. As illustrative example calculations for the $(SN)_x$-$(\overset{SN}{\underset{H}{|}})_x$ two-component mixed polymer are presented. Spikes and dips in the density of states of the mixed system are found already for 3 mol. per cent hydrogen concentration.

The resolvent method developed previously for the *ab initio* self-consistent field linear-combination of atomic-orbitals (SCF LCAO) treatment of a cluster of impurities embedded in a one-dimensional periodic polymer is extended for the *ab initio* SCF LCAO computation of the surface and chemisorption states of a three-dimensional semi-infinite crystal. The way of solving the matrix equations which are derived is outlined.

1. INTRODUCTION

The paper presented here is concerned with problems which arise when the full crystalline symmetry is broken or more generally, the periodicity of a system is disturbed. This is the case for random substitutional alloys, for a large class of polymers which play an important role in technical applications (plastics, highly conducting polymers) or in biology (biopolymers like DNA, RNA and proteins) and, of course, at surfaces. In the last years a vast amount of experimental information has been gathered both on the different properties of the highly conducting polymers and on surface states. To interpret these different physical and chemical properties new theoretical techniques are needed.

The electronic structure of systems with an underlying symmetry is relatively well understood. The assumption of a general symmetry operation in connection with periodic (Born-von Kármán) boundary conditions makes it possible to block-diagonalize in any one-electron (like Hartree-Fock) approximation the cyclic hypermatrix and to reduce the problem to the size of the unit cell [1]. When the symmetry is broken blockdiagonalization is no longer possible and one has to find other ways to treat the arising aperiodic system.

In the first part the application of the coherent potential approximation (CPA) method [2] to aperiodic polymers using a k- and energy-dependent self-energy will be discussed. As an example the calculation of the density of states of the valence band of $(SN)_x$ with different hydrogen impurity concentrations [3] is presented. In the second part it is shown how the resolvent method for the calculation of the extra states of a cluster of impurities embedded in an one-dimensional periodic system [4] can be generalized for the calculation of the surface states of a semi-infinite three-dimensional crystal in the Hartree-Fock (ab initio SCF LCAO) level [5]. This method in its matrix formulation of the problem facilitates, as we shall see, very much numerical calculations.

2. CPA CALCULATION OF THE $(SN)_x$ - $(\overset{SN}{\underset{H}{|}})_x$ MIXED SYSTEM

Highly conducting polymers like $(SN)_x$ (poly-sulfurnitride) have raised considerable interest in the last ten years. $(SN)_x$ is a highly anisotropic metal at higher temperatures and becomes superconductive below 0.26°K [6].

Several authors performed band structure calculations for this quasi one-dimensional system (both semiempirical [7] and minimal bases ab initio [8] ones). There are also in the literature non-self-consistent OPW and LCAO calculations for the three-dimensional system [9]. Recently we have executed also an ab initio double ζ band structure calculation [10] for the linear chain which has given rather good agreement with experiment for the effective electronic mass and density of states at the Fermi level.

Recently at IBM (San Jose) five to ten mol. per cent hydrogen was found in $(SN)_x$ [11]. The position of the hydrogen impurities is unknown (as it is of the Br_2, J_2 and JCl molecules with which $(SN)_x$ was modified also [12]), but most probably the H atoms bind to the N atoms. In this way they change the hybridization state of the N atoms and the number of π-electrons in the partially filled band of $(SN)_x$ (in a $\diagdown S{=}N\diagdown$ unit there are 3π-electrons, while in a $\diagdown S\diagup\overset{N}{\underset{H}{|}}\diagdown$ unit there are 4).

To determine theoretically the shift of the Fermi level and the change in the density of states when $(SN)_x$ contains hydrogen impurities we applied CPA, using a newly developed method which does not neglect the k-dependence of the self-energy.

2.1. Method

The basis assumption of the CPA method is [2] that one substitutes the average of the multicomponent system by an effective medium determined so that the average fluctuation through the medium is zero. This can be achieved by replacing each site in the crystal but one+ with an unknown coherent potential. One embeds then at the reference site an A or a B component (in a simple case of an A, B two-component compositionally disordered system) with the probability f and 1-f, respectively. We solve then the problem of this single impurity embedded in the effective medium described by the coherent potential. The coherent potential on the other hand is determined by the self-consistency requirement that the average scattering (or fluctuation) from the chosen reference site is also zero.

To formulate mathematically the method++ we can start from the Dyson equation of the single particle Green's function G of the disordered system, which is in the one-band case

$$G = G^0 + G^0 \Delta G \tag{2.1}$$

where G^0 is the Green's function of the unperturbed perfectly periodic system which is in its Fourier transformed form

$$G^0(\bar{k},E) = [E - \varepsilon^A(\bar{k})]^{-1} \ , \ G^0(E) = \Omega^{-1} \sum_k [E - \varepsilon^A(\bar{k})]^{-1} \tag{2.2}$$

(Ω is the volume of the crystal). Further the deviation from the perfectly periodic system A $\Delta(\bar{k})$ can be written in the case of a single site diagonal perturbation [2] as+++

$$\Delta(\bar{k}) = \varepsilon^B(\bar{k}) - \varepsilon^A(\bar{k}) \tag{2.3}$$

+ In this single site approximation of CPA it is assumed that after a scattering form this site no repeated scattering occurs before a scattering from another site has taken place.

++ We write here down only the basic equations, for their derivation see [2].

+++ Since system A and B and also the effective medium are periodic one can represent Δ also in the $\bar{k}$ space. Working with a tight-binding Hamiltonian and neglecting its off-diagonal elements it is easy to show that Eq.(2.3) is the Fourier transform of the more familiar $\Delta_i = \varepsilon_i^B - \varepsilon_i^A$ expression (here, i stands for the site characterized by $\bar{R}_i$).

Eq. (2.1) (which can be derived [2] from the perturbation expansion

$$G = G^0 + G^0 \Delta G^0 + G^0 \Delta G^0 \Delta G^0 + \ldots \tag{2.4}$$

of G) can be written also in the form

$$G = G^0 + G^0 T G^0 \tag{2.5}$$

where the one-dimensional scattering matrix T is defined as

$$T = \Delta(1 - \Delta G^0)^{-1} \tag{2.6}$$

In the case of CPA we can define a Green's function G_e for the effective medium again through the Dyson equation

$$G_e = G^0 + G^0 \sum G_e \tag{2.7}$$

(which defines also the self-energy $\sum$). Solving (2.7) for G^0 we obtain

$$G^0 = G_e(1 + \sum G_e)^{-1} \tag{2.8}$$

Substituting this expression of G^0 into the Dyson equation (2.4) of the exact Green's function of the system one can write

$$G = G^0(1 + \Delta G) = G_e(1 + \sum G_e)^{-1} (1 + \Delta G) \tag{2.9}$$

which leads after rearrangement to the equation

$$G = G_e + G_e \Delta_e G \tag{2.10}$$

where $\Delta_e = \Delta - \sum$ [2]. Thus at the unperturbed sites ($\Delta = 0$) of the original system $\Delta_e = -\sum$ and of the perturbed sites $\Delta_e = \Delta - \sum$. Using the definition (2.6) of T we can now write for its average (by substituting instead of Δ Δ_e and instead of G^0 G_e)

$$\langle T\rangle = (1 - f)(-\sum)[1 + \sum G_e]^{-1} + f(\Delta - \sum)[1 - (\Delta - \sum)G_e]^{-1} \tag{2.11}$$

where f is the probability of the perturbed sites.

According to the fundamental approximation of CPA the average fluctuation from the effective medium is zero,

$$\langle T\rangle = 0 \tag{2.12}$$

Putting eq. (2.11) equal to zero one obtains after some manipulations the CPA equation [2]:

$$\sum(\bar{k},E) = f\Delta(\bar{k})/\{1 + [\sum(\bar{k},E) - \Delta(\bar{k})]\ G_e(E)\} \tag{2.13}$$

With (2.2) we obtain from (2.7) for G_e

$$G_e(\bar{k},E) = [E - \varepsilon^A(\bar{k}) - \sum(\bar{k},E)]^{-1} \qquad (2.14a)$$

and finally

$$G_e(E) = \Omega^{-1}\sum_k G_e(\bar{k},E) = \Omega^{-1}\sum[E - \varepsilon^A(\bar{k}) - \sum(\bar{k},E)]^{-1} \qquad (2.14b)$$

In the case when the $\bar{k}$ dependence of $\sum$ can be neglected $G_e(E)$ is simply [15]

$$G_e(E) = G^0(E - \sum) \qquad (2.15)$$

Eq. (2.15) can be applied, however, only in cases when the density of states curves of the periodic systems A and B have the same shape, only one curve is shifted with respect to the other by a constant $\sum$ value. In the case of substitutionally disordered polymers due to the complexity of their subunits we cannot expect, however that this condition will be fullfilled.

Returning to the general formalism with $\bar{k}$-dependent self-energy we can write for the spectral density [15] and for the average density of states per molecule, respectively,

$$A(\bar{k},E) = -\pi^{-1} \operatorname{Im} G_e(\bar{k},E + i0), \qquad (2.16)$$

$$\rho(E) = \Omega^{-1} \sum_{\bar{k}} A(\bar{k},E) = -\pi^{-1} \operatorname{Im} G_e(E + i0) \qquad (2.17)$$

Using further the Kramers-Kronig [16] relation one obtains

$$G_e(E) = \int_{-\infty}^{\infty} \rho(E')/(E - E')dE' \qquad (2.18)$$

Finally using the rule

$$\frac{\alpha_1 + i\beta_1}{\alpha_2 + i\beta_2} = \frac{\alpha_1\alpha_2 + \beta_1\beta_2}{\alpha_2^2 + \beta_2^2} + \frac{\alpha_2\beta_1 - \alpha_1\beta_2}{\alpha_2^2 + \beta_2^2}\, i$$

for the ratio of two complex numbers we can write

$$\operatorname{Im} G_e(\bar{k},E) = \operatorname{Im} \sum(\bar{k},E)\{[E - \varepsilon^A(\bar{k}) - \operatorname{Re} \sum(\bar{k},E)]^2 + [\operatorname{Im} \sum(\bar{k},E)]^2\}^{-1} \qquad (2.19)$$

Equations (2.13) - (2.19) are the expressions which can be used for an actual calculation with a $\bar{k}$-dependent self-energy. (In the case of a linear aperiodic polymer we can write instead of the vector $\bar{k}$ everywhere in these equations the scalar k). One starts the iterative procedure with a guess for $G_e(E)^+$, one obtains the first values for $\sum(k,E)$ from (2.13) for different k and E values. Substituting this into (2.19) and the resulting Im $G_e(k,E)$ values into (2.16) and (2.17), respectively, one obtains the first approximation for $\rho(E)$. Putting finally this into (2.18) one gets the next approximation for $G_e(E)$. This numerical procedure (which only recently was worked out [3] for the treatment of an aperiodic polymer with k-dependent self-energy) has to be repeated until self consistency is reached in $\rho(E)$.

2.2. Application to the $(SN)_x$-System with Hydrogen Impurities

The above described form of CPA has been applied for the treatment of hydrogen impurities in $(SN)_x$ using as input the ab initio minimal basis band structures of the periodic $(SN)_x$ and $(\overset{SN}{\underset{H}{|}})_x$ chains [17]. From these the densities of states $\rho(E)$ of the two periodic chains have been computed using the method of Delhalle [18] (see Fig. 2.1)

As we can see from Fig. 2.1 the shape of the two density of states curves is very different. Our first attempts to apply CPA to the mixed system with a constant (k-independent) $\sum$ have therefore failed. As next step we applied the above described procedure with a k- and E-dependent $\sum$, $\sum(k,E)$ [3].

In the actual calculations 93 different E_i-values and k_i points $(0 \leq k_i \leq \pi/a)$ were chosen for which $\sum(k_i,E_i)$ was computed in every iteration step. To reach self-consistency between 24 $(f = 0.3)$ and 70 $(f = 0.5)$ iteration steps were needed using the SCF criterion

$$|\rho_e^{(n)}(E_i) - \rho_e^{(n-1)}(E_i)| \leq 10^{-3}\ [\text{a.u.mol.}]^{-1}\ ,\ \forall E_i \tag{2.21}$$

Finally, after self-consistency was reached the position of the Fermi level was determined for each value of f by numerical integration using the relation

$$\int_{\varepsilon_{min}}^{\varepsilon_F} \rho^{(SCF)}(E)dE = 1 + f \tag{2.22}$$

+ Taking either the Green's function of pure $A(\sum = 0,\ G_e(E)=G^0(E))$ or the virtual crystal Green's function $(\sum(k) = f\Delta(k)$ which one obtains in a good approximation from (2.13) if $\Delta(k)$ is small) computing it by (2.18) with the virtual crystal density of states $\rho_{vc}(E')$ which belongs to the energy band

$$\varepsilon_{vc}(k) = (1 - f)\varepsilon^A(k) + f\varepsilon^B(k) \tag{2.20}$$

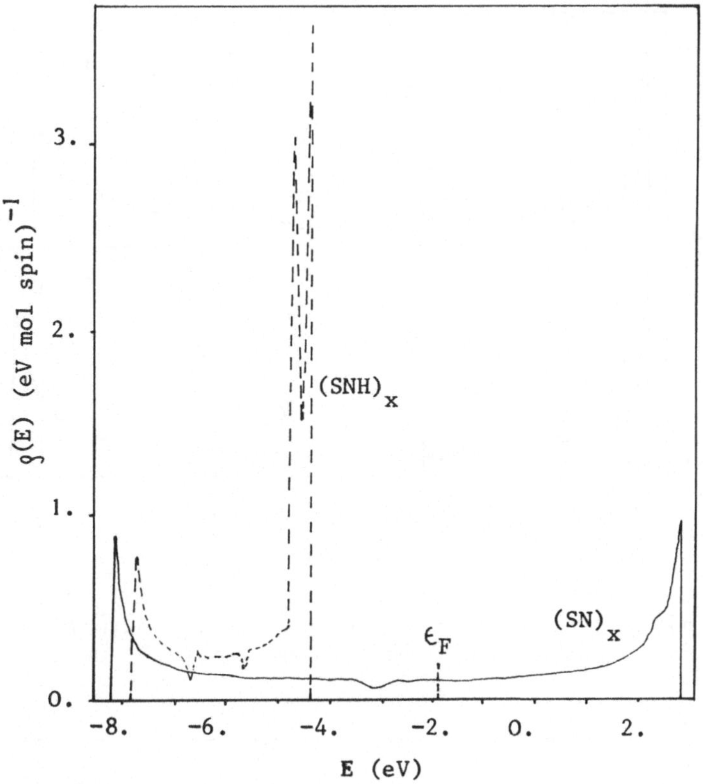

Fig. 2.1 : The density of states curves of the pure $(SN)_x$ and $(\overset{SN}{\underset{H}{|}})_x$ systems in $(eV.\ mol.\ spin)^{-1}$ units.

The calculations for a given f value have taken between 100 and 200 seconds on a CYBER 172 computer (for further details of the numerical calculations see [3]).

Figs. 2.2 and 2.3 present the density of states curves of the mixed system for different hydrogen concentrations f obtained with the aid of the described procedure. Further density of states curves for f= 0.05, 0.07, 0.20 and 0.40 can be found in [3].

The most interesting feature of the $\rho_e(E)$ curves in the Figs. is their complicated structure with spikes and dips which in many cases go to zero producing gaps in the band of the mixed system. In this connection it should be emphasized that there are no zero dips in the $\rho(E)$ curves of the two periodic systems (see Fig. 2.1) and therefore their occurrence (already at f = 0.03) is a genuine effect of the aperiodicity. Such structures are well known from computer experiments on linear chains [2] and from cluster-CPA

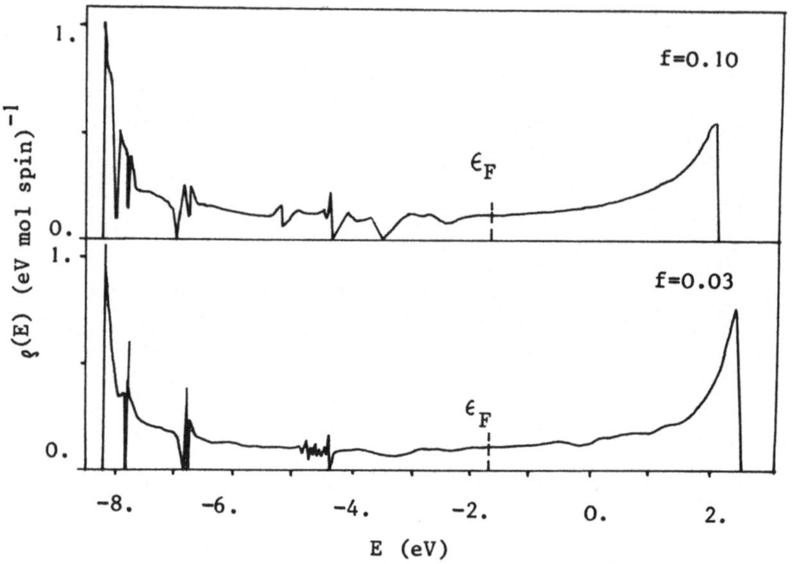

Fig. 2.2 : The density of states curves of the $(SN)_x$ and $(\underset{\mathrm{H}}{\mathrm{SN}})_x$ mixed system obtained in the CPA approximation with f = 0.03, and 0.10 in (eV. mol. spin)$^{-1}$ units.

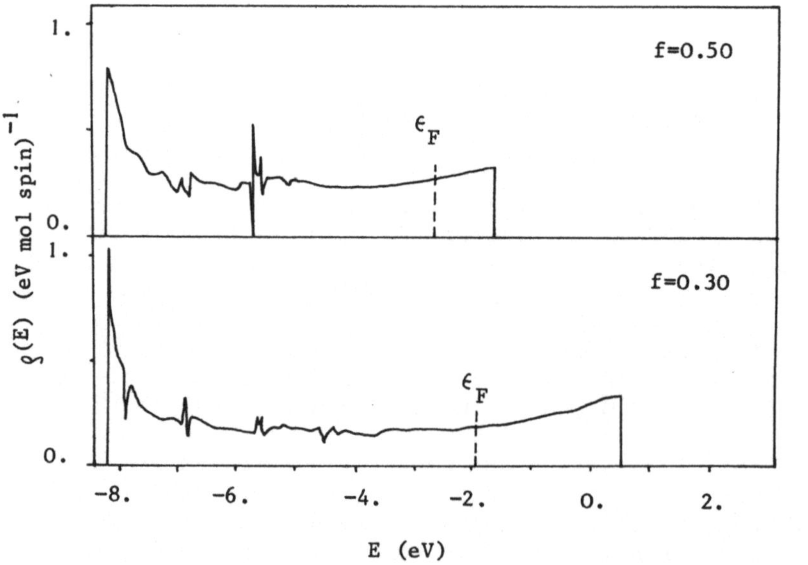

Fig. 2.3 : The density of states curves of the $(SN)_x$ and $(\underset{\mathrm{H}}{\mathrm{SN}})_x$ mixed system obtained in the CPA approximation with f = 0.30 and f = 0.50 in (eV. mol. spin)$^{-1}$ units.

calculations [2] (in which the impurities are extended over more sites), but could not be obtained until now with the one-site CPA with a k-independent $\sum$.

Looking at the positions of these spikes and dips it is easy to recognize that they always occur at such E values at which the $\rho(E)$ curves of the periodic systems have a large curvature (compare Fig. 2.2 and2.3 with Fig. 2.1). Therefore probably the small gaps occurring in the band of the mixed system are not due to Anderson localization, but to the fact that levels of the original periodic systems which lie in a high curvature region of $\rho(E)$ are more sensitive to perturbations due to the other component, than the other levels.

From Fig.2.1 one can see that the $(\overset{\mathrm{SN}}{\underset{\mathrm{H}}{|}})_x$ periodic hain has a much narrower valence band than $(\mathrm{SN})_x$ (with corresponding large peaks in the density of states). The Fermi level of $(\mathrm{SN})_x$ lies at - 1.09eV, while the upper limit of the completely filled valence band of $(\overset{\mathrm{SN}}{\underset{\mathrm{H}}{|}})_x$ is at -4.38eV. One would expect on the basis of the rather large differences in the density of states curves of the two systems that in the mixed system already a small percentage of $(\overset{\mathrm{SN}}{\underset{\mathrm{H}}{|}})_x$ would have a comparatively large influence on the density of states curve of pure $(\mathrm{SN})_x$.

This expectation if fullfilled as one can see from Fig. 2.2 where the density of states curves obtained in the CPA approximation for the mixed system with 10% (or less) hydrogen are shown. At very low hydrogen concentration (f = 0.03) already new peaks in the density of states curve start to develop in the region between - 4.4eV and - 8.0 eV. At higher f values (see Fig. 2.3) peaks of $(\mathrm{SN})_x$ and $(\overset{\mathrm{SN}}{\underset{\mathrm{H}}{|}})_x$ in the region between -7 and -8eV fuse to one broader peak. On the other hand in consequence of the very high but extremely narrow peaks of $(\overset{\mathrm{SN}}{\underset{\mathrm{H}}{|}})_x$ between -4.8 and -4.4eV new peaks develop but they are shifted to the region around -5.6eV (see especially the curve belonging to f = 0.50). In all these cases one can see a clear demonstration for the fact that the CPA method gives essentially different results than the simple virtual crystal approximation.

The position of the Fermi level of the mixed system is not a sensitive function of f at low concentrations (ε_F = - 1.9eV for f = 0.00, ε_F = - 1.6eV for f = 0.03 and 0.10). At high concentrations, of course, its position shifts towards lower energies (ε_F = - 2.1eV at f = 0.30 and ε_F = - 2.6eV at f = 0.50). One should mention that the density of states at the Fermi level monotonously increases with the increase of f ($\rho(E)$ = 0.10, 0.11, 0.13, 0.18 and 0.26 at f = 0.00, 0.03, 0.10, 0.30 and 0.50, respectively).

Finally one should point out that by doping $(\mathrm{SN})_x$ with hydrogen one could expect to find the theoretically obtained spikes and dips in the $\rho(E)$ curve of the mixed system by photoelectron spectroscopy (for the complications which can occur by the interpretation of these spectra see, however, [3]). Further, since according to the BCS theory of superconductivity the transition temperature T_c depends exponentially on the density of states at the Fermi level [19],

$$T_c = 1.14\ \Theta_D \exp\{-\frac{1}{\rho(\varepsilon_F)V}\} \tag{2.23}$$

(here Θ_D) is the Debye temperature and V is the BCS electron-electron interaction parameter), one would expect that due to the increase of $\rho(\varepsilon_F)$ with the concentration of hydrogen, the transition temperature would increase also with increased H doping.

3. HARTREE-FOCK THEORY OF SURFACE STATES OF THREE-DIMENSIONAL CRYSTALS

A self-consistent-field theory of the surface states of three-dimensional crystals is already published in several papers. All these works apply, however, either to a model Hamiltonian [20],[21] or a general operator formalism [22], [23] which is not very suitable for immediate applications. The ab initio SCF LCAO method for the calculation of the surface states of a semi-infinite three-dimensional crystal in the Hartree-Fock level [5] which will be presented now is a generalization of the resolvent method which was derived for the calculation of the extra states due to a cluster of impurities embedded in an one-dimensional periodic system [4]. The matrix formulation of the problem has the advantage that numerical applications to the calculation of surface states of real systems are more feasible.

The way we will proceed to derive the final matrix equations is sketched in Fig. 3.1. We start from the problem of a cluster of impurities embedded in an infinite one-dimensional periodic chain (Fig.3.1a) [4]; as special case of the cluster of impurities we can intersect the loop at some point, then we are faced with the problem of a finite linear chain (Fig.3.1b) and the resolvent method can provide the surface states belonging to the two ends. The case of a half-infinite chain (Fig.3.1c) yields then the final problem of the surface of a semi-infinite three-dimensional crystal (Fig.3.1d) if one thinks of the linear chain of cells with one end as embedded in a three-dimensional crystal and repeated periodically in the other two dimensions an infinite number of times ($N_1 \to \infty$, but one end, $N_2 \to \infty$, $N_3 \to \infty$).

3.1. Problem of the Finite Linear Chain

Let us assume that we have a three-dimensional periodic crystal with $2N_1 + 1$, $2N_2 + 1$, and $2N_3 + 1$ unit cells in each direction and

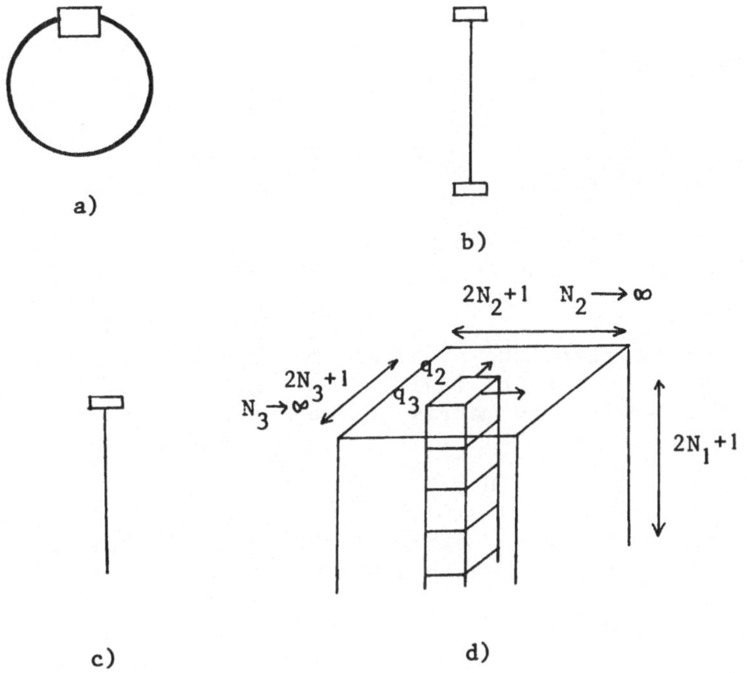

Fig. 3.1 : The generalization of the problem of one perturbation in a linear chain to the problem of the surface of a 3-dimensional crystal.

m orbitals in the unit cell. Let us further assume that though N_2 and $N_3 \to \infty$, N_1 does not. Let us consider now a linear chain of cells in the first direction. The position of this in the crystal chain should be characterized by the integers q_2 and q_3 ($\overline{R} = q_2 a_2 \overline{i} + q_3 a_3 \overline{j}$, where a_2 and a_3, and $\overline{i}$ and $\overline{j}$, are the elementary translations and unit vectors in the second and third directions, respectively). In Fig. 3.1d we show this chain in the crystal. Introducing periodic boundary conditions the coefficients of the Bloch orbitals of the electrons delocalized in the chain can be calculated with the aid of the matrix equation [1]

$$\underline{\underline{F}}_0(k_1,q_2,q_3)\underline{d}_i(k_1,q_2,q_3) = \varepsilon_i(k_1,q_2,q_3)\underline{\underline{S}}_0(k_1,q_2,q_3)\underline{d}_i(k_1,q_2,q_3) \tag{3.1}$$

where

$$\underline{\underline{F}}_0(k_1,q_2,q_3) = \sum_{q_1=-\infty}^{\infty} \underline{\underline{F}}_0(q_1,q_2,q_3)e^{ik_1q_1a_1} \tag{3.2a}$$

$$\underline{\underline{S}}_0(k_1,q_2,q_3) = \sum_{q_1=-\infty}^{\infty} \underline{\underline{F}}_0(q_1,q_2,q_3)e^{ik_1q_1a_1} \tag{3.2b}$$

with

$$[\underline{\underline{F}}_0(q_1,q_2,q_3)]_{f,g} = \langle X_f^{(0,q_2,q_3)}|\hat{F}^P|X_g^{(q_1,q_2,q_3)}\rangle \quad , \tag{3.3a}$$

$$[\underline{\underline{S}}_0(q_1,q_2,q_3)]_{f,g} = \langle X_f^{(0,q_2,q_3)}|X_g^{(q_1,q_2,q_3)}\rangle \quad . \tag{3.3b}$$

$\hat{F}^P$ is the Fock operator of the periodic linear chain and $X_f^{(0,q_2,q_3)}$ is the f-th orbital in the cell characterized by the integers $0,q_2,q_3$.

If we now intersect the chain at the crystal surface ($q_1 = 0$), we obtain instead of the original loop a finite chain. We can write then the Hamiltonian matrix of the whole finite chain in the form

$$\underline{\underline{H}} = \underline{\underline{H}}_0^P + \underline{\underline{H}}' \tag{3.4}$$

where $\underline{\underline{H}}_0^P$ is the Hamiltonian matrix of the original periodic chain (loop) and the matrix H' (which gives the deviation from the periodic case) is defined (see also [4]) in the first-neighbor interactions approximation by

$$\underline{\underline{H}}' = \begin{bmatrix} \tilde{\underline{\underline{F}}}(0) & \tilde{\underline{\underline{F}}}(1) & \ldots\underline{0}\ldots & . & \underline{\underline{F}}_0(2N_1) - \underline{\underline{F}}(1)^{tr} \\ \tilde{\underline{\underline{F}}}(1) & 0 & & & \\ \vdots & & \vdots & & \vdots \\ \underline{\underline{0}} & & \ldots\underline{\underline{0}}\ldots & & \underline{\underline{0}} \\ \vdots & & \vdots & & \vdots \\ & & & \underline{\underline{0}} & \tilde{\underline{\underline{F}}}(1)^{tr} \\ \underline{\underline{F}}_0(2N_1)^{tr} - \underline{\underline{F}}(1) & & \ldots\underline{\underline{0}}\ldots & \tilde{\underline{\underline{F}}}(1) & \tilde{\underline{\underline{F}}}(0) \end{bmatrix} \tag{3.5}$$

Here

$$\tilde{\underline{\underline{F}}}(0) = \underline{\underline{F}}_0(0) - \underline{\underline{F}}(0) \ , \ \tilde{\underline{\underline{F}}}(1) = \underline{\underline{F}}_0(1) - \underline{\underline{F}}(1) \tag{3.6}$$

The matrices $\underline{\underline{F}}(0)$ and $\underline{\underline{F}}(1)$ occur in the hypermatrix $\underline{\underline{H}}_0^P$ which can be written in the first-neighbor interactions approximation as

---

+ While $\underline{\underline{H}}_0(k_1,q_2,q_3)$ is a mxm matrix, $\underline{\underline{H}}_0^P$ and $\underline{\underline{H}}'$ are $m(2N_1 + 1) \times m(2N_1 + 1)$ hypermatrices.

$$\underline{\underline{H}}_0^P = \begin{bmatrix} \underline{\underline{F}}(0) & \underline{\underline{F}}(1) & \underline{\underline{0}} & \cdots & & & \underline{\underline{F}}(1)^{tr} \\ \underline{\underline{F}}(1)^{tr} & \underline{\underline{F}}(0) & \underline{\underline{F}}(1) & & & & \\ \underline{\underline{0}} & \underline{\underline{F}}(1) & \underline{\underline{F}}(0) & & & & \\ \vdots & & & \ddots & & & \\ & & & & \underline{\underline{F}}(1)^{tr} & \underline{\underline{F}}(0) & \underline{\underline{F}}(1) \\ \underline{\underline{F}}(1) & & & \cdots & (\underline{\underline{0}}) & \underline{\underline{F}}(1)^{tr} & \underline{\underline{F}}(0) \end{bmatrix} \tag{3.7}$$

From (3.7) it is clear that the elements of the matrix $\underline{\underline{F}}(0)$ (suppressing the indices q_2 and q_3),

$$[\underline{\underline{F}}(0)]_{f,g} = \langle X_f^{(0)}|\hat{F}^P|X_g^{(0)}\rangle = \langle X_f^{(q_1)}|\hat{F}^P|X_g^{(q_1)}\rangle \tag{3.8}$$

give interactions within the unit cell, while the matrix $\underline{\underline{F}}(1)$ provides the first-neighbor interactions. The matrix $\underline{\underline{F}}_0(0) \neq \underline{\underline{F}}(0)$ corresponds to the end of the chain, and its elements are defined as

$$[\underline{\underline{F}}_0(0)]_{f,g} = \langle \tilde{X}_f^{(0)}|\hat{F}^S|\tilde{X}_g^{(0)}\rangle \tag{3.9}$$

Here $\tilde{X}_f^{(0)}$ can be different from $X_f^{(0)}$ due to chemisorption in the case if the end cell has a larger number of orbitals ($\tilde{m} = m + m_{chem}$, where m_{chem} is the number of orbitals in the chemisorbed molecule) than the bulk cells. Further generally $\underline{\underline{F}}^S \neq \underline{\underline{F}}^P$, due both to the change of the potential at the chain end due to possible changes in the geometry in the end cell. Correspondingly the elements

$$[\underline{\underline{F}}_0(1)]_{f,g} = \langle \tilde{X}_f^{(0)}|F^S|X_g^{(1)}\rangle \tag{3.10}$$

of the matrix $\underline{\underline{F}}_0(1)$ give the interactions between orbitals centered in the end cell and in the next-neighbor cell, respectively.

In full analogy to (3.4) we can write

$$\underline{\underline{S}} = \underline{\underline{S}}_0^P + \underline{\underline{S}}' \tag{3.11}$$

where the hypermatrix $\underline{\underline{S}}_0^P$ can be given by (3.7) if we substitute in it the corresponding overlap matrices $\underline{\underline{S}}(0)$ and $\underline{\underline{S}}(1)$, respectively. The deviation matrix $\underline{\underline{S}}'$ has (in the first-neighbor interactions approximation) again the form of (3.5) and we can write it down if we substitute into (3.5), instead of $\underline{\underline{F}}(0)$ and $\underline{\underline{F}}(1)$, the corresponding overlap matrices

$$\tilde{\underline{\underline{S}}}(0) = \underline{\underline{S}}_0(0) - \underline{\underline{S}}(0)\ ,\ \tilde{\underline{\underline{S}}}(1) = \underline{\underline{S}}_0(1) - \underline{\underline{S}}(1) \tag{3.12}$$

In Eq. (3.12) the elements of $\underline{\underline{S}}_0(0)$ and $\underline{\underline{S}}_0(1)$ can be easily obtained if we substitute into (3.9) and (3.10) instead of $\hat{F}^S$ the unity operator $\hat{1}$.+ Further it should be mentioned that in the first-neighbor interactions approximation, the matrix $\underline{\underline{F}}_0(2N_1)$ and the corresponding matrix $\underline{\underline{S}}_0(2N_1)$ (which give the interactions between the orbitals belonging to the two different end cells of the chain) should be neglected.

With the help of these definitions we can formulate the problem of our finite open chain as

$$\underline{\underline{H}}\,\underline{c} = (\underline{\underline{H}}_0^P + \underline{\underline{H}}')\underline{c} = \lambda\underline{\underline{S}}\,c = \lambda(\underline{\underline{S}}_0^P + \underline{\underline{S}}')\underline{c} \tag{3.13}$$

Let us multiply this equation from the left by $\underline{\underline{U}}_1^+$ (the unitary matrix which block-diagonalizes $\underline{\underline{H}}_0^P$ and $\underline{\underline{S}}_0^P$). By inserting $\underline{\underline{U}}_1\underline{\underline{U}}_1^+ = 1$ and collecting the periodic matrices on the left-hand side, we obtain

$$\underline{\underline{U}}_1^+\underline{\underline{H}}_0^P\underline{\underline{U}}_1\underline{\underline{U}}_1^+\underline{c} - \lambda\underline{\underline{U}}_1^+\underline{\underline{S}}_0\underline{\underline{U}}_1\underline{\underline{U}}_1^+\underline{c} = -\,\underline{\underline{U}}_1^+\underline{\underline{H}}'\underline{c} - \lambda\underline{\underline{U}}_1^+\underline{\underline{S}}'\underline{c}\ , \tag{3.14}$$

or by introducing the notations

$$\underline{\underline{H}}_0^{BD} = \underline{\underline{U}}_1^+\underline{\underline{H}}_0^P\underline{\underline{U}}\ ,\quad \underline{\underline{S}}_0^{BD} = \underline{\underline{U}}_1^+\underline{\underline{S}}_0^P\underline{\underline{U}}_1\ ,$$

we can write

$$(\underline{\underline{H}}_0^{BD} - \lambda\underline{\underline{S}}_0^{BD})\underline{\underline{U}}_1^+\underline{c} = -\,\underline{\underline{U}}_1^+(\underline{\underline{H}}' - \lambda\underline{\underline{S}}')\underline{c} \tag{3.15}$$

The blocks of the unitary matrix $\underline{U}_1$ are defined [1] as

$$[\underline{\underline{U}}_1]_{pq} = \frac{1}{(2N_1 + 1)^{1/2}} \exp\{\frac{2\pi ipq}{2N_1 + 1}\} \tag{3.16}$$

If we introduce the resolvent matrix

$$\underline{\underline{Z}}(q_2,q_3) \equiv \underline{\underline{U}}_1[\underline{\underline{H}}_0^{BD}(q_2,q_3) - \lambda\underline{\underline{S}}_0^{BD}(q_2,q_3)]^{-1}\underline{\underline{U}}_1^+ \tag{3.17}$$

and multiply Eq. (3.16) from the left by

$$\underline{\underline{U}}_1(\underline{\underline{H}}_0^{BD} - \lambda\underline{\underline{S}}_0^{BD})^{-1}$$

we obtain the equation

$$\underline{c}(q_2,q_3) = -\,\underline{\underline{Z}}(q_2,q_3)[\underline{\underline{H}}'(q_2,q_3) - \lambda\underline{\underline{S}}'(q_2,q_3)]\underline{c}(q_2,q_3). \tag{3.18}$$

+ If the geometry of the end cell remains unchanged and there is no chemisorption, $\underline{\underline{S}}_0(0) = \underline{\underline{S}}(0)$, $\underline{\underline{S}}_0(1) = \underline{\underline{S}}(1)$, and therefore $\tilde{\underline{\underline{S}}}(0) = \tilde{\underline{\underline{S}}}(1) = 0$.

3.2. Half-Infinite Linear Chain

In the case of a half-infinite chain ($N_1 \to \infty$, but one end), we can repeat the preceding derivation, only now our deviation matrix H' is defined as

$$\underline{\underline{H}}' = \begin{bmatrix} \underline{\underline{0}} & -\underline{\underline{F}}(1) & \underline{\underline{0}} & \cdots & \underline{\underline{0}} & \cdots \\ -\underline{\underline{F}}(1)^{tr} & \underline{\underline{F}}(0) & \underline{\underline{F}}(1) & & & \\ \underline{\underline{0}} & \underline{\underline{F}}(1)^{tr} & \underline{\underline{0}} & \ddots & & \\ \vdots & & & & & \\ \underline{\underline{0}} & & & & \underline{\underline{0}} & \\ \vdots & & & & & \ddots \end{bmatrix} \tag{3.19}$$

and $\underline{\underline{S}}'$ has a similar expression. Substituting these expressions for $\underline{\underline{H}}'$ and $\underline{\underline{S}}'$ into (3.19) we obtain the following equations for the determination of the subvectors (having only m components) of $\underline{c}$:

$$-\underline{c}_{-1} = \underline{\underline{M}}_{-1,-1}\underline{c}_{-1} + \underline{\underline{M}}_{-1,0}\underline{c}_0 + \underline{\underline{M}}_{1,1}\underline{c}_1 ,$$

$$-\underline{c}_0 = \underline{\underline{M}}_{0,-1}\underline{c}_{-1} + \underline{\underline{M}}_{0,0}\underline{c}_0 + \underline{\underline{M}}_{0,1}\underline{c}_1 ,$$

$$-\underline{c}_1 = \underline{\underline{M}}_{1,-1}\underline{c}_{-1} + \underline{\underline{M}}_{1,0}\underline{c}_0 + \underline{\underline{M}}_{1,1}\underline{c}_1 ,$$

$$-c_i = \underline{\underline{M}}_{i,-1}\underline{c}_{-1} + \underline{\underline{M}}_{i,0}\underline{c}_0 + \underline{\underline{M}}_{i,1}\underline{c}_1 \quad (i = 2,3,\ldots). \tag{3.20}$$

By performing the matrix multiplications in (3.18) it is easy to show that the matrices M_{q-1}, $M_{q,0}$ and $M_{q,1}$ ($q = -1,0,1$) are defined as

$$\underline{\underline{M}}_{q,-1} = \underline{\underline{Z}}_{q,0}\,[-\underline{\underline{F}}(1) + \lambda\underline{\underline{S}}(1)]^{tr} , \tag{3.21a}$$

$$\underline{\underline{M}}_{q,0} = \underline{\underline{Z}}_{q,-1}\,[-\underline{\underline{F}}(1) + \lambda\underline{\underline{S}}(1)] + \underline{\underline{Z}}_{q,0}[\tilde{\underline{\underline{F}}}(0) - \lambda\tilde{\underline{\underline{S}}}(0)] + \underline{\underline{Z}}_{q,1}[\tilde{\underline{\underline{F}}}(1) - \lambda\tilde{\underline{\underline{S}}}(1)]^{tr} , \tag{3.21b}$$

$$\underline{\underline{M}}_{q,1} = \underline{\underline{Z}}_{q,0}[\tilde{\underline{\underline{F}}}(1) - \lambda\tilde{\underline{\underline{S}}}(1)] \quad (q = -1,0,1). \tag{3.21c}$$

In the system of Eqs. (3.20) one has to take into account that the part of the LCAO wave function, which has coefficients given by the subvector $\underline{c}_{-1}$, does not correspond to any physical reality because in a semifinite chain the electron belonging to the surface states of the end cannot get to the other end of the chain (which is in infinity) or to the space outside the chain (we are discussing here always bound electrons). To rule out the occurrence of these unphysical parts of the wave functions and also the occurrence of the unphysical solutions (spurious states) of the system of Eqs. (3.20), one has to leave out the first equation and take into account only the second and third terms in the subsequent equations.[+] In this way we can rewrite our Eqs. (3.20) as

$$-\underline{c}_0 = \underline{\underline{M}}_{0,0}\underline{c}_0 + \underline{\underline{M}}_{0,1}\underline{c}_1 ,$$

$$-\underline{c}_1 = \underline{\underline{M}}_{1,0}\underline{c}_0 + \underline{\underline{M}}_{1,1}\underline{c}_1 ,$$

$$-\underline{c}_i = \underline{\underline{M}}_{i,0}\underline{c}_0 + \underline{\underline{M}}_{i,1}\underline{c}_1 \quad (i = 2,3,\ldots). \tag{3.22}$$

The definition of the matrices $\underline{\underline{M}}_{0,0}$, $\underline{\underline{M}}_{0,1}$ etc., remains however, also in this case the same (see Eqs. (3.21)). If we want to solve Eqs. (3.22) only for $\underline{c}_0$ and $\underline{c}_1$ we obtain the following matrix equation :

$$\underline{\underline{M}}(\lambda)\begin{bmatrix}\underline{c}_0\\ \underline{c}_1\end{bmatrix} = \begin{bmatrix}\underline{\underline{M}}_{0,0}+\underline{\underline{1}} & \underline{\underline{M}}_{0,1}\\ \underline{\underline{M}}_{1,0} & \underline{\underline{M}}_{1,1}+\underline{\underline{1}}\end{bmatrix}\begin{bmatrix}\underline{c}_0\\ \underline{c}_1\end{bmatrix} = \underline{\underline{0}} \tag{3.23}$$

3.3. Surface States of a Three-Dimensional Crystal

Let us remember that the half-infinite chain treated in the previous section can be periodically repeated in the other two dimensions an infinite number of times ($N_2 \to \infty$, $N_3 \to \infty$; see Fig. 3.1d). If we introduce periodic boundary conditions in these two other dimensions, all the matrices $\underline{\underline{M}}_{q,0}$ and $\underline{\underline{M}}_{q,1}$ ($q = 0,1,2,3,\ldots$) occurring in Eqs. (3.22) will become cyclic hypermatrices. This can be easily proved [5]. The matrices $\underline{\underline{Z}}_{p_1,s}$ are defined as

$$\underline{\underline{Z}}_{p_1,s} = \frac{1}{\omega_1}\int_{\omega_1}\frac{\exp\{ia_1(p_1 - s)k_1\}}{\underline{\underline{F}}_0(k_1) - \lambda\underline{\underline{S}}_0(k_1)}\,dk_1 \tag{3.24}$$

(see Eq. (3.13) of [4], where $\underline{\underline{F}}_0(k_1)$ and $\underline{\underline{S}}_0(k_1)$ are again cyclic hypermatrices due to the periodicity in dimensions 2 and 3. The matrices $\underline{\underline{Z}}_{p_1,s}$ are also cyclic hypermatrices, because the inverse of a cyclic hypermatrix is also a cyclic hypermatrix.

+ Instead of this intuitive argument one can also give a more rigorous derivation of this procedure with the aid of projection-operator technique (see also Ref. 23).

Performing the matrix multiplications in (3.24) and knowing that all the four matrices $\underline{\underline{M}}_{0,0}$, $\underline{\underline{M}}_{0,1}$, $\underline{\underline{M}}_{1,0}$ and $\underline{\underline{M}}_{1,1}$ are cyclic hypermatrices, we can write

$$\underline{\underline{U}}^{+}_{2,3}(\underline{\underline{M}}_{0,0} + \underline{\underline{1}})\underline{\underline{U}}_{2,3}\underline{\underline{U}}^{+}_{2,3}\underline{c}_0 + \underline{\underline{U}}^{+}_{2,3}\underline{\underline{M}}_{0,1}\underline{\underline{U}}_{2,3}\underline{\underline{U}}^{+}_{2,3}\underline{c}_1 = \underline{\underline{0}} \tag{3.25}$$

$$\underline{\underline{U}}^{+}_{2,3}\underline{\underline{M}}_{1,0}\underline{\underline{U}}_{2,3}\underline{\underline{U}}^{+}_{2,3}\underline{c}_0 + \underline{\underline{U}}^{+}_{2,3}(\underline{\underline{M}}_{1,1} + \underline{\underline{1}})\underline{\underline{U}}_{2,3}\underline{\underline{U}}^{+}_{2,3}\underline{c}_1 = \underline{\underline{0}} , \tag{3.26}$$

where the unitary matrix $\underline{\underline{U}}_{2,3}$ which blockdiagonalizes the matrices $\underline{\underline{M}}_{q,0}$ and $\underline{\underline{M}}_{q,1}$ (q = 0,1) is defined through its blocks as (if we assume $N_1 = N_2 = N_3 = N$) [1]

$$(\underline{\underline{U}}_{2,3})_{p_2,p_3;q_2,q_3} = (2N + 1)^{-1}\exp\{i2\pi(p_2q_2 + p_3q_3)(2N + 1)^{-1}\}\underline{\underline{1}} . \tag{3.27}$$

Introducing the notations

$$\underline{\underline{M}}^{BD}_{q,s} = \underline{\underline{U}}^{+}_{2,3}\underline{\underline{M}}_{q,s}\underline{\underline{U}}_{2,3} , \quad \underline{\tilde{D}}_q = \underline{\underline{U}}^{+}_{2,3}\underline{c}_q \qquad (q,s = 0,1) \tag{3.28}$$

we can rewrite Eqs. (3.25) and (3.26) as

$$(\underline{\underline{M}}^{BD}_{0,0} + \underline{\underline{1}})\underline{\tilde{D}}_0 + \underline{\underline{M}}^{BD}_{0,1}\underline{\tilde{D}}_1 = 0 \tag{3.29a}$$

$$\underline{\underline{M}}^{BD}_{1,0}\cdot\underline{\tilde{D}}_0 + (\underline{\underline{M}}^{BD}_{1,1} + \underline{\underline{1}})\underline{\tilde{D}}_1 = \underline{0} . \tag{3.29b}$$

Taking into account the definitions (3.21) of the $\underline{\underline{M}}_{q,0}$ and $\underline{\underline{M}}_{q,1}$ (q = 0,1) matrices we can write for instance $\underline{\underline{M}}^{BD}_{1,1}$ as

$$\begin{aligned}\underline{\underline{M}}^{BD}_{1,1} &= \underline{\underline{U}}^{+}_{2,3}\underline{\underline{M}}_{1,1}\underline{\underline{U}}_{2,3} = \\ &= \underline{\underline{U}}^{+}_{2,3}\underline{\underline{Z}}_{1,0}\underline{\underline{U}}_{2,3}\underline{\underline{U}}^{+}_{2,3}[\underline{\underline{\tilde{F}}}(1) - \lambda\underline{\underline{\tilde{S}}}(1)]\underline{\underline{U}}_{2,3} .\end{aligned} \tag{3.30}$$

Since the product of blockdiagonal matrices is again a blockdiagonal matrix, the matrix equations (3.29) can be reduced to such matrix equations which have only the order of the number of orbitals in the unit cell (m) [1]. If we assume further that $N_2 = N_3 \to \infty$ and introduce the continuous variables

$$k_2 = 2\pi q_2/(2N + 1)a_2 \quad , \quad k_3 = 2\pi q_3/(2N + 1)a_3 \tag{3.31}$$

we can write

$$[\underline{\underline{M}}_{0,0}(k_2,k_3) + \underline{\underline{1}}]\underline{\tilde{d}}_0(k_2,k_3) + \underline{\underline{M}}_{0,1}(k_2,k_3)\underline{\tilde{d}}_1(k_2,k_3) = \underline{0} \tag{3.32a}$$

$$\underline{\underline{M}}_{1,0}(k_2,k_3)\underline{\tilde{d}}_0(k_2,k_3) + [\underline{\underline{M}}_{1,1}(k_2,k_3) + \underline{\underline{1}}]\underline{\tilde{d}}_1(k_2,k_3) = \underline{0} . \tag{3.32b}$$

With the notation $\overline{k}' = k_2\overline{i} + k_3\overline{j}$, we can write Eqs. (3.32) in the form of the hypermatrix equation

$$\underline{\underline{M}}(\lambda(\overline{k}'),\overline{k}')\underline{\tilde{d}}(\overline{k}') = \begin{bmatrix} \underline{\underline{M}}_{0,0}(\overline{k}') + \underline{\underline{1}} & \underline{\underline{M}}_{0,1}(\overline{k}') \\ \underline{\underline{M}}_{1,0}(\overline{k}') & \underline{\underline{M}}_{1,1}(\overline{k}') + \underline{\underline{1}} \end{bmatrix} \begin{bmatrix} \tilde{d}_0(\overline{k}') \\ \tilde{d}_1(\overline{k}') \end{bmatrix} = \underline{0} \tag{3.33}$$

From the criterion of the existence of the non-vanishing solution for the vectors $\underline{d}(\overline{k}')$ one obtains the equation

$$\det[\underline{\underline{M}}(\lambda(\overline{k}'),\overline{k}')] = 0 \tag{3.34}$$

for the determination of the 2m surface energy bands $\lambda_i(\overline{k}')$.

To form the matrix $\underline{\underline{M}}(\overline{k}')$ one has to write down the individual matrices

$$\underline{\underline{M}}_{0,0}(\overline{k}'),\ \underline{\underline{M}}_{0,1}(\overline{k}'),\ \underline{\underline{M}}_{1,0}(\overline{k}')$$

and $\underline{\underline{M}}_{1,1}(\overline{k}')$, respectively. The last of these, $\underline{\underline{M}}_{1,1}(\overline{k}')$, is if we write it out in more detail,

$$\underline{\underline{M}}_{1,1}(\overline{k}') = \underline{\underline{Z}}_{1,0}(\overline{k}')[\tilde{\underline{\underline{F}}}(1,\overline{k}') - \lambda(\overline{k}')\tilde{\underline{\underline{S}}}(1,\overline{k}')] \tag{3.35}$$

where, as one can show [1], [24]

$$\underline{\underline{Z}}_{1,0}(\overline{k}') = \sum_{q_2,q_3=-\infty}^{\infty} \exp\{i(k_2q_2a_2 + k_3q_3a_3)\}\underline{\underline{Z}}_{1,0}(q_2,q_3) \tag{3.36}$$

with (see Eq. (3.24) in the case of $p_1 = 1$, $s = 0$)

$$\underline{\underline{Z}}_{1,0}(q_2,q_3) = \frac{1}{\omega_1}\int_{\omega_1} \frac{\exp\{ia_1(1-0)k_1\}}{\underline{\underline{F}}_0(k_1,q_2,q_3) - \lambda(\overline{k}')\underline{\underline{S}}_0(k_1,q_2,q_3)}\,dk_1 \tag{3.37}$$

Further $\underline{\underline{F}}_0(k_1,q_2,q_3)$ and $\underline{\underline{S}}_0(k_1,q_2,q_3)$ were defined by Eq. (3.2),

$$\tilde{\underline{\underline{F}}}(1,\overline{k}') = \sum_{q_2,q_3=-\infty}^{\infty} \exp\{i(k_2q_2q_2 + k_3q_3q_3)\}[\underline{\underline{F}}_0(1,q_2,q_3) - \underline{\underline{F}}(1,q_2,q_3)] \tag{3.38}$$

and analogous expressions are valid for $\tilde{\underline{\underline{F}}}(0,\overline{k}')$, $\tilde{\underline{\underline{S}}}(1,\overline{k}')$, and $\tilde{\underline{\underline{S}}}(0,\overline{k}')$, respectively.

3.4. Self-Consistent-Field Procedure for the Calculation of Surface States

To be able to calculate the surface states in a SCF way we need the definitions of the elements of the matrices $\underline{\underline{F}}(q_1,q_2,q_3) = \underline{\underline{F}}(\bar{q})$, $\underline{\underline{F}}_0(0,q_2,q_3)$, and $F_0(1,q_2,q_3)$, respectively. The elements of the matrix $\underline{\underline{F}}(\bar{q})$ belonging to the periodic problem are [1, 24] (see also [4])

$$[\underline{\underline{F}}(\bar{q})]_{f,g} = \langle X_f^{\bar{0}}| - \frac{1}{2}\Delta - \sum_{\bar{q}_1}\sum_{\alpha}^{M_0} \frac{{}^0Z_\alpha}{|\bar{r} - \bar{r}_\alpha^{\bar{q}_1}|}|X_g^{\bar{q}}\rangle + \sum_{\bar{q}_1,\bar{q}_2}\sum_{u,v=1}^{m} p(\bar{q}_1-\bar{q}_2)_{u,v}$$

$$(\langle X_f^{\bar{0}}X_u^{\bar{q}_1}|X_g^{\bar{q}}X_v^{\bar{q}_2}\rangle - \frac{1}{2}\langle X_f^{\bar{0}}X_u^{\bar{q}_1}|X_v^{\bar{q}_2}X_g^{\bar{q}}\rangle), \tag{3.39}$$

where the elements of the generalized-charge-bond-order matrix are defined as [1,4]

$$p(\bar{q}_1 - \bar{q}_2)_{u,v} = \frac{1}{\omega}\int_\omega \sum_f^{n^\star} \exp[i\bar{k}(\bar{R}_{q_1} - \bar{R}_{q_2})]\, d^\star(\bar{k})_{f,u} d(\bar{k})_{f,v} d\bar{k} \tag{3.40}$$

(ω is the volume of the first Brillouin zone in the $\bar{k}$ space). In the case of the first-neighbor interactions approximation used here in the formulation of the problem, we have to substitute for $\bar{q}$ either $(0,q_2,q_3)$ or $(1,q_2,q_3)$.

The elements of the additional matrix $\underline{\underline{F}}_0(0,q_2,q_3)$ are given by

$$[\underline{\underline{F}}_0(0,q_2,q_3)]_{f,g} = \langle\tilde{X}_f^{\bar{0}}| - \frac{1}{2}\Delta - \sum_{\alpha=1}^{M}\sum_{q_5,q_6} \frac{Z_\alpha}{|\bar{r} - \bar{r}_\alpha^{(0,q_5,q_6)}|} -$$

$$- \sum_{\alpha=1}^{M_0}\sum_{q_5,q_6} \frac{{}^0Z_\alpha}{|\bar{r} - \bar{r}_\alpha^{(1,q_5,q_6)}|}|\tilde{X}_g^{(0,q_2,q_3)}\rangle + \sum_{q_4,q_7=0}^{1}\ \sum_{\substack{q_5,q_6,\\ q_8,q_9=-\infty}}^{\infty}\ \sum_{u,v=1}^{\tilde{m}}$$

$$\tilde{p}(q_4,q_5,q_6;q_7,q_8,q_9)_{u,v}\cdot(\langle\tilde{X}_f^{\bar{0}}\overline{X}_u^{(q_4,q_5,q_6)}|\tilde{X}_g^{(0,q_2,q_3)}\overline{X}_v^{(q_7,q_8,q_9)}\rangle$$

$$-\frac{1}{2}\langle\ \ \rangle_{exch}), \tag{3.41}$$

where $\tilde{X}_f^0 = \tilde{X}_f^{(0,q_2,q_3)}$ is the fth atomic orbital in the surface cell characterized by q_2, q_3,

$$\overline{X}_u^{(q_4,q_5,q_6)} = \begin{cases} \tilde{X}_u^{(q_4,q_5,q_6)} & \text{if } q_4 = 0 \\ X_u^{(q_4,q_5,q_6)} & \text{if } q_4 = 1 \end{cases} \tag{3.42}$$

${}^0Z_\alpha$ and M_0 is the nuclear charge of the αth nucleus and the number of nuclei, respectively, in the bulk cell, while Z_α and M are the same in the surface cell (which may be constituted from a pure surface cell and a chemisorbed molecule). Correspondingly, as mentioned before, $\tilde{m} = m + m_{chem}$, where m_{chem} is the number of orbitals in the chemisorbed molecule. Finally,

$$\tilde{P}(q_4,q_5,q_6;q_7,q_8,q_9)_{u,v} = \sum_{j=1}^{\tilde{n}^\star} \frac{1}{\omega_{23}} \int_{\omega_{23}} dk_2 dk_3 \, d^\star_{f,q_4,u}(k_2,k_3)$$

$$\tilde{d}_{j,q_7,v}(k_2,k_3) \exp\{i[k_2 a_2(q_5 - q_8) + k_3 a_3 (q_6 - q_9)]\}$$

$$(q_4 = 0,\ q_7 = 0 \text{ or } 1),\ \tilde{n}^\star = \frac{1}{2}[n^\star(q_4) + n^\star(q_7)] \tag{3.43}$$

We can immediately obtain also the elements of the matrix $\underline{\underline{F}}(1,q_2,q_3)$ if we substitute in (3.41) everywhere the orbital $X_g(1,q_2,q_3)$ instead of $\tilde{X}_g(0,q_2,q_3)$.

After having all the necessary expressions for the <u>ab initio</u> SCF LCAO treatment of the surface (and chemisorption states) of a three-dimensional crystal, one can proceed in the following way :

(i) One has to solve first the SCF problem of the three-dimensional periodic infinite crystal.

(ii) Next one has to obtain the $\lambda_i(\overline{k}')$ solutions of Eq. (3.34). One promising way to do it is, as van der Avoird et al.[22] have suggested it, to find the zero eigenvalues $E_i(\lambda(\overline{k}')) = 0$ of the eigenvalue problem of the Hermitian matrix $\underline{\underline{M}}^H$ ($\underline{\underline{M}}(\lambda)$ is not Hermitian) :

$$\underline{\underline{M}}^H(\lambda(\overline{k}'))\underline{d}_j(\lambda(\overline{k}')) = [\underline{\underline{H}}'(\overline{k}') - \lambda(\overline{k}')\underline{\underline{S}}'(\overline{k}')] .$$

$$.\underline{\underline{M}}(\lambda(\overline{k}')\underline{d}_j(\lambda(\overline{k}')) = E_j(\lambda(\overline{k}')\underline{d}_j(\lambda(\overline{k}')). \tag{3.44}$$

where

$$\underline{\underline{H}}'(\overline{k}') = \begin{bmatrix} \tilde{\underline{\underline{F}}}(0,k_2,k_3) & \tilde{\underline{\underline{F}}}(1,k_2,k_3) \\ \tilde{\underline{\underline{F}}}(1,k_2,k_3) & \underline{\underline{0}} \end{bmatrix} \tag{3.45}$$

and $\underline{\underline{S}}'(\overline{k}')$ has an analogous expression.[+] These zeros can be calculated by an elegant algorithm proposed by van der Avoird et al.[22] because the matrix $\underline{\underline{M}}^H$ and its eigenvalues $E_j(\lambda(\overline{k}'))$ have some special properties which were proved in the case of a cluster of impurity embedded in a one-dimensional chain [13]. These properties lead to the following theorem :

The number of λ_0's for which $E_j(\lambda_0) = 0$ is in a given interval $[\lambda_1 < \lambda_2]$ equal to

$$n_0(\lambda_1,\lambda_2) = n_+(\lambda_2) - n_+(\lambda_1) + \sum_{\substack{\overline{k} \\ \lambda_1 < \varepsilon_{\overline{k}} < \lambda_2}} d^0_{\overline{k}} \tag{3.46}$$

when $n_+(\lambda_2)$ and $n_+(\lambda_1)$ are the number of positive eigenvalues of $\underline{\underline{M}}^H(\lambda_2)$ and $\underline{\underline{M}}^H(\lambda_1)$, respectively, and $d^{(0)}_{\overline{k}}$ is the degeneracy of the unperturbed energies $\varepsilon_{\overline{k}}$. The last summation in (3.46) just counts the number of poles $\lambda = \varepsilon_{\overline{k}}$ in the given interval $[\lambda_1, \lambda_2]$.

Using this theorem, the roots λ_0 for which $E_j(\lambda_0) = 0$ can be found by repeated bisection of the interval, until the required accuracy is reached. The fastest way of calculating the number of positive eigenvalues $n_+(\lambda)$ of $\underline{\underline{M}}^H(\lambda)$ at a given point λ is probably to bring $\underline{\underline{M}}^H$ into upper triangular form by the Gauss elimination process and to count the number of positive diagonal elements. This number equals $n_+(\lambda)$ [22].

(iii) After having found the roots $\lambda_i(\overline{k}')$, one can substitute these back into Eq. (3.33) and solve that for the different vectors $\underline{\tilde{d}}_{i,0}(\overline{k}')$ and $\underline{\tilde{d}}_{i,1}(\overline{k}')$, respectively.

(iv) With the aid of the vectors $\underline{\tilde{d}}_{i,0}(\overline{k}')$ and $\underline{\tilde{d}}_{i,1}(\overline{k}')$, one can form again the matrices $\underline{\underline{F}}_0(0;q_2,q_3)$ and $\underline{\underline{F}}(1;q_2,q_3)$, and with the latter ones the matrix $\underline{\underline{M}}(\lambda(\overline{k}'))$. In this way the whole procedure can be repeated until self-consistency is reached.

Using the van der Avoird algorithm in step (ii) instead of solving (3.34) has an advantage of practical importance : one can calculate also those energy eigenvalues which lie within the bulk bands of the crystal and thus coincide with a pole in the resolvent. In such a case one or more eigenvalues go to infinity but some $E_j(\lambda)$ eigenvalues can go to zero and are found by formula (3.46). These solutions cannot be obtained by solving (3.34) directly. As it can be seen from (3.43) in every cycle of a SCF procedure <u>all</u> eigenvectors d_j, $j = 1,\ldots,n$, must be known for the construction of the charge-bond-order matrix and therefore <u>all</u> λ_j, i.e., also those which coincide with a pole.

+ One can show on the basis of the definition of $\underline{\underline{M}}(\lambda(\overline{k}'))$ [see Eqs. (3.21b), (3.21c), (3.33), and (3.35)-(3.38)] that $\underline{\underline{M}}^H(\lambda(\overline{k}')) = (\underline{\underline{H}}' - \lambda\underline{\underline{S}}')\underline{\underline{M}}$ is Hermitian.

Until now we did not discuss the problem of bulk distortion. If we treat the problem in the first-neighbors interactions approximation as it was done here we can calculate the surface and chemisorption states in a self-consistent way but we cannot treat the effect of these states on the bulk bands. If we include, however, more neighbors in the description of the interactions between perturbed and unperturbed cells, then matrix elements of the form

$$<X_f^B \,|\hat{F}|\, X_g^B > \qquad (3.47)$$

will occur where X_f^B and X_g^B are pure bulk states but the Fock operator $\hat{F}$ contains also F^S the contribution of the perturbation (surface). Matrix elements of the form (3.47) give interactions between bulk states in the presence of a perturbation. The problem of calculating bulk distortion as well as other questions related to localization of the surface states and surface exciton states will be the subject of future work.

One can hope that after coding the described method one can obtain, in the near future, surface states for three-dimensional crystals in an ab initio SCF way. Comparing the results of these planned calculations with those of cluster calculations and of two-dimensional (layer orbital) calculations on simple metals (like Be crystal), one will be able on the one hand to decide better when these approximations are justified. On the other hand the results of these calculations will give also better answers on the following questions :

(i) How many neighbors perpendicular to the surface have to be taken into account in different types of systems (metals, covalently bound crystals, etc.) to describe correctly the surface and chemisorption states ;

(ii) How much are the surface states localized or delocalized in these systems ; and

(iii) How much effect may the (usually experimentally not well-known) possible changes have in the geometry of the surface cells as compared to the bulk cells on the surface states. All these studies will serve of course the better understanding of the surface properties and in final analysis of the catalytic properties of different solids.

ACKNOWLEDGEMENTS

The authors wish to express their gratitude to Professors T.C. Collins and F. Martino and to Dr. S. Suhai for very useful discussions. They are further indebted to the German Research Council (Deutsche Forschungsgemeinschaft, Project no. La 371/1) and to the Fonds of Chemical Industry (Fond der Chemischen Industrie, BRD) for the financial support of this research.

REFERENCES

1. G. Del Re, J. Ladik and G. Biczó, Phys. Rev. 155, 967 (1967).

2. For a review of this method see : A.J. Elliot, J.A. Krumhansl and P.L. Leath, Rev. Mod. Phys. 46, 465 (1974). F. Martino in "Quantum Theory of Polymers", J.-M. André, J. Delhalle and J. Ladik, Eds., D. Reidel Publ. Co., Dordrecht-Boston, p. 169 (1978).

3. M. Seel, T.C. Collins, F. Martino, D.K. Rai and J. Ladik, Phys. Rev. B (submitted).

4. J. Ladik and M. Seel, Phys. Rev. B13, 5338 (1976).

5. J. Ladik, Phys. Rev. B17, 1663 (1978).

6. For a review see : H.P. Geserich and L. Pintschovius, Adv. Solid State Phys. 16, 65 (1976).

7. H. Kamimura, A.J. Grant, F. Levy, A.D. Yoffe, C.D. Pitt, Solid State Commun. 17, 49 (1975) ; D.E. Parry, J.M. Thomas, J. Phys. C3, L45 (1975) ; A. Zunger, J. Chem. Phys. 63, 4854 (1975) ; S. Suhai, M. Kertész, J. Phys. C9, L347 (1976) ; N.T. Rajan and L.M. Falicov, Phys. Rev. B12, 1240 (1975).

8. C. Merkel and J. Ladik, Phys. Lett. 56A, 395 (1976) ; M. Kertész, J. Koller, A. Ažman and S. Suhai, Phys. Lett. 55A, 107 (1975).

9. W.E. Rudge and P.M. Grant, Phys. Rev. Lett. 35, 1799 (1975) ; W.Y. Ching, J.G. Harrison and C.C. Lin, Phys. Rev. B15, 5975 (1977) ; I.P. Batra, S. Ciraci and W.E. Rudge, Phys. Rev. B15, 5858 (1977).

10. S. Suhai and J. Ladik, Solid State Commun. 22, 227 (1977).

11. B.L. Györffy and J.S. Faulkner (personal communication).

12. W.D. Gill, W. Bludan, R.H. Geiss, P.M. Grant, R.L. Greene, J.J. Mayerle and G.B. Street, Phys. Rev. Lett. 38, 1305 (1977).

13. M. Seel, "Theoretische Untersuchungen zur Aperiodizität und zum Korrelationsproblem in Polymeren und Molekülkristallen" (Theoretical Investigations of Aperiodicity and Correlation in Polymers and Molecular Crystals) Thesis, University Erlangen-Nürnberg 1978.

14. See for instance : W.L. McCubbin in "Quantum Theory of Polymers" p. 185 (1978).

15. P. Soven, Phys. Rev. 156, 809 (1967) ; S. Kirkpatrick, B. Velický, H. Ehrenreich, Phys. Rev. B1, 3250 (1970).

16. See for instance : J. Callaway "Quantum Theory of the Solid State", Academic Press, New York-London, p. 517 (1974).

17. S. Suhai (unpublished results).

18. J. Delhalle, Bull. Soc. Chim. Belg. 84, 135 (1975).

19. See for instance : G. Rickayzen, "Theory of Superconductivity" Interscience-Wiley, New York-London (1965).

20. P.W. Anderson, Phys. Rev. 124, 41 (1961).

21. T.B. Grimley, Proc. Phys. Soc. Lond. 90, 751 (1966); 92, 776 (1967); Ber. Bunsen Ges. 75, 1003 (1971) ; M.J. Kelly, Surf. Sci. 43, 587 (1974).

22. A. van der Avoird, S.P. Liebmann, and J.M. Fassaert, Phys. Rev. B10, 1230 (1974).

23. J. Koutecký, Progr. Surf. and Membr. Sci. 11, 1 (1976).

24. J. Ladik, in "Electronic Structure of Polymers and Molecular Crystals", J.-M. André and J. Ladik, Eds., Plenum Press, New York, p. 23 (1975).

ON THE LECTURES WHOSE FULL TEXT DOES NOT APPEAR IN THIS BOOK

Electronic structure at metallic surfaces (5 lectures) by P. Soven (Univ. of Pennsylvania).

An excellent introductory reference to the entire subject of surface states can be found in the article "Electron States at Clean Surfaces" by F. Forstmann, which appears in the book "Photoemission and the Electronic Properties of Surfaces" edited by B. Feuerbacher, B. Fitton and R.F. Willis (John Wiley and Sons, (1978)).

Theory of Low Energy Electron Diffraction and Photoemission (5 lectures) by J. Pendry (Daresbury Laboratory).

The topics covered are discussed at length in :
"Advances in Leed Theory" published in "Electron Diffraction (1977)" (1977), Eds. P.J. Dobson, J.B. Pendry and C.J. Humphreys (Institute of Physics, Bristol, (1978)).
and
"Electron Emission from Solids" published in "Photoemission and the Electronic Properties of Surfaces", Eds. B. Feuerbacher, B. Fitton and R.F. Willis (John Wiley and Sons, Chichester, (1978)).

Superconductivity in Amorphous Metals (3 lectures) by G. Bergmann (KFA - Jülich).

This subject is thoroughly discussed in the review article "Amorphous Metals and their Superconductivity" by G. Bergmann Physics Reports 27C, (1976).

LIST OF LECTURERS

Bergmann G.	KFA - Jülich GmbH, Institut 10 - Supraleitung und Tieftemperaturphysik, Postfach 1913, D-5170 Jülich 1	W.-Germany
Berko S.	Department of Physics, Brandeis University, Waltham, Mass. 02154	U.S.A.
Edwards D.M.	Imperial College of Science and Technology, Department of Mathematics, Huxley Building Queen's Gate, London SW7 2BZ	England
Evans R.	University of Bristol, H.H. Wills Physics Lab., Royal Fort, Tyndall Avenue, Bristol BS8 1TL	England
Gunnarsson O.	Institute of Theoretical Physics, Chalmers, Fack, S-402 20 Göteborg 5	Sweden
Güntherodt H.J.	Institut für Physik der Univ. Basel, Exper. Physik der kond. Materie, Klingelbergstrasse 82, CH-4056 Basel	Switzerland
Györffy B.L.	University of Bristol, H.H. Wills Physics Lab., Royal Fort, Tyndall Avenue, Bristol BS8 1TL	England
Neddermeyer H.	Ruhr-Universität Bochum, Institut für Experimentalphysik, Postfach 1021 48, D-4630 Bochum	W.-Germany
Pendry J.	Daresbury Laboratory, Science Research Council, Daresbury, Warrington WA4 4AD	England
Soven P.	University of Pennsylvania, Department of Physics, Philadelphia, Pa. 19104	U.S.A.

LIST OF PARTICIPANTS

Ackermann K.P.	University of Basel	Switzerland
Bansil A.	Northeastern Univ., Boston	U.S.A.
Bilir N.	Middle East Techn. Univ., Ankara	Turkey
Campagna M.	K.F.A. - Jülich	W.-Germany
Coutinho S.	Imperial College, London	England
Daams J.M.	University of Toronto	Canada
Dalmai G.	I.S.E.N. Lille	France
De Meyer G.	University of Gent	Belgium
Djafari-Rouhani B.	I.S.E.N. Lille	France
Durham P.J.	University of Bristol	England
Echenique P.M.	University of Barcelona	Spain
Fiermans L.	University of Gent	Belgium
Funnemann D.	Westfälische-Wilhelms Univ., Münster	W.-Germany
Ganachaud J.-P.	Ecole Nat. Sup. de Méca-nique, Nantes	France
Giuliano E.S.	Università di Messina	Italy
Greuter F.	Swiss Fed. Inst. of Techn., Zürich	Switzerland
Gubler U.M.	University of Basel	Switzerland
Hague C.F.	Université Paris 6	France
Heimann P.	Universität München	W.-Germany
Hoogewijs R.	University of Gent	Belgium
Humberg H.	Universität Münster	W.-Germany
Jones R.O.	K.F.A. - Jülich	W.-Germany
Krieg J.	University of Basel	Switzerland
Kuhlmann E.	K.F.A. - Jülich	W.-Germany
Ladik J.	Univ. Erlangen-Nürnberg	W.-Germany
Lähdeniemi M.	University of Turku	Finland
Lambrecht W.	University of Gent	Belgium
Liard M.E.	University of Basel	Switzerland
Maca F.	Czech.Acad. of Sciences, Prague	Czechoslovakia
Manninen M.J.	University of Helsinki	Finland
Mijnarends P.E.	N.E.R.F., Petten	The Netherlands
Monnier R.	E.T.H., Zürich	Switzerland
Moore I.D.	Science Res. Council, Daresbury	England

Neve J.	Royal Inst. of Technol., Stockholm	Sweden
Oelhafen P.	University of Basel	Switzerland
Oxinos G.	University of Salford	England
Parussel M.	University of Erlangen-Nürnberg	W.-Germany
Pessa M.	Tampere Univ. of Technol.	Finland
Pindor A.J.	Instytut Fizyki PAN, Warszawa	Poland
Pinski F.J.	State Univ. of New York, Stony Brook	U.S.A.
Rotthier R.	University of Gent	Belgium
Rundgren J.	Royal Inst. of Technol., Stockholm	Sweden
Scheire L.	University of Gent	Belgium
Singh V.	S.U.N.Y.A., Albany	U.S.A.
Staunton J.	University of Bristol	England
Stocks G.M.	Oak Ridge Nat. Lab.	U.S.A.
Szulkin M.	Pedagogical Univ., Czestochowa	Poland
Temmerman W.	University of Bristol	England
Ulehla M.	Oak Ridge Nat. Lab.	U.S.A.
Van Steenberge P.	University of Gent	Belgium
Verhuyck K.	University of Gent	Belgium
Verschelde H.	University of Gent	Belgium
Weinberger P.	Techn. Univ. Wien, Vienna	Austria
Wells G.	Brookhaven Nat. Laboratory	U.S.A.
Werner K.	K.F.A. - Jülich	W.-Germany
Westrin J.P.	Univ. of Stockholm	Sweden
Zagorski A.	Technical University,Warsaw	Poland
Zangwill A.	University of Pennsylvania, Philadelphia	U.S.A.
Zwicknagl G.	K.F.A. - Jülich	W.-Germany

SUBJECT INDEX